AF714771

Springer Complexity

Springer Complexity is an interdisciplinary program publishing the best research and academic-level teaching on both fundamental and applied aspects of complex systems—cutting across all traditional disciplines of the natural and life sciences, engineering, economics, medicine, neuroscience, social and computer science.

Complex Systems are systems that comprise many interacting parts with the ability to generate a new quality of macroscopic collective behavior the manifestations of which are the spontaneous formation of distinctive temporal, spatial or functional structures. Models of such systems can be successfully mapped onto quite diverse "real-life" situations like the climate, the coherent emission of light from lasers, chemical reaction–diffusion systems, biological cellular networks, the dynamics of stock markets and of the internet, earthquake statistics and prediction, freeway traffic, the human brain, or the formation of opinions in social systems, to name just some of the popular applications.

Although their scope and methodologies overlap somewhat, one can distinguish the following main concepts and tools: self-organization, nonlinear dynamics, synergetics, turbulence, dynamical systems, catastrophes, instabilities, stochastic processes, chaos, graphs and networks, cellular automata, adaptive systems, genetic algorithms and computational intelligence.

The three major book publication platforms of the Springer Complexity program are the monograph series "Understanding Complex Systems" focusing on the various applications of complexity, and the "Springer Series in Synergetics", which is devoted to the quantitative theoretical and methodological foundations, and the "SpringerBriefs in Complexity" which are concise and topical working reports, case-studies, surveys, essays and lecture notes of relevance to the field. In addition to the books in these two core series, the program also incorporates individual titles ranging from textbooks to major reference works.

Editorial and Programme Advisory Board

Henry Abarbanel, Institute for Nonlinear Science, University of California, San Diego, USA

Dan Braha, New England Complex Systems, Institute and University ofMassachusetts, Dartmouth, USA

Péter Érdi, Center for Complex Systems Studies, Kalamazoo College, USA and Hungarian Academy of Sciences, Budapest, Hungary

Karl Friston, Institute of Cognitive Neuroscience, University College London, London, UK

Hermann Haken, Center of Synergetics, University of Stuttgart, Stuttgart, Germany

Viktor Jirsa, Centre National de la Recherche Scientifique (CNRS), Université de la Méditerranée, Marseille, France

Janusz Kacprzyk, System Research, Polish Academy of Sciences, Warsaw, Poland

Kunihiko Kaneko, Research Center for Complex Systems Biology, The University of Tokyo, Tokyo, Japan

Scott Kelso, Center for Complex Systems and Brain Sciences, Florida Atlantic University, Boca Raton, USA

Markus Kirkilionis, Mathematics Institute and Centre for Complex Systems, University of Warwick, Coventry, UK

Jürgen Kurths, Nonlinear Dynamics Group, University of Potsdam, Potsdam, Germany

Andrzej Nowak, Department of Psychology, Warsaw University, Poland

Hassan Qudrat-Ullah, School of Administrative Studies, York University, Canada

Linda Reichl, Center for Complex Quantum Systems, University of Texas, Austin, USA

Peter Schuster, Theoretical Chemistry and Structural Biology, University of Vienna, Vienna, Austria

Frank Schweitzer, System Design, ETH Zürich, Zürich, Switzerland

Didier Sornette, Entrepreneurial Risk, ETH Zürich, Zürich, Switzerland

Stefan Thurner, Section for Science of Complex Systems, Medical University of Vienna, Vienna, Austria

Understanding Complex Systems

Founding Editor: Scott Kelso

Future scientific and technological developments in many fields will necessarily depend upon coming to grips with complex systems. Such systems are complex in both their composition—typically many different kinds of components interacting simultaneously and nonlinearly with each other and their environments on multiple levels—and in the rich diversity of behavior of which they are capable.

The Springer Series in Understanding Complex Systems series (UCS) promotes new strategies and paradigms for understanding and realizing applications of complex systems research in a wide variety of fields and endeavors. UCS is explicitly transdisciplinary. It has three main goals: First, to elaborate the concepts, methods and tools of complex systems at all levels of description and in all scientific fields, especially newly emerging areas within the life, social, behavioral, economic, neuro- and cognitive sciences (and derivatives thereof); second, to encourage novel applications of these ideas in various fields of engineering and computation such as robotics, nano-technology and informatics; third, to provide a single forum within which commonalities and differences in the workings of complex systems may be discerned, hence leading to deeper insight and understanding.

UCS will publish monographs, lecture notes and selected edited contributions aimed at communicating new findings to a large multidisciplinary audience.

More information about this series at http://www.springer.com/series/5394

Christopher Schlick • Bruno Demissie

Product Development Projects

Dynamics and Emergent Complexity

Springer

Christopher Schlick
Institute of Industrial Engineering
and Ergonomics
RWTH Aachen University
Aachen, Germany

Fraunhofer Institute for Communication,
Information Processing &
Ergonomics FKIE
Wachtberg, Germany

Bruno Demissie
Fraunhofer Institute for Communication,
Information Processing &
Ergonomics FKIE
Wachtberg, Germany

ISSN 1860-0832 ISSN 1860-0840 (electronic)
Understanding Complex Systems
ISBN 978-3-319-37354-6 ISBN 978-3-319-21717-8 (eBook)
DOI 10.1007/978-3-319-21717-8

Springer Cham Heidelberg New York Dordrecht London
© Springer International Publishing Switzerland 2016
Softcover re-print of the Hardcover 1st edition 2016
This work is subject to copyright. All rights are reserved by the Publisher, whether the whole or part of the material is concerned, specifically the rights of translation, reprinting, reuse of illustrations, recitation, broadcasting, reproduction on microfilms or in any other physical way, and transmission or information storage and retrieval, electronic adaptation, computer software, or by similar or dissimilar methodology now known or hereafter developed.
The use of general descriptive names, registered names, trademarks, service marks, etc. in this publication does not imply, even in the absence of a specific statement, that such names are exempt from the relevant protective laws and regulations and therefore free for general use.
The publisher, the authors and the editors are safe to assume that the advice and information in this book are believed to be true and accurate at the date of publication. Neither the publisher nor the authors or the editors give a warranty, express or implied, with respect to the material contained herein or for any errors or omissions that may have been made.

Printed on acid-free paper

Springer International Publishing AG Switzerland is part of Springer Science+Business Media (www.springer.com)

Acknowledgments

The development of the concepts, models, and methods presented in this book began in conversations on emergent complexity in sociotechnical systems, talks that started over a decade ago with colleagues at RWTH Aachen University and the Fraunhofer Institute for Communication, Information Processing and Ergonomics (FKIE). The early research findings were centered on the evaluation of interacting processes with discrete states and were expressed in a series of papers which were published under the umbrella of the IEEE Systems, Man, and Cybernetics Society. To support industrial companies in developing new products and help them manage associated product development projects more effectively, we shifted our focus to concurrent interactive processes with continuous states and devoted the entirety of our efforts to the development of innovative model-driven approaches for analyzing cooperative work and evaluating emergent complexity in these open organizational systems. We preferred statistical models based on Work Transformation Matrices, because they can account for unpredictable performance fluctuations and also make it possible to apply powerful estimation methods to analyze, predict, and evaluate project dynamics. These research results were regularly presented to the Design Structure Matrix Community—an interdisciplinary group consisting of researchers, practitioners, and tool developers with a common interest in matrix representations of products, processes, and organizational structures—at their annual international conferences. We thank these colleagues for the inspiring discussions and the open-minded and supportive climate at the conferences, which fostered the creativity necessary to write this book. The basic research results presented here on the modeling and simulation of cooperative work with vector autoregression models were supported by grants from the German Research Foundation (Deutsche Forschungsgemeinschaft DFG, Normalverfahren SCHL 1805/3-1 and SCHL 1805/3-3). In addition, the theoretical work on linear dynamical systems and criterion-based model selection was supported by the German Research Foundation under the Cluster of Excellence "Integrative Production Technology for High-Wage Countries." C.M. Schlick would like to thank the German Research Foundation for its kind support over all the years. Dr. Andreas Kräußling and

Prof. Eric Beutner deserve acknowledgment for reviewing and correcting the mathematical formulas related to the vector autoregressive representations in an early phase of the book project. Furthermore, Dr. Ursu deserves special recognition for providing us with the original Matlab routines that Ursu and Duchesne (2009) used in their seminal paper to fit a model to quarterly seasonally unadjusted West German income and consumption data. Finally, C.M. Schlick would like to thank his spouse, Maike Schlick, who unfailingly supported and encouraged him. Without her this book would never have seen the light of day.

Contents

Chapter 1
Introduction

1.1 Concurrent Engineering Approach and Product Development Flow

Industrial companies operating today persistently face strong competition and must adapt to rapid technological progress and fast-changing customer needs. Under these conditions, if companies want to gain a competitive advantage in global markets, they must be able to successfully develop innovative products and effectively manage the associated product development (PD) projects. To shorten time-to-market and lower development/production costs, PD projects often undergo concurrent engineering (CE). In their landmark report, Winner et al. (1988) define CE as "a systematic approach to the integrated, concurrent design of products and their related processes, including manufacture and support. This approach is intended to cause the developers, from the outset, to consider all elements of the product life cycle from conception through disposal, including quality, cost, schedule, and user requirements." A large-scale vehicle development project in the automotive industry offers a good example. In the late development stage, such a project involves hundreds of engineers collaborating in dozens of CE teams. The CE teams are usually structured according to the subsystems of the product to be developed (e.g. body-in-white, powertrain, interior systems, electronics etc.) and are coordinated by systems integration and management teams of responsible engineers who know the entire product (see e.g. Midler and Navarre 2007). Under an integrated approach to concurrent design of products and processes, multi-disciplinary teams are formed to develop recommendable configurations of the intended subsystems. These configurations should satisfy all constraints and incorporate the different types of technical expertise and methodological approaches to problem-solving needed in order for a parallel execution of work processes to be successful (Molina et al. 1995). The constraints and requirements imposed on the design by the various engineering disciplines (engineering design, production engineering, control engineering etc.) are discussed by the subject-matter experts

© Springer International Publishing Switzerland 2016

C. Schlick, B. Demissie, *Product Development Projects*, Understanding Complex Systems, DOI 10.1007/978-3-319-21717-8_1

in team meetings and are mapped onto specific design parameters in a process of intensive collaboration. To avoid unnecessary system integration problems, in CE the teamwork usually follows a continuous integration rhythm with regular team meetings that are typically held at intervals of just a few weeks. Additional team meetings, e.g. to solve time-critical or quality-critical problems and to find sound compromises for conflicting constraints that have arisen during the design process, are held as needed.

The subject-matter experts in the multi-disciplinary teams differ in terms of how they use engineering methods to solve design problems and satisfy constraints, and thus generate a variety of possible solutions from different technological perspectives and by applying different fields of knowledge (Eversheim and Schuh 2005). PD projects (and this is not limited to large-scale automotive innovation) thus involve a considerable amount of creative work in the sense of Rohmert (1983) and can display highly informative but also highly complex and difficult-to-manage patterns of project dynamics. These patterns are partly the result of pure creative thought concerning the relevant aspects of the product and its related processes from a specific technological perspective, and partly the result of systematic engineering design. The latter is inherently iterative because of the cyclic interdependencies between facets of the design problem and because loops of analysis, synthesis and evaluation are fundamental to the design process (Braha and Maimon 1997). Furthermore, the aim of the design process is to generate knowledge that reduces uncertainty and increases design maturity by incrementally defining design parameters and concretizing the product on both the functional and physical level (Feldhusen et al. 2013). During the iteration process the functions of the product and its constituent components are fully integrated into an overall systems architecture that involves numerous interfaces between mechanical, electrical and electronic modules. The product development flow (Reinertsen 2009) has frequent and sometimes irregular iterations due to the availability of new or updated information on generalized functions, physical functions, geometric/topological entities, etc. It is also important for the developer's workflows to be organized in such a way as to ensure effective and flexible forms of cooperation. For our purposes, this means that mutual agreements on design (sub)goals are developed within teams and that mutually compatible action plans are derived on the basis of those goals and are executed with a high level of agreement and without dividing the teams (Luczak et al. 2003; Mühlfelder 2003). In this context, effective cooperation between developers and production engineers is essential in order to satisfy all constraints that are necessary to ensure an efficient and timely production ramp-up (sensu Stahl et al. 2000; Mütze-Niewöhner and Luczak 2003). Another important aspect of collaboration within PD involves sharing knowledge resources within and between teams, and creating additional resources through continuous interaction between subject-matter experts (Durst and Kabel 2001; Luczak et al. 2000). Concurrent development tasks are therefore both highly variable and strongly dependent on each other and on elements of "surprise," i.e. on seemingly erratic but profoundly creative activities that are essential to design work. As Shalizi (2006) and Nicolis and Nicolis (2007) point out, this coexistence of variability and dependency is typical of complex systems in a variety of domains. In this sense, PD projects employing long-range, cross-functional coordination mechanisms can be regarded as one of the most authentic prototypes of complex sociotechnical systems. They provide inspiration for raising new issues and perspectives, and encourage further research on the sources and mechanisms of complexity.

Clearly, the higher-level cognitive processes underlying human capacities for reasoning, decision-making and problem-solving represent one of the most important sources of complexity in PD projects. The complexity of accessing the knowledge resources on the product and processes in question and the large number of interlinked technical documents involved usually overwhelm the information-processing capacity of any project manager, team leader or team member, and require many assumptions to be made about design ranges, optimal operating points and other important variables of the system under development (sensu Loch and Terwiesch 2007). During the iteration process, some assumptions invariably turn out to be wrong and must be reexamined, others turn out to be too vague and must be refined, and a few might turn out to be too rigid and must be relaxed. Furthermore, design errors of various kinds are unavoidable in such a distributed and open organizational system. Accordingly, detecting, identifying and correcting errors is an essential part of cooperative work in PD. Although most errors are processed fairly quickly and successfully in the form of ad-hoc design changes, there is an inevitable disparity between analysis and synthesis efforts, which, from the perspective of the product development flow, can lead to significant performance variability. Borrowing a statistical concept from physics, we can also speak of unpredictable performance fluctuations. These fluctuations can be found at every organizational level and are an irreducible feature of the participating organizational units' knowledge-generating and knowledge-sharing efforts, which are intended to reduce uncertainty. By definition, they are not random. Developers in the current work context often believe them to be random because the intricate mechanisms in which events unfold in time and manifest in teams are too complex to be understood by individual reasoning and conjecture. In addition to representing the negative side effects of the limited capacity for processing information, and incorporating fundamental mechanisms of error correction in PD, unpredictable performance fluctuations can be seen as essential components of creative thinking. As such, they are basic ingredients of success that should not be limited in their reach and capacity to benefit the product development flow. Nevertheless, due to the high level of individualization they often make it hard to predict and control the project as a whole, as a large body of individual knowledge concerning the development history is required to develop effective managerial interventions, and progress toward a stable design solution can differ significantly from the expected (unimpaired) process (Huberman and Wilkinson 2005). Depending on the kind and intensity of cooperative relationships, some of the development teams can fall into "vicious cycles" of multiple revisions, which entail significant unplanned and unwanted effort, as well as long delays (Huberman and Wilkinson 2005). Moreover, the revision cycles can be reinforced, and a fatal pattern of organizational dynamics termed "design churns" (Yassine et al. 2003) or "problem-solving oscillations" (Terwiesch et al. 2002; Mihm et al. 2003; Mihm and Loch 2006) can emerge. In this case, the progress of the project irregularly oscillates between being on, ahead of, or behind schedule. Ultimately, the project must be abandoned entirely to break the cycle. The design churn effect was analyzed by Terwiesch et al. (2002) and Yassine et al. (2003) in PD projects in the automotive industry. Aptly summarizing the problem, an anonymous product development manager at an automobile manufacturer commented, "We just churn and chase our tails until someone says they won't be able to make the launch date" (Yassine et al. 2003). According to the literature review by Mihm and Loch (2006), design churns occur not only in the automotive industry but also in large development projects across different domains.

Design churns are an intriguing example of emergent (or "self-generated") complexity in PD projects. They can produce disastrous results, painful financial losses and a great deal of frustration for all stakeholders concerned. The emergence is strong, in the sense that the only way to reliably anticipate critical patterns in the product development flow is to analyze the distant past of each particular instance of task processing and to have access to a large body of (mostly explanatory) knowledge on the prior history of the interacting processes (Chalmers 2002). To cope with this kind of emergent complexity, a deeper understanding of the interrelationships between unpredictable performance fluctuations and project dynamics is needed, as are new methods for analyzing and evaluating quantitative complexity.

1.2 Goals and Structure of this Book

This book continues the tradition of research works produced by RWTH Aachen University's Institute of Industrial Engineering and Ergonomics on the organizational and ergonomic analysis of work processes in concurrent product and process development. To gain a detailed understanding of the interrelationships between unpredictable performance fluctuations and the dynamics of concurrent development in an open organizational system, the book's first goal is to present different mathematical models of cooperative work on the basis of the theory of stochastic processes. To promote the development of new methods for analyzing and evaluating quantitative complexity, its second goal is to introduce an information-theoretical complexity measure that is underpinned by a convincing complexity theory from the field of theoretical physics (Grassberger 1986; Bialek et al. 2001) and therefore makes it possible to quantify strong emergence in terms of mutual information between past and future histories of interacting processes. The complexity measure is a key invariant of generalized stochastic processes; as such, its formulation makes the same contribution to theory as the famous entropy rate, which was discovered and popularized much earlier. To model cooperative work in an open PD environment, we focus on the development tasks and their interactions in the product development flow and assume that additional dependencies related to the design product and design problem domains (see Summers and Shah 2010) were integrated into a joint model. This approach is supported by the practical complexity definition put forward by Tatikonda and Rosenthal (2000), who define project complexity as the nature, quantity and magnitude of organizational subtasks and subtask interactions within a certain project. The challenge is that, even if the breakdown and dependency structures of the development tasks, the rate of processing, the variability in processing and the rules of interaction are given, it is difficult to anticipate the performance of the project as a whole. Self-reinforcing feedback processes can exacerbate performance fluctuations and generate effects that cannot simply be reduced to properties of the constituent components of the product development flow. Instead, these phenomena emerge from higher-order interactions and can be considered properties of the organization as a whole (Huberman and Wilkinson 2005). The combined approach of providing mathematical models of cooperative work in an open PD environment as well as information-theoretic complexity measures builds on our previous work on project simulation and the management of dynamic complexity (see e.g. Schlick et al. 2007, 2008, 2009, 2012, 2013a, b, 2014, 2015; Petz et al. 2015; Tackenberg et al. 2009, 2010). However, it differs from said work in terms of how discrete-time, continuous-state models with different internal configurations for gaining a complete description of the work processes are formulated, and in that the parameterized models are consistently validated in different validation studies. Furthermore, we present results of mathematical analyses of emergent complexity and formulate closed-form solutions of different strengths that can be used to identify the variables that are fundamental in shaping complexity, making it possible to carry out a theory-driven model selection and to optimize the organization design of the project at hand.

The structure of this book is derived from a theoretical framework on the dynamic complexity analysis of the product development flow: Chapter 2 lays the foundations for deterministic and stochastic modeling of cooperative work in PD projects, and formulates the state equations that are needed to model the time evolution of the amount of work done in the iteration process. We follow the principle of successive refinement and begin by formulating a simple linear first-order difference equation based on the seminal work that Smith and Eppinger (1997) did on Work Transformation Matrices. These matrices can be regarded as a task-oriented variant of the popular Design Structure Matrix (see, e.g. Steward 1981; Lindemann et al. 2009; and Eppinger and Browning 2012) and as such can be easily interpreted in terms of structure and parameters. Following Huberman and Wilkinson (2005), the deterministic model formulation is generalized towards the theory of stochastic processes. This means performance variability in the iteration process is represented by continuous-valued random variables and interpreted as unpredictable performance fluctuations. Viewing organizational systems in this way also allows us to apply powerful estimation methods to analyze, predict and evaluate project dynamics. To model cooperative work in PD projects, we consider two classes of models. The basic class comprises vector autoregression models of finite order, which can capture the typical cooperative processing of the development tasks with short iteration length. For a first-order model, a transformation into the spectral basis is performed in order to uncover the essential dynamic mechanisms. An augmented state-space formulation also makes it possible to represent more complex autoregressive processes with periodically correlated components. These processes incorporate a hierarchical coordination structure and can therefore also be used to simulate the long-term effects of intentionally withholding design information to improve the implementation of the product architecture. In addition to autoregression models, Chapter 2 considers the theoretically interesting class of linear dynamical systems with additive Gaussian noise. With this class of models, the state of the project is only partially observable. In other words, regular observations of the work processes required to complete a particular development activity are correlated with the state of the project, but are insufficient to precisely determine this state. In this sense, a kind of hidden-state process of cooperative development is distinguished from the observation process in the product development flow. As the term suggests, the state variables cannot be directly observed and must be inferred through a causal model from other variables that are measured directly. This fundamental uncertainty in the project state and its evolution can lead to a non-negligible degree of long-term correlations between development activities and can therefore significantly increase emergent complexity. In addition to the mathematical models, we also introduce least squares and maximum likelihood estimation methods to show how the independent parameters can be efficiently estimated from time series of task processing. These methods can be very useful for applying the stochastic models in different project phases. The estimation methods also help achieve a deeper understanding of the interrelationships between performance variability and project dynamics. A case study was carried out in a German industrial company to validate the models with field data. The validation results are presented and discussed in detail in separate chapters. Chapter 3 provides a project management-based review of various complexity frameworks, theories and measures that have been developed in organizational theory, systematic engineering design and basic scientific research on complex systems. We analyze an information-theoretic quantity—called the effective measure complexity—in detail because of its outstanding construct validity and conceptual advantages for evaluating emergent complexity in the field of application. As mentioned above, the measure also stands out as a key invariant of stochastic processes. The stochastic models developed in Chapter 2 also make it possible to calculate the effective measure complexity and present closed-form solutions with different numbers of independent parameters. These solutions are derived and discussed in Chapter 4. The calculations are carried out in considerable detail for both model classes. We use different coordinate systems to formulate solutions of different expressiveness and with different structural richness. Simplified polynomial-based solutions for first-order models representing the processing of two and three development tasks in the spectral basis are presented to clarify the complicated interrelationships between the individual parameters. We also put upper and lower bounds on the effective measure complexity for first-order models, so as to support the interpretation of emergent complexity in the sense of the measure. For the broader class of linear dynamical systems, we calculate an explicit solution of the effective measure complexity as an original contribution to the theory of linear stochastic systems. Because of the complicated internal structure of this solution, we also present implicit formulations that are much easier to interpret and can be directly applied. Following a comprehensive and unified treatment of emergent complexity in PD projects based on mathematical models of cooperative work and application of information-theoretic methods, Chapter 5 focuses on the validity of selected closed-form complexity solutions that were obtained for vector autoregression models as the basic model class. In terms of methodology, we follow classic validity theory and evaluate the criterion-related validity. In the validation studies we investigate project organization forms in which the developers directly cooperate and also analyze work processes that have periodically correlated components due to a hierarchical coordination structure. Furthermore, in the strict sense of the concept of criterion-related validity we investigate whether it is possible to obtain a valid formulation of a model selection criterion that is based on the effective measure complexity and can be used to identify an optimal model order within the class of vector autoregression models from data. By means of this additional theoretical contribution our aim is to focus not exclusively on PD environments in the validation studies but also to find a criterion with universal reach in model selection tasks. This complexity-based criterion is the subject of the first validation study in Section 5.1. To formulate the criterion, we follow the principles of model selection developed by Li and Xie (1996) and generalize their solution to vector-valued autoregressive processes. The corresponding validation studies are based on two Monte Carlo experiments. These experiments have the same overall objective of comparing the accuracy of the complexity-based criterion with standard model selection criteria like the Akaike information criterion and the Schwarz-Bayes criterion. It is hypothesized that model selection based on the effective measure complexity makes it possible to select the true model order with a high degree of accuracy and that the probabilities for the selected model orders are not significantly different from the distribution obtained under the alternative criteria. The parametric models evaluated in the Monte Carlo experiments are not only derived from field data from the cited industrial company, but are also synthetically generated in order to allow a systematic comparison of the different criteria. Sections 5.2 and 5.3 provide a more practical explanation of the theoretical considerations of emergent complexity by using applied examples of optimizing project organization. To conduct these validation studies, we systematically manipulate different independent variables related to project organization forms in additional Monte Carlo experiments to see how the levels of these variables affect emergent complexity and whether it is possible to derive meaningful and useful organizational design recommendations. The study in Chapter 5.2 has the objective of designing cooperative work with minimal emergent complexity by selecting the optimal staffing of three concurrent engineering teams using developers with different levels of productivity in a simulated PD project. We hypothesize that for large productivity differences, "productivity balancing" at the team level minimizes emergent complexity. Productivity balancing is a self-developed concept for systematically designing interactions between humans, tasks and products that views performance fluctuations as an opportunity to innovate and learn (Schlick et al. 2009). The objective of the final study in Section 5.3 is to optimize the period for minimal emergent complexity in which information on integration and tests of components is deliberately withheld by subsystem-level teams and not directly released to component-level teams. This kind of non-cooperative behavior in a multi-level hierarchical coordination structure aims at improving solution maturity and reducing coordination efforts. In both studies, we formulate and solve constrained and unconstrained optimization problems. For constrained optimization, we consider the total amount of work done in all tasks during the iteration process as the major constraint. Chapter 6 draws the main conclusions and provides a brief outlook for future research.

1.3 Notation

Throughout this book we will use the following mathematical notation: $A_{:i}$ denotes the i-th column of the matrix A. A^{T} is the transpose. We use the normal font style for the superscript T in order to discriminate the transpose from the variable T indicating the time instant $T \in \mathbb{N}$ in the work processes. A^{*} denotes the conjugate of A. The conjugate matrix is obtained by taking the complex conjugate of each element of A. The inverse of A is denoted by A^{-1}. The elements of a matrix are either written as sub-scripted, *non-bold lower-case* letters, e.g. a_{ij}, or are indexed by i and j as $A_{[\![i,j]\!]}$. The index form stems from the notation of the Mathematica® modeling and simulation environment. Although quite unusual, additional operations on matrices and vectors begin with the capital letter of the operation, and the argument is written in square brackets, e.g. Arg[.], E[.], Cov[.], Corr[.], Det[.], Diag[.], Exp[.], Eig[.], Erf[.], SVD[.], Tr[.], Total[.], Var[.], Vol[.]. This representation style is also derived from the Mathematica® modeling and simulation environment. Similarly, the linear algebraic product of matrices, vectors or vector/matrices is written explicitly, for instance as $A \cdot A = A^2$ for the product of two matrices, and $\{a\} \cdot A$ for the multiplication of a scalar a with a matrix A. This rule is only violated if the terms grow too long and their meaning is clear from the context, e.g. in Sections 2.7, 2.9 and 4.2. An identity matrix of size n is denoted by the symbol I_n. A zero column vector with n components is denoted by 0_n. A continuous-type or discrete-type random state variable is denoted by a Latin capital letter, e.g. X. An observed value (realization) of a random state variable is indicated by a lower-case letter symbol, e.g. x. A random variable that represents unpredictable fluctuations is denoted by a lower-case Greek letter, e.g. ε. The symbol $\sim$ means that a random variable is distributed according to a certain probability distribution, e.g. $\varepsilon \sim \mathcal{N}(\mu, \Sigma)$. A multivariate Gaussian distribution with location (mean) μ and covariance matrix Σ is written as $\mathcal{N}(\mu, \Sigma)$. The corresponding probability density function with parameter vector $\theta = (\mu, \Sigma)$ is denoted by $f_\theta[x] = \mathcal{N}(x; \mu, \Sigma)$. Equations that use or generate a time series include a time index for the state variables and the fluctuations, e.g. x_t or ε_t. The complete stochastic state process is written as $\{X_t\}$. Finite process fragments $(X_{t_1}, X_{t_1+1}, \ldots, X_{t_2})$ from time step $t_1 \in \mathbb{Z}$ to $t_2 \in \mathbb{Z}$ are written as $X_{t_1}^{t_2}$. Similarly, the term $x_{t_1}^{t_2} = (x_{t_1}, x_{t_1+1}, \ldots, x_{t_2})$ denotes the sequence of states that was observed across the same interval of time steps.

References

Bialek, W., Nemenman, I., Tishby, N.: Predictability, complexity and learning. Neural Comput. **13** (1), 2409–2463 (2001)

Braha, D., Maimon, O.: The design process: Properties, paradigms, and structure. IEEE Trans. Syst. Man Cybern. **27**(2), 146–166 (1997)

Chalmers, D.J.: Strong and weak emergence. In: Clayton, P., Davies, P. (eds.) The Re-emergence of Emergence, pp. 244–256. Oxford University Press, Oxford (2002)

Durst, R., Kabel, D.: Cross-functional teams in a concurrent engineering environment—principles, model, and methods. In: Beyerlein, M., Johnson, D., Beyerlein, S. (eds.) Virtual teams, pp. 163–210. Elsevier Science Ltd., London (2001)

Eppinger, S.D., Browning, T.: Design structure matrix methods and applications. MIT Press, Cambridge, MA (2012)

Eversheim, W., Schuh, G.: Integrierte Produkt- und Prozessgestaltung. Springer, Berlin (2005) (in German)

Feldhusen, J., Grote, K.-H., Nagarajah, A., Pahl, G., Beitz, W., Wartzack, S.: Vorgehen bei einzelnen Schritten des Produktentstehungsprozesses (in German). In: Feldhusen, J., Grote, K.-H. (eds.) Pahl/Beitz Konstruktionslehre—Methoden und Anwendung erfolgreicher Produktentwicklung, 8th edn, pp. 291–410. Springer, Berlin (2013)

Grassberger, P.: Toward a quantitative theory of self-generated complexity. Int. J. Theor. Phys. **25** (9), 907–938 (1986)

Huberman, B.A., Wilkinson, D.M.: Performance variability and project dynamics. Comput. Math. Organ. Theor. **11**(4), 307–332 (2005)

Li, L., Xie, Z.: Model selection and order determination for time series by information between the past and the future. J. Time Ser. Anal. **17**(1), 65–84 (1996)

Lindemann, U., Maurer, M., Braun, T.: Structural complexity management. An approach for the field of product design. Springer, Berlin (2009)

Loch, C.H., Terwiesch, C.: Coordination and information exchange. In: Loch, C., Kavadias, S. (eds.) Handbook of New Product Development Management, pp. 315–343. Butterworth Heinemann, Amsterdam (2007)

Luczak, H., Wimmer, R., Kabel, D., Durst, R. (2000). What engineers do learn from team effectiveness models: An investigation of applicability and utility of team effectiveness models in production systems. In: Proceedings of the 7th International Conference Human Aspects of Advanced Manufacturing: Agility and Automation, HAAMAHA, pp. 1–8, (2000)

Luczak, H., Mühlfelder, M., Schmidt, L.: Group task analysis and design of computer supoorted cooperative work. In: Hollnagel, E. (ed.) Handbook of Cognitive Task Design, pp. 99–127. Lawrence Erlbaum Associates, Mahwah, NJ (2003)

Midler, C., Navarre, C.: Project management in the automotive industry. In: Morris, P.W.G., Pinto, J.K. (eds.) The Wiley Guide to Managing Projects, pp. 1368–1388. Wiley, New York, NY (2007)

Mihm, J., Loch, C.: Spiraling out of control: Problem-solving dynamics in complex distributed engineering projects. In: Braha, D., Minai, A.A., Bar-Yam, Y. (eds.) Complex Engineered Systems: Science Meets Technology, pp. 141–158. Springer, Berlin (2006)

Mihm, J., Loch, C., Huchzermeier, A.: Problem-solving oscillations in complex engineering. Manag. Sci. **46**(6), 733–750 (2003)

Molina, A., Al-Ashaab, A., Ellis, T.I.A., Young, R.I.M., Bell, R.: A review of computer aided simultaneous engineering systems. Res. Eng. Design **7**(1), 38–63 (1995)

Mühlfelder, M.: Das kollektive Handlungsfeld—Ein psychologisches Konzept zur Modellierung interpersonal koordinierten Handelns (in German). Ph.D. Thesis, Europa-Universität Flensburg, (2003)

Mütze-Niewöhner, S., Luczak, H.: Prospective job design and evaluation in early stages of production system design. In: Proceedings of the Seventh International Symposium on Human Factors in Organizational Design and Management, ODAM 2003, pp. 323–328, (2003)

Nicolis, G., Nicolis, C.: Foundations of Complex Systems—Nonlinear Dynamics, Statistical Physics, Information and Prediction. World Scientific, Singapore (2007)

Petz, A., Schneider, S., Duckwitz, S., Schlick, C.M.: Modeling and simulation of service systems with design structure and domain mapping matrices. J. Mod. Proj. Manag. **3**(3), 65–71 (2015)

Reinertsen, D.G.: The Principles of Product Development Flow—Second Generation Lean Product Development. Celeritas Publishing, Redondo Beach, CA (2009)

Rohmert, W.: Formen menschlicher Arbeit (in German). In: Rohmert, W., Rutenfranz, J. (eds.) Praktische Arbeitsphysiologie. Georg Thieme Verlag, Stuttgart, New York (1983)

Schlick, C.M., Beutner, E., Duckwitz, S., Licht, T.: A complexity measure for new product development projects. Proceedings of the 19th International Engineering Management Conference, pp. 143–150, (2007)

Schlick, C.M., Duckwitz, S., Gärtner, T., Schmidt, T.: A complexity measure for concurrent engineering projects based on the DSM. In: Proceedings of the 10th International DSM Conference, pp. 219–230, (2008)

Schlick, C.M., Duckwitz, S., Gärtner, T., Tackenberg, S.: Optimization of concurrent engineering projects using an information-theoretic complexity metric. Proceedings of the 11th International DSM Conference, pp. 53–64, (2009).

Schlick, C.M., Schneider, S., Duckwitz, S.: Modeling of cooperative work in concurrent engineering projects based on extended work transformation matrices with hidden state variables. In: Proceedings of the 14th International Dependency and Structure Modeling Conference, DSM 2012, pp. 411–422, (2012).

Schlick, C.M., Schneider, S., Duckwitz, S.: A universal complexity criterion for model selection in dynamic models of cooperative work based on the DSM. In: Proceedings of the 15th International Dependency and Structure Modeling Conference, DSM 2013, pp. 99–105, (2013a).

Schlick, C.M., Duckwitz, S., Schneider, S.: Project dynamics and emergent complexity. Comput. Math. Organ. Theor. **19**(4), 480–515 (2013b)

Schlick, C.M., Schneider, S., Duckwitz, S.: Estimation of work transformation matrices for large-scale concurrent engineering projects. In: Proceedings of the 16th International Dependency and Structure Modeling Conference, DSM 2014, pp. 211–221, (2014).

Schlick, C.M., Schneider, S., Duckwitz, S.: Estimation of work transformation matrices for large-scale concurrent engineering projects. J. Mod. Proj. Manag. **3**(3), 73–79 (2015)

Shalizi, C.R.: Methods and techniques of complex systems science: An overview. In: Deisboeck, T.S., Kresh, J.Y. (eds.) Complex Systems Science in Biomedicine, pp. 33–114. Springer, New York (2006)

Smith, R.P., Eppinger, S.D.: Identifying controlling features of engineering design iteration. Manag. Sci. **43**(3), 276–293 (1997)

Stahl, J., Mütze, S., Luczak, H.: A method for job design in concurrent engineering. Hum. Fact. Ergon. Manuf. **10**(3), 291–307 (2000)

Steward, D.V.: The design structure system: A method for managing the design of complex systems. IEEE Trans. Eng. Manag. **28**(3), 71–74 (1981)

Summers, J.D., Shah, J.J.: Mechanical engineering design complexity metrics: Size, coupling, and solvability. J Mech. Des. **132**(2), 1–11 (2010)

Tackenberg, S., Duckwitz, S., Kausch, B., Schlick, C.M., Karahancer, S.: Organizational simulation of complex process engineering projects in the chemical industry. J. Universal Comp. Sci. **15**(9), 1746–1765 (2009)

Tackenberg, S., Duckwitz, S., Schlick, C.M.: Activity- and actor-oriented simulation approach for the management of development projects. Int. J. Comp Aided Eng. Technol. **2**(4), 414–435 (2010)

Tatikonda, M.V., Rosenthal, S.R.: Technology novelty, Project complexity and product development project execution success. IEEE Trans. Eng. Manag. **47**, 74–87 (2000)

Terwiesch, C., Loch, C.H., De Meyer, A.: Exchanging preliminary information in concurrent engineering: Alternative coordination strategies. Organ. Sci. **13**(4), 402–419 (2002)

Winner, R.I., Pennell, J.P., Bertrand, H.E., Slusarczuk, M.M.: The role of concurrent engineering in weapons system acquisition. IDA-Report R-338, Institute for Defense Analyses; Alexandria, VA, (1988)

Yassine, A., Joglekar, N., Eppinger, S.D., Whitney, D.: Information hiding in product development: The design churn effect. Res. Eng. Des. **14**(3), 145–161 (2003)

Chapter 2
Mathematical Models of Cooperative Work in Product Development Projects

The Merriam-Webster dictionary defines the word "project," a fundamental term in industrial engineering and engineering management, as "a planned piece of work that has a specific purpose (such as to develop a new product or introduce a new manufacturing technology) and that usually requires a lot of time." The word stems from the Middle English *projecte*, from the Medieval Latin *projectum*, and from the Latin neuter form of *projectus*, past participle of *proicere* to 'throw forward', from pro- + jacere 'to throw'. Cambridge Dictionaries Online defines a project as a "piece of planned work or an activity that is finished over a period of time and intended to achieve a particular aim." It defines project management as "the activity of organizing and controlling a project."

Based on these definitions it is clear that project management originated when people started to plan, organize and control pieces of work that had a specific purpose and were intended to achieve a particular aim. Examples from ancient history include the pyramids of Egypt, the Great Wall of China, the temple in Jerusalem and the Angkor Wat complex of temples in Cambodia. Modern examples include product development (PD) projects, construction projects, reorganization projects, and many more. Although humans have been pursuing projects for thousands of years, published literature on project management methodologies is a comparatively new phenomenon. The early work of Henry Gantt, published in the beginning of the twentieth century, dealt with the scheduling of activities in the job shop. His Gantt chart was subsequently adopted for scheduling project activities. Its simplicity and excellent visualization make it a popular tool even today, a century after its invention.

Following the advent of tools and techniques for operations research in the mid twentieth century, mathematical models were developed to support project management. Early tools like the Critical Path Method (CPM; see Kelley and Walker 1959) were based on purely deterministic techniques. By focusing on the longest sequence of activities connecting the start of a project to its end, the CPM method identifies the critical activities and the slack of non-critical activities. Each critical

© Springer International Publishing Switzerland 2016
C. Schlick, B. Demissie, *Product Development Projects*, Understanding Complex Systems, DOI 10.1007/978-3-319-21717-8_2

path problem can be converted into a linear programming problem and solved by standard mathematical methods. This also permits the inclusion of time–cost relationships (Salmon 1962). Stochastic tools like the Program Evaluation and Review Technique (PERT; see Fazar 1959), also developed in the mid twentieth century, were designed to deal with the uncertain nature of projects and the associated risk. The development of powerful, inexpensive computers in the 1970s encouraged not only the development of CPM- and PERT-based software but also the use of computers to allocate project work among participants by means of the Work Breakdown Structure model (see e.g. Shtub et al. 2004) and the sharing of project information between stakeholders with the help of distributed databases and computer networks. Computer technology made it possible to collect, save, retrieve and analyze large volumes of data and to support the management of the project scope (the work that is to be done), the product scope (features, functions and components of the product to be developed), the management of resources needed for the project (availability of resources, resource required to perform the work), and to provide cost information. Modern project management software supports decision-making by offering scheduling algorithms, resource allocation and resource-leveling algorithms, budgeting and cost-management algorithms, as well as monitoring and control techniques. Computing power also made it possible to analyze risk with Monte Carlo simulations. Early simulations focused on project scheduling risks and sought to estimate the probability that a project activity would be on the critical path and the probability that the project would be finished by a given date. System dynamics models were used to simulate the behavior of projects by modeling cause-effect relationships within feedback loops (see e.g. Sterman 2000). These continuous time, continuous state-space models supported the analysis of different strategies and their impact on project duration, cost and performances (see e.g. Williams 2002).

Published research (Cooper 2011) highlights the critical success drivers in PD projects:

- A unique and superior product—a differentiated product that delivers unique benefits and a compelling value proposition to the customer or user.
- Building in the voice of the customer—a market-driven and customer-focused development process.
- Doing the homework and front-end loading the project.
- Getting product and project definitions early on—and avoiding scope creep and unstable specs.
- Spiral development—build, test, obtain feedback, and revise—putting something in front of the customer early and often.
- The world product—a global product (global concept, locally tailored) targeted at international markets.
- A well-conceived, properly executed launch with a solid marketing plan at the heart of the launch.
- Speed counts—accelerate development projects, but not at the expense of quality of execution.

These eight critical success drivers are centered around one aspect—effective cooperative work within and between the development teams. As already pointed out in the introductory chapter, the challenge is that even when key aspects like the work breakdown structure, task scheduling, integration rhythm, available resources for processing the tasks and organizational structures are clearly defined, due to the intrinsic performance variability it is usually very difficult to ensure a focused and fast development process that leads to a high-quality product. Depending on the structure and intensity of interactions in the development process, self-reinforcing feedback processes can cause continual revisions and lead to significant additional work and long delays (Huberman and Wilkinson 2005) or to the cited design churns and problem-solving oscillations as fatal patterns of project dynamics (Yassine et al. 2003; Mihm et al. 2003; Mihm and Loch 2006), in which the project irregularly oscillates between being on, ahead of, or behind schedule. These success-critical phenomena emerge from higher-order interactions and must be addressed at the level of PD project organization as a whole (Huberman and Wilkinson 2005).

The following chapters present different mathematical models of cooperative work in an open PD environment. These models are based on structuring the PD process into phases and integrating the (concurrent) design and engineering activities into a coherent and comprehensive PD "funnel" with clearly defined process elements (subphases). A corresponding PD funnel model developed by Hauser (2008) at the MIT Center for Innovation in Product Development (CIPD) illustrates the main phases in Fig. 2.1. The funnel model adopts the stages of opportunity identification and idea generation, concept development and selection, design and engineering, testing and launch used by Urban and Hauser (1993) in their classic textbook. The key management concepts are that it is much less expensive to screen products in the early phases than in the later phases and that each phase can improve product functionality, product quality and positioning in the market so that the likelihood of

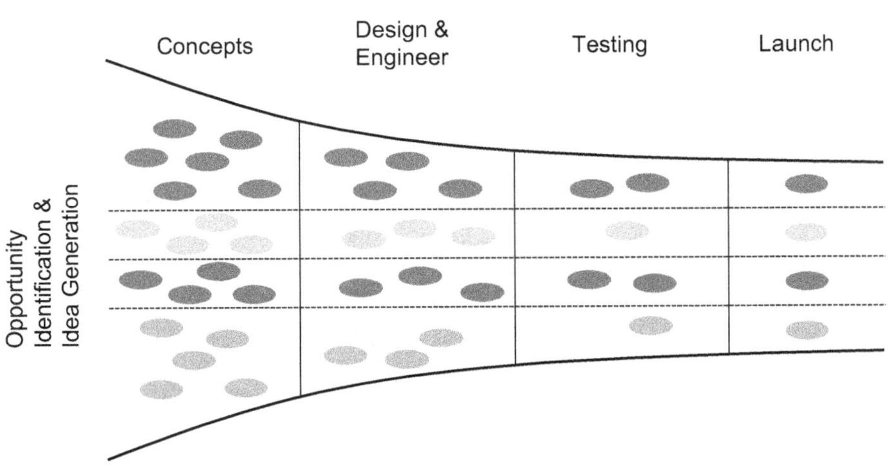

Fig. 2.1 Product development funnel model according to Hauser (2008)

success increases (Urban and Hauser 1993). The funnel model in Fig. 2.1 also illustrates the concept of pipeline management, in which multiple, parallel sets of projects move through the company's development funnel.

In Hauser's product development funnel model and similar models (see e.g. Wheelwright and Clark 1992; McGrath 1996; Katz 2011), a particular set of components in the work breakdown structure of a single system has to be processed entirely in parallel and, theoretically, no task is processed independently of the others, since, to arrive at completion, all tasks regularly require information on the state of system functions or components under development. In this sense, the tasks are fully interdependent; for instance the tasks for the design of the main functions and basic components of a new product in the design and engineering phase (Fig. 2.1). Following Puranam et al. (2011), we say that two tasks are interdependent when, within the project, the value generated by performing one of them differs according to whether or not the other task is performed or not. Given the number, strength and structure of the interdependencies, these phases are often critical to project success.

2.1 Deterministic Formulation

To analyze the interrelationships between project dynamics and emergent complexity explicitly and from a special complexity-theoretical perspective, a dynamic model of cooperative work in an open product development environment has to be formulated and the independent parameters have to be defined. We begin with the deterministic formulation of a continuous-state, discrete time model based on the seminal work of Smith and Eppinger (1997), according to which a distinct phase of the PD project life cycle with p parallel and interacting development tasks can be modeled by a linear first-order difference equation as

$$x_t = A_0 \cdot x_{t-1} \quad t = 1, \ldots, T . \tag{1}$$

In PD it is generally desirable to process tasks in parallel so as to reduce the overall development time and to get the product to market earlier (Cooper 2011). The above state equation is also termed a linear homogeneous recurrence relation. The p-dimensional state vector $x_t \in \mathbb{R}^p$ represents the work remaining for all p tasks at time step t. Smith and Eppinger (1997) simply speak of the "work vector" x_t. It is assumed that the work vector is observed (or estimated) by the project manager at equally spaced time instants and therefore that time can be indexed by the discrete variable t. The amount of work remaining in the phase can be measured by the time left to finalize a specific design, by the definable labor units required to complete a particular development activity or component of the work breakdown structure, by the number of engineering drawings requiring completion before design release, by the number of engineering design studies required before design release, or by the number of issues that still need to be addressed/resolved before design release (Yassine et al. 2003). The matrix $A_0 = (a_{ij})$ is a dynamical operator for the iteration over all p tasks, also called the "work transformation matrix" (WTM). The WTM is a square real matrix of dimension $p \times p$, i.e. $A_0 \in \mathbb{R}^{p \times p}$. The WTM can be regarded as a task-oriented variant of the popular design structure matrix (Steward 1981), which is often used in industry and academia to analyze and optimize complex products. It enables the project manager to model, visualize and evaluate the dependencies between the development tasks and to derive suggestions for improvement or reorganization. It is clear that not only the tasks to be processed but also the structure of the product (in terms of an envisioned physical and functional solution) and the formalized design problem are important in meeting the project goals and satisfying the functional and nonfunctional requirements. However, for the sake of simplicity, in the following we focus on the tasks and their interactions and assume that additional dependencies from other domains were integrated into a joint work transformation model. Given a distinct phase of a PD project, it is assumed that the WTM does not vary with time, and that the state equation is autonomous.

In this book, we use the improved WTM concept of Yassine et al. (2003) and Huberman and Wilkinson (2005). Hence, the diagonal elements a_{ii} $(i = 1 \ldots p)$ account for developers' different productivity levels when processing tasks. This contrast with the original WTM model by Smith and Eppinger (1997), in which tasks are processed at the same rate. The diagonal elements a_{ii} are defined as autonomous task processing rates (Huberman and Wilkinson 2005, who also speak of autonomous task completion rates). They indicate the ratio of work left incomplete after and before an iteration over task i, under the assumption that the tasks are processed independently of the others. Therefore, the autonomous task processing rates must be nonnegative real numbers $(a_{ii} \in \mathbb{R}^{+})$. The off-diagonal elements $a_{ij} \in \mathbb{R}\,(i \neq j)$, however, model the informational coupling between tasks and indicate the intensity and nature of cooperative relationships between developers. Depending on their values, they have different meanings: (1) if $a_{ij} = 0$, work carried out on task j has no direct effect on task i; (2) if $a_{ij} > 0$, work on task j slows down the processing of task i, and one unit of work on task j at time step t generates a_{ij} units of extra work on task i at time step $t + 1$; (3) if $a_{ij} < 0$, work on task j accelerates the processing of task i, and one unit of work on task j reduces the work on task i by a_{ij} units at time step $t + 1$. The only limitation on the use of negative entries is that negative values of the work remaining in the state vector x_t are not permitted at any time instant. In practice, many off-diagonal elements must be expected to be nonnegative, because PD projects usually require intensive cooperation, leading to additional work. For instance, Klein et al. (2003) analyzed the design of the Boeing 767 and found that half of the engineering labor budget was spent on redoing work because the original work did not yield satisfactory results. Roughly 25%–30% of the design decisions required reworking, and in some instances up to 15 iterations had to be done to reach a stable design state.

Following Smith and Eppinger (1997), whose concept lies behind the formulation of state Eq. 1, this book focuses on the design and engineering phase of Hauser's product development funnel model (see Fig. 2.1), where a particular set of components in the work breakdown structure related to detailed design and engineering development of the new product, along with simple product tests and the development of a production plan, have to be processed completely in parallel and are fully interdependent.

In general, we do not recommend modeling cooperative work on the basis of linear recurrence relations in the preliminary phases of opportunity identification, idea generation, concept generation and concept evaluation. This is because the work processes related to this "fuzzy front end," to use Katz's terminology (Katz 2011), are often weakly structured and highly nonlinear. However, this recommendation does not affect the applicability of the model to concurrent product and process design. In fact, it is sometimes also possible to build models with a similar structure, estimate their parameters and use the parametric representations to predict performance for earlier subphases of interest covering the conceptual system design. This will be demonstrated in Section 2.5 for a PD project executed at a small industrial company in Germany that develops mechanical and electronic sensor components for the automotive industry (Schlick et al. 2008, 2012). Furthermore, one can also build corresponding models for later subphases of interest, e.g. when launching the product.

In our way of structuring time, the initial time step $t = 0$ indicates the beginning of the detailed design or engineering development phase with fully interdependent tasks. For instance, if detailed design is modeled by state Eq. 1, the initial time step usually denotes the start of the set of activities that are carried out concurrently to describe a product through solid modeling and drawings so that the external dimensions are specified and the materials, packaging, test and reliability requirements are met. The end of the project phase of interest is indicated by time instant $T \in \mathbb{N}$. Decomposing the whole process into distinct phases with fully interdependent tasks may seem rather inconvenient, but does not limit the generality of the approach. To model and analyze a generalized product development process architecture in which the development tasks can have arbitrary serial and parallel interconnections, we simply have to decompose the whole process into serially concatenated subphases of fully interdependent parallel subtasks. This can be done by splitting the overlapping tasks in such a way that the partial overlaps can be assigned to the subphases in a one-to-one relation. For two overlapping tasks, for instance, in which the first tasks leads to the second, three serially concatenated subphases are required to model the process. In the first subphase only the processing of the non-overlapping portion of the first task is modeled. In the second subphase the simultaneous processing of the overlapping portions of both tasks is represented, while the third subphase captures only the processing of the portion of the second task that does not overlap with the first. The corresponding autonomous task processing rates and the coupling strengths then have to be redistributed (Murthy et al. 2012). The serially concatenated subphases can be modeled by separate WTMs in conjunction with separate initial states that are executed subphase-by-subphase (Smith and Eppinger 1998). Separate initial states unambiguously define the interfaces between phases because of the linear first-order recurrence relation. In the case of higher-order interrelations in task processing, the states have to be recoded in order to define the interfaces (see Section 2.4). As shown by Murthy et al. (2012), Smith and Eppinger's (1997) classic WTM model can be extended towards this kind of generalized product development process architecture. Their results can easily be applied to the improved WTM concept of Yassine et al. (2003) and Huberman and Wilkinson (2005) and the above model formulation generalized to formulate extended work transformation matrices and initial work vectors. For small PD projects, the time index T can also cover the total time spent to complete the project (see validation studies in Sections 5.2 and 5.3). Moreover, for PD projects that undergo major reorganization, separate initial states and WTMs can be defined. The analysis would then apply separately to each reorganized phase of the project (Huberman and Wilkinson 2005).

In the literature (see e.g. Smith and Eppinger 1997; Yassine et al. 2003; Huberman and Wilkinson 2005; and Schlick et al. 2007) the initial state x_0 is often taken to be a vector of ones, i.e.

$$x_0 = \begin{pmatrix} 1 \\ \vdots \\ 1 \end{pmatrix}, \tag{2}$$

which defines a relative scale that measures how much work still has to be done on the development tasks. Using this scale, it is very easy to model overlapping tasks by setting the work remaining for the corresponding vector components to values less than one (see the parameter example in Eq. 73). Alternatively, the project manager can define an absolute scale and assign other nonnegative values to vector components of x_0 to indicate the absolute number of work units needed to complete the tasks. The scale that measures how much work is left then determines the autonomous task processing rates and the off-diagonal elements of the WTM A_0, and can therefore have significant effects on performance evaluation. The transformation between relative and absolute scales can be carried out using the linear transformation $x'_t = W \cdot x_t$, where $W = \left(w_{ij}\right)$ is a diagonal matrix of the absolute work units $w_{ii} \in \mathbb{R}^+$ related to task i (cf. diagonal matrix of task times introduced by Smith and Eppinger 1997, 1998). Depending on the intensity and nature of the cooperative relationships between developers, the linear transformation can produce different values for the total time spent to complete the project phase (see Eq. 5). It is therefore important to choose the scale carefully and to use it consistently to calculate the dependent and independent parameters (cf. Section 2.4).

Owing to the fact that the cooperative task processing modeled by the WTM A_0 in conjunction with state Eq. 1 is a linear iteration, the work remaining x_t at time step t can also be expressed as a multiple of the initial state x_0:

$$x_t = A_0^t \cdot x_0. \tag{3}$$

From the theory of linear dynamic systems we know that the rate and nature of convergence of the task processing are determined by the eigenmodes of the

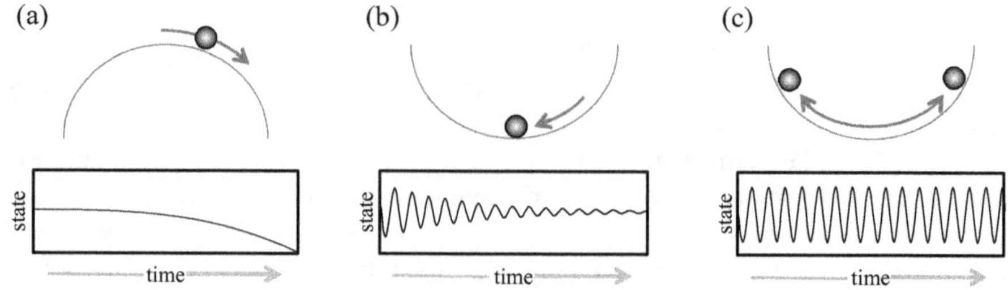

Fig. 2.2 Types of system equilibria. (**a**) Unstable equilibrium: the state vector rapidly moves away from the equilibrium point when the system is perturbed. (**b**) Asymptotically stable equilibrium: the state vector returns to the original equilibrium point when perturbed. (**c**) Stable equilibrium: the perturbed state vector oscillates interminably around the equilibrium point ("no resistance") (adopted from Boots 2009)

dynamical operator A_0. Following Smith and Eppinger (1997), we use the term "design mode" $\phi_i = (\lambda_i(A_0), \vartheta_i(A_0))$ to refer to an eigenvalue $\lambda_i(A_0)$ inherent to A_0 associated to its eigenvector $\vartheta_i(A_0)$ $(1 \leq i \leq p)$. Strictly speaking, there are an infinite number of eigenvectors associated to each eigenvalue of a dynamical operator. Because any scalar multiple of an eigenvector is still an eigenvector, an infinite family of eigenvectors exists for each eigenvalue. However, these vectors are all proportional to each other. In this sense, each design mode ϕ_i has both temporal (eigenvalue) and structure-organizational (scalar multiple of eigenvector) characteristics. Every dynamical operator A_0 has exactly p eigenvalues, which are not necessarily distinct. Another term used for the eigenvalues is characteristic values. Eigenvectors corresponding to distinct eigenvalues are linearly independent.

The characteristics of the design modes also determine the stability of the process. Stability is an important property of dynamical systems in general, not just PD projects. Stability is usually defined in terms of equilibrium points (see e.g. Luenberger 1979 or Hinrichsen and Pritchard 2005). According to Luenberger (1979), if all solutions of the linear system from Eq. 1 that start out near an equilibrium state x_e of work remaining stay near or converge to x_e, the state is called stable or asymptotically stable respectively. This is illustrated in Fig. 2.2 in a one-dimensional state representation of a simple dynamical system. The illustration also clearly shows that the notion of asymptotic stability is stronger than stability.

The origin $\overline{x} = 0$ is always a singular point of the vector field $x \rightarrow A_0 \cdot x$ on $\mathbb{R}^p$ and therefore an equilibrium point of the linear homogenous recurrence relation given by Eq. (1). A linear homogeneous recurrence relation is internally stable if its dynamical operator is stable in the sense of Lyapunov (Hinrichsen and Pritchard 2005). A square real matrix is said to be asymptotically stable in the sense of Lyapunov if and only if for an operator A_0 and any positive semi-definite matrix C there exists a positive-definite symmetric matrix Σ satisfying the following Lyapunov equation (see e.g. Halanay and Rasvan 2000; Hinrichsen and Pritchard 2005 or Siddiqi 2010):

$$\Sigma - A_0 \cdot \Sigma \cdot A_0^{\mathrm{T}} = C. \tag{4}$$

For the first-order linear autoregressive model that will be introduced in the next Section 2.2, A_0 is the cited WTM, Σ is the steady-state state covariance matrix (Eq. 245) and C is the covariance matrix of the unpredictable performance fluctuations (Eq. 9). According to Siddiqi (2010), the Lyapunov equation can be interpreted as satisfied for a linear autoregressive model if, for a given observation covariance, there exists a legitimate belief distribution in which the predicted belief over project state is equivalent to the previous belief over project state, that is, if there exists an equilibrium point of the distribution.

In the following, we use the convention of listing the eigenvalues of the design modes in order of decreasing magnitude ($|\lambda_1(A_0)| \geq |\lambda_2(A_0)| \geq \ldots$). For the matrix A_0 with these eigenvalues, we define the spectral radius as the greatest-magnitude eigenvalue and denote it by $\rho(A_0) = \max|\lambda_i|$. An eigenvalue corresponding to $\max|\lambda_i|$ (that is, λ_1) is called the dominant eigenvalue (Gentle 2007).

The Lyapunov equation (Eq. 4) holds for the linear homogenous recurrence relation given by Eq. (1) if and only if the spectral radius is less than or equal to one, i.e. $\rho(A_0) \leq 1$ (Hinrichsen and Pritchard 2005). Recall that a matrix M is positive semidefinite if and only if it holds that $v^{\mathrm{T}} \cdot M \cdot v \geq 0$ for all non-zero column vectors v of $m = \mathrm{Dim}[M]$ real numbers. Let λ be a left eigenvalue of A_0 and ϑ_l a corresponding normalized eigenvector satisfying $\vartheta_l^{\mathrm{T}} \cdot A_0 = \{\lambda\} \cdot \vartheta_l^{\mathrm{T}}$. Then the Lyapunov equation can be written as

$$\begin{aligned} \vartheta_l^{\mathrm{T}} \cdot C \cdot \vartheta_l &= \vartheta_l^{\mathrm{T}} \cdot \left(\Sigma - A_0 \cdot \Sigma \cdot A_0^{\mathrm{T}}\right) \cdot \vartheta_l \\ &= \vartheta_l^{\mathrm{T}} \cdot \Sigma \cdot \vartheta_l - \vartheta_l^{\mathrm{T}} \cdot \{\lambda\} \cdot \Sigma \cdot \{\lambda\} \cdot \vartheta_l \\ &= \vartheta_l^{\mathrm{T}} \cdot \Sigma \cdot \vartheta_l \cdot \{1 - \lambda^2\}. \end{aligned}$$

Since the matrices Σ and C are positive-definite symmetric matrices, it holds that $\vartheta_l^{\mathrm{T}} \cdot \Sigma \cdot \vartheta_l \geq 0$ and $\vartheta_l^{\mathrm{T}} \cdot C \cdot \vartheta_l \geq 0$. It follows that $\left(1 - \lambda^2\right) \geq 0$. Therefore, the Lyapunov criterion from Eq. 4 is satisfied if the spectral radius is less than or equal to one and it holds that $|\lambda| \leq 1$ (Siddiqi 2010).

A modeled PD project phase is said to be asymptotically stable if and only if the spectral radius is less than 1: that is, $\rho(A_0) < 1$. In this case, irrespective of the initial state x_0 the work remaining converges to the zero vector, meaning that all tasks are fully completed. If it holds that $\rho(A_0) = 1$, the project phase is stable but not necessarily asymptotically stable. Asymptotic stability can be relevant for stochastic model formulations in which unpredictable performance fluctuations are represented by continuous-type random variables. The integration into the state equation can be done either by means of additive (Schlick et al. 2007) or multiplicative fluctuations (Huberman and Wilkinson 2005). We will return to this point in the next Section 2.2. In the case of stochastic model formulations, for a stable project phase the work remaining x_t would eventually oscillate around x_0 indefinitely. As already stated and illustrated in Fig. 2.2, the notion of asymptotic stability is stronger than stability. For the first design mode ϕ_1 with the dominant eigenvalue λ_1 the equation $|\lambda_1| = 1$ determines the boundary between stable and asymptotically stable regimes. If the project phase is neither stable nor asymptotically stable and $|\lambda_1(A_0)| > 1$, it is said to be unstable. If it is unstable, a redesign of tasks and their interactions is necessary, because the work remaining then exceeds all given limits.

Unfortunately, even if the project phase modeled is asymptotically stable, theoretically an infinite number of iterations are necessary to reach the final state where zero work remains for all tasks. Therefore, project managers have to specify an additional stopping criterion. In the following we use a simple one-dimensional parametric criterion $\delta \in]0;1[$ indicating that the goal has been reached if the work remaining is at most 100δ percent for all p tasks. According to Huberman and Wilkinson (2005), the zero vector represents a theoretically optimal solution, and the values of the state vector are an abstract measure of the amount of work left to be done before a task's solution is optimal. Formally speaking, with an initial state x_0, a WTM A_0 and a stopping criterion δ, the total time T_δ spent to complete the project phase can be determined by the equation (cf. Eq. 3):

$$\begin{aligned} T_\delta &= \min_t \left\{ \max_i x_t^{(i)} \leq \delta \right\}_{t=0}^{T_{max}} \\ &= \min_t \left\{ \max_i \left[A_0^t \cdot x_0 \right]^{(i)} \leq \delta \right\}_{t=0}^{T_{max}} . \end{aligned} \tag{5}$$

$x_t^{(i)}$ denotes the i-th component of the work vector at time step t and $\left[A_0^t \cdot x_0\right]^{(i)}$ the i-th component of the product $A_0^t \cdot x_0$. The time index is expressed explicitly in the $\min\{.\}$ function for greater clarity. The maximum time step T_{max} must be set to a value that is sufficiently large to satisfy the specified stopping criterion. We can also generalize the closed-form solutions derived in the seminal work of Smith and Eppinger (1997, 1998) and express the total amount of work done in all p tasks during the iteration process across the time interval T_δ as

$$\begin{aligned} \sum_{t=0}^{T_\delta} x_t &= \sum_{t=0}^{T_\delta} \left(A_0^t \cdot x_0 \right) \\ &= \left(\sum_{t=0}^{T_\delta} A_0^t \right) \cdot x_0 \\ &= \left(I_p - A_0 \right)^{-1} \cdot \left(I_p - A_0^{T_\delta + 1} \right) \cdot x_0 . \end{aligned} \tag{6}$$

The above solution is based on the Neumann series generated by A_0 (Bronstein et al. 2000); I_p denotes the $p \times p$ identity matrix. The total amount of work done over all tasks, x_{tot}, can easily be calculated as a scalar indicator of the total effort involved in completing the deliverables in the modeled project phase by summing over the components of the cumulated work vectors. We have

$$x_{tot} = \text{Total}\left[\left(I_p - A_0\right)^{-1} \cdot \left(I_p - A_0^{T_\delta+1}\right) \cdot x_0\right]. \tag{7}$$

The function Total[...] computes the sum of the argument vector's components. The formulation of cooperative work processes as a linear recurrence relation therefore makes it easy to take extra work into account and precisely determine the total effort involved. The effort-centered approach can also cope with deliverables that do not meet all the original requirements or that have a quality problem and therefore need to be reworked (Smith and Eppinger 1997, 1998). This is not to be confused with changes of scope, where separate state variables have to be defined and a dedicated scope change management system should be used. In the limit $T_\delta \rightarrow \infty$ for an asymptotically stable project phase, we have:

$$\lim_{T_\delta \rightarrow \infty} \sum_{t=0}^{T_\delta} x_t = \left(I_p - A_0\right)^{-1} \cdot x_0 .$$

2.2 Stochastic Formulation in Original State Space

In their seminal paper on performance variability and project dynamics, Huberman and Wilkinson (2005) showed how to model cooperative work in PD projects based on the theory of stochastic processes, and how to apply formal methods of statistics to analyze, predict and evaluate the dynamics of open organizational systems. An open organizational system is a sociotechnical system in which humans continuously interact with each other and with their work environment. These interactions usually take the form of goal-directed information exchange within and through the system boundary and lead to a kind of self-organization, since patterns of coordination can emerge that convey new properties, such as oscillations or pace-setting. Furthermore, there is a regular supply of energy and matter from the environment. In the work presented here, we follow the basic ideas of Huberman and Wilkinson and formulate a stochastic model of cooperative work based on the theory of Gauss–Markov processes (Cover and Thomas 1991; Papoulis and Pillai 2002). However, we do not incorporate "multiplicative noise" to represent performance variability as Huberman and Wilkinson do, but rather assume that the effects of performance fluctuations on work remaining are cumulative. Clearly, there are subtle conceptual differences between the two approaches, but they are beyond the scope of this book, which is in this context to carry out a first validation study of the formulated stochastic model based on field data from an industrial PD project (Section 2.5) and to analyze the interrelationships between projects dynamics and emergent complexity explicitly in analytical and numerical studies (Chapters 3, 4 and 5).

Our model generalizes the first-order difference equation (Eq. 1) according to Smith and Eppinger (1997) to a deterministic random process $\{X_t\}$ (Puri 2010) with state equation

$$X_t = A_0 \cdot X_{t-1} + \varepsilon_t \quad t = 1, \ldots, T. \tag{8}$$

In this first-order linear autoregressive model, the multivariate random variable X_t represents the measured (or estimated) work remaining at time step t of the project phase under consideration. A_0 is the cited WTM. The random vector ε_t is used to model unpredictable performance fluctuations. In terms of state estimation ε_t can also be interpreted as an "error vector." Each component in the error vector indicates a specific "error bar" in the sense of a reduced mathematical representation of the performance variability when processing the corresponding development task.

In PD projects there are many factors shaping performance variability. Although we do not know their exact number or distribution, the central limit theorem tells us that, to a large degree, the sum of independently and identically distributed factors can be represented by a Gaussian distribution $\mathcal{N}(x; \mu, C)$ with location $\mu = E[\varepsilon_t]$ and covariance $C = E\left[(\varepsilon_t - \mu)(\varepsilon_t - \mu)^{\mathrm{T}}\right]$. The location is often simply termed "mean." We assume that the performance fluctuations are independent of the work remaining and therefore that the location and covariance do not depend on the time index. Hence, we can also write ε in place of ε_t in the following definitions of the entries of the covariance matrix.

The covariance matrix C is a square matrix of size p, whose entry $C_{[[i,j]]}$ in the, (i,j)-th position is the covariance between the i-th element $\varepsilon^{(i)}$ and the j-th element $\varepsilon^{(j)}$ of the random vector ε, i.e.

$$\begin{aligned} C_{[[i,j]]} &= \operatorname{Cov}\left[\varepsilon^{(i)}, \varepsilon^{(j)}\right] \\ &= E\left[\left(\varepsilon^{(i)} - \mu^{(i)}\right)\left(\varepsilon^{(j)} - \mu^{(j)}\right)\right]. \end{aligned} \tag{9}$$

C is symmetric by definition and also positive-semidefinite (Lancaster and Tismenetsky 1985). We assume that C has full rank. The diagonal elements $C_{[[i,i]]}$ represent the scalar-valued variances c_{ii}^2 of vector components $\varepsilon^{(i)}$ (i.e. performance fluctuations in work tasks i):

$$\begin{aligned} c_{ii}^2 &= \operatorname{Var}\left[\varepsilon^{(i)}\right] \\ &= E\left[\left(\varepsilon^{(i)} - \mu^{(i)}\right)^2\right]. \end{aligned} \tag{10}$$

The square root of the scalar-valued variance c_{ii}^2 is the well-known standard deviation c_{ii}. The off-diagonal elements $C_{[[i,j]]}$ $(i \neq j)$ represent the scalar-valued covariances and can be factorized as

$$\rho_{ij} c_{ii} c_{jj} = \rho_{ij} \sqrt{\operatorname{Var}[\varepsilon^{(i)}] \operatorname{Var}[\varepsilon^{(j)}]} \quad (i \neq j), \tag{11}$$

where the first factor is Pearson's famous product–moment coefficient

$$\rho_{ij} := \mathrm{Corr}\left[\varepsilon^{(i)}, \varepsilon^{(j)}\right] = \frac{\mathrm{Cov}\left[\varepsilon^{(i)}, \varepsilon^{(j)}\right]}{\sqrt{\mathrm{Var}\left[\varepsilon^{(i)}\right]\mathrm{Var}\left[\varepsilon^{(j)}\right]}}. \tag{12}$$

The Pearson correlation ρ_{ij} is $+1$ in the case of a perfect positive linear relationship (correlation) and -1 in the case of a perfect negative linear relationship (anticorrelation). It has values between -1 and 1 in all other cases, indicating the degree of linear dependence between the variables.

In the developed autoregression model of cooperative work it is assumed that the performance fluctuations have no systematic component and that $\mu = 0_p = (\,0\ 0\ \ldots\ 0\,)^{\mathrm{T}}$. We imposed no additional a priori constraints on the covariance matrix C. Hence, the fluctuations

$$\varepsilon_t \sim \mathcal{N}(0_p, C)$$

in the state equation can be expressed explicitly by a Gaussian probability density function $f[x] = \mathcal{N}(x; 0_p, C)$ (*pdf*, see e.g. Puri 2010) as

$$\mathcal{N}(x; 0_p, C) = \frac{1}{(2\pi)^{p/2}(\mathrm{Det}[C])^{1/2}} \mathrm{Exp}\left[-\frac{1}{2}x^{\mathrm{T}} \cdot C^{-1} \cdot x\right]. \tag{13}$$

The covariance matrix C can be written in vector form as

$$C = \mathrm{Cov}[\varepsilon_t, \varepsilon_t] = E\left[\varepsilon_t \varepsilon_t^{\mathrm{T}}\right].$$

In terms of basic geometric concepts, the covariance matrix C can be visualized in the prediction error space spanned by the components of Δx through the concentration ellipsoid (see e.g. Bronstein et al. 2000)

$$\Delta x^{\mathrm{T}} \cdot C^{-1} \cdot \Delta x = \kappa.$$

The constant κ determines the size of the p-dimensional region enclosed by the ellipsoid surface. By setting the value of κ we can define the probability that the prediction error will fall inside the ellipsoid. Figure 2.3 shows an illustrative example of the concentration ellipsis in the two dimensional case. The size and orientation of the concentration ellipsoid depend on the eigenvalues $\lambda_i(C)$ and eigenvectors $k_i(C)$ of the covariance matrix C $(i = 1, \ldots, p)$. We can determine the eigenvalues and eigenvectors by solving the well-known eigenvalue problem

$$C \cdot k_i(C) = \lambda_i(C) \cdot k_i(C).$$

Because C is symmetric by definition, the eigenvectors are mutually orthogonal. The mutually orthogonal eigenvectors point in the directions of the principal axis of the concentration ellipsoid, and the eigenvalues determine the length of the semiaxis $\sqrt{\kappa/\lambda_i(C)}$. The concentration ellipsoid containing 68.3%, 95.4% and

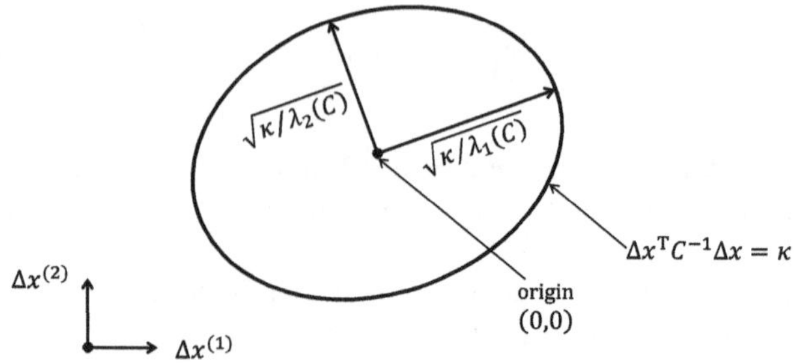

Fig. 2.3 Concentration ellipse in the two-dimensional prediction error space spanned by the components of Δx (adopted from Oispuu 2014)

99.73% (1σ-, 2σ- and 3σ-ellipse) of normally distributed prediction errors are defined by constants $\kappa_1 = 2.30$, $\kappa_2 = 6.17$ and $\kappa_3 = 11.80$, respectively (Bronstein et al. 2000).

In the following we assume that the performance fluctuations are uncorrelated from time step to time step and that it holds for all time steps $\{\mu, \nu\} \in \mathbb{Z}$ that

$$E\left[\varepsilon_\mu \varepsilon_\nu^{\mathrm{T}}\right] = \left\{\delta_{\mu\nu}\right\} \cdot C.$$

$\delta_{\mu\nu}$ is the Kronecker delta which is defined as

$$\delta_{\mu\nu} = \begin{cases} 1 & \mu = \nu \\ 0 & \mu \neq \nu. \end{cases} \tag{14}$$

If the covariance matrix is a nonzero scalar multiple of the identity matrix I_p, that is $C = \left\{\sigma^2\right\} \cdot I_p$, we speak of isotropic fluctuations, and the variance σ^2 represents the overall strength ($\sigma^2 \in \mathbb{R}^+$). In spite of the stochastic task processing, it is assumed in the following that perfect initial conditions exist, and that the components of the initial state vector according to state Eq. 8 are positive real numbers and not random variables. This assumption is justified by the fact that in most projects the initial state x_0 represents the planned amount of work at the beginning of a given project phase (cf. Eq. 2), which is predefined by the project manager. In this case a real valued parameter vector $\theta = \begin{bmatrix} x_0 & A_0 & C \end{bmatrix}$ is sufficient to parameterize the model. Alternatively, the initial state vector X_0 can be assumed to be a Gaussian random vector with location μ_0 and covariance C_0. In Section 2.9 we will present a stochastic model formulation with hidden state variables that can cover this case under a more general theoretical framework. When this alternative formulation is used, the parameter vector must be extended, and becomes $\theta = \begin{bmatrix} \mu_0 & C_0 & A_0 & C \end{bmatrix}$. A graphical representation of the first-order autoregression model is shown in Fig. 2.4 in the form of a dynamic Bayesian network (see e.g. Gharahmani 2001). In a dynamic Bayesian network the random state variables are related to each other over adjacent time steps and are drawn as nodes of the graph. At any point in time t, the value of a state variable can be calculated from the internal regressors and the immediate prior value (time step $t-1$). The directed arcs represent conditional

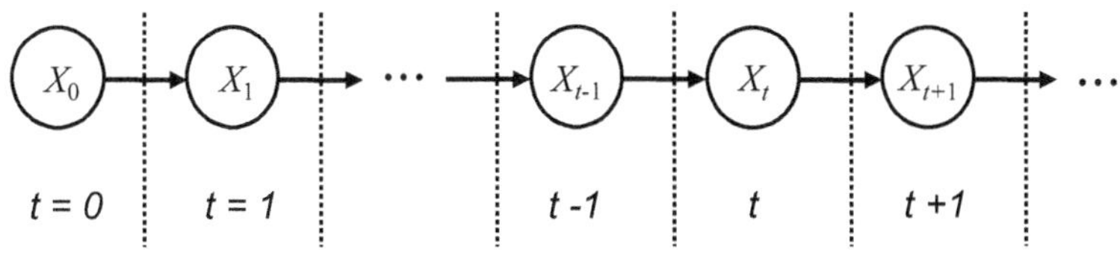

Fig. 2.4 Graphical representation of the first-order autoregression model in the form of a dynamic Bayesian network. The nodes in the graph represent the random state variables of the stochastic process. The directed arcs encode conditional dependencies between the variables

dependencies between the variables. Exogenous inputs to the model are not considered in the following.

It is not difficult to see that the process $\{X_t\}$ can be decomposed into a deterministic and stochastic part as

$$X_t = A_0^t \cdot x_0 + \sum_{v=1}^{t} A_0^{t-v} \cdot \varepsilon_v \qquad (t \geq 1)\,.$$

The deterministic part represents the mean vectors (Eq. 3)

$$E[X_t] = A_0^t \cdot x_0$$

of work remaining, which evolve unperturbed. As shown in the previous Section 2.1, for an arbitrary project phase with predefined initial state x_0 and WTM A_0, we can derive the following closed-form solution to the expected total amount of work done in all p tasks during the iteration process until the stopping criterion δ is satisfied:

$$\begin{aligned} E\left[\sum_{t=0}^{T_\delta} X_t\right] &= \sum_{t=0}^{T_\delta} E[X_t] \\ &= \sum_{t=0}^{T_\delta} \left(A_0^t \cdot x_0\right) \\ &= \left(\sum_{t=0}^{T_\delta} A_0^t\right) \cdot x_0 \\ &= \left(I_p - A_0\right)^{-1} \cdot \left(I_p - A_0^{T_\delta+1}\right) \cdot x_0\,. \end{aligned} \tag{15}$$

To obtain the above solution we have assumed that the decision on whether the stopping criterion is satisfied (see Eq. 5) is based on the mean vectors $E[X_t]$ of work remaining at time step t and not on individual instances $\{x_t\}$ of the stochastic process $\{X_t\}$. The duration T_δ of the project phase is then determined entirely by the deterministic part of the process. The expected total amount of work x_{tot} done over all tasks until the stopping criterion is satisfied is estimated by:

$$x_{tot} = \text{Total}\left[\left(I_p - A_0\right)^{-1} \cdot \left(I_p - A_0^{T_\delta+1}\right) \cdot x_0\right]. \tag{16}$$

In the limit $T_\delta \to \infty$ for an asymptotically stable project phase, we have the expected total amount of work done for all p tasks:

$$\lim_{T_\delta \to \infty} E\left[\sum_{t=0}^{T_\delta} X_t\right] = \left(I_p - A_0\right)^{-1} \cdot x_0. \tag{17}$$

In addition to the deterministic evolution of the mean work remaining the stochastic part of the process $\{X_t\}$ represents the accumulated unpredictable performance fluctuations. The formulation of the linear model means that the variances and covariances of the vector components of the fluctuations are independent of the work remaining. In view of an information processing system, the process $\{X_t\}$ satisfies the Markov property. The Markov property describes a special kind of "memorylessness" in the sense that conditional on the present state x_t of the modeled project, its future $\{X_{t+1}, \ldots\}$ and past $\{X_1, \ldots, X_{t-1}\}$ are rendered independent:

$$f_\theta[x_{t+1}|x_0, \ldots, x_t] = f_\theta[x_{t+1}|x_t] \quad \forall t \geq 0. \tag{18}$$

$f_\theta[x_{t+1}|x_0, \ldots, x_t]$ denotes the conditional *pdf* of vector X_{t+1}, given the sequence of vectors $X_0, \ldots, X_t$ (Papoulis and Pillai 2002).

From the decomposition of the process into a deterministic and a stochastic part it is evident that the *pdf* of the current state X_t is Gaussian with location $A_0^t x_0$ and covariance $\Sigma_{v=1}^t A_0^{t-v}\, C\left(A_0^{\mathrm{T}}\right)^{t-v}$, that is

$$X_t \sim \mathcal{N}\left(A_0^t \cdot x_0, \sum_{v=1}^{t} A_0^{t-v} \cdot C \cdot \left(A_0^{\mathrm{T}}\right)^{t-v}\right).$$

The density function $f_\theta[x_t]$ of X_t can be written explicitly as (Puri 2010)

$$f_\theta[x_t] = \frac{1}{(2\pi)^{p/2} (\text{Det}[\Sigma_t])^{1/2}} \text{Exp}\left[-\frac{1}{2}\left(x_t - A_0^t \cdot x_0\right)^{\mathrm{T}} \cdot \Sigma_t^{-1} \cdot \left(x_t - A_0^t \cdot x_0\right)\right], \tag{19}$$

where

$$\Sigma_t = \sum_{v=1}^{t} A_0^{t-v} \cdot C \cdot \left(A_0^{\mathrm{T}}\right)^{t-v} = \sum_{v=0}^{t-1} A_0^{v} \cdot C \cdot \left(A_0^{\mathrm{T}}\right)^{v}.$$

The conditional density of state X_{t+1} given state $X_t = x_t$ (Eq. 18) is

$$f_\theta[x_{t+1}|x_t] = \frac{1}{(2\pi)^{p/2}(\mathrm{Det}[C])^{1/2}}\mathrm{Exp}\left[-\frac{1}{2}(x_{t+1} - A_0 \cdot x_t)^{\mathrm{T}} \cdot C^{-1} \cdot (x_{t+1} - A_0 \cdot x_t)\right]. \tag{20}$$

Next, we turn our attention to the total time T_δ spent to complete the project phase described by the stochastic formulation in state space. Owing to the fact that the state vector is a random variable, the total time T_δ is not uniquely determined by Eq. 5—we can only consider the probability that the project phase ends at a certain time instant. As we will see below, it is possible to derive a recursive procedure for calculating the time-dependent probabilities, but there seems to be no obvious way to obtain a closed form solution. This motivates to use a model for the *pdf* of the duration of the process. A reasonable approach is to model the duration by a log-normal distribution because it possesses many advantageous properties (see Baker and Trietsch 2009; Trietsch et al. 2012) and a high external validity in the given application area. As the log-normal distribution shares many properties with the generalized Rayleigh distribution its density function can also be used to effectively model skewed execution times. We will return to this point in the Monte Carlo studies of Sections 5.2 and 5.3. Without loss of generality let us assume that $x_t = \max_i x_t^{(i)}$ is the distinct largest component of the previously introduced state vector. The probability for the process to be finished at time instant $t = 1$, i.e. that the work remaining for all tasks is smaller than or equal to δ, can be obtained as follows: Observing that X_1 is a Gaussian random variable with distribution $\mathcal{N}(A_0 \cdot x_0, C)$ the total probability that the distinct largest component x_1 is less than or equal to δ is given by

$$\begin{aligned} P_1 &= \int_{-\infty}^{\delta} \mathcal{N}(x_1; A_0 \cdot x_0, C) dx_1 \\ &= \frac{1}{2}\left(1 + \mathrm{Erf}\left[\left(\frac{\delta - (A_0 \cdot x_0)}{\sqrt{2C}}\right)\right]\right). \end{aligned}$$

The function Erf[...] denotes the Gauss error function (see e.g. Puri 2010). We used a simplified notation to indicate that the error function only refers to the mean and variance corresponding to the distinct largest component of the state vector X_1. In the field of detection theory, P_1 is known as the probability of missed detection of a signal X_1 with mean $A_0 \cdot x_0$ embedded in Gaussian noise with covariance C and a detection threshold equal to δ.

On the other hand, the total probability that the stochastic process exceeds the threshold δ for all tasks and propagates to the second time step is $\overline{P}_1 = 1 - P_1$, see Fig. 2.5. The *pdf* for the work remaining at the first time step and exceeding the threshold is given by

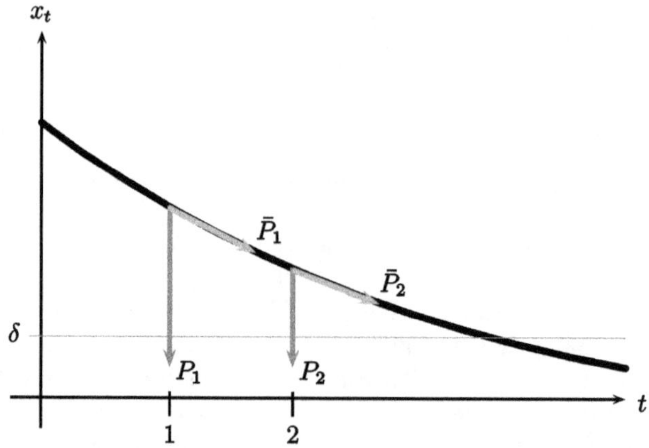

Fig. 2.5 Graphical illustration of the recursive procedure for calculating the probabilities that the project phase ends at a certain time instant T_δ: the total probability that the process exceeds the threshold determined by the stopping criterion δ at time instant $t = 1$ is given by $\overline{P}_1$; this probability is propagated to the second time step to calculate the total probability $\overline{P}_2$. The subsequent probabilities P_t for the process to finish at time instant t can be computed iteratively based on Eq. 21

$$\overline{f}_\theta[x_1] = \begin{cases} \dfrac{1}{\overline{P}_1}\dfrac{1}{\sqrt{2\pi C}}\, e^{-(x_1-(A_0\cdot x_0))^2/2C} & x_1 \geq \delta \\ 0 & x_1 < \delta\,. \end{cases}$$

As we can see, $\overline{f}_\theta[x_t]$ is no longer Gaussian. The *pdf* for the propagated state at the second time instant is not Gaussian either and can be computed as

$$\begin{aligned} f_\theta[x_2] &= \int_{-\infty}^{\delta} f_\theta[x_2|x_1]\overline{f}_\theta[x_1]dx_1 \\ &= \frac{1}{\overline{P}_1}\frac{1}{2\sqrt{1+A_0^2}}\, e^{-\left(x_2-\left(A_0^2\cdot x_0\right)\right)^2/2\left(1+A_0^2\right)C}\left(1+\mathrm{Erf}\left[\left(\frac{(1+A_0^2)\delta - A_0(x_0+x_2)}{\sqrt{2(1+A_0^2)C}}\right)\right]\right). \end{aligned}$$

Then, the total probability that the process terminates at the second time step is given by the integral

$$\begin{aligned} P_2 &= \overline{P}_1 \int_{-\infty}^{\delta} f_\theta[x_2]dx_2 \\ &= (1-P_1)\int_{-\infty}^{\delta} f_\theta[x_2]dx_2. \end{aligned}$$

Unfortunately, this integral cannot be solved analytically, and we can only evaluate the probability numerically. The subsequent probabilities P_{t+1} for the project to finish at time instant $t+1$ are then computed iteratively as

$$\begin{aligned} P_{t+1} &= \overline{P}_t \int_{-\infty}^{\delta} \int f_\theta[x_{t+1}|x_t] \overline{f}_\theta[x_t] dx_t dx_{t+1} \\ &= (1 - P_t) \int_{-\infty}^{\delta} \int f_\theta[x_{t+1}|x_t] \overline{f}_\theta[x_t] dx_t dx_{t+1}. \end{aligned} \tag{21}$$

Alternatively, we could simulate the stochastic task processing and compute the probabilities in a Monte Carlo study.

At first glance, the chosen memoryless perturbation mechanism may appear to over-simplify the problem. However, the correlations ρ_{ij} between performance fluctuations between tasks i and j can strongly influence the course of the project not only at single time steps but also on long time scales and therefore lead to unexpected stateful behavior. This is the case if the correlations are reinforced through the informational coupling between the development tasks. To reinforce the correlations, the covariance matrix C must have nonzero off-diagonal elements: in other words, the fluctuations must be nonisotropic. Depending on the structure of the dynamical operator A_0, the correlations ρ_{ij} can significantly excite the design modes and lead to unexpected effects of emergent complexity, such as the cited problem-solving oscillations in the preasymptotic range of development projects (Mihm and Loch 2006; Schlick et al. 2008). We will return to the interesting phenomenon of excitation of design modes in Section 4.1.2, where the interrelationships between project dynamics and emergent complexity are analyzed in detail in the spectral basis.

Following the theoretical considerations of system stability from Section 2.1, the first-order linear autoregressive model defined in Eq. 8 is asymptotically stable in the sense of Lyapunov (Eq. 4) if and only if the spectral radius of the dynamical operator A_0 is strictly less than one, i.e. $\rho(A_0) < 1$, and the matrix in Eq. 4 is positive definite. In contrast to the deterministic model formulation, an autoregression model with Gaussian performance fluctuations without drift and unit spectral radius $\rho(A_0) = 1$ would steadily move away from the equilibrium state x_e and therefore not be stable (Papoulis and Pillai 2002; Siddiqi 2010). If $\rho(A_0) = 1$, the autoregressive process is said to be marginally stable (Halanay and Rasvan 2000).

In extension of state Eq. 8, we can formulate a model of cooperative work on the basis of a forcing matrix K in conjunction with a random variable η_t whose covariance matrix does not indicate correlations between vector components (i.e. work tasks) and is therefore diagonal. To do so, the covariance matrix C is decomposed into eigenvectors and eigenvalues through an eigendecomposition:

$$C = K \cdot \Lambda_K \cdot K^{-1}, \tag{22}$$

where

$$K \cdot K^{\mathrm{T}} = I_p \quad \text{and} \quad K^{-1} = K^{\mathrm{T}}.$$

Because the covariance matrix C is symmetric by definition, the forcing matrix K resulting from the eigendecomposition has mutually orthogonal column vectors $k_i(C) = K_{:i}$ and is therefore orthogonal. These vectors are the eigenvectors of C. Λ_K is simply a diagonal matrix with the eigenvalues $\lambda_i(C)$ along the principal diagonal. The associated state equation is

$$X_t = A_0 \cdot X_{t-1} + K \cdot \eta_t, \tag{23}$$

with

$$\eta_t \sim \mathcal{N}(0_p, \Lambda_K) \tag{24}$$

and

$$\Lambda_K = \mathrm{Diag}[\lambda_i(C)] \quad 1 \leq i \leq p. \tag{25}$$

According to the above equation, the eigenvalues $\lambda_i(C)$ of the decomposed covariance matrix C can be interpreted as the variances of the performance fluctuations along the rotated axes of the identified eigenvectors $k_i(C)$. Following our terminology we will use the term "performance fluctuation mode," $\Psi_i = (\lambda_i(C), k_i(C))$, to refer to an eigenvalue $\lambda_i(C)$ of C along with its eigenvector $k_i(C)$ $(1 \leq i \leq p)$.

Finally, we analyze the properties of the generated stochastic process $\{X_t\}$ in steady state and derive a closed-form expression for its joint *pdf*. Under the assumption of asymptotic stability $(\rho(A_0) < 1)$, the *pdf* of the stochastic process commutes in the long-term evolution $t \to \infty$ in a distribution which is invariant under a shift of the origin. It follows that the stationary behavior is characterized by a stable distribution for state variable X_t with mean (also termed location)

$$\mu(t \to \infty) = 0_p \tag{26}$$

and covariance Σ that satisfies the Lyapunov criterion from Eq. 4 in the sense that the famous Lypunov equation

$$\Sigma = A_0 \cdot \Sigma \cdot A_0^{\mathrm{T}} + C \tag{27}$$

for the steady-state covariance matrix Σ, the dynamical operator A_0 and the covariance matrix C of the performance fluctuations is fulfilled. The closed-form solution of the steady-state covariance can be written as a simple matrix power series (see Eq. 245), which we will introduce and dicuss in Section 4.1.1.

Given the Markov property the joint *pdf* of the process can be factorized as follows:

$$f_\theta[x_{t_1},\ldots,x_{t_2}] = f_\theta[x_{t_1}] \prod_{t=t_1+1}^{t_2} f_\theta[x_t|x_{t-1}]\,. \tag{28}$$

If the initial state and transition probabilities are Gaussians,

$$f_\theta[x_{t_1}] = \mathcal{N}\left(x_{t_1}; 0_p, \Sigma\right) \tag{29}$$

$$f_\theta[x_t|x_{t-1}] = \mathcal{N}(x_t; A_0 \cdot x_{t-1}, C), \tag{30}$$

then the joint *pdf* in steady state reads:

$$f_\theta[x_{t_1},\ldots,x_{t_2}] = c_x \mathrm{Exp}\left[-\frac{1}{2}x_{t_1}^{\mathrm{T}} \cdot \Sigma^{-1} \cdot x_{t_1} - \frac{1}{2}\sum_{t=t_1+1}^{t_2} (x_t - A_0 \cdot x_{t-1})^{\mathrm{T}} \cdot C^{-1} \cdot (x_t - A_0 \cdot x_{t-1})\right] \tag{31}$$

with the normalization constant

$$c_x = (2\pi)^{-\frac{\Delta tq}{2}} (\mathrm{Det}\,\Sigma)^{-\frac{1}{2}} (\mathrm{Det}\,C)^{-\frac{\Delta t-1}{2}} \tag{32}$$

and $\Delta t = t_2 - t_1 + 1$ the number of time steps. We can write the joint *pdf* in a more compact form:

$$f_\theta[x_{t_1},\ldots,x_{t_2}] = c_x \mathrm{Exp}\left[-\frac{1}{2}\boldsymbol{x}^{\mathrm{T}} \cdot \mathcal{C}_2 \cdot \boldsymbol{x}\right],$$

where $\boldsymbol{x}$ is a large column vector containing all states from time step t_1 to t_2, $\boldsymbol{x}^{\mathrm{T}} = \left(x_{t_1}^{\mathrm{T}} \cdots x_{t_2}^{\mathrm{T}}\right)$. The elements of the matrix $\mathcal{C}_2$ are:

$$\mathcal{C}_2 = \begin{pmatrix} \Sigma^{-1} + A_0^{\mathrm{T}} \cdot C^{-1} \cdot A_0 & -A_0^{\mathrm{T}} \cdot C^{-1} & & & 0 \\ -C^{-1} \cdot A_0 & C^{-1} + A_0^{\mathrm{T}} \cdot C^{-1} \cdot A_0 & -A_0^{\mathrm{T}} \cdot C^{-1} & & \\ & \ddots & \ddots & \ddots & \\ & & -C^{-1} \cdot A_0 & C^{-1} + A_0^{\mathrm{T}} \cdot C^{-1} \cdot A_0 & -A_0^{\mathrm{T}} \cdot C^{-1} \\ 0 & & & -C^{-1} \cdot A_0 & C^{-1} \end{pmatrix}. \tag{33}$$

For the inverse of $\mathcal{C}_2$, which is the covariance of the joint *pdf*, we find the following closed form:

$$\mathcal{C}_x = \mathcal{C}_2^{-1} = \begin{pmatrix} \Sigma & \Sigma \cdot A_0^{\mathrm{T}} & \Sigma\left(A_0^{\mathrm{T}}\right)^2 & \cdots & \Sigma\left(A_0^{\mathrm{T}}\right)^{\Delta t-1} \\ A_0 \cdot \Sigma & \Sigma & \Sigma \cdot A_0^{\mathrm{T}} & \ddots & \Sigma\left(A_0^{\mathrm{T}}\right)^{\Delta t-2} \\ \vdots & \ddots & \ddots & \ddots & \vdots \\ \vdots & & \ddots & \Sigma & \Sigma \cdot A_0^{\mathrm{T}} \\ A_0^{\Delta t-1} \cdot \Sigma & \cdots & \cdots & A_0 \cdot \Sigma & \Sigma \end{pmatrix}. \tag{34}$$

The above form for the covariance can be easily verified by showing that the proposed inverse leads to the desired result, i.e. that it holds that $\mathcal{C}_2 \cdot \mathcal{C}_x = I_p$ and $\mathcal{C}_X \cdot \mathcal{C}_2 = I_p$. Because the matrices are symmetric, the second equation follows from the first one. That means we only have to prove the first one:

$$\begin{aligned} \mathcal{C}_x \cdot \mathcal{C}_2 &= \mathcal{C}_x^{\mathrm{T}} \cdot \mathcal{C}_2^{\mathrm{T}} \\ &= (\mathcal{C}_2 \cdot \mathcal{C}_x)^{\mathrm{T}} \\ &= I_p^{\mathrm{T}} = I_p. \end{aligned}$$

For example, the $(1,1)$-block of $\mathcal{C}_2 \cdot \mathcal{C}_x$ is

$$(\Sigma^{-1} + A_0^{\mathrm{T}} \cdot C^{-1} \cdot A_0)\Sigma - A_0^{\mathrm{T}} \cdot C^{-1} \cdot A_0 \cdot \Sigma = I_p.$$

The $(1,2)$-block is

$$\begin{aligned} (\Sigma^{-1} + A_0^{\mathrm{T}} \cdot C^{-1} \cdot A_0)\Sigma \cdot A_0^{\mathrm{T}} - A_0^{\mathrm{T}} \cdot C^{-1} \cdot \Sigma &= A_0^{\mathrm{T}} + A_0^{\mathrm{T}} \cdot C^{-1} \cdot A_0 \cdot \Sigma \cdot A_0^{\mathrm{T}} - A_0^{\mathrm{T}} \cdot C^{-1} \cdot \Sigma \\ &= A_0^{\mathrm{T}} + A_0^{\mathrm{T}} \cdot C^{-1}(\Sigma - C) - A_0^{\mathrm{T}} \cdot C^{-1} \cdot \Sigma \\ &= A_0^{\mathrm{T}} \cdot C^{-1} \cdot \Sigma - A_0^{\mathrm{T}} \cdot C^{-1} \cdot \Sigma \\ &= 0. \end{aligned}$$

The other $(1,j)$-blocks have just an additonal factor $\left(A_0^{\mathrm{T}}\right)^{j-2}$ and thus also yield zero. The $(2,1)$-block is easily computed as

$$-C^{-1} \cdot A_0 \cdot \Sigma + (C^{-1} + A_0^{\mathrm{T}} \cdot C^{-1} \cdot A_0)A_0 \cdot \Sigma - A_0^{\mathrm{T}} \cdot C^{-1} \cdot A_0^2 \cdot \Sigma = 0.$$

Then the second diagonal block is

$$\begin{aligned} &-C^{-1}A_0 \cdot \Sigma \cdot A_0^{\mathrm{T}} + (C^{-1} + A_0^{\mathrm{T}} \cdot C^{-1} \cdot A_0)\Sigma - A_0^{\mathrm{T}} \cdot C^{-1} \cdot A_0 \cdot \Sigma \\ &= -C^{-1}(\Sigma - C) + C^{-1} \cdot \Sigma = I_p. \end{aligned}$$

The other diagonal blocks (except for the last one) are computed in the same way. For the last diagonal block we have

$$-C^{-1} \cdot A_0 \cdot \Sigma \cdot A_0^{\mathrm{T}} + C^{-1} \cdot \Sigma = -C^{-1}(\Sigma - C) + C^{-1} \cdot \Sigma = I_p.$$

All other blocks can easily be computed in a similar way and will yield zero.

The introduced stochastic models of cooperative work in PD projects are quite closely related to the dynamical model of product development on complex directed networks that was introduced by Braha and Bar-Yam (2007). However, there are some important differences: (1) the autoregression models are defined over a continuous range of state values and can therefore represent different kinds of cooperative relationships as well as precedence relations (e.g. overlapping);

(2) each task is unequally influenced by other tasks; (3) correlations ρ_{ij} between performance fluctuations among tasks i and j can be captured.

2.3 Stochastic Formulation in Spectral Basis

In order to analyze explicitly the intricate interrelationships between project dynamics and emergent complexity in later chapters (see Chapter 3 in conjunction with Chapters 4 and 5), we use the spectral basis and spectral methods (Neumaier and Schneider 2001). Spectral methods are based on the eigenvalues of matrix representations of systems and can therefore capture global information on structure and dynamics. Where a system's matrix representation is in the form of a static design structure matrix of system components (see e.g., Eppinger and Browning 2012 and Section 3.3), a spectral analysis of these dependency structures related to architectural elements and interfaces can be used to detect modules and assess modularity (Sarkar et al. 2013, Sarkar and Dong 2014). Facets of structural complexity can then be evaluated on a global level. We focus on persistent components of task-based, dynamic complexity in the following and carry out a spectral analysis of the WTM A_0 as dynamical operator of state Eq. 8. To carry out the transformation of the state-space coordinates, A_0 is diagonalized through an eigendecomposition (cf. Eq. 22) as

$$A_0 = S \cdot \Lambda_S \cdot S^{-1}, \tag{35}$$

with

$$\Lambda_S = \mathrm{Diag}[\lambda_i(A_0)] \quad 1 \le i \le p. \tag{36}$$

The eigenvectors $\vartheta_i(A_0) = S_{:i}$ of the design modes ϕ_i of A_0 are the column vectors of S resulting from the eigendecomposition $(i = 1 \ldots p)$. However, because A_0 must not be symmetric, the eigenvectors are in general not mutually orthogonal, and their elements can be complex numbers. The diagonal matrix Λ_s stores the ordered eigenvalues $\lambda_i(A_0)$ along the principal diagonal. The dynamical operator A_0 can always be diagonalized if it has p distinct eigenvalues and therefore also p linearly independent eigenvectors. On the other hand, if the operator has eigenvalues of multiplicity greater than one a diagonalization can still be carried out if p linearly independent eigenvectors can be found (Puri 2010). Otherwise, the above linear transformation does not lead to the diagonal form but to the so-called Jordan canonical form (see e.g. Puri 2010).

It is easy to analyze the stability of the modeled PD project in the spectral basis. According to Section 2.1, the autoregressive model defined in Eq. 8 is asymptotically stable in the sense of Lyapunov (Eq. 4) if and only if it holds for the spectral radius of the dynamical operator A_0 that $\rho(A_0) < 1$. Based on Eq. 35 we can conclude that the limit

$$\lim_{k\to\infty} A_0^k = \lim_{k\to\infty} S \cdot \Lambda_S^k \cdot S^{-1}$$
$$= S \cdot \left(\lim_{k\to\infty} \Lambda_S^k \right) \cdot S^{-1}$$

is a null matrix, since it is evident that the entries of Λ_S^k along the principal diagonal are just the eigenvalues raised to power k, which converge to zero when $\rho(A_0) < 1$. Hence, we have

$$\lim_{k\to\infty} \Lambda_S^k = \begin{pmatrix} \lambda_1^k & 0 & \cdots \\ 0 & \lambda_2^k & \cdots \\ \vdots & \vdots & \ddots \end{pmatrix} = 0.$$

If $\Lambda_S = I_p$, the project is said to be marginally stable but not asymptotically stable, because the work remaining would steadily move away from the equilibrium state x_e.

In the spectral basis, the dynamic model from Eq. 8 can be represented by the state vector X_t and the vector ε_t of unpredictable performance fluctuations as simple linear combinations, as follows:

$$\begin{aligned} X_t &= S \cdot X_t' \\ &= \sum_{i=1}^{p} X_t'^{(i)} \cdot \vartheta_i(A_0) \end{aligned} \tag{37}$$

and

$$\begin{aligned} \varepsilon_t &= S \cdot \varepsilon_t' \\ &= \sum_{i=1}^{p} \varepsilon_t'^{(i)} \cdot \vartheta_i(A_0), \end{aligned} \tag{38}$$

with coefficient vectors

$$X_t' = \begin{pmatrix} X_t'^{(1)} \\ \vdots \\ X_t'^{(p)} \end{pmatrix}$$

and

$$\varepsilon_t' = \begin{pmatrix} \varepsilon_t'^{(1)} \\ \vdots \\ \varepsilon_t'^{(p)} \end{pmatrix}.$$

For the initial state, we have

$$x_0 = S \cdot x_0'.$$

We obtain the transformed stochastic process $\{X_t'\}$ that is generated by the coefficient vectors on the basis of the state equation

$$X_t' = \Lambda_S \cdot X_{t-1}' + \varepsilon_t' \quad t = 1, \ldots, T, \tag{39}$$

with

$$\varepsilon_t' \sim \mathcal{N}\left(0_p, C'\right) \tag{40}$$

and

$$C' = S^{-1} \cdot C \cdot \left(\left[S^{\mathrm{T}}\right]^*\right)^{-1}. \tag{41}$$

The transformed covariance matrix $C' = E\left[\varepsilon_t'\left[\varepsilon_t'^{\mathrm{T}}\right]^*\right]$ is also positive-semidefinite.

When we substitute the eigendecomposition of C according to Eq. 22 in Eq. 41 we have

$$\begin{aligned} C' &= S^{-1} \cdot K \cdot \Lambda_K \cdot K^{-1} \cdot \left(\left[S^{\mathrm{T}}\right]^*\right)^{-1} \\ &= S^{-1} \cdot K \cdot \Lambda_K \cdot \left(\left[S^{\mathrm{T}}\right]^* \cdot K\right)^{-1} \\ &= S^{-1} \cdot \left(K^{\mathrm{T}}\right)^{\mathrm{T}} \cdot \Lambda_K \cdot \left(\left[S^{\mathrm{T}}\right]^* \cdot K\right)^{-1} \\ &= S^{-1} \cdot \left(K^{\mathrm{T}}\right)^{-1} \cdot \Lambda_K \cdot \left(\left[S^{\mathrm{T}}\right]^* \cdot K\right)^{-1} \\ &= \left(K^{\mathrm{T}} \cdot S\right)^{-1} \cdot \Lambda_K \cdot \left(\left[S^{\mathrm{T}}\right]^* \cdot K\right)^{-1}. \end{aligned} \tag{42}$$

Let $\left(K^{\mathrm{T}} \cdot S\right)^{-1} = (d_{ij})$ and $\left(\left[S^{\mathrm{T}}\right]^* \cdot K\right)^{-1} = (e_{ij}), \quad 1 \leq i, j \leq p$. On the basis of the matrix elements, we can derive simple formulas for the diagonal and off-diagonal elements of C', which are needed in Chapter 4 to calculate the EMC in an expressive closed form (see Eq. 262 in conjunction with Eqs. 260 and 261). The diagonal elements are

$$C'_{[[i,i]]} = \sum_{n=1}^{p} d_{in} \lambda_n(C) e_{ni}$$

and the off-diagonal elements are

$$C'_{[[i,j]]} = \sum_{n=1}^{p} d_{in} \lambda_n(C) e_{nj}.$$

Hence, the correlations ρ'_{ij} in the spectral basis are

$$\rho'_{ij} := \frac{C'_{[[i,j]]}}{\sqrt{C'_{[[i,i]]} C'_{[[j,j]]}}} = \frac{\Sigma_{n=1}^{p} d_{in} \lambda_n(C) e_{nj}}{\sqrt{\left(\Sigma_{n=1}^{p} d_{in} \lambda_n(C) e_{ni}\right)\left(\Sigma_{n=1}^{p} d_{jn} \lambda_n(C) e_{nj}\right)}}. \tag{43}$$

Interestingly, the correlations ρ'_{ij} can be interpreted in a geometrical framework (de Cock 2002). Let $\epsilon'^{(i)}_t$ be the row vector of the normalized (by $1/\sqrt{t}$) sequence of samples that were drawn from the i-th vector component of the multivariate distribution of the random variable ε'_t representing the performance fluctuations in the spectral basis, that is

$$\epsilon'^{(i)}_t := \frac{1}{\sqrt{t}}\left(\varepsilon'^{(i)}_0, \ldots, \varepsilon'^{(i)}_{t-1}\right).$$

The correlations ρ'_{ij} between the i-th and j-th components of ε'_t are defined as the angle between the vectors $\epsilon'^{(i)}_t$ and $\epsilon'^{(j)}_t$ for $t \to \infty$:

$$\rho'_{ij} = \lim_{t \to \infty} \cos\left(\epsilon'^{(i)}_t \angle \epsilon'^{(j)}_t\right). \tag{44}$$

In an analogous way, let $x'^{(i)}_t$ be the row vector of the normalized sequence of samples that were drawn from the i-th vector component of the multivariate distribution of the transformed state variable X'_t, that is

$$x'^{(i)}_t := \frac{1}{\sqrt{t}}\left(x'^{(i)}_0, \ldots, x'^{(i)}_{t-1}\right).$$

The correlations between the i-th and j-th components of X'_t in the steady state ($t \to \infty$) of the stochastic process are the reinforced correlations ρ'_{ij} between noise components. The reinforcement factor is $1/\left(1 - \lambda_i(A_0)\overline{\lambda_j(A_0)}\right)$, and there holds (Neumaier and Schneider 2001)

$$\frac{1}{1 - \lambda_i(A_0)\overline{\lambda_j(A_0)}} \rho'_{ij} = \lim_{t \to \infty} \cos\left(x'^{(i)}_t \angle x'^{(j)}_t\right). \tag{45}$$

In the above equation the terms $\overline{\lambda_j(A_0)}$ denote the complex conjugates of the eigenvalues. As mentioned earlier, this interesting reinforcement phenomenon will be discussed again in Chapter 4.

If A_0 is a symmetric matrix, we can also obtain expressive vector calculus forms of both the diagonal (representing the variances along the rotated coordinate axes in the spectral basis) and off-diagonal (representing the correlations ρ'_{ij} between them) elements of C', as shown in the following steps. This is because the eigenvectors $\vartheta_i(A_0)$ are mutually orthogonal and have only real components. However, if A_0 is not symmetric, the eigenvectors are not orthogonal, and the following simplifications are impossible.

If A_0 is symmetric, all p components of the eigenvectors are real, and Eq. 41 can be rewritten as

$$C' = S^{-1} \cdot C \cdot \left(S^{\mathrm{T}}\right)^{-1}.$$

The eigenvectors $\vartheta_i(A_0) = S_{:i}$ are mutually orthogonal. The normalized eigenvectors are denoted by $\tilde{\vartheta}_i \left(\left\|\tilde{\vartheta}_i\right\| = 1\right)$, the corresponding orthonormal matrix by $S_\perp$, and the mutually orthogonal (but not normalized) column vectors of the forcing matrix K by $k_i = K_{:i}$. We can write

$$C' = (S_\perp \cdot N)^{-1} \cdot C \cdot \left((S_\perp \cdot N)^{\mathrm{T}}\right)^{-1}$$

with

$$N = \mathrm{Diag}[\|\vartheta_i(A_0)\|], \quad 1 \le i \le p.$$

Because N is diagonal, the transpose can be written as

$$C' = (S_\perp \cdot N)^{-1} \cdot C \cdot \left(N \cdot S_\perp^{\mathrm{T}}\right)^{-1},$$

and the inverse can be factorized:

$$C' = N^{-1} \cdot S_\perp^{-1} \cdot C \cdot \left(S_\perp^{\mathrm{T}}\right)^{-1} \cdot N^{-1}.$$

For orthonormal matrices, $S_\perp^{-1} = S_\perp^{\mathrm{T}}$, and we have

$$\begin{aligned} C' &= N^{-1} \cdot S_\perp^{\mathrm{T}} \cdot C \cdot S_\perp \cdot N^{-1} \\ &= N^{-1} \cdot S_\perp^{\mathrm{T}} \cdot K \cdot \Lambda_K \cdot K^{\mathrm{T}} \cdot S_\perp \cdot N^{-1} \\ &= N^{-1} \cdot \left(S_\perp^{\mathrm{T}} \cdot K\right) \cdot \Lambda_K \cdot \left(S_\perp^{\mathrm{T}} \cdot K\right)^{\mathrm{T}} \cdot N^{-1} \end{aligned} \tag{46}$$

with

$$S_{\perp}^{\mathrm{T}} \cdot K = \begin{pmatrix} \tilde{\vartheta}_1 \cdot k_1 & \tilde{\vartheta}_1 \cdot k_2 & \cdots \\ \tilde{\vartheta}_2 \cdot k_1 & \tilde{\vartheta}_2 \cdot k_2 & \cdots \\ \cdots & \cdots & \ddots \end{pmatrix}$$

$$\left(S_{\perp}^{\mathrm{T}} \cdot K\right)^{\mathrm{T}} = \begin{pmatrix} \tilde{\vartheta}_1 \cdot k_1 & \tilde{\vartheta}_2 \cdot k_1 & \cdots \\ \tilde{\vartheta}_1 \cdot k_2 & \tilde{\vartheta}_2 \cdot k_2 & \cdots \\ \cdots & \cdots & \ddots \end{pmatrix}$$

$$\left(S_{\perp}^{\mathrm{T}} \cdot K\right) \cdot \Lambda_K = \begin{pmatrix} \tilde{\vartheta}_1 \cdot k_1\{\lambda_1(C)\} & \tilde{\vartheta}_1 \cdot k_2\{\lambda_2(C)\} & \cdots \\ \tilde{\vartheta}_2 \cdot k_1\{\lambda_1(C)\} & \tilde{\vartheta}_2 \cdot k_2\{\lambda_2(C)\} & \cdots \\ \cdots & \cdots & \ddots \end{pmatrix}.$$

Now we can formulate a simple geometric relationship for the diagonal elements of C'

$$C'_{[[i,i]]} = \frac{1}{\|\vartheta_i\|^2} \sum_{n=1}^{p} \lambda_n(C) \left(\tilde{\vartheta}_i \cdot k_n\right)^2,$$

as well as for the off-diagonal elements

$$C'_{[[i,j]]} = \frac{1}{\|\vartheta_i\| \|\vartheta_j\|} \sum_{n=1}^{p} \lambda_n(C) \left(\tilde{\vartheta}_i \cdot k_n\right) \left(\tilde{\vartheta}_j \cdot k_n\right).$$

The closed-form solution for the correlations ρ'_{ij} embedded in C' is:

$$\begin{aligned} \rho'_{ij} &= \frac{\|\vartheta_i\| \|\vartheta_j\|}{\|\vartheta_i\| \|\vartheta_j\|} \cdot \frac{\Sigma_{n=1}^{p} \lambda_n(C) \left(\tilde{\vartheta}_i \cdot k_n\right) \left(\tilde{\vartheta}_j \cdot k_n\right)}{\sqrt{\left(\Sigma_{n=1}^{p} \lambda_n(C) \left(\tilde{\vartheta}_i \cdot k_n\right)^2\right) \left(\Sigma_{n=1}^{p} \lambda_n(C) \left(\tilde{\vartheta}_j \cdot k_n\right)^2\right)}} \\ &= \frac{\Sigma_{n=1}^{p} \lambda_n(C) \left(\tilde{\vartheta}_i \cdot k_n\right) \left(\tilde{\vartheta}_j \cdot k_n\right)}{\sqrt{\left(\Sigma_{n=1}^{p} \lambda_n(C) \left(\tilde{\vartheta}_i \cdot k_n\right)^2\right) \left(\Sigma_{n=1}^{p} \lambda_n(C) \left(\tilde{\vartheta}_j \cdot k_n\right)^2\right)}}. \end{aligned} \tag{47}$$

When we analyze Eqs. 46 and 47, it is not difficult to see that the correlations ρ'_{ij} in the spectral basis are zero, in either of the following cases.

(i) The column vectors of the forcing matrix K and the column vectors of the transformation matrix S are pairwise collinear, and $S = K \cdot \Lambda_{SK}$ ($\Lambda_{SK} = \mathrm{Diag}[c_i], c_i \in \mathbb{R}$).

(ii) The forcing matrix K is equal to the identity matrix I_p, and the fluctuations are isotropic with overall strength $\sigma^2 \in \mathbb{R}^+$, that is

$$\Lambda_K = \mathrm{Diag}[\lambda(C)] = \{\sigma^2\} \cdot I_p.$$

In case (i) the off-diagonal elements $C'_{[[i,j]]}$ are zero, because $\tilde{\vartheta}_i \cdot k_{n'} \neq 0$ for only one column vector $k_{n'}$ with index n' that is aligned with $\tilde{\vartheta}_{n'}$, while for this index $\tilde{\vartheta}_i \cdot k_{n'} = 0$ for all $j \neq i$, because the vectors are mutually orthogonal (keeping in mind that A_0 is supposed to be symmetric). Therefore,

$$\sum_{n=1}^{p} \lambda_n(C) \left(\tilde{\vartheta}_i \cdot k_n\right) \left(\tilde{\vartheta}_j \cdot k_n\right) = 0,$$

from which $C'_{[[i,j]]} = 0$ and $\rho'_{ij} = 0$ follow.

In case (ii) we can substitute $K = I_p$, as well as $\Lambda_K = \{\sigma^2\} \cdot I_p$ in Eq. 46:

$$\begin{aligned} C' &= N^{-1} \cdot \left(S_\perp^{\mathrm{T}} \cdot I_p\right) \cdot \{\sigma^2\} \cdot I_p \cdot \left(S_\perp^{\mathrm{T}} \cdot I_p\right)^{\mathrm{T}} \cdot N^{-1} \\ &= \{\sigma^2\} \cdot N^{-1} \cdot S_\perp^{\mathrm{T}} \cdot S_\perp \cdot N^{-1} \\ &= \{\sigma^2\} \cdot \left(N^{-1}\right)^2. \end{aligned}$$

Since N is a diagonal matrix, again $C'_{[[i,j]]} = 0$ and $\rho'_{ij} = 0$ follow.

Interestingly, under the assumption of isotropic fluctuations, that is $C = \{\sigma^2\} \cdot I_p$ or $K = I_p$ and $\Lambda_K = \{\sigma^2\} \cdot I_p$, only the dynamic part of the iteration process and not the fluctuations are relevant for evaluating emergent complexity (see Eqs. 250 and 251). This also holds for a dynamical operator A_0 that is not symmetric.

We can also write the state equation (39) in the spectral basis component-wise (Neumaier and Schneider 2001). For each vector component the eigenvalue $\lambda_i = \lambda_i(A_0)$ is the dynamical operator of the scalar state equation with coefficient $X'^{(i)}_t$:

$$X'^{(i)}_t = \lambda_i X'^{(i)}_{t-1} + \varepsilon'^{(i)}_t \quad t = 1, \ldots, T. \tag{48}$$

It is important to note that in the scalar state equations the fluctuations are correlated and that for the expectations it holds for all time steps $\mu \in \mathbb{Z}$ that

$$\begin{aligned} E\left[\varepsilon'^{(i)}_\mu \left[\varepsilon'^{(j)}_t\right]^*\right] &= \delta_{\mu t} C'_{[[i,j]]} \\ &= \delta_{\mu t} \rho'_{ij} C'_{[[i,i]]} C'_{[[j,j]]}, \end{aligned} \tag{49}$$

where $\rho'_{ij}(k, l = 1, \ldots, p)$ is the correlation coefficient defined in Eq. 43 and $\delta_{\mu t}$ is the Kronecker delta (Eq. 14). From Eq. 48 it follows that for $\mathrm{Arg}[\lambda_i] \neq 0$ the expectations

$$E\left[X_t'^{(i)}\right] = \lambda_i E\left[X_{t-1}'^{(i)}\right]$$

describe spirals in the Gaussian plane represented by

$$\begin{aligned} E\left[X_{t+\tau}'^{(i)}\right] &= \lambda_i^\tau E\left[X_t'^{(i)}\right] \\ &= e^{-\frac{\tau}{\tau_i}} e^{(\mathrm{Arg}[\lambda_i])i\tau} E\left[X_t'^{(i)}\right] \end{aligned} \tag{50}$$

with damping time scales (Neumaier and Schneider 2001)

$$\tau_i := -\frac{1}{\log|\lambda_i|} \tag{51}$$

and periods

$$T_i := \frac{2\pi}{|\mathrm{Arg}[\lambda_i]|}. \tag{52}$$

The function Arg[λ_i] denotes the argument of the possibly complex eigenvalue $\lambda_i = a_i + ib_i$, which can be computed as

$$\mathrm{Arg}[\lambda_i] = \tan^{-1}\left(\frac{a_i}{b_i}\right).$$

We use the convention that $-\pi \le \mathrm{Arg}[\lambda_i] \le \pi$ to ensure that a pair of complex conjugate eigenvalues of the dynamical operator is associated with a single period of the described spiral. For a stable stochastic process $\{X_t\}$ all eigenvalues must be less than one in magnitude and therefore the damping time scale τ_i is positive and bounded. If the eigenvalue λ_i of interest has a nonzero imaginary part or is real but negative, the period T_i is also bounded. In this case we can consider the scalar stochastic process $\{X_t'^{(i)}\}$ as a stochastically driven damped oscillator (Neumaier and Schneider 2001). The period of the damped oscillator is minimal if λ_i is real and negative and we have $T_i = 2$ (or $|\mathrm{Arg}[\lambda_i]| = \pi$). This period is equivalent to the famous Nyquist frequency which plays an important role in Fourier analysis and system theory (see e.g. Puri 2010). In contrast, if the eigenvalue λ_i is real and positive, then the period T_i grows over all limits ($T_i \to \infty$) and the scalar-valued system spirals indefinitely around the expectation value. In this case, the scalar stochastic process $\{X_t'^{(i)}\}$ can be regarded as a stochastically driven relaxator (Neumaier and Schneider 2001). Therefore, the linear combinations according to Eqs. 37 and 38 decompose the vector autoregression process $\{X_t\}$ generated by state Eq. 8 into linear combinations of damped oscillators and relaxators with oscillation and relaxation modes $\vartheta_i(A_0)$ that operate on damping time scales τ_i and have periods T_i.

Where a stochastic process is asymptotically stable in the meaning of Lyapunov (Eq. 4), the time evolution of the scalar states $\{X_t'^{(i)}\}$ of the damped oscillators and relaxators combined in a linear way can be represented in the complex plane in the form of three schematic diagrams. These schematic diagrams are shown in Fig. 2.6.

Since the work transformation matrix A_0 is real, its eigenvalues and eigenvectors can be written as complex conjugate pairs. Based on these pairs we can also decompose the introduced linear autoregression model (Eq. 8) into real rather than complex design modes (see e.g. Neumaier and Schneider 2001 or Hinrichsen and Pritchard 2005). In fact, for any complex eigenvalue $\lambda_i = a_i + ib_i$ the scalar process $\{X_t'^{(i)}\}$ defined by state Eq. 48 can be expressed as a real bivariate autoregressive process of first order

$$\begin{pmatrix} \mathrm{Re}\left[X_t'^{(i)}\right] \\ \mathrm{Im}\left[X_t'^{(i)}\right] \end{pmatrix} = \begin{pmatrix} a_i & -b_i \\ b_i & a_i \end{pmatrix} \begin{pmatrix} \mathrm{Re}\left[X_{t-1}'^{(i)}\right] \\ \mathrm{Im}\left[X_{t-1}'^{(i)}\right] \end{pmatrix} + \varepsilon_t'^{(i)} \tag{53}$$

with the bivariate random vectors

$$\varepsilon_t'^{(i)} = \begin{pmatrix} \mathrm{Re}\left[\varepsilon_t^{(i)}\right] \\ \mathrm{Im}\left[\varepsilon_t^{(i)}\right] \end{pmatrix} = \frac{1}{2} \begin{pmatrix} \varepsilon_t'^{(i)} + \varepsilon_t^{(i)} \\ i\left(\varepsilon_t'^{(i)} - \varepsilon_t^{(i)}\right) \end{pmatrix}.$$

In the above equation the components indexed by i' are conjugates of the components indexed by i. From the definition of the correlated fluctuations in Eq. 49, the covariance matrix can be written as (Neumaier and Schneider 2001):

$$\begin{aligned}
&E\left[\varepsilon_\mu'^{(k)}\left[\varepsilon_\nu'^{(l)}\right]\right] \\
&= \frac{\delta_{\mu\nu}}{4} \begin{pmatrix} C'_{[\![k,l]\!]} + C'_{[\![k',l']\!]} + C'_{[\![k',l]\!]} + C'_{[\![k,l']\!]} & -i\left(-C'_{[\![k,l]\!]} + C'_{[\![k',l']\!]} - C'_{[\![k',l]\!]} + C'_{[\![k,l']\!]}\right) \\ -i\left(-C'_{[\![k,l]\!]} + C'_{[\![k',l']\!]} - C'_{[\![k',l]\!]} + C'_{[\![k,l']\!]}\right) & C'_{[\![k,l]\!]} + C'_{[\![k',l']\!]} - C'_{[\![k',l]\!]} - C'_{[\![k,l']\!]} \end{pmatrix} \\
&= \frac{\delta_{\mu\nu}}{2} \begin{pmatrix} \mathrm{Re}\left[C'_{[\![k,l]\!]}\right] + \mathrm{Re}\left[C'_{[\![k',l]\!]}\right] & \mathrm{Im}\left[C'_{[\![l,k]\!]}\right] + \mathrm{Im}\left[C'_{[\![l,k']\!]}\right] \\ \mathrm{Im}\left[C'_{[\![k,l]\!]}\right] + \mathrm{Im}\left[C'_{[\![k',l]\!]}\right] & \mathrm{Re}\left[C'_{[\![k,l]\!]}\right] - \mathrm{Re}\left[C'_{[\![k',l]\!]}\right] \end{pmatrix} \\
&= \frac{\delta_{\mu\nu}}{2} \begin{pmatrix} \mathrm{Re}\left[\rho'_{kl}C'_{[\![k,k]\!]}C'_{[\![l,l]\!]}\right] + \mathrm{Re}\left[\rho'_{k'l}C'_{[\![k',k']\!]}C'_{[\![l,l]\!]}\right] & \mathrm{Im}\left[\rho'_{lk}C'_{[\![l,l]\!]}C'_{[\![k,k]\!]}\right] + \mathrm{Im}\left[\rho'_{lk'}C'_{[\![l,l]\!]}C'_{[\![k',k']\!]}\right] \\ \mathrm{Im}\left[\rho'_{kl}C'_{[\![k,k]\!]}C'_{[\![l,l]\!]}\right] + \mathrm{Im}\left[\rho'_{k'l}C'_{[\![k',k']\!]}C'_{[\![l,l]\!]}\right] & \mathrm{Re}\left[\rho'_{kl}C'_{[\![k,k]\!]}C'_{[\![l,l]\!]}\right] - \mathrm{Re}\left[\rho'_{k'l}C'_{[\![k',k']\!]}C'_{[\![l,l]\!]}\right] \end{pmatrix}.
\end{aligned}$$

The eigenmodes of the decomposed process are given by the real and imaginary parts of the eigenvector $\vartheta_i(A_0) = S_{:i}$.

For small PD projects with only $p = 2$ tasks, we can obtain simple analytical solutions for the eigenvalues, eigenvectors and correlation coefficient in the spectral basis. To do so, we use the following parametric representation of the WTM and the covariance matrix of the fluctuations in the original state space coordinates:

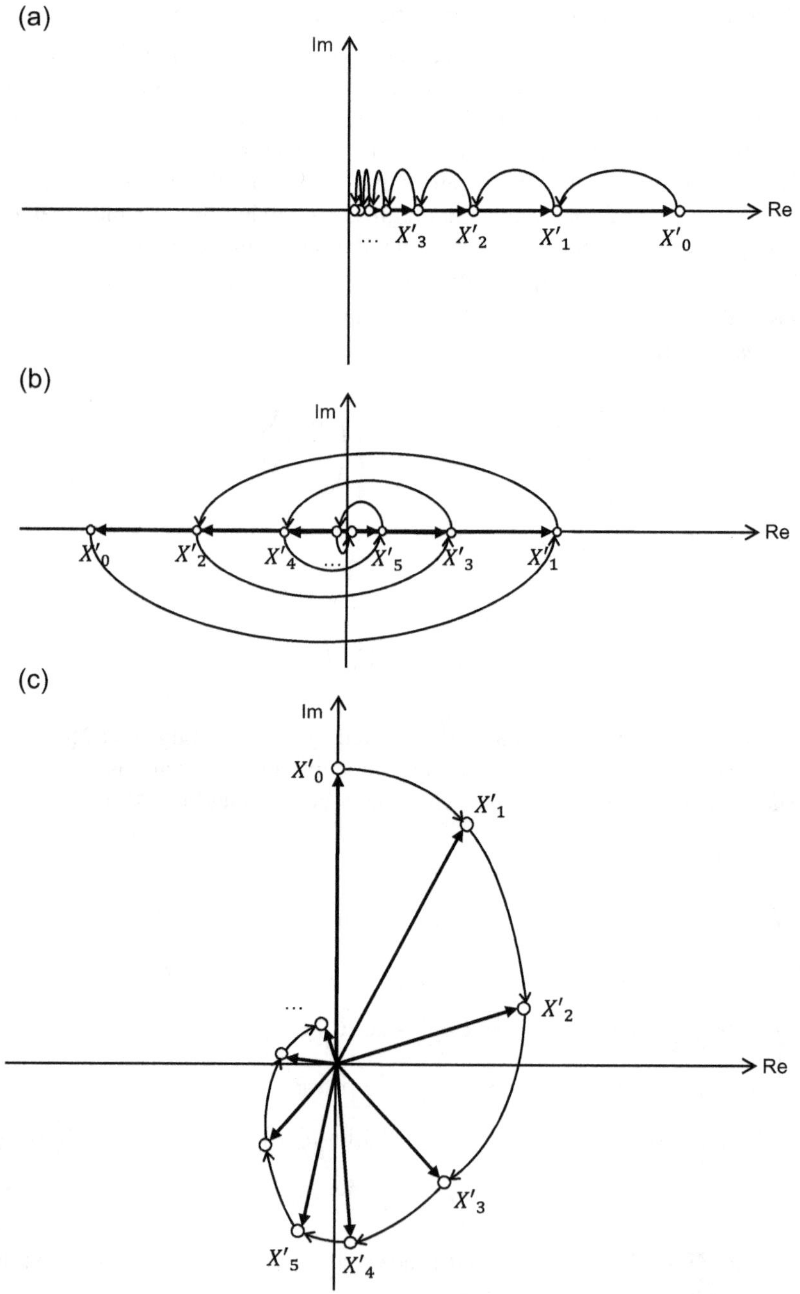

Fig. 2.6 Schematic diagrams of the time evolution of the scalar states $X'^{(i)}_t$ of relaxators and damped oscillators in the complex plane. The superscript (i) was dropped for greater clarity. The initial value is therefore represented by X'_0. The thick arrows represent the scalar states. The thin arrows indicate the transitions between states. (**a**) Relaxator: λ_i is real and $\lambda_i > 0$. (**b**) Stochastically driven damped oscillator with minimum period $T_i = 2$: λ_i is real and $\lambda_i < 0$. (**c**) Stochastically driven damped oscillator with period $T_i > 2$: λ_i is complex (adopted from von Storch et al. 1995)

$$A_0 = \begin{pmatrix} a_{11} & a_{12} \\ a_{21} & a_{11} + \Delta a \end{pmatrix} \tag{54}$$

$$C = \begin{pmatrix} \sigma_{11}^2 & \rho\sigma_{11}\sigma_{22} \\ \rho\sigma_{11}\sigma_{22} & \sigma_{22}^2 \end{pmatrix}, \tag{55}$$

where $\{a_{11}, \sigma_{11}, \sigma_{22}\} \in \mathbb{R}^+$, $\{\Delta a, a_{12}, a_{21}\} \in \mathbb{R}$ and $\rho \in [-1; 1]$. Note that the parametric representation of the autonomous task processing rates a_{11} and $a_{22} = a_{11} + \Delta a$ through the rate difference Δa is slightly different than in the applied example which will be presented Section 2.5. The rate difference is introduced in order to obtain solutions that are easier to interpret.

The above parametric representation leads to the eigenvalues

$$\lambda_1 = \frac{1}{2}(2a_{11} + \Delta a - \sqrt{g}) \tag{56}$$

$$\lambda_2 = \frac{1}{2}(2a_{11} + \Delta a + \sqrt{g}), \tag{57}$$

the infinite families ($\{c_1, c_2\} \in \mathbb{R}$) of eigenvectors

$$\vartheta_1 = \{c_1\} \cdot \begin{pmatrix} -\dfrac{\Delta a + \sqrt{g}}{2a_{21}} \\ 1 \end{pmatrix}$$

$$\vartheta_2 = \{c_2\} \cdot \begin{pmatrix} -\dfrac{\Delta a - \sqrt{g}}{2a_{21}} \\ 1 \end{pmatrix},$$

and therefore to the matrix

$$S = \begin{pmatrix} -\dfrac{\Delta a + \sqrt{g}}{2a_{21}} & -\dfrac{\Delta a - \sqrt{g}}{2a_{21}} \\ 1 & 1 \end{pmatrix}$$

for the basis transformation. In order to obtain an instance of an autoregressive model that is asymptotically stable in the sense of Lyapunov (Eq. 4), the spectral radius $\rho(A_0)$ must be less than one and therefore it must hold that $1/2(2a_{11} + \Delta a + \sqrt{g}) < 1$.

In the spectral basis, the transformed variances $\sigma'^2_{11} = C'_{[1,1]}$ and $\sigma'^2_{22} = C'_{[2,2]}$ are given by

$$\sigma'^2_{11} = \frac{4a_{21}^2\sigma_{11}^2 + 4\rho(\Delta a - \mathrm{Re}[\sqrt{g}]a_{12}\sigma_{11}\sigma_{22}) + (\mathrm{Abs}[g] + \Delta a(\Delta a - 2\mathrm{Re}[\sqrt{g}]))\sigma_{22}^2}{4\mathrm{Abs}[g]}$$
$$\sigma'^2_{22} = \frac{4a_{21}^2\sigma_{11}^2 + 4\rho(\Delta a + \mathrm{Re}[\sqrt{g}]a_{12}\sigma_{11}\sigma_{22}) + \mathrm{Abs}[g] + \Delta a(\Delta a + 2\mathrm{Re}[\sqrt{g}])\sigma_{22}^2}{4\mathrm{Abs}[g]}.$$

The correlation coefficient ρ' can be expressed in the spectral basis as

$$\rho' = \frac{(-4a_{21}^2\sigma_{11}^2 - 4\rho a_{21}(\Delta a - \mathrm{Re}[\sqrt{g}] + \sqrt{g}))\sigma_{11}\sigma_{22} + (\mathrm{Abs}[g] - \Delta a(\Delta a + 2i\mathrm{Im}[\sqrt{g}]))\sigma_{22}^2}{\sigma'_{11}\sigma'_{22}}.$$

In the equations above we have used the coupling constant g which is defined as

$$g := \Delta a^2 + 4a_{12}a_{21}.$$

$\sqrt{g}$ is the distance between the eigenvalues, see Eqs. 56 and 57. If all entries of the WTM A_0 are nonnegative, the correlation coefficient ρ' can be further simplified and we have

$$\rho' = \frac{\sigma_{22}(a_{12}\sigma_{22} - \Delta a\rho\sigma_{11}) - a_{21}\sigma_{11}^2}{\sqrt{(a_{21}^2\sigma_{11}^4 + 2\Delta a\rho a_{21}\sigma_{11}^3\sigma_{22} + (\Delta a^2 + 2(1-2\rho^2)a_{12}a_{21})\sigma_{11}^2\sigma_{22}^2 - 2\Delta a\rho a_{12}\sigma_{11}\sigma_{22}^3 + a_{12}^2a_{22}^4)}}.$$

Furthermore, if not only all the entries of the WTM are nonnegative, but also the correlation coefficient of the fluctuations in the original state space coordinates is zero, i.e. $\rho = 0$, we arrive at the most simple parametric form:

$$\rho' = \frac{a_{12}\sigma_{22}^2 - a_{21}\sigma_{11}^2}{\sqrt{(a_{21}\sigma_{11}^2)^2 + (\Delta a^2 + 2a_{12}a_{21})\sigma_{11}^2\sigma_{22}^2 + (a_{12}\sigma_{22}^2)^2}}.$$

It is evident that the transformed correlation coefficient ρ' is zero if the fluctuations are isotropic and the WTM is symmetric.

2.4 Formulation of Higher-Order Models, Least Squares Parameter Estimation and Model Selection

Neumaier and Schneider (2001) term the stochastic process that is generated by the introduced linear autoregression models (Eqs. 8 and 39) a vector autoregressive process of order 1, abbreviated as VAR(1) process, with zero mean. A VAR

(1) process with zero mean is the least complex stochastic process in its class. In the original state-space coordinates, a VAR(1) model can be easily generalized by increasing the regression order and therefore the correlation length of the process. When the regression order is increased not only the present state of the development project is considered to make accurate predictions of its future course but also states that lay in the finite past. The corresponding vector autoregression model of order n, abbreviated as VAR(n) model, is defined by the extended state equation:

$$X_t = \sum_{i=0}^{n-1} A_i \cdot X_{t-i-1} + \varepsilon_t. \tag{58}$$

Similar to a VAR(1) model, $\{X_t\}$ is a time series of p-dimensional state vectors representing the work remaining observed at equal space time instants t. The probability density function of the vector ε_t was already given in Eq. 13 and it holds that $\varepsilon_t \sim \mathcal{N}(0_p, C)$. One could also include a p-dimensional parameter vector ω of intercept terms to allow for a nonzero mean of the time series (see Neumaier and Schneider 2001 or Lütkepohl 2005). From a theoretical point of view, this systematic shift or drift of work remaining is not essential in the model formulation and is consequently ignored in the following analysis. Like the first-order model (see Fig. 2.4), Fig. 2.7 shows a graphical representation of the second-order autoregression model in the form of a dynamic Bayesian network. By generalizing the graphical model shown, it becomes clear that the *VAR*(n) model's memory only reaches back n time steps into the past because its state is the weighted value of the last n observations (Rudary 2009).

The VAR(n) model is one of the most flexible models for the analysis of multivariate time series. This model has proven especially useful for describing the dynamic behavior of economic time series and for forecasting. Neumaier and Schneider (2001), Franses and Paap (2004), Lütkepohl (2005) and others developed efficient and robust methods to estimate the order of the model, the values of its parameters, spectral information and confidence regions based on empirically acquired time series. We will describe the parameter estimation methods for vector autoregression models based on the material from Neumaier and Schneider (2001). The description of the criteria for model selection is based on the textbooks by Burnham and Anderson (2002) and Lütkepohl (2005). All methods are standard and

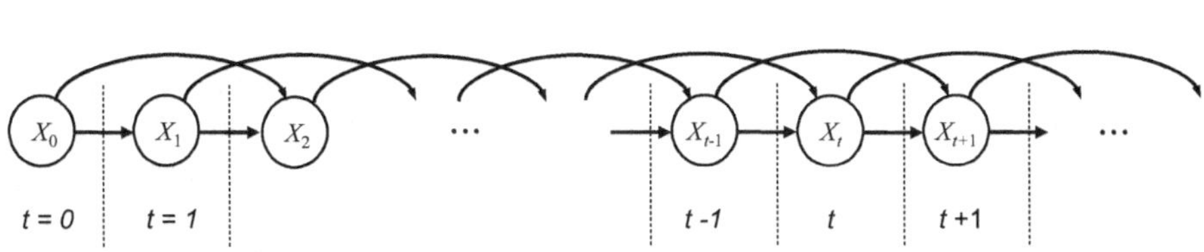

Fig. 2.7 Graphical representation of the second-order autoregression model in the form of a dynamic Bayesian network. The nodes in the graph represent the random state variables of the stochastic process. The directed arcs represent conditional dependencies between the variables

can, in part, be found in many other textbooks, e.g. Box and Jenkins (1976), Marple (1987), Kay (1988), Soderstrom and Stoica (1989) and Stoica and Moses (1997).

An important result in the theory of stochastic dynamical systems is that every linear autoregression model of finite order generating a process $\{X_t\}_{1-n}^{T} = (X_{1-n}, \ldots, X_T)$ of p-dimensional state vectors can be rewritten as a first-order model based on the state equation

$$\tilde{X}_t = \tilde{A} \cdot \tilde{X}_{t-1} + \tilde{\varepsilon}_t \quad t = 1, \ldots, T, \tag{59}$$

where $\tilde{X}_t$ is the augmented state vector

$$\tilde{X}_t = \begin{pmatrix} X_t \\ X_{t-1} \\ \vdots \\ X_{t-n+1} \end{pmatrix} \in \mathbb{R}^{np}, \tag{60}$$

$\tilde{\varepsilon}_t$ is the augmented noise vector

$$\tilde{\varepsilon}_t = \begin{pmatrix} \varepsilon_t \\ 0 \\ \vdots \\ 0 \end{pmatrix} \tag{61}$$

and $\tilde{A}$ is the extended dynamical operator

$$\tilde{A} = \begin{pmatrix} A_0 & A_1 & \cdots & A_{n-2} & A_{n-1} \\ I_p & 0 & \cdots & 0 & 0 \\ 0 & I_p & \cdots & 0 & 0 \\ & & & & \\ 0 & 0 & \ddots & 0 & 0 \\ 0 & 0 & \ldots & I_p & 0 \end{pmatrix} \in \mathbb{R}^{np \times np}. \tag{62}$$

This order reduction by state-space augmentation makes it possible to develop relatively simple parameter estimation algorithms (see literature review in Neumaier and Schneider 2001). However, the only problem with this approach is that the augmented state vector $\tilde{X}_t$ has vector components that are also included in the previous state vector $\tilde{X}_{t-1}$ and therefore the past and future are not completely shielded in information-theoretic terms, given the present state. To be able to apply the complexity measures that will be presented in Sections 4.1.1, 4.1.2 and 4.1.3 in different closed forms directly to the higher-order model in order to evaluate emergent complexity in PD projects, one has to find a state representation with disjoint vector components. This can be easily done as will be shown in Section 4.1.6.

If process order n is known in advance, various techniques exist to estimate coefficient matrices $A_0, \ldots, A_{n-1}$ in terms of multiple work transformation matrices and the covariance matrix C from a time series of empirically acquired state vectors x_t (see Neumaier and Schneider 2001). The most prominent techniques are based on the well-known Maximum Likelihood Estimation (MLE, see for instance Brockwell and Davis 1991). For a fixed series of data and an underlying parameterized model, MLE picks the value of parameters that maximize the probability of generating the observations or minimize the one-step prediction error. It is usually assumed that the fluctuations are uncorrelated from time step to time step and that it holds for all time steps $\{\mu, \nu\} \in \mathbb{Z}$ that $E\left[\varepsilon_\mu \varepsilon_\nu^{\mathrm{T}}\right] = \left\{\delta_{\mu\nu}\right\} \cdot C$. Since the covariance matrix C depends on the regression parameters, the likelihood must be maximized in several iterations. The iterative estimates are not only asymptotically efficient as the sample size grows to infinity, they also show good small-sample performance (Brockwell and Davis 1991). An alternative iterative MLE technique with good small-sample performance that can also deal with "hidden" (not directly observable) state variables will be introduced in Section 2.10. Although efficiency and robustness are advantageous properties for estimation techniques, the iterative procedure is computationally quite demanding and slows the parameter estimation for large projects significantly. A much simpler technique is based on the classic least squares estimation (Neumaier and Schneider 2001). Our analyses have shown that in the application area of project management least-square estimation is comparatively accurate and robust. The least-square estimates can also be used as a bootstrap for subsequent maximum likelihood iterations. Furthermore, least-square estimation can be intuitively extended by information-theoretic or Bayesian model selection techniques. This reveals interesting theoretical connections between selecting a predictive model based on universal criteria and evaluating its statistical or information-theoretical complexity. We will return to this important point in Sections 3.2.2 and 3.2.4.

According to Neumaier and Schneider (2001) the least squares estimates for a VAR(n) process are most conveniently derived when the process $\{X_t\}_{1-n}^{T} = (X_{1-n}, \ldots, X_T)$ generated by state Eq. 58 is cast in the classic linear regression form

$$\begin{aligned} X_t &= A \cdot U_t + \varepsilon_t \quad t = 1, \ldots, T \\ \varepsilon_t &\sim \mathcal{N}\left(0_p, C\right), \end{aligned} \tag{63}$$

with the coefficient matrix

$$A = \begin{pmatrix} A_0 & \cdots & A_{n-1} \end{pmatrix}$$

and predictors

$$U_t = \begin{pmatrix} X_{t-1} \\ \vdots \\ X_{t-n} \end{pmatrix} \in \mathbb{R}^{n_p}, \quad n_p = pn.$$

The key insight in the derivation of the least squares estimates of the independent parameters is to consider the casted regression model as a model with fixed predictors U_t. Clearly, this is only an approximation, because for the multivariate time series the U_t are realizations of a random variable. However, the above definition of the predictors implies that the assumption of fixed predictors leads to treating

$$U_1 = \begin{pmatrix} X_0 \\ \vdots \\ X_{1-n} \end{pmatrix}$$

as a vector of fixed initial states. Since the relative effect of the initial state vanishes as the length T of the time series approaches infinity, using corresponding parameter estimates for the regression model in the autoregressive model formulation can be expected to be asymptotically correct (Neumaier and Schneider 2001). In fact, when using a linear regression model, the least squares principle leads to the best unbiased parameter estimates. We define the following data-related matrices obtained from the observations x_t and u_t of the random vectors X_t and U_t:

$$U = \sum_{t=1}^{T} u_t u_t^{\mathrm{T}}$$
$$V = \sum_{t=1}^{T} x_t x_t^{\mathrm{T}}$$
$$W = \sum_{t=1}^{T} x_t u_t^{\mathrm{T}}.$$

The least square estimate for the coefficient matrix A can be written as the matrix product

$$\hat{A} = W \cdot U^{-1}. \tag{64}$$

The corresponding estimate for the covariance matrix is given by

$$\hat{C} = \frac{1}{T - n_p} \sum_{t=1}^{T} \left(x_t - \hat{A} \cdot u_t\right)\left(x_t - \hat{A} \cdot u_t\right)^{\mathrm{T}}. \tag{65}$$

The leading factor $1/(T - n_p)$ is used to adjust the degrees of freedom of the covariance matrix. In terms of a multivariate regression, the estimated covariance matrix can be interpreted as a point estimate of the inherent one-step prediction error.

Alternatively, the estimate of the covariance matrix can be expressed as

$$\hat{C} = \frac{1}{T - n_p}\left(V - W \cdot U^{-1} \cdot W^{\mathrm{T}}\right).$$

This estimate is proportional to the Schur complement of the composed matrix

$$\begin{pmatrix} U & W^{\mathrm{T}} \\ W & V \end{pmatrix} = \sum_{t=1}^{T} \begin{pmatrix} u_t \\ x_t \end{pmatrix} \cdot \begin{pmatrix} u_t \\ x_t \end{pmatrix}^{T} = K^{\mathrm{T}} \cdot K,$$

where the aggregated data matrix K is defined as

$$K = \begin{pmatrix} u_1^{\mathrm{T}} & x_1^{\mathrm{T}} \\ \vdots & \vdots \\ u_T^{\mathrm{T}} & x_T^{\mathrm{T}} \end{pmatrix},$$

and therefore assured to be positive semidefinite. Neumaier and Schneider (2001) have shown that a QR factorization of the data matrix K as

$$K = Q \cdot R$$

with an orthogonal matrix Q and an upper triangular matrix

$$R = \begin{pmatrix} R_{11} & R_{12} \\ 0 & R_{22} \end{pmatrix}$$

allows the development of an efficient procedure to compute the parameter estimates numerically. Therefore, the above Schur complement is rewritten as

$$\begin{pmatrix} U & W^{\mathrm{T}} \\ W & V \end{pmatrix} = R^{\mathrm{T}} \cdot R = \begin{pmatrix} R_{11}^{\mathrm{T}} \cdot R_{11} & R_{11}^{\mathrm{T}} \cdot R_{12} \\ R_{12}^{\mathrm{T}} \cdot R_{11} & R_{12}^{\mathrm{T}} \cdot R_{12} + R_{22}^{\mathrm{T}} \cdot R_{22} \end{pmatrix}.$$

Based on the rewritten Schur complement, the following least squares estimates are obtained:

$$\hat{A} = \left(R_{11}^{-1} \cdot R_{12}\right)^{\mathrm{T}}$$
$$\hat{C} = \frac{1}{T - n_p}\left(R_{22}^{\mathrm{T}} \cdot R_{22}\right).$$

The estimate for the initial state is simply

$$\widehat{U}_1 = \begin{pmatrix} x_0 \\ \vdots \\ x_{1-n} \end{pmatrix}.$$

As an alternative to the QR factorization, the matrix R can be obtained from a Cholesky decomposition. Furthermore, regularization schemes can be used to reduce effects of noise. More details are available in Neumaier and Schneider (2001).

If not only a single time series of empirically acquired state vectors x_t is given but multiple realizations of the stochastic process had been acquired in N independent measurement trials, the additional time series are simply appended as additional $N - 1$ blocks of rows in the regression Eq. 63. Similarly, the predictors are extended by additional row blocks. The initial state is determined by averaging over all initial data points.

If not only the coefficient matrices and the covariance matrix of the VAR(n) process have to be estimated from data but also the model order, a good trade-off between the predictive accuracy gained by increasing the number of independent parameters and the danger of overfitting the model has to be found. Overfitting means in our context of project management that the model is fitted to unpredictable performance fluctuations instead of the implicit or explicit rules of cooperative work that are necessary for the functioning of the project. In order to find an optimal solution to the trade-off in terms of a universal principle, Rissanen's (1989, 2007) minimum description length principle aims at selecting the model with the briefest recording of all attribute information–not only the likelihood of a fixed series of data and an underlying parameterized model. This integrative view provides a natural safeguard against overfitting as it defines a method to reduce the part of the data that looks like noise by using a more elaborate—but in the sense of Occam's Razor not unnecessary complex—model. We will come back to this important model selection principle in Section 3.2.2 when we discuss the stochastic complexity of a generative model within a simple parametric model class compris-ing of distributions indexed by a specific parameter set. In the following, we take a more pragmatic view and select the order of an autoregression model on the basis of the standard selection criteria of Akaike (Final Prediction Error Criterion as well as Information Criterion, see Akaike 1971, 1973 and 1974) and Schwarz (Schwarz-Bayes Criterion, see Schwarz 1978, also termed Bayesian information criterion). These criteria had been validated in many scientific studies (see e.g. Lütkepohl 1985 and 2005) and can also be calculated easily. For regular parametric distribu-tion families of dimension p, the simplified two-stage minimum description length criterion takes the form of the Schwarz-Bayes Criterion (Eq. 71) and therefore penalizes for stationary data model complexity with the same factor as the Schwarz criterion (Hansen and Yu 2001). By the simplified two-stage minimum description length criterion, we mean an encoding scheme in which the description length for the best-fitting member of the model class is calculated in the first stage and then the description length of data based on the parameterized probability distribution is determined (see Section 3.2.2). However, it is important to point out that Schwarz's Bayesian approximation only holds if the number of parameters is kept fixed and the number of observations goes to infinity. If the number of parameters is either infinite or grows with the number of observations, then model selection based on minimum description length can lead to quite different results (see Grünwald 2007). A detailed comparison of these (and other) criteria can be found in Section 4.3 of Lütkepohl (2005) and Lütkepohl (1985).

Akaike's (1971) Final Prediction Error (FPE) is one of the historically most significant criteria and provides a generalized measure of model quality by simulating the situation where a parameterized model is tested on a different dataset. Clearly, the approximating model with smallest FPE is favored. The FPE is defined for a VAR(n) model according to Eq. 58 as

$$FPE(n) = \ln\left(\left(\frac{T+n_p}{T-n_p}\right)^p \mathrm{Det}\left[\widehat{\Sigma}_{(n)}\right]\right),$$

where

$$\widehat{\Sigma}_{(n)} := \frac{\hat{\Delta}_{(n)}}{T} \tag{66}$$

is a measure for the not biased corrected variance of the estimator. $\widehat{\Sigma}_{(n)}$ can be interpreted as the one-step prediction error of order n (Lütkepohl 2005). The determinant $\mathrm{Det}\left[\widehat{\Sigma}_{(n)}\right]$ is a scalar measure for the not biased corrected variance. $\hat{\Delta}_{(n)}$ denotes the not biased corrected estimate of the accumulated variances and covariances for the nth order model that was fitted to the time series of task processing in the project and it holds that

$$\begin{aligned}\hat{\Delta}_{(n)} &= (T-n_p)\hat{C}\\ &= R_{22}^{\mathrm{T}} \cdot R_{22}.\end{aligned}$$

In the chosen model formulation with Gaussian random variables, the weighted least squares estimation is equivalent to the maximum likelihood estimation. Note that the above definition of the FPE is only valid for an autoregression model without a parameter vector of intercept terms. If intercept terms are included, the model-related product n_p in the denominator and numerator of the first factor has to be corrected by $pn+1$ (Lütkepohl 2005). It is evident that the FPE can also be expressed as the sum

$$FPE(n) = \ln \mathrm{Det}\left[\widehat{\Sigma}_{(n)}\right] + p \ln\frac{T+n_p}{T-n_p},$$

which is the form of representation most frequently found in textbooks. In this form, the second summand $p\ln\left((T+n_p)/(T-n_p)\right)$ can be interpreted as a penalty term penalizing project models that are unnecessarily complex in the sense of Occam's Razor.

Based on the FPE, Akaike (1973, 1974) later developed an information-theoretic criterion that the literature calls the Akaike Information Criterion (AIC). The AIC is the most widely known and used criterion in statistical model selection among scientists and practitioners. To develop the AIC, Akaike proposed an information-theoretic framework wherein the estimation of the model parameters and model selection could be simultaneously accomplished. In fact, the AIC is an asymptotically unbiased estimator of the expected relative Kullback–Leibler distance, which represents the amount of information lost when we use approximating models g_i within a family G to approximate another model f. The AIC chooses the model with the smallest expected Kullback–Leibler distance. Model f may be the "true" model or not. If f itself is only a poor approximation of the true generative mechanisms, the AIC selects the relatively best among the poor models. Details about the Kullback–Leibler divergence and its underlying information-theory principles can be found in Cover and Thomas (1991). For a VAR(n) model, the AIC is defined as (Lütkepohl 2005):

$$AIC(n) = \ln \mathrm{Det}\left[\widehat{\Sigma}_{(n)}\right] + \frac{2}{T}k. \tag{67}$$

The variable k in the above definition of the criterion represents the effective number of parameters of the approximating model. In many publications, not only Lütkepohl's (2005) standard textbook on multiple time series analysis, the effective number of parameters is determined by counting the freely estimated parameters in the coefficient matrices $A_0, \ldots, A_{n-1}$ and can therefore, for practical applications in project management, be expressed as

$$k = np^2.$$

The same expression is used in the popular ARfit toolbox. The ARfit toolbox was published by Schneider and Neumaier (2001) in a paper accompanying their theoretical considerations and later revised to include multiple time series of empirically acquired state vectors x_t. We will use this toolbox in the numerical project modeling and simulation example in the following Section 2.5. The above approximation works well in many areas of application and leads to highly accurate results in model selection. However, it is important to note that the original criterion developed by Akaike (1973) determines the number of effective parameters based not only on the coefficient matrices $A_0, \ldots, A_{n-1}$ but also on the covariance matrix C of the inherent one-step prediction error. From a theoretical point of view (Cavanaugh and Neath 1999), we have to count the number of functionally independent parameters in the parameter vector θ which must be estimated. Hence, the effective number of parameters of the approximating model must be

$$k = np^2 + \frac{p(p+1)}{2} \tag{68}$$

in order to consider the functionally independent autonomous task processing rates a_{ii}, the informational couplings between tasks $a_{ij}(i \neq j)$, the scalar-valued variances c_{ii}^2 of performance fluctuations in work tasks i (Eq. 11) and the correlations $\rho_{ij}(i \neq j)$ between performance fluctuations (Eq. 12). The increased number of effective parameters can also be found in the paper by Hurvich and Tsai (1993) and the textbook by Burnham and Anderson (2002). In the numerical example of project modeling and simulation that will be presented in Section 2.5 we will use both expressions to penalize models that are unnecessarily complex and to compare the selected model orders. It turns out that both approaches lead to the same model selection decisions. Additional Monte Carlo studies have shown that using the smaller number of effective parameters does not significantly influence the model selection accuracy in use cases of practical interest.

The heuristic to estimate the effective number of parameters by counting the functionally independent parameters in the parameter vector θ which must be estimated can easily be generalized to other model classes, such as linear dynamical systems. This more complex class of models with latent state variables is introduced and discussed in Section 2.9.

Following the previous interpretation of the penalty term $2k/T$, AIC penalizes model complexity with a factor that scales linearly in the number of effective parameters and inversely in the number of observations in the joint ensemble.

When the effective number of parameters is expressed by np^2, it holds that (Lütkepohl 2005):

$$\ln FPE(n) = AIC(n) + 2\frac{p}{T} + o\left(T^{-2}\right).$$

The third term $o\left(T^{-2}\right)$ denotes an arbitrary sequence indexed by T that remains bounded when multiplied by T^{-2}. As the second term $2p/T$ does not depend on the model order n, $AIC(n)$ and $AIC(n) + 2p/T$ indicate their minimum for the same value of n. Hence, $\ln FPE(n)$ and $AIC(n)$ differ essentially by a term of order $o\left(T^{-2}\right)$. Due to this property the difference between $\ln FPE(n)$ and $AIC(n)$ tends rapidly towards zero for $T \to \infty$ and $FPE(n)$ and $AIC(n)$ are asymptotically equivalent.

If the sample size is small with respect to the number of estimated parameters, Burnham and Anderson (2002) recommend a version of AIC, which was developed by Hurvich and Tsai (1993) and is defined as follows:

$$AIC_c(n) = \ln \operatorname{Det}\left[\widehat{\Sigma}_{(n)}\right] + \frac{2b}{T}k, \tag{69}$$

where the scale factor b in the penalty term is given as

$$b = \frac{T}{T - (np + p + 1)}.$$

A simulation study of Hurvich and Tsai (1993) shows that AIC_c has superior bias properties and strongly outperforms AIC for vector autoregressive model selection in small samples. Burnham and Anderson (2002) advocate the use of AIC_c when it holds that

$$\frac{T}{k} < 40. \tag{70}$$

To make a decision about using AIC_c instead of AIC, one must consider the value of k for the highest-dimensional model in the set of candidate models (Burnham and Anderson 2002). It is evident that if the ratio T/k is sufficiently large, AIC and AIC_c are similar and will strongly tend to select the same model order. In a given analysis, either AIC or AIC_c must be used consistently (Burnham and Anderson 2002).

Finally, the Schwarz-Bayes Criterion is introduced and evaluated. This criterion is also one of the most widely known and used tools in statistical model selection. Other terms for this used in the literature are the Bayesian information criterion and the Schwarz information criterion. In view of the Bayesian methodology used to derive the criterion it is abbreviated as BIC in the following (note that Sawa (1978) developed a model selection criterion derived from a Bayesian modification of the AIC criterion, which is also sometimes called the Bayesian information criterion in the literature and some commercial software packages, see, e.g., Beal (2007), but should not be confused with the Schwarz-Bayes criterion). Schwarz derived the BIC to provide an asymptotic approximation to a transformation of the Bayesian posterior probability of an approximating model. In settings with a large sample size, the model favored by this criterion ideally corresponds to the approximating model, which is most probable a posteriori and therefore rendered most plausible by the observed data. This is indicated by minimum scores. The calculation of the BIC is based on the empirical logarithmic likelihood function and does not require the specification of prior distributions. To briefly explain Schwarz's deductive approach, we can refer to Bayes factors (Raftery 1996). They are the Bayesian analogues of the famous likelihood ratio tests (see e.g. Honerkamp 2002). Under the assumption that two approximating models g_1 and g_2 are regarded as equally probable a priori, the Bayes factor $l(g_1, g_2)$ represents the ratio of the posterior probabilities of the models. Which model is most probable a posteriori is determined by whether the Bayes factor $l(g_1, g_2)$ is greater or is less than one. Within a specific class of nested models, model selection based on BIC under certain conditions is very similar to model selection through Bayes factors as BIC provides a close approximation to the Bayes factor when the prior over the parameters is the unit information prior. The term unit information prior refers to a multivariate normal prior with its mean at the maximum likelihood estimate and variance equal to the expected information matrix for one observation (Kass and Wasserman 1995). For a VAR(n) model the BIC is defined as

$$BIC(n) = \ln \text{Det}\left[\widehat{\Sigma}_{(n)}\right] + \frac{\ln T}{T} k. \tag{71}$$

In a similar manner to the Akaike Information Criterion, the variable k represents the effective number of parameters of the approximating model (Eq. 68).

Under all three selection criteria, the order n_{opt} of the VAR(n) model is considered to be the optimal one if it is assigned minimum scores, that is

$$n_{opt} = \arg\min_n \begin{cases} FPE(n) \\ AIC(n) \\ BIC(n)\ . \end{cases} \tag{72}$$

A substantial advantage in using information-theoretic or Bayesian criteria is that they are also valid for nonnested models and can therefore also be used to evaluate sinusoidal performance curves with given amplitudes. Traditional likelihood ratio tests are defined only for nested models such as vector autoregression models of finite order, and this represents another substantial limitation on the use of hypothesis testing in model selection (sensu Burnham and Anderson 2002). Vector autoregression models are nested in the sense that a VAR(n_1) model can be considered as a special case of a VAR(n_2) model if it holds that $n_1 < n_2$. This is a direct consequence of the formulation of the state equation (Eq. 58). It is important to note that for *FPE* and *AIC* the estimate $\hat{n} = n_{opt}$ is inconsistent under the assumption that the maximum order is larger than the maximum evaluated order n_{max}. Furthermore, it can be shown that, under quite general conditions, the limiting probability for underestimating the model order is zero for both criteria and the probability of overfitting $P_{overfit}$ is a nonnegative constant.

Hence, for FPE and AIC it holds that

$$P_{overfit} \xrightarrow{T\to\infty} \begin{cases} 0 & for\ FPE \\ constant > 0 & for\ AIC \end{cases}$$

(Stoica and Selen 2004; Lütkepohl 2005). In other words, *FPE* and *AIC* tend to overestimate the true autoregression order.

In contrast, Schwarz's *BIC*(n) is strongly consistent for any dimension of the state space. For this criterion it can be shown that, under the assumption that the data-generating process belongs to the considered model class, the autoregression order is consistently selected and the probability of correct selection

$$P_{correctselection} \to 1$$

as $T \to \infty$ (Stoica and Selen 2004; Lütkepohl 2005).

2.5 Product Development Project Example from Industry and Model Validation

To demonstrate the state-space concept introduced and validate the developed models of cooperative work in PD projects with field data, a detailed analysis of the course of PD projects was carried out at a small industrial company in Germany (Schlick et al. 2008, 2012). The company develops mechanical and electronic sensor components for the automotive industry. We investigated task processing by a team of three engineers in a multiproject setting comprising projects A, B and C. Project A was the research focus, comprising 10 partially overlapping development tasks covering all project phases—from conceptual design of the particular sensor to product documentation for the customer. In addition, the workloads of two concurrent smaller projects, B and C, were acquired. The acquired time data of all three projects were very accurate, because the company used a barcode-based labor time system: an engineer in this company who starts processing a development task has to indicate this with a manual barcode scan on a predefined task identification sheet. When the task was finished, the task identifier was scanned again. The recorded "time-on-development-task" had a resolution of 1 min and was used as an estimator for the values of the components of the introduced state variable X_t representing the work remaining. Let $y_t^{(i)}$ be the recorded time-on-task for task i at time step t. The estimated mean work remaining for the i-th component of the state variable is $x_t^{(i)} = 1 - y_t^{(i)} / y_{t_{max}}^{(i)}$, where t_{max} represents the time step in which the processing of task i was completed. The time scale is weeks.

Figure 2.8 shows the obtained time series for the initial five tasks of project A and the first two tasks of project C. Figure 2.8 also shows the complete work remaining in project B, a "fast-track project." Its detailed work breakdown structure was not considered and only the accumulated time data was analyzed (Schlick et al. 2012).

The validation study included interviews and informal discussions with management and engineers to understand the development projects in detail. We found that the development of sensor technologies is a good subject for PD project modeling because tasks are largely processed in parallel and frequent iterations occur.

For simplicity, we focus on the first two overlapping development tasks of project A, "conceptual sensor design" (task 1) and "design of circuit diagrams" (task 2), and model only their overlapping range (cf. Fig. 2.8). Concerning the left bound of this range, the conceptual sensor design had reached a completion level of 39.84% when the design of the circuit diagram began. Therefore, the estimated work remaining at the initial time step is:

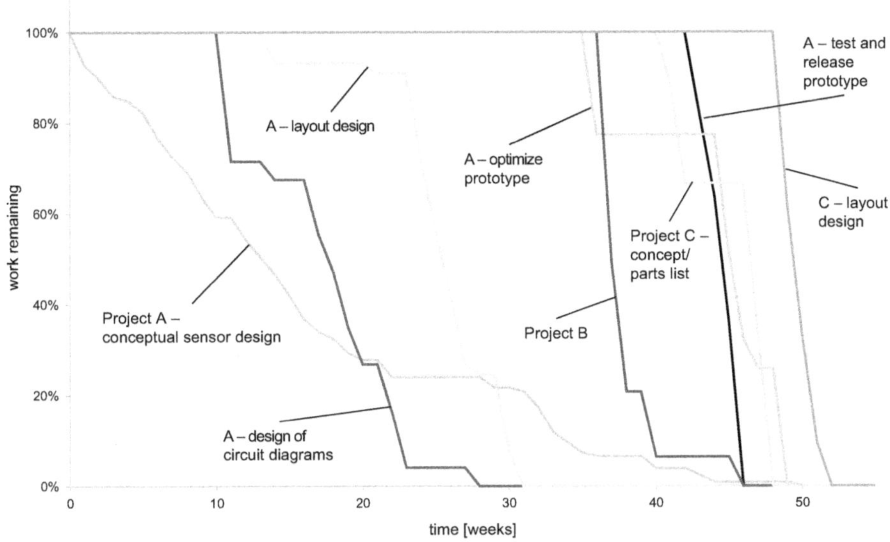

Fig. 2.8 Time series of work remaining in three real product development projects. The data were acquired in a small industrial company in Germany. Only the first five tasks of project A and the first two tasks of project C are shown. Project B was a "fast-track project." Its work breakdown structure is not considered and only the complete work remaining is shown. The time scale is weeks

$$\hat{x}_0 = \begin{pmatrix} x_0^{(1)} = 0.6016 \\ x_0^{(2)} = 1.0000 \end{pmatrix}. \tag{73}$$

The least square method developed by Neumaier and Schneider (2001) was used to estimate the additional parameters of the VAR(n) process (see Section 2.4). The maximum model order to be considered in model selection was set to $n_{max} = 6$. The estimation algorithm and model selection procedures were implemented in Mathematica® based on the Matlab® source code of the ARfit toolbox. The estimation results were also verified through the ARfit toolbox in the original Matlab® simulation environment. Note that the fast algorithm introduced by Neumaier and Schneider (2001) in Section 4.2 of their paper was used to improve computational efficiency. The fast algorithm does not require separate QR factorizations (see Section 2.4) for each approximating autoregression model, but instead only one factorization for the most complex model of order $n_{max} = 6$. Due to this simplification, the least square estimates are slightly less accurate and therefore only approximate order selection criteria for lower order models can be obtained.

We start by presenting and discussing the parameter estimates for a vector autoregression model of first order. This model with least independent parameters was selected due to Schwarz' Bayesian Criterion (Section 2.4). Afterwards, the parameters of a second-order model that is favored by Akaike's criteria will be presented.

The model selection procedure showed that Schwarz' criterion is minimal for a VAR model of first order and we have $BIC(n_{opt} = 1) = -15.74$ for this model. If the number of effective parameters is determined based not only on the coefficient matrices $A_0, \ldots, A_{n-1}$ but also on the covariance matrix C and we set $k = np^2 + n(n+1)/2$, a first-order VAR model is also selected.

Due to the cited design of the selection algorithm for fast processing, the first $n_{max} - 1$ data points of the mean work remaining are ignored in the implementation of the ARfit toolbox. Although this approach does not affect the accuracy of results asymptotically, for short time series such as in our industrial case study it can constitute a significant loss of information. Therefore, the least square fitting was repeated with $n_{max} = 1$. The estimated WTM A_0 for this model is given by

$$\hat{A}_0 = \begin{pmatrix} a_{11} = 0.9406 & a_{12} = -0.0017 \\ a_{21} = 0.0085 & a_{22} = 0.8720 \end{pmatrix}. \tag{74}$$

The estimated covariance matrix $\hat{C}$ of the normally distributed random variable ε_t is given the representation (cf. Eq. 55)

$$\varepsilon_t \sim \mathcal{N}\left(\begin{pmatrix} 0 \\ 0 \end{pmatrix}, \begin{pmatrix} \sigma_{11}^2 = (0.0135)^2 & \rho\sigma_{11}\sigma_{22} = -0.38 \cdot 0.0135 \cdot 0.0416 \\ \rho\sigma_{11}\sigma_{22} & \sigma_{22}^2 = (0.0416)^2 \end{pmatrix}\right). \tag{75}$$

In the above probability distribution, the variable $\rho \epsilon [-1; 1]$ represents Pearson's correlation coefficient in the original state-space coordinates. We can also rewrite the formulation of the covariance matrix and assume that the standard deviation $\sigma_{ii} = c_{ii} = \sqrt{C_{[i,i]}}$ of the fluctuations is proportional with the proportionality constant s_i to the autonomous task processing rate a_{ii}:

$$\varepsilon_t \sim \mathcal{N}\left(\begin{pmatrix} 0 \\ 0 \end{pmatrix}, \begin{pmatrix} (s_1 a_{11})^2 = (0.0144 \cdot 0.9406)^2 & \rho(s_1 a_{11})(s_2 a_{22}) \\ \rho(s_1 a_{11})(s_2 a_{22}) & (s_2 a_{22})^2 = (0.0477 \cdot 0.8720)^2 \end{pmatrix}\right).$$

Figure 2.9 shows the list plots of work remaining in the overlapping range of tasks 1 and 2 over 50 weeks. The figure not only presents the empirical time series of task processing which had been acquired in the industrial company but also the results of a Monte Carlo simulation based on the parameterized model of the focused project phase (see state Eq. 8 in conjunction with parameter estimates from Eqs. 73, 74 and 75). A total of 1,000 separate and independent simulation runs were performed.

Concerning the field data, the fact that the conceptual design of the sensor was processed entirely in parallel with the design of the circuit diagram and continued after the circuit diagram was finished is of particular interest (Fig. 2.8). More than ten iterations are necessary to reach a stable conceptual design state after week 35. In the list plot of Fig. 2.9, the simulated task processing is represented by the means and 95% confidence intervals of work remaining. The estimated stopping criterion

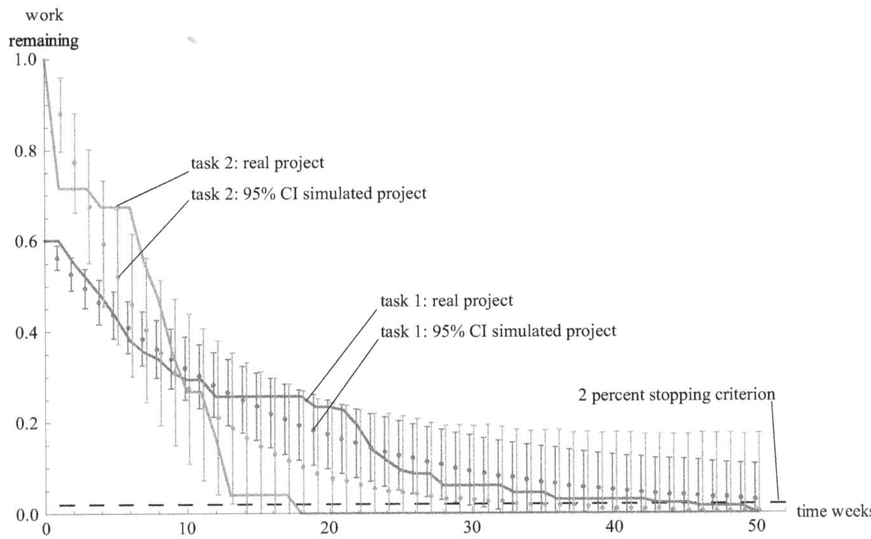

Fig. 2.9 List plot of work remaining in the real product development project. The data were acquired in a small industrial company in Germany. Only the overlapping range of the first two tasks, "conceptual sensor design" (task 1) and "design of circuit diagram" (task 2), is shown. The plot also shows means of simulated time series of task processing as note points and 95% confidence intervals as *error bars*. The Monte Carlo simulation was based on state Eq. 3 in conjunction with the least square estimates of the independent parameters according to Eqs. 73, 74 and 75. A total of 1000 separate and independent runs were calculated. Note points have been offset to distinguish the *error bars*. The stopping criterion of 2% is marked by a dashed line at the *bottom* of the plot

of $\delta = 0.02$ is plotted as a dashed line at the bottom of the chart. According to Fig. 2.9, 49 out of 50 confidence intervals of the simulated work remaining of task 1 include the empirical data points from the real project before the stopping criterion is met. Only the confidence interval computed for week 1 is a little too small. Furthermore, the center of the confidence interval calculated for week 50 is approximately 2% and therefore the simulated processing of task 1 will—in line with the definition of the stopping criterion—be completed at week 50 on average. The task completion is therefore accurately predicted. In contrast, when the processing of task 2 is compared against the field data, the goodness-of-fit is significantly lower, and only 28 out of the 31 confidence intervals cover the empirical data points before the stopping criterion is met. Moreover, the quite abrupt completion of task 2 in week 18 is poorly predicted by the smoothly decaying means of the autoregression model of first order (Schlick et al. 2008, 2012). On average task 2 is predicted to complete in week 31, which is 13 weeks later than the real time point. However, the predictive accuracy in singular cases can be much better because of the large variance of the fluctuation variable ε_t in the second dimension covering task 2.

The root-mean-square deviation (RMSD) between the work remaining in task 1 predicted by the 1000 simulation runs and the empirical data is $\text{RMSD}_{\text{task1}} = 0.046$. For task 2, the deviation is more than twice that value: $\text{RMSD}_{\text{task2}} = 0.107$. Regarding the established scientific standards of organizational simulation (see Rouse and Boff 2005), the total deviation is low, and this confirms the validity of the model.

In total, a parameter vector $\theta_1 = \left[x_0^{(1)} \quad x_0^{(2)} \quad a_{11} \quad a_{12} \quad a_{21} \quad a_{22} \quad \sigma_{11} \quad \sigma_{22} \quad \rho \quad \delta\right]$ with 10 components is necessary for modeling task 1 and 2 in this phase of the project based on a VAR(1) model.

If the alternative formulation with a forcing matrix K in the original state-space coordinates is used for project modeling (Eq. 23), we have in addition to the initial state x_0 (Eq. 73) and A_0 (Eq. 74) the estimated independent parameters

$$\widehat{K} = \begin{pmatrix} -0.1355 & 0.9908 \\ 0.9908 & 0.1355 \end{pmatrix}$$

and

$$\eta_t \sim \mathcal{N}\left(\begin{pmatrix} 0 \\ 0 \end{pmatrix}, \begin{pmatrix} \lambda_1(C) = (0.00176)^2 & 0 \\ 0 & \lambda_2(C) = (0.00015)^2 \end{pmatrix}\right).$$

The parameter vector is $\theta_2 = \left[x_0^{(1)} \, x_0^{(2)} \, a_{11} \, a_{12} \, a_{21} \, a_{22} \, k_{11} \, k_{12} \, \lambda_1(C) \, \lambda_2(C) \, \delta\right]$.

For the sake of completeness, the estimated independent parameters in the spectral basis (Eqs. 39 and 40) are presented as well. The eigendecomposition of the dynamical operator A_0 according to Eq. 35 leads to the matrix of eigenvectors:

$$S = \begin{pmatrix} 0.9924 & 0.0247 \\ 0.1228 & 0.9997 \end{pmatrix}.$$

The estimate of the initial state $\hat{x}'_0 = S^{-1} \cdot \hat{x}_0$ is given by

$$\hat{x}'_0 = \begin{pmatrix} x_0'^{(1)} = 0.5830 \\ x_0'^{(2)} = 0.9287 \end{pmatrix},$$

the transformed dynamical operator is given by

$$\hat{\Lambda}_S = \begin{pmatrix} \lambda_1(A_0) = 0.9404 & 0 \\ 0 & \lambda_2(A_0) = 0.8722 \end{pmatrix},$$

and the estimated covariance $\hat{C'}$ of the normally distributed performance fluctuations is represented by

$$\varepsilon'_t \sim \mathcal{N}\left(\begin{pmatrix} 0 \\ 0 \end{pmatrix}, \begin{pmatrix} \sigma'^2_{11} = (0.0141)^2 & \rho'\sigma'_{11}\sigma'_{22} = -0.48 \cdot 0.0141 \cdot 0.0424 \\ \rho'\sigma'_{11}\sigma'_{22} & \sigma'^2_{22} = (0.0424)^2 \end{pmatrix}\right).$$

Interestingly, the basis transformation slightly reinforces variances $\sigma'^2_{11} = C'_{[1,1]}$ and $\sigma'^2_{22} = C'_{[2,2]}$ and correlation coefficient ρ'_{22}. The corresponding parameter vector is $\theta_3 = \left[x'^{(1)}_0 \quad x'^{(2)}_0 \quad \lambda_1(A_0) \quad \lambda_2(A_0) \quad \sigma'_{11} \quad \sigma'_{22} \quad \rho' \quad \delta'\right]$.

It is important to note that in contrast to Schwarz' criterion, Akaike's FPE and AIC model selection criteria lead to a result, in which minimum scores were assigned to a model of second-order, $n_{opt} = 2$, not to a first-order autoregression model. In this case we have $FPE(n_{opt} = 2) = -16.06$ and $AIC(n_{opt} = 2) = -12.45$. A second-order model is also selected under the FPE and AIC criteria, if the larger number of effective parameters is considered and it holds that $k = np^2 + n(n+1)/2$. As before, the model selection results are invariant in the heuristics to count the effective number of parameters. As the AIC_c version of Akaike's Information Criterion has superior bias properties for small samples, it assigns minimum scores to a model of second order under both counting heuristics. Hence, the model selection results for these criteria are consistent and robust. In contrast to model selection based on the BIC criterion the least square fitting was not repeated with a maximum model order of $n_{max} = 2$ to make use of more data points from the beginning of the project phase, as it leads to inconsistent results. Under this condition a first-order model is assigned minimum scores. However, the difference to the second-order model is very small. This inconsistency seems to be a negative consequence of the preferred design of the selection algorithm for fast processing.

For the second-order model the estimated work remaining at the initial two time steps is:

$$\hat{x}_0 = \begin{pmatrix} 0.6016 \\ 1.0000 \end{pmatrix}$$
$$\hat{x}_1 = \begin{pmatrix} 0.6016 \\ 0.7154 \end{pmatrix}. \tag{76}$$

The estimated WTMs $\hat{A}_0$ and $\hat{A}_1$ are given by

$$\hat{A}_0 = \begin{pmatrix} 1.1884 & -0.1476 \\ 0.0470 & 1.1496 \end{pmatrix} \tag{77}$$

$$\hat{A}_1 = \begin{pmatrix} -0.2418 & 0.1344 \\ -0.0554 & -0.2622 \end{pmatrix}. \tag{78}$$

The estimated covariance matrix $\hat{C}$ is given by the representation

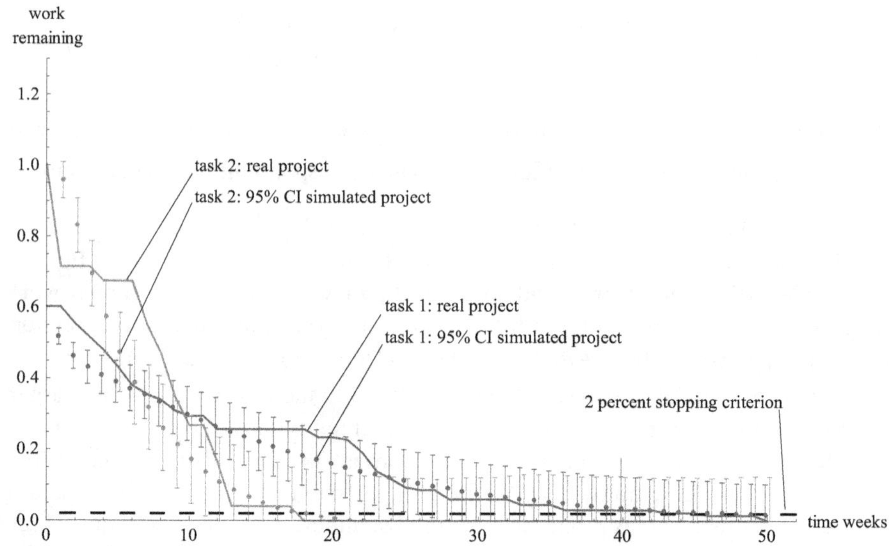

Fig. 2.10 List plot of work remaining in the real and simulated product development projects. As in Fig. 2.9, only the overlapping range of the first two tasks is shown. The means of simulated time series of task processing are shown as note points and 95% confidence intervals appear as error bars. The Monte Carlo simulation was based on state Eq. 58. The selected order of the autoregression model was $n_{opt} = 2$. The least square estimates of the independent parameters are given in Eqs. 73, 76, 77, 78 and 79. A total of 1000 separate and independent runs were calculated. Note points have been offset to distinguish the error bars. The stopping criterion of 2% is marked by a dashed line at the bottom of the plot

$$\varepsilon_t \sim \mathcal{N}\left(\begin{pmatrix} 0 \\ 0 \end{pmatrix}, \begin{pmatrix} (0.0116)^2 & -0.013 \cdot 0.0116 \cdot 0.0257 \\ -0.013 \cdot 0.0116 \cdot 0.0257 & (0.0257)^2 \end{pmatrix}\right). \quad (79)$$

Figure 2.10 shows the corresponding list plots of real and simulated work remaining in the overlapping range of tasks 1 and 2 over 50 weeks.

The simulation model is based on state Eq. 58 in conjunction with parameter estimates from Eqs. 73, 76, 77, 78 and 79. As before, 1,000 separate and independent simulation runs were calculated. The RMSD between the work remaining in task 1 predicted by the project simulations and the field data is $\text{RMSD}_{\text{task1}} = 0.053$. For task 2, the deviation is almost twice that value: $\text{RMSD}_{\text{task2}} = 0.103$. For both tasks the root-mean-square deviations for the second-order model are similar to the first-order model. Although the deviations are similar, significant qualitative differences between the two models exist and can be clearly seen in Fig. 2.10 between the first and fourth week of task 1 and the fifth and eighth week of task 2, in which the confidence intervals related to the second-order model do not include the data points from the real project, and an "undershoot" effect can be observed for both tasks. In particular the confidence intervals computed for weeks 1, 2, 4, 5, 6, 7, 8 and 9 of task 2 demonstrate low goodness-of-fit indices. In this context, it is important to note that the center of the confidence interval that was calculated for week 18 for task 2 is approximately 2%, which means that on average the simulated processing of task 2 will be completed by this point in time. Hence, in contrast to the first-order autoregression model, the rapidly decaying means accurately predict the relatively abrupt completion of task 2. The goodness-of-fit for task 1 is comparable to the first-order model, and 46 of the 50 confidence intervals cover the empirical data points before the stopping criterion is satisfied. In spite of these apparent similarities in terms of goodness-of-fit, the center of the confidence interval calculated for week 47 is approximately 2% and therefore the simulated processing of task 1 will, on average, be completed in the same week and not in week 50 as it is in the real development project. In conclusion, we can say that the parameterized first- and second-order VAR models lead to a similar overall predictive accuracy. The first-order model has the advantage that it is not only a simpler representation but also predicts the completion of task 1 with high accuracy. Conversely, the second-order model can predict the completion week of task 2 with high accuracy, while the first-order model leads to an entirely unacceptable prediction error of 13 weeks on average. However, attaining the high predictive accuracy of the second-order model concerning the completion time comes at the cost of having a significant deviation of the means at the beginning of the processing of both tasks. Therefore, on a phenomenological level of project management and schedule control, it is impossible to find a low-order model with optimal properties. In Section 2.9 we will introduce with linear dynamical system models an advanced approach based on hidden state variables that makes it possible to reach a higher predictive accuracy for both tasks without an unnecessarily complex internal configuration.

2.6 Stochastic Formulation with Periodically Correlated Processes

An extension of the introduced autoregressive approach to modeling cooperative work in PD projects that is especially interesting from theoretical and practical perspectives is to formulate a so-called "periodic vector autoregressive" (PVAR) stochastic process (Franses and Paap 2004; Ursu and Duchesne 2009). In principle, a PVAR model can capture the dynamic processing of the development tasks with short iteration length, and the long-term effects of inadvertent information hiding due to the asynchronous exchange of engineering design data. According to the previous validation study, short iterations for a given amount of work are necessary to process and disseminate component-level design information within CE teams and to develop the corresponding system components. Short iterations are also necessary if the scope of predictability for the development project is small and only a few stable assumptions can be made about the design ranges or physical functions of the product under development. A corresponding simulation study is

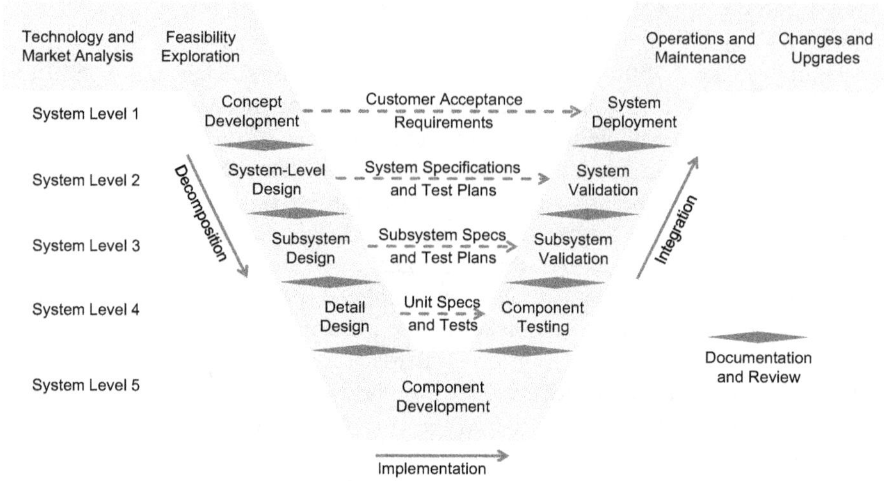

Fig. 2.11 V-model of the systems engineering process (Eppinger and Browning 2012, adapted from the U.S. Department of Transportation)

presented in Section 5.1. However, designing complex products that include mechanical, electrical and software components not only requires detailed design and implementation of units, but also a definition of the product architecture. According to Eppinger and Browning (2012), defining product architecture often involves breaking the product down into smaller elements such as subsystems, modules, components and parts, and then defining their interfaces. These elements must be integrated at different levels of abstraction and aggregation so that they can work together efficiently and thus perform the intended function of the system as a whole and achieve the desired levels of performance, robustness and reliability. In Fig. 2.11, the traditional V-model of the systems engineering process illustrates the main activities involved in developing complex products. The downward-facing side of the V represents design and decomposition activities, and the upward-facing side shows the complementary component-level-to-system-level integration and testing activities. The upward-facing side also includes model integration and data aggregation activities. The basic activities involve developing the components to perform the single or multiple physical functions that will ensure the product can function as a whole. Following the same logic of decomposition, implementation, and integration, it is also possible to design and develop an ultra-large-scale "system of systems" (see, e.g. Maier 1998).

To meet their specification, projects that develop complex products therefore need frequent iterations between teams doing detailed design and component development work, and between teams at the system, subsystem, and module levels. At every level of system design, the problem-solving processes are organized in such a way that, by definition, the tasks are cooperatively processed and that information on functional, topological, and geometrical entities and the progress of the work is not hidden but freely accessible and widely disseminated. An additional long-term effect, however, can occur in large-scale projects with a hierarchical coordination structure and globally distributed engineering service providers. This is because teams working at the system, subsystem or module level may withhold, for a limited period, certain pieces of information on the design, integration and testing of these entities (Yassine et al. 2003). Between the releases, new information is hidden (kept secret) and work in the subordinate teams is based on the old knowledge. Such a hold-and-release policy is typical for PD projects in the automotive and aerospace industries. This kind of uncooperative behavior is justified by the desire to improve the implementation of the product architecture through better system-level design and validation and thus release only those designs that have a sufficient level of maturity. This can significantly reduce the overall amount of coordination required by the teams in the development project. A deterministic model capable of capturing both cooperative and noncooperative task processing was developed by Yassine et al. (2003). In their seminal paper, a time-variant (nonautonomous) state equation was formulated and validated based on simulation runs. We build directly upon their results in the following. However, the PVAR approach we have developed can also account for unpredictable performance fluctuations (Schlick et al. 2011, 2014) and provides a firm basis for analytical and numerical complexity evaluations (see Section 5.3).

To develop a PVAR model of cooperative work in PD projects with a multi-level hierarchical coordination structure, we assume that a small amount of finished work related to concept development and systems engineering accumulates over the short iterations by the corresponding team, and that this work is released to the component development level only at time steps $ns (n \in \mathbb{N})$ with fixed, predetermined period $s \geq 2$. At all other time steps, $ns + v (n = 0, 1, \ldots ; v = 1, \ldots, s - 1)$, the tasks are processed by short iterations, and the detailed design and development information is freely accessible within teams and widely disseminated between teams. As a consequence of the hold-and-release policy, certain components, their constituent parts or interfaces regularly need to be reworked, which leads to additional work in the component development teams. For the sake of simplicity, we start by modeling a hierarchical coordination structure with two levels: System Level 2, in line with Fig. 2.11, which represents system-level design and integration activities; and System Level 5, which focuses on component development (and, to a certain extent, integration testing) in the mathematical formulation of the basic PVAR model. An additional subsystem level will be integrated into the model after defining the state equation and the independent parameters.

Under the assumption of a hierarchical coordination structure with two levels, the stochastic difference equation 8 can be generalized to a process $\{X_{ns+v}\}$ with periodically correlated components. The state equation is

$$X_{ns+v} = \Phi_1(v) \cdot X_{ns+v-1} + \varepsilon_{ns+v}, \tag{80}$$

where the index n indicates the long time scale with period s, and v the short time scale. $X_t = (X_t(1), \ldots, X_t(d))^{\mathrm{T}}$ is a $d \times 1$ single random vector encoding the state of the project at time step $t = ns + v$ ($n = 0, 1, \ldots; v = 1, \ldots, s$). The leading components of the state vector X_t represent the work remaining of the $p^{\mathrm{C}} \in \mathbb{N}$ component-level and $p^{\mathrm{S}} \in \mathbb{N}$ system-level tasks that are processed on the short time scale. As already mentioned in Section 2.1, the amount of work remaining can be measured by the time left to finalize a specific design, by the definable labor units required to complete a particular development activity or component of the work breakdown structure, by the number of engineering drawings requiring completion before design release, by the number of engineering design studies required before design release or by the number of issues that still need to be addressed/resolved before design release (Yassine et al. 2003). For the component-level and system-level tasks that are processed on the short time scale, the work transformation can be captured by a combined WTM A_0 as

$$A_0 = \begin{pmatrix} A_0^{\mathrm{C}} & A_0^{\mathrm{SC}} \\ A_0^{\mathrm{CS}} & A_0^{\mathrm{S}} \end{pmatrix}.$$

In the combined WTM A_0, the submatrix A_0^{C} of size $p^{\mathrm{C}} \times p^{\mathrm{C}}$ is the dynamic operator for the cooperative processing of component-level tasks. The $p^{\mathrm{S}} \times p^{\mathrm{S}}$ submatrix A_0^{S} refers to system-level tasks in an analogous manner. The $p^{\mathrm{C}} \times p^{\mathrm{S}}$ submatrix A_0^{SC} determines the fraction of work remaining created by system-level tasks for the corresponding component-level tasks in each short iteration, whereas the $p^{\mathrm{S}} \times p^{\mathrm{C}}$ submatrix A_0^{CS} determines the fraction of work remaining created by component-level tasks for the system-level tasks. Moreover, the substates $(X_t(1), \ldots, X_t(p^{\mathrm{C}} + p^{\mathrm{S}}))$ have to be augmented by other p^{S} substates to account for the periodic hold-and-release policy of system-level design information. The augmented p^{S} substates do not represent the work remaining as the leading states do, but represent the amount of finished work on the system level that has accumulated over the short iterations. The finished work remains hidden for the component-level teams until it is released at time step ns. Through the PVAR model formulation, the finished work can be placed in a hold state. The associated $p^{\mathrm{S}} \times p^{\mathrm{S}}$ submatrix A_0^{SH} covers the fraction of work that has been generated by the system-level tasks in each iteration at time step $v = 1, \ldots, s-1$ and has been finished and put in a hold state to reduce the amount of coordination needed. After release, additional work is generated for the component-level tasks. This work is calculated based on the WTM A_0^{HC}. This WTM is of size $p^{\mathrm{C}} \times p^{\mathrm{S}}$. There, $d = p^{\mathrm{C}} + 2p^{\mathrm{S}}$ holds.

The periodically correlated work processes in the project are represented by the time evolution of the state vector X_{ns+v} based on the autoregressive coefficients $\Phi_1(v)$. The two time scales correspond to indices n and v. The long time scale is indexed by n. In seasonal macroeconomic models, for instance, n indicates the year that the acquired samples of the time series refer to (e.g. income and consumption). However, in large-scale projects the release period is much shorter and covers typically intervals of four to eight weeks. On the other hand, the short time scale is indexed by v. On this scale, the iterations usually occur on a daily or weekly basis (see Section 2.5). In the terminology of macroeconomic models, v indicates a specific "season" of the "year." Furthermore, the length s of the period between releases of hidden information about system integration and testing ("number of seasons within year") has to be specified by the project manager. For a predetermined period length s, the random vector X_{ns+v} contains the realization of work remaining during the vth iteration over all component-level and system-level tasks at the release period n and the amount of finished work on the system level that is ready to be released to component-level tasks in period $n+1$. According to the previous analysis the state vector can be separated into three substate vectors and defined as

$$X_t = \begin{pmatrix} X_t^{\mathrm{C}} \\ X_t^{\mathrm{S}} \\ X_t^{\mathrm{H}} \end{pmatrix}$$

with

$$X_t^{\mathrm{C}} = \begin{pmatrix} X_t(1) \\ \vdots \\ X_t(p^{\mathrm{C}}) \end{pmatrix}$$
$$X_t^{\mathrm{S}} = \begin{pmatrix} X_t(p^{\mathrm{C}}+1) \\ \vdots \\ X_t(p^{\mathrm{C}}+p^{\mathrm{S}}) \end{pmatrix}$$
$$X_t^{\mathrm{H}} = \begin{pmatrix} X_t(p^{\mathrm{C}}+p^{\mathrm{S}}+1) \\ \vdots \\ X_t(p^{\mathrm{C}}+2p^{\mathrm{S}}) \end{pmatrix}.$$

Furthermore, the task processing on the long and short time scales has to be modeled, and for this aim, two independent dynamical operators are introduced (Yassine et al. 2003). These operators correspond to the autoregressive coefficients $\Phi_1(v)$ in state equation 80. The release of hidden information over s time steps is modeled by the first dynamical operator $\Phi_1(s)$. It is assumed that the release occurs at the end of the period, that is at relative time step $v = s$. The operator $\Phi_1(s)$ can be composed of the previously defined submatrices as

$$\Phi_1(s) = \begin{pmatrix} A_0^{\mathrm{C}} & A_0^{\mathrm{SC}} & A_0^{\mathrm{HC}} \\ A_0^{\mathrm{CS}} & A_0^{\mathrm{S}} & 0 \\ 0 & 0 & \{\varepsilon\} \cdot I_{p^{\mathrm{S}}} \end{pmatrix}. \tag{81}$$

In the above equation the ε-symbol denotes an arbitrarily small positive quantity. The definition of positive interactions between the augmented p^{S} substates is necessary for explicitly evaluating the emergent complexity of the periodically correlated work processes on the basis of the chosen information-theoretic measure EMC, which is presented and discussed in Sections 3.2.4 and Chapter 4. EMC simply scales linearly with ε. For practical purposes, it is recommended to calculate with $\varepsilon = 10^{-4}$. By doing so, the finished work after release is set back to a nonzero but negligible amount in terms of productivity.

The task processing in the $v = 1, \ldots, s-1$ iterations before release is modeled on the basis of a second dynamical operator $\Phi_1(1)$. In contrast to macroeconomic models, it is assumed that the cooperative processing of the development tasks by short iterations without withholding information follows a regime in which the kind and intensity of interactions does not change before the end of period s and therefore the autoregressive coefficients are constant, that is $\Phi_1(1) = \ldots = \Phi_1(s-1)$. No other dynamical operators are needed to capture project dynamics within the period. $\Phi_1(1)$ can be composed of the previously defined submatrices in an analogous manner as

$$\Phi_1(1) = \begin{pmatrix} A_0^{\mathrm{C}} & A_0^{\mathrm{SC}} & 0 \\ A_0^{\mathrm{CS}} & A_0^{\mathrm{S}} & 0 \\ 0 & A_0^{\mathrm{SH}} & \{1-\varepsilon\} \cdot I_{ps} \end{pmatrix}. \tag{82}$$

By the same reasoning, it is assumed that the covariance of the noise vector ε_{ns+v} representing unpredictable performance fluctuations when processing the development tasks cooperatively by short iterations is constant for all $v = 1, \ldots, s-1$ iterations before the end of period s and no other unpredictable effects influence the cooperative work within the period.

The evolution of the process state X_t over t time steps can therefore be expressed by two dynamical operators $\Phi_1(1)$ and $\Phi_1(s)$ in conjunction with two noise vectors ε_1 and ε_s in the form of a recurrence relation with switching regime as

$$X_t = \begin{cases} \Phi_1(1) \cdot X_{t-1} + \varepsilon_1 & \text{for } t = ns + v \\ \Phi_1(s) \cdot X_{t-1} + \varepsilon_s & \text{for } t = (n+1)s \end{cases}$$

with $n = 0, 1, \ldots$ and $v = 1, \ldots, s-1$.

The noise vectors ε_1 and ε_s correspond to zero-mean white noise and have zero means and covariances C_1 and C_s, respectively:

$$\begin{aligned} \varepsilon_1 &\sim \mathcal{N}(0_d, C_1) \\ \varepsilon_s &\sim \mathcal{N}(0_d, C_s). \end{aligned} \tag{83}$$

Hence, the combined error process $\{\varepsilon_{ns+v}\}$ can also be expressed by a zero-mean periodic white noise process. $\{\varepsilon_{ns+v}\}$ is composed of $d \times 1$ random vectors, such that $E[\varepsilon_{ns+v}] = 0_{\mathrm{d}}$ and $E\left[\varepsilon_{ns+v}\, \varepsilon_{ns+v}^{\mathrm{T}}\right] = C_1$ for $v = 1, \ldots, s-1$, and $E[\varepsilon_{ns+v}] = 0_d$ and $E\left[\varepsilon_{ns+v}\, \varepsilon_{ns+v}^{\mathrm{T}}\right] = C_s$ for $v = s$. It is assumed that the covariance matrices C_1 and C_s are not singular.

If all parallel tasks are initially to be completed in full and no finished work exists in hold state, the initial state is simply

$$x_0 = \begin{pmatrix} \begin{pmatrix} 1 \\ \vdots \\ 1 \end{pmatrix} \\ \begin{pmatrix} 1 \\ \vdots \\ 1 \end{pmatrix} \\ \begin{pmatrix} 0 \\ \vdots \\ 0 \end{pmatrix} \end{pmatrix},$$

where the first row vector of ones has size $|A_0^{\mathrm{C}}|$, the second row vector of ones has size $|A_0^{\mathrm{S}}|$ and the third vector of zeros has size $|A_0^{\mathrm{SH}}|$. The inner brackets of the vector notation are there to clarify the structure of the substates and do not have any mathematical relevance.

Following the modeling concept with distinct dynamical operators and combined work transformation matrices, we can also easily model multi-level hierarchical coordination structures. Consider for instance a project in which the system-level and component-level design teams do not cooperate directly and an intermediate organizational level exists where a highly specialized team of systems engineers designs subsystems and carries out subsystem integration testing. The subsystems integrate the components in the same manner as the whole system integrates the subsystems, and therefore they improve modularity, adaptability and testability of the product. This three-level decomposition means that the development team responsible for subsystem design and integration hide a certain amount of finished work, accumulate it and then later release it to the subordinate component-level team. It also means that, at the system-design level a dedicated organizational unit hides, accumulates and releases a certain amount of its finished work to the subsystem level to improve the implementation of the product architecture as a whole. It is reasonable to assume that these high-level coordination processes occur on a very large time scale. We represent this time scale by the variable s'. To simplify the definition of the switching regime of the recurrence relation (see below), we assume it holds that $s' = 2s$. The basic mechanisms of inadvertent information hiding and releasing remain the same. To model this three-level hierarchical coordination structure, the previous substates $(X_t(1), \ldots, X_t(p^{\mathrm{C}} + p^{\mathrm{S}} + p^{\mathrm{S}}))$ have to be augmented by additional $p^{\mathrm{SS}} + p^{\mathrm{SS}}$ substates to account for the extra periodic correlations that are generated by the tasks at the subsystem level. Following hierarchical system decomposition, the work remaining for subsystem tasks is stored in the components in the upper-middle section of the state vector, i.e. between vector components related to the component-level and system-level tasks. The amount of finished work for subsystems placed in a hold state is indicated by the lower-middle components of the state vector. We have:

$$X_t = \begin{pmatrix} X_t^{\mathrm{C}} \\ X_t^{\mathrm{SS}} \\ X_t^{\mathrm{S}} \\ X_t^{\mathrm{SSH}} \\ X_t^{\mathrm{SH}} \end{pmatrix}$$

with

$$X_t^{\mathrm{C}} = \begin{pmatrix} X_t(1) \\ \vdots \\ X_t(p^{\mathrm{C}}) \end{pmatrix}$$
$$X_t^{\mathrm{SS}} = \begin{pmatrix} X_t(p^{\mathrm{C}}+1) \\ \vdots \\ X_t(p^{\mathrm{C}}+p^{\mathrm{SS}}) \end{pmatrix}$$
$$X_t^{\mathrm{S}} = \begin{pmatrix} X_t(p^{\mathrm{C}}+p^{\mathrm{SS}}+1) \\ \vdots \\ X_t(p^{\mathrm{C}}+p^{\mathrm{SS}}+p^{\mathrm{S}}) \end{pmatrix}$$
$$X_t^{\mathrm{SSH}} = \begin{pmatrix} X_t(p^{\mathrm{C}}+p^{\mathrm{SS}}+p^{\mathrm{S}}+1) \\ \vdots \\ X_t(p^{\mathrm{C}}+2p^{\mathrm{SS}}+p^{\mathrm{S}}) \end{pmatrix}$$
$$X_t^{\mathrm{SH}} = \begin{pmatrix} X_t(p^{\mathrm{C}}+2p^{\mathrm{SS}}+p^{\mathrm{S}}+1) \\ \vdots \\ X_t(p^{\mathrm{C}}+2p^{\mathrm{SS}}+2p^{\mathrm{S}}) \end{pmatrix}.$$

The indices "C", "SS" and "S" in the state variables above refer to component-level, subsystem-level and system-level tasks, respectively. The three time scales of information exchange mean that we also have to formulate three dynamical operators. We define them as follows:

$$\Phi_1(1) = \begin{pmatrix} A_0^{\mathrm{C}} & A_0^{\mathrm{SSC}} & 0 & 0 & 0 \\ A_0^{\mathrm{CSS}} & A_0^{\mathrm{SS}} & A_0^{\mathrm{SSS}} & 0 & 0 \\ 0 & A_0^{\mathrm{SSS}} & A_0^{\mathrm{S}} & 0 & 0 \\ 0 & A_0^{\mathrm{SSH}} & 0 & \{1-\varepsilon\}\cdot I_{p^{\mathrm{SS}}} & 0 \\ 0 & 0 & A_0^{\mathrm{SH}} & 0 & \{1-\varepsilon\}\cdot I_{p^{\mathrm{S}}} \end{pmatrix}$$
$$\Phi_1(s) = \begin{pmatrix} A_0^{\mathrm{C}} & A_0^{\mathrm{SSC}} & 0 & A_0^{\mathrm{HC}} & 0 \\ A_0^{\mathrm{CSS}} & A_0^{\mathrm{SS}} & A_0^{\mathrm{SSS}'} & 0 & 0 \\ 0 & A_0^{\mathrm{SSS}} & A_0^{\mathrm{S}} & 0 & 0 \\ 0 & 0 & 0 & \{\varepsilon\}\cdot I_{p^{\mathrm{SS}}} & 0 \\ 0 & 0 & 0 & 0 & \{1-\varepsilon\}\cdot I_{p^{\mathrm{S}}} \end{pmatrix}$$
$$\Phi_1(s') = \begin{pmatrix} A_0^{\mathrm{C}} & A_0^{\mathrm{SSC}} & 0 & A_0^{\mathrm{HC}} & 0 \\ A_0^{\mathrm{CSS}} & A_0^{\mathrm{SS}} & A_0^{\mathrm{SSS}'} & 0 & A_0^{\mathrm{HSS}} \\ 0 & A_0^{\mathrm{SSS}} & A_0^{\mathrm{S}} & 0 & 0 \\ 0 & 0 & 0 & \{\varepsilon\}\cdot I_{p^{\mathrm{SS}}} & 0 \\ 0 & 0 & 0 & 0 & \{\varepsilon\}\cdot I_{p^{\mathrm{S}}} \end{pmatrix}.$$

In this representation with nine distinct work transformation matrices, the matrix A_0^{SSH} determines the fraction of finished work that is generated by the subsystem-level tasks in each short iteration and is accumulated through the associated sub-states $(X_t(p^{\mathrm{C}}+p^{\mathrm{SS}}+p^{\mathrm{S}}+1),\ldots,X_t(p^{\mathrm{C}}+2p^{\mathrm{SS}}+p^{\mathrm{S}}))$. The release of this work is indicated by A_0^{HC}, as before. Similarly, the matrix A_0^{SH} is needed to compute the fraction of finished work that was processed by the system-level team in each short iteration. The accumulated work is stored in substates $(X_t(p^{\mathrm{C}}+2p^{\mathrm{SS}}+p^{\mathrm{S}}+1),\ldots,X_t(p^{\mathrm{C}}+2p^{\mathrm{SS}}+2p^{\mathrm{S}}))$. However, the release of the accumulated finished work now occurs at time steps $2(n+1)s$, and not at $(n+1)s$ as was the case for the two-level hierarchical coordination structure that we developed previously. This can also be seen in the above definition of the dynamical operator $\Phi_1(s')$. The variable A_0^{HSS} is related to the work that the subsystem design and integration team has put in a hold state and is transferred to the system-level at time steps $(n+1)s$. The matrices A_0^{C}, A_0^{CSS}, A_0^{SSC}, A_0^{SS}, A_0^{SSS}, $A_0^{\mathrm{SSS'}}$ and A_0^{S} have the same meaning as before. They describe the short-cycle work transformation processes on the three distinct levels, and the coupling between levels. According to the above definitions of the dynamical operators, for a given amount of work, the task processing at the system and component levels is decoupled and only "mediated" through the design activities at the subsystem level. It is also quite easy to develop additional operator representations of periodically correlated work processes that describe fully or partially synchronized release processes at the system and subsystem levels. We will leave this as an exercise for the interested reader.

The evolution of the process state X_t under the developed regime of the three-level hierarchical coordination structure can obviously be expressed by three dynamical operators $\Phi_1(1)$, $\Phi_1(s)$ and $\Phi_1(s')$ in conjunction with three noise vectors ε_1, ε_s, and $\varepsilon_{s'}$ as

$$X_t=\begin{cases}\Phi_1(1)\cdot X_{t-1}+\varepsilon_1 & \text{for } t=ns+v\\ \Phi_1(s)\cdot X_{t-1}+\varepsilon_s & \text{for } t=(2n+1)s\\ \Phi_1(s')\cdot X_{t-1}+\varepsilon_{s'} & \text{for } t=2(n+1)s\end{cases}$$

with $n=0,1,\ldots$ and $v=1,\ldots,s-1$.

The noise vectors ε_1, ε_s and $\varepsilon_{s'}$ correspond to zero-mean white noise as before:

$$\begin{gathered}\varepsilon_1\sim\mathcal{N}(0_d,C_1)\\ \varepsilon_s\sim\mathcal{N}(0_d,C_s)\\ \varepsilon_{s'}\sim\mathcal{N}(0_d,C_{s'}).\end{gathered}$$

An important result from the theory of stochastic processes (Franses and Paap 2004; Ursu and Duchesne 2009) is that the PVAR process with two time scales in Eq. 80 also offers a compact representation as a VAR(1) model:

$$\Phi_0^* \cdot X_n^* = \Phi_1^* \cdot X_{n-1}^* + \varepsilon_n^*, \tag{84}$$

where $X_n^* = \left(X_{ns+s}^{\mathrm{T}}, X_{ns+s-1}^{\mathrm{T}}, \ldots, X_{ns+1}^{\mathrm{T}}\right)^{\mathrm{T}}$ and $\varepsilon_n^* = \left(\varepsilon_{ns+s}^{\mathrm{T}}, \varepsilon_{ns+s-1}^{\mathrm{T}}, \ldots, \varepsilon_{ns+1}^{\mathrm{T}}\right)^{\mathrm{T}}$ are $ds \times 1$ state and error vectors, respectively. The matrix Φ_0^* and the autoregressive coefficient Φ_1^* are given by the nonsingular matrices

$$\Phi_0^* = \begin{pmatrix} I_d & -\Phi_1(s) & 0 & \cdots & 0 & 0 \\ 0 & I_d & -\Phi_1(s-1) & \cdots & 0 & 0 \\ \vdots & & & \ddots & & \vdots \\ 0 & 0 & 0 & \cdots & I_d & -\Phi_1(2) \\ 0 & 0 & 0 & \cdots & 0 & I_d \end{pmatrix} = \begin{pmatrix} I_d & -\Phi_1(s) & 0 & \cdots & 0 & 0 \\ 0 & I_d & -\Phi_1(1) & \cdots & 0 & 0 \\ \vdots & & & \ddots & & \vdots \\ 0 & 0 & 0 & \cdots & I_d & -\Phi_1(1) \\ 0 & 0 & 0 & \cdots & 0 & I_d \end{pmatrix} \tag{85}$$

and

$$\Phi_1^* = \begin{pmatrix} 0 & 0 & \cdots & 0 \\ 0 & 0 & & 0 \\ \vdots & & \ddots & \vdots \\ \Phi_1(1) & 0 & \cdots & 0 \end{pmatrix}. \tag{86}$$

The matrices Φ_0^* and Φ_1^* are both of size $ds \times ds$. Similar representations can also be developed for PVAR processes with multiple time scales.

The process $\{\varepsilon_n^*\}$ corresponds to a zero-mean periodic white noise process as before, with $E\left[\varepsilon_n^*\right] = 0_{ds}$ and $E\left[\varepsilon_n^* \, \varepsilon_n^{*\mathrm{T}}\right] = C^*$. The covariance matrix C^* is not singular. We assume that the vectors $\varepsilon_{ns+s}^{\mathrm{T}}, \varepsilon_{ns+s-1}^{\mathrm{T}}, \ldots$ are temporally uncorrelated and that C^* can be expressed by

$$C^* = \begin{pmatrix} C_s & 0 & 0 & 0 \\ 0 & C_1 & 0 & 0 \\ 0 & 0 & \ddots & 0 \\ 0 & 0 & 0 & C_1 \end{pmatrix}. \tag{87}$$

If all parallel tasks are initially to be completed in full, the initial state in the VAR (1) representation is

$$x_0^* = \begin{pmatrix} x_0 \\ \begin{pmatrix} 0 \\ \vdots \\ 0 \end{pmatrix} \\ \vdots \\ \begin{pmatrix} 0 \\ \vdots \\ 0 \end{pmatrix} \end{pmatrix}, \tag{88}$$

where the row vectors of zeros following the original initial state x_0 have size $|x_0|$. A total of $s - 1$ row vectors of zeros must be appended to the original initial state for a complete state representation.

It is clear that the VAR(1) model according to Eq. 84 can be rewritten as

$$X_n^* = \left(\Phi_0^*\right)^{-1} \cdot \Phi_1^* \cdot X_{n-1}^* + \left(\Phi_0^*\right)^{-1} \cdot \varepsilon_n^*. \tag{89}$$

The matrix Φ_0^* can be easily inverted:

$$\begin{aligned}(\Phi_0^*)^{-1} &= \begin{pmatrix} I_d & \Phi_1(s) & \Phi_1(s-1)\Phi_1(s-2) & \dots & \Phi_1(s-1)\Phi_1(s-2)\dots\Phi_1(3) & \Phi_1(s)\Phi_1(s-1)\dots\Phi_1(2) \\ 0 & I_d & \Phi_1(s-1) & \dots & \Phi_1(s-2)\dots\Phi_1(3) & \Phi_1(s-1)\dots\Phi_1(2) \\ \vdots & & & \ddots & & \vdots \\ 0 & 0 & 0 & \dots & I_d & \Phi_1(2) \\ 0 & 0 & 0 & \dots & 0 & I_d \end{pmatrix} \\ &= \begin{pmatrix} I_d & \Phi_1(s) & \Phi_1(1)^2 & \cdots & \Phi_1(1)^{s-3} & \Phi_1(s)\Phi_1(1)^{s-2} \\ 0 & I_d & \Phi_1(1) & \cdots & \Phi_1(1)^{s-4} & \Phi_1(1)^{s-2} \\ \vdots & & & \ddots & & \vdots \\ 0 & 0 & 0 & \cdots & I_d & \Phi_1(1) \\ 0 & 0 & 0 & \cdots & 0 & I_d \end{pmatrix}. \end{aligned} \tag{90}$$

The most convenient VAR(1) representation is therefore

$$X_n^* = A_0^* \cdot X_{n-1}^* + \varepsilon_n^\star,$$

with the combined dynamical operator

$$A_0^* = \left(\Phi_0^*\right)^{-1} \cdot \Phi_1^* = \begin{pmatrix} \Phi_1(s)(\Phi_1(1))^{s-1} & 0 & 0 & \cdots & 0 \\ (\Phi_1(1))^{s-1} & 0 & 0 & \cdots & 0 \\ (\Phi_1(1))^{s-2} & 0 & 0 & \cdots & 0 \\ \vdots & \vdots & 0 & \ddots & \vdots \\ \Phi_1(1) & 0 & 0 & \cdots & 0 \end{pmatrix}$$

and the transformed vector $\varepsilon_n^\star = \left(\Phi_0^*\right)^{-1} \cdot \varepsilon_n^*$ with covariance

$$C^{\star} = \left(\Phi_0^*\right)^{-1} \cdot C^* \cdot \left(\Phi_0^*\right)^{-\mathrm{T}}.$$

Analogous to Sections 2.1 and 2.2, a closed-form solution to the expected total amount of work done during the iteration process until the stopping criterion δ is satisfied can be calculated across $T_\delta = n_\delta s$ time steps for an arbitrary process with predefined initial state x_0^* and combined dynamical operator A_0^* as

$$\begin{aligned} E\left[\sum_{i=0}^{n_\delta} X_i^*\right] &= \sum_{i=0}^{n_\delta} E\left[X_i^*\right] \\ &= \sum_{i=0}^{n_\delta} \left(\left(A_0^*\right)^i \cdot x_0^*\right) \\ &= \left(\sum_{i=0}^{n_\delta} \left(A_0^*\right)^i\right) \cdot x_0^* \\ &= \left(I_p - A_0^*\right)^{-1} \cdot \left(I_p - \left(A_0^*\right)^{n_\delta+1}\right) \cdot x_0^*. \end{aligned}$$

The expected total amount of work x_{tot} done over all tasks across $T_\delta = n_\delta s$ time steps can be estimated by:

$$x_{tot} = \mathrm{Total}\left[\left(I_p - A_0^*\right)^{-1} \cdot \left(I_p - \left(A_0^*\right)^{n_\delta+1}\right) \cdot x_0^*\right].$$

In the limit $n_\delta \to \infty$, for an asymptotically stable project phase with periodically correlated work processes the expected total amount of work done over all tasks we have:

$$\lim_{n_\delta \to \infty} E\left[\sum_{i=0}^{n_\delta} X_i^*\right] = \left(I_p - A_0^*\right)^{-1} \cdot x_0^*. \tag{91}$$

To evaluate explicitly the intricate interrelationships between project dynamics and emergent complexity, the stochastic process must satisfy the criterion of strict stationarity. We will return to this point in Section 4.1. A strictly stationary process has a joint probability density that is invariant with shifting the origin, and therefore the locus and variance do not change over time. It is clear that the periodic autoregression model in Eq. 80 is a non-stationary model as the variance and autocovariances take different values in different time steps ("seasons"). In order to facilitate the analysis of stationarity, the introduced time-invariant representations as VAR(1) models (cf. Eq. 84) have to be considered. Using general properties of these models (see e.g. Brockwell and Davis 1991), it follows that the stochastic process $\{X_t^*\}$ is stationary if

$$\mathrm{Det}\left[\Phi_0^* - \Phi_1^* z\right] \neq 0$$

for all $z \in \mathbb{C}$ satisfying the condition $|z| \leq 1$ (Ursu and Duchesne 2009). The characteristic equation can be simplified to

$$\begin{aligned}\mathrm{Det}\left[\Phi_0^* - \Phi_1^* z\right] &= \mathrm{Det}\left[I_{ds} - \left(\left(\Phi_0^*\right)^{-1} \cdot \Phi_1^*\right) z\right] \\ &= \mathrm{Det}\left[I_d - \left(\prod_{k=0}^{s-1} \Phi_1(s-k)\right) z\right] \\ &= \mathrm{Det}\left[I_d - \Phi_1(s) \cdot (\Phi_1(1))^{s-1} z\right] \neq 0\end{aligned}$$

for all z such that $|z| \leq 1$. Hence, the process $\{X_t^*\}$ is stationary if the eigenvalues of the matrix product $\Phi_1(s) \cdot (\Phi_1(1))^{s-1}$ are all strictly smaller than one in modulus (Franses and Paap 2004; Ursu and Duchesne 2009).

2.7 Extended Least Squares Parameter Estimation and Model Selection

The autoregressive coefficients and the error covariances of a PVAR model of arbitrary order (not necessarily limited to a first-order autoregressive process for each season as previously formulated) without linear constraints on the independent parameters can be calculated efficiently on the basis of standard least-squares or maximum-likelihood estimation techniques from textbooks (see, e.g. Brockwell and Davis 1991; Franses and Paap 2004; Lütkepohl 2005). To apply these techniques, one only has to bring the introduced linear recurrence relations into a standard regression form and then execute the (usually iterative) estimation procedures. However, in the developed model formulation we had to pose the constraint that some entries of the dynamical operators $\Phi_1(1)$ (Eq. 81) and $\Phi_1(s)$ (Eq. 82) must be zero in order to be able to model the typical hold-and-release policy of design information in a PD project with periodically correlated work processes or must be equal to ε or $(1-\varepsilon)$ for an analytical evaluation of emergent complexity in later chapters. Furthermore, some coefficients of the work transformation submatrices are linear dependent. Consequently, we cannot use the standard estimation techniques but instead have to use an extended algorithm developed by Ursu and Duchesne (2009) that is able to carry out least-squares estimation with linear constraints on the regression parameters. For the given model formulation with zero-mean periodic white noise the least squares estimators are equivalent to maximum likelihood estimators. In the following sections we will present closed-form solutions of the estimators of different strength based on the original work of Ursu and Duchesne (2009). To allow the interested reader to follow the accompanying proofs in the original material, we will use a similar notation to the developers of the algorithm. We start by presenting a convenient regression form and then proceed with the specification of the least square estimators for the full unconstraint case and the more complex restricted case.

In principle, the asymptotic properties of the least square estimators could be derived from generalized results for time series based on the multivariate representation from Eq. 84 (see, e.g. Brockwell and Davis 1991; Lütkepohl 2005). However, to estimate the statistical properties of the autoregressive coefficient matrices for each release period v of the formulated model of a PD project with periodically correlated work processes, the multivariate stochastic process generated by state Eq. 84 needs to be inverted. This operation seems to be unnecessarily complex in the multivariate setting. Instead, it is more efficient to use the individual PVAR components directly in parameter estimation. Consider the sequence of aggregated random variables $\{X_{ns+v}, 0 \leq ns + v < Ts\}$ representing the task processing in the PD project over T time steps, $t = 0, 1, \ldots, T-1$. At each time step the $v = 1, \ldots, s$ short iterations of the development teams without purposefully withholding design information are aggregated and therefore the states in between the n long iterations in which the release of hidden information occurs are combined. Hence, we have a total sample size equal to $n' = Ts$. For a convenient state representation let

$$Z(v) = \left(X_v, X_{s+v}, \ldots, X_{(T-1)s+v}\right) \tag{92}$$

$$E(v) = \left(\varepsilon_v, \varepsilon_{s+v}, \ldots, \varepsilon_{(T-1)s+v}\right) \tag{93}$$

$$\boldsymbol{X}(v) = \left(\boldsymbol{X}_0(v), \ldots, \boldsymbol{X}_{T-1}(v)\right) \tag{94}$$

all be $d \times T$ random matrices, where

$$\boldsymbol{X}_t(v) = X_{ts+v-1}$$

denotes the $d \times 1$ random vectors of work remaining of the component-level and system-level tasks at time steps $t = 0, 1, \ldots, T-1$. Utilizing these aggregated variables, the PVAR model from Eq. 80 can be recast and written in the following convenient regression form:

$$Z(v) = B(v) \cdot \boldsymbol{X}(v) + E(v), \quad v = 1, \ldots, s. \tag{95}$$

The independent parameters of the regression model are collected in the $d \times d$ parameter matrix $B(v)$. Using the definitions of the dynamical operators $\Phi_1(1)$ and $\Phi_1(s)$, the parameter matrix can be defined as:

$$B(v) = \begin{cases} \Phi_1(1) & \text{for } v = 1, \ldots, s-1 \\ \Phi_1(s) & \text{for } v = s\,. \end{cases}$$

Since for all $v = 1, \ldots, s-1$ the regression equation 95 contains the same unknown regression parameters, it is convenient to concatenate them into one equation

$$Z_c = \Phi_1(1) \cdot \boldsymbol{X}_c + E_c, \tag{96}$$

where

$$Z_c = (Z(1), \ldots, Z(s-1)) \tag{97}$$
$$\boldsymbol{X}_c = (\boldsymbol{X}(1), \ldots, \boldsymbol{X}(s-1)) \tag{98}$$
$$E_c = (E(1), \ldots, E(s-1)). \tag{99}$$

Convenient vector representations of the regression equations 95 and 96 can be obtained by using the Kronecker product $\otimes$. A vectorization of the dependent variables based on the Kronecker product leads to

$$z_1 = vec[Z_c] = \left(\boldsymbol{X}_c^{\mathrm{T}} \otimes I_d\right) \cdot \beta(1) + vec[E_c] \tag{100}$$

and

$$z_2 = vec[Z(s)] = \left(\boldsymbol{X}^{\mathrm{T}}(s) \otimes I_d\right) \cdot \beta(s) + vec[E(s)]. \tag{101}$$

The vectors of the regression coefficients are given by

$$\beta(v) = vec[\Phi_1(v)] \quad v = 1, \ldots, s\,.$$

In general, the vector operator $vec[A]$ represents the vector obtained by stacking the columns of the matrix A onto each other. We can also combine both regression equations in one large equation:

$$z = \begin{bmatrix} z_1 \\ z_2 \end{bmatrix} = \begin{bmatrix} \boldsymbol{X}_c^{\mathrm{T}} \otimes I_d & 0 \\ 0 & \boldsymbol{X}^{\mathrm{T}}(s) \otimes I_d \end{bmatrix} \cdot \begin{bmatrix} \beta(1) \\ \beta(s) \end{bmatrix} + e. \tag{102}$$

The combined error vector is given by

$$e = \begin{bmatrix} vec[E_c] \\ vec[E(s)] \end{bmatrix}.$$

The vector of regression coefficients

$$\beta = \begin{bmatrix} \beta(1) \\ \beta(s) \end{bmatrix}$$

contains by parts the same elements (see the definitions of the matrices in Eqs. 81 and 82), i.e. they are linear dependent. Furthermore, many elements of β are known to be zero. This can be expressed by the following linear relation:

$$\beta = R \cdot \xi + b. \tag{103}$$

The vector ξ represents a $K \times 1$ vector of unknown regression parameters in the restricted case. It is evident that the parameter setting

$$R = I_{2d^2} \quad \text{and} \quad b = 0$$

represents the full unconstraint case, in which no constraints are imposed on the entries of the dynamical operators $\Phi_1(1)$ and $\Phi_1(s)$. Through the specification of the entries in R and b, additional linear constraints can be imposed on the parameters for each release period v. If, for instance, a matrix entry in $\Phi_1(1)$ must be zero, the corresponding row vector of R and vector component of b are set to zero. Through this encoding the null entries in the dynamical operators are ignored in least squares estimation and only the non-zero informational couplings between tasks are determined. Secondly, if the n-th component of β is related to the m-th component of the irreducible regression vector ξ then we set the element $R_{[[n,m]]} = 1$.

When we convert the linear relation from Eq. 103 into a vector representation that is similar to Eq. 101, we arrive at the following expression for z:

$$\begin{aligned} z &= \left(\boldsymbol{X}^{\mathrm{T}} \otimes I_d\right) \cdot \beta + e \\ &= \left(\boldsymbol{X}^{\mathrm{T}} \otimes I_d\right) \cdot (R \cdot \xi + b) + e, \end{aligned} \tag{104}$$

where

$$\boldsymbol{X}^{\mathrm{T}} = \begin{bmatrix} \boldsymbol{X}_c^{\mathrm{T}} & 0 \\ 0 & \boldsymbol{X}^{\mathrm{T}}(s) \end{bmatrix}.$$

The least squares estimators of the parameter vector ξ are calculated by minimizing the generalized least squares criterion:

$$\mathfrak{J}_g(\xi) = e^{\mathrm{T}} (I_T \otimes C_e)^{-1} e. \tag{105}$$

The matrix C_e represents the covariance matrix of the combined error vector e, that is $C_e = E[e \cdot e^{\mathrm{T}}]$. It can be easily composed of the individual covariance matrices C_1 and C_s.

In the unrestricted case, an equivalent representation of the least squares estimators based on the above generalized least squares criterion $\mathfrak{J}_g(\xi)$ can be obtained by minimizing the ordinary least squares:

$$\mathfrak{J}(\beta) = e^{\mathrm{T}} \cdot e.$$

A similar result holds for VAR models, see Schneider and Neumaier (2001) and Lütkepohl (2005). To obtain the ordinary least squares estimators, the function $\mathfrak{J}(\beta)$ is differentiated with respect to each "vectorized" dynamical operator $\Phi_1(v)$:

$$\frac{\delta \Im(\beta)}{\delta vec[\Phi_1(v)]} = -2\sum_{t=0}^{T-1}(X_{ts+v-1} \otimes \varepsilon_{ts+v}), \quad v = 1, \ldots, s.$$

Setting the derivatives to zero yields the following system of equations for a given release period v:

$$\sum_{t=0}^{T-1}(\boldsymbol{X}_t(v) \otimes \varepsilon_{ts+v}) = 0_{d^2}.$$

In the above equation 0_{d^2} is the $d^2 \times 1$ null vector. Since the fluctuations can be expressed as $\varepsilon_{ns+v} = X_{ns+v} - \left(\boldsymbol{X}_n^{\mathrm{T}}(v) \otimes I_d\right)\beta(v)$, the normal equations for each short iteration v are given by

$$\sum_{t=0}^{T-1}(\boldsymbol{X}_t(v) \otimes X_{ts+v}) = \left(\sum_{t=0}^{T-1}\left(\boldsymbol{X}_t(v) \cdot \boldsymbol{X}_t^{\mathrm{T}}(v) \otimes I_d\right)\right)\beta(v).$$

Hence, the desired least squares estimators $\hat{\beta}(v)$ satisfy the relation:

$$\hat{\beta}(v) = \left(\left(\boldsymbol{X}(v) \cdot \boldsymbol{X}^{\mathrm{T}}(v)\right)^{-1}\boldsymbol{X}(v) \otimes I_d\right)z(v). \tag{106}$$

The estimated residuals are given by the difference:

$$\hat{\varepsilon}_{ns+v} = X_{ns+v} - \left(\boldsymbol{X}_n^{\mathrm{T}}(v) \otimes I_d\right)\hat{\beta}(v). \tag{107}$$

In the above estimators the independent variables $\boldsymbol{X}(v)$ and X_{ns+v}, respectively denote the time series of empirically acquired state vectors as single realizations of the periodically correlated work processes in the PD project. If multiple realizations of the PVAR process had been acquired in N independent trials as opposed to merely a single time series is being given, the additional time series are simply appended as additional $N-1$ blocks of rows in the regression Eq. 95. Similarly, the predictors are extended by additional row blocks. The initial state is determined by averaging over all initial state vectors.

Solving the least squares problem directly, Ursu and Duchesne (2009) give the following alternative equation for the least squares estimators:

$$\hat{B}(v) = Z(v) \cdot \boldsymbol{X}^{\mathrm{T}}(v)\left(\boldsymbol{X}(v) \cdot \boldsymbol{X}^{\mathrm{T}}(v)\right)^{-1}. \tag{108}$$

Based on the above relation one can also express the difference between estimator $\hat{B}(v)$ and $B(v)$ as:

$$\hat{B}(v) - B(v) = \left\{\frac{1}{T}\right\} \cdot E(v) \cdot \boldsymbol{X}^{\mathrm{T}}(v) \left(\left\{\frac{1}{T}\right\} \cdot \boldsymbol{X}(v) \cdot \boldsymbol{X}^{\mathrm{T}}(v) \right)^{-1}.$$

Noting that for the sum over T time steps it holds that

$$\sum_{t=0}^{T-1} vec\left[\varepsilon_{ns+v} \cdot X_n^{\mathrm{T}}(v)\right] = vec\left[E(v) \cdot X^{\mathrm{T}}(v)\right],$$

it follows for the convergence in distribution (symbol "$\xrightarrow{d}$") that

$$\left\{\frac{1}{\sqrt{T}}\right\} \cdot vec\left[E(v) \cdot \boldsymbol{X}^{\mathrm{T}}(v)\right] \xrightarrow{d} \mathcal{N}(0_{d^2}, \Omega(v) \otimes C(v))$$

and for the convergence in probability (symbol "$\xrightarrow{p}$") that

$$\left\{\frac{1}{T}\right\} \cdot vec\left[E(v) \cdot \boldsymbol{X}^{\mathrm{T}}(v)\right] \xrightarrow{p} 0_{d^2}.$$

The function $\mathcal{N}(0_{d^2}, \Omega(v) \otimes C(v))$ denotes the d^2-variate Gaussian distribution with location 0_{d^2} and covariance $\Omega(v) \otimes C(v)$. The *pdf* of this distribution is given in Eq. 13. $\Omega(v)$ denotes the $d \times d$ covariance matrix of the aggregated random vector $\boldsymbol{X}_t(v)$, see Ursu and Duchesne (2009).

After the derivation of the least square estimators for the full unconstraint case, we proceed with the restricted case, i.e. the case in which additional linear constraints must be satisfied. If the parameters satisfy the linear constraint in Eq. 95, the least squares estimators of $\xi(v)$ minimize the generalized criterion $\mathfrak{I}_g(\xi)$ (Eq. 105). It is evident that the generalized criterion is not equivalent to the ordinary least squares criterion $\mathfrak{I}(\beta)$ (see e.g. Lütkepohl 2005). Rearranging the regression Eq. 104 leads to the following relation for the combined error vector:

$$e = z - \left(\boldsymbol{X}^{\mathrm{T}} \otimes I_d\right)(R \cdot \xi + b).$$

This relation is sufficient to derive the asymptotic properties of the least squares estimator of ξ under linear constraints. Owing to limited space we will not present the stepwise derivation of ξ in this book but only cite the result from the original work by Ursu and Duchesne (2009):

$$\hat{\xi} = \left(R^{\mathrm{T}}\left(\boldsymbol{X} \cdot \boldsymbol{X}^{\mathrm{T}} \otimes C_e^{-1}\right)R\right)^{-1} R^{\mathrm{T}}\left(\boldsymbol{X} \otimes C_e^{-1}\right)\left(z - \left(\boldsymbol{X}^{\mathrm{T}} \otimes I_d\right)b\right).$$

Ursu and Duchesne (2009) show that the estimator $\hat{\xi}$ is consistent for ξ and asymptotically follows a Gaussian distribution:

$$\left\{\sqrt{T}\right\} \cdot (\hat{\xi} - \xi) \xrightarrow{d} \left(0_K, \left(R^{\mathrm{T}}\left(\Omega \otimes C_e^{-1}\right)R\right)^{-1}\right).$$

However, the estimator $\hat{\xi}$ is unfeasible in almost all practical applications in project management because it relies on the (usually) unknown covariance matrix C_e. Instead, a consistent estimator $\tilde{C}_e$ of the covariance matrix C_e can be used and we have the alternative representation:

$$\hat{\hat{\xi}} = \left(R^{\mathrm{T}}\left(X \cdot X^{\mathrm{T}} \otimes \tilde{C}_e^{-1}\right)R\right)^{-1} R^{\mathrm{T}}\left(X \otimes \tilde{C}_e^{-1}\right)\left(z - \left(X^{\mathrm{T}} \otimes I_d\right)b\right). \tag{109}$$

According to Ursu and Duchesne (2009) good candidate consistent estimators are given by the unconstrained least squares estimators:

$$\tilde{C}_e = \left\{\frac{1}{T-d}\right\} \cdot \left(Z - \hat{B} \cdot X\right)\left(Z - \hat{B} \cdot X\right)^{\mathrm{T}}. \tag{110}$$

In the above equation $\hat{B}$ denotes the least squares estimators from Eq. 108, which were obtained for the full unconstraint case. The resulting estimator of β is given by

$$\hat{\hat{\beta}} = R \cdot \hat{\hat{\xi}} + b. \tag{111}$$

Its asymptotic distribution is Gaussian:

$$\left\{\sqrt{T}\right\} \cdot \left(\hat{\hat{\beta}} - \beta\right) \xrightarrow{d} \mathcal{N}\left(0_{2d^2}, R\left(R^{\mathrm{T}}\left(\Omega \otimes \tilde{C}_e^{-1}\right)R\right)^{-1} R^{\mathrm{T}}\right).$$

The detailed proof of the above results can be found in Ursu and Duchesne (2009). It follows lines of reasoning similar to the proof in Lütkepohl (2005). However, Lütkepohl (2005) established the asymptotic properties of least squares estimators only for VAR models in which the model parameters satisfy linear constraints according to Eq. 103 and he did not generalize his results to PVAR models.

For applied studies in project management and schedule management/control, it can also be of interest not only to estimate the coefficient matrices $\Phi_1(v)$ and the error covariance matrices $C(v)$ of the PVAR process from time series data based on the introduced model formulation, but to follow a fully data-driven approach and also include the regression order for each iteration v and the corresponding multiple dynamical operators in a combined estimation procedure. Similar to model selection in the class of VAR(n) models from the previous Section 2.4, in a fully data-driven approach a good trade-off must be found between the predictive accuracy gained by increasing the number of independent parameters and the danger of overfitting the model to performance fluctuations and not persistent patterns of cooperation. We start by formulating an extended model of periodically correlated work processes with iteration-dependent correlation lengths and then proceed with solving the more subtle problem of selecting the "right" regression order for each iteration. To incorporate iteration-dependent correlation lengths into a PVAR process, the state Eq. 80 has to be extended towards multiple interacting autoregression models (Ursu and Duchesne 2009):

$$X_{ns+v} = \sum_{k=1}^{n(v)} \Phi_k(v) \cdot X_{ns+v-1} + \varepsilon_{ns+v}. \tag{112}$$

The variable $n(v)$ denotes the autoregressive model order at iteration v of the work process and $\Phi_k(v)$ represents the multiple dynamical operators holding for that period of time. $n(v)$ must be smaller than the the length s of the period between releases of hidden information. Both the autoregressive model order $n(v)$ and the dynamical operators $\Phi_k(v),\ k = 1, \ldots, n(v)$, are the model coefficients during iteration $v = 1, \ldots, s$. Therefore, the regression order of the extended PVAR model is not just a non-negative integer as for the VAR(n) model, but an s-tuple $(n(1), \ldots, n(s))$ of multiple regression orders in which the vector components determine the regression order for the individual iteration $v = 1, \ldots, s$.

Similar to the previous model formulation, the combined error process $\{\varepsilon_{ns+v}\}$ corresponds to a zero-mean periodic white noise process. $\{\varepsilon_{ns+v}\}$ is composed of $d \times 1$ random vectors, such that $E[\varepsilon_{ns+v}] = 0_d$ and $E[\varepsilon_{ns+v}\, \varepsilon_{ns+v}{}^{\mathrm{T}}] = C(v)$ for $v = 1, \ldots, s$. It is assumed that the covariance matrices $C(v)$ for the iterations are not singular.

Following the same procedure as before, we can develop a generalized state representation for the extended PVAR model:

$$Z(v) = \left(X_v, X_{s+v}, \ldots, X_{(T-1)s+v}\right) \tag{113}$$

$$E(v) = \left(\varepsilon_v, \varepsilon_{s+v}, \ldots, \varepsilon_{(T-1)s+v}\right) \tag{114}$$

$$\boldsymbol{X}(v) = (\boldsymbol{x}_0(v), \ldots, \boldsymbol{x}_{T-1}(v)). \tag{115}$$

In the generalized state representation $Z(v)$ and $E(v)$ are the same $d \times T$ random matrices as before. By contrast, $\boldsymbol{X}(v)$ is a $\left(d^2 n(v)\right) \times T$ matrix, where the entries

$$\boldsymbol{x}_t(v) = \left(X_{ts+v-1}^{\mathrm{T}}, \ldots, X_{ts+v-n(v)}^{\mathrm{T}}\right)^{\mathrm{T}}$$

denote the $\left(d^2 n(v)\right) \times 1$ random vectors of work remaining of the component-level and system-level tasks at time steps $t = 0, 1, \ldots, T-1$. It is evident that the full PVAR model can be rewritten as $Z(v) = B(v) \cdot \boldsymbol{X}(v) + E(v),\ v = 1, \ldots, s$. This regression form was already introduced in Eq. 95.

The dynamical operators of the full PVAR model are collected for each short iteration in the extended $d \times (dn(v))$ parameter matrix $B(v)$. The parameter matrix is defined as

$$B(v) = \left(\Phi_1(v), \ldots, \Phi_{n(v)}(v)\right). \tag{116}$$

It is important to note that the generalized state representation according to Eqs. 113–116 is in principle sufficient to estimate the independent parameters in the full unconstraint case, and the least squares estimators for the parameter matrix (Eq. 106) and the error covariance (Eq. 107) can be directly applied. To make the estimation procedure fully operational, the parameters simply have to be stacked into the parameter vector

$$\beta(v) = \left(vec^{\mathrm{T}}[\Phi_1(1)], \ldots, vec^{\mathrm{T}}\left[\Phi_{n(v)}(v)\right]\right)$$

of dimension $d^2n(v) \times 1$.

If a least square estimation with linear constraints on the parameters of the dynamical operators needs to be carried out, we have to define an extended $\left(d^2n(v)\right) \times K(v)$ matrix $R(v)$ of rank $K(v)$ and an extended $\left(d^2n(v)\right) \times 1$ vector $b(v)$ to satisfy the linear relation given by Eq. 103. Similar as before, the vector $\xi(v)$ represents a $K(v) \times 1$ vector of unknown regression parameters. The parameter setting $R(v) = I_{d^2n(v)}$ and $b(v) = 0$ reflects the full unconstraint case. If certain matrix entries in $\Phi_k(v)$ must be zero, the corresponding row vectors of $R(v)$ and vector components of $b(v)$ have to be set to zero. Such a coherent theoretical framework for constraint satisfaction allows us to use the feasible least squares estimator from Eq. 109 directly. A more complicated estimation relation is not necessary. According to Ursu and Duchesne (2009) good candidate consistent estimators for the error covariance matrix $\tilde{C}(v)$ are also given by the unconstrained least squares estimators from Eq. 110.

If the s-tuple $(n(1), \ldots, n(s))$ of regression orders holding for the individual short iterations $v = 1, \ldots, s$ has to be estimated from time series data for an unconstraint or constraint model in a fully data-driven approach, the cited trade-off between the predictive accuracy gained by increasing the regression order and the danger of overfitting the model can be resolved in a similar fashion as in the previous Section 2.4 by using the standard selection criteria of Akaike (1971, 1973 and 1974) and Schwarz (1978). This is due to the fact that PVAR processes do not constitute a model class in their own right, but can be expressed as basic vector autoregressive processes (see Eq. 89). In this chapter we focus on the Schwarz-Bayes Criterion (cf. Eq. 71) because within the scope of this book it has the same consequences for regression order selection as the (simplified two-stage) minimum description length criterion (Hansen and Yu 2001), which in turn is well grounded in Rissanen's theory of minimum description length that will be presented and discussed in the complexity-theoretic Section 3.2.2. Generalizing the fundamental ideas of McLeod (1994) on diagnostic checking periodic autoregression models, Ursu and Duchesne (2009) introduce a heuristic approach in which Schwarz's BIC criterion is decomposed to obtain separate selection criteria for each short iteration $v = 1, \ldots, s$. For the unconstrained model, they define the cumulative criterion

$$BIC = \sum_{v=1}^{s} BIC_{n(v)}(v) \tag{117}$$

and the iteration dependent criteria

$$BIC_{n(v)}(v) = \ln \mathrm{Det}\left[\widehat{\Sigma}_{(n(v))}(v)\right] + \frac{\ln T}{T} k(v), \tag{118}$$

where

$$k(v) = n(v)\,p^2.$$

For each separate criterion, the variable $\widehat{\Sigma}_{(n(v))}(v)$ denotes the not bias corrected least squares estimate of $C(v)$ (cf Eq. 107) for the approximating autoregression model of order $n(v)$. $\widehat{\Sigma}_{(n(v))}(v)$ can also be interpreted as the one-step prediction error resulting from the separate autoregression model with parameter matrix $(\Phi_1(v), \ldots, \Phi_{n(v)}(v))$. Similarly to the definition of the Akaike information criterion presented in Section 2.5, the last factor $k(v)$ represents the effective number of parameters for a given iteration v. In the paper of Ursu and Duchesne (2009) the effective number of parameters is estimated by counting only the freely estimated parameters in the coefficient matrices. As already pointed out in Section 2.5, from a theoretical point of view, all functionally independent parameters in the parameter vector θ which must be estimated have to be considered and therefore the effective number of parameters of the approximating model is

$$k(v) = n(v)\,p^2 + \frac{p(p+1)}{2}. \tag{119}$$

For a constraint model, the reduced effective number of parameters

$$k(v) = n(v)\,p^2 - \Upsilon(v) + \frac{p(p+1)}{2} \tag{120}$$

must be used (cf. Songsiri et al. 2010). $\Upsilon(v)$ denotes the total number of constraints that are posed in iteration v on the work transformation submatrices to model the hold-and-release policy of design information (in the form of 0 or 1 matrix entries) and to allow an analytical evaluation of emergent complexity (in the form of ε or $(1-\varepsilon)$ matrix entries). A calculation of the cumulative criterion from Eq. 117 must also take account of the coefficients of the work transformation submatrices that are linear dependent.

The heuristic to estimate the effective number of parameters by counting the functionally independent parameters in the parameter vector can easily be generalized to other model classes, such as linear dynamical systems, which are introduced and discussed in Section 2.9. For practical purposes, the regression order can be varied systematically in the range of $n(v) \in \{1, \ldots, s\}$ at each iteration v and the one-step-ahead prediction error is evaluated using the criterion from Eq. 118.

The regression order $n_{opt}(v)$ of the generative model holding at a given short iteration v is considered to be the optimal one if it is assigned minimum scores, that is

$$n_{opt}(v) = \arg\min_{n(v)} BIC_{n(v)}(v). \tag{121}$$

The optimum tuple n_{opt} of regression orders for the extended PVAR model is given by:

$$n_{opt} = \left(n_{opt}(1), \ldots, n_{opt}(s)\right). \tag{122}$$

Ursu and Duchesne (2009) used the introduced model selection criteria to fit a PVAR model to quarterly seasonally unadjusted West German income and consumption data for the years 1960–1987 and found the autoregressive orders $(2, 1, 3, 1)$ to be the optimal ones. The same data were also analyzed by Lütkepohl (2005) based on the classic PVAR model formulation from Section 2.4. Using the BIC selection criterion according to Eq. 71, Lütkepohl (2005) obtained a minimum score for a VAR(5) model.

After a model for the full unconstraint case has been fitted to data from a PD project on the basis of the above two-step procedure, it is sometimes also possible to reduce the number of independent parameters by setting selected entries in the dynamical operators $\Phi_1(v), \ldots, \Phi_{n(v)}(v)$ to zero. Ursu and Duchesne (2009) use a straightforward selection heuristic in which the standard errors of the individual regression coefficients are evaluated: If the absolute value of the t-statistic of the given autoregressive parameter is less than one, the corresponding parameter is set to zero. The t-statistic is computed as the value of the least squares estimator divided by its standard error. In a third step these additionally identified constraints on the parameters are defined in the form of the linear relationship (Eq. 103) and the parameters are re-estimated using the feasible estimators from Eq. 109 in conjunction with the consistent estimators from Eq. 110. The effectiveness of this kind of heuristic parameter reduction was also demonstrated by the authors on the basis of the quarterly seasonally unadjusted West German income and consumption data. They were able to reduce the number of independent parameters from 28 for the full unconstraint case to only 22 for a PVAR model with 6 null regression coefficients.

2.8 Simulation Study of Periodically Correlated Work Processes, and Validation of Estimation and Model Selection Procedures

Having comprehensively analyzed least squares estimation for applied studies with and without linear constraints on the entries of the dynamical operators, and having defined the cumulative and iteration-dependent Schwarz's criteria for model selection in periodic vector autoregression models, we now focus on the accuracy of the methods for estimating the independent parameters of the constraint model according to the two-level formulation from Eq. 80 (in conjunction with Eqs. 81, 82 and 83) in a fully data-driven approach (Schlick et al. 2014). By fully data-driven approach, we mean an applied study where the work transformation matrices (A_0^{C}, $A_0^{\mathrm{CS}}, A_0^{\mathrm{SC}}, A_0^{\mathrm{S}}, A_0^{\mathrm{SH}}, A_0^{\mathrm{HC}}$), the covariance matrices (C_1, C_s) and the initial state vector have to be estimated from data, as does the regression order for each short iteration v. Although the model formulation according to Eq. 80 assumes that a first-order autoregression is sufficient for capturing the essential dynamics of project phases with periodically correlated work processes, we have to verify whether or not larger regression orders are needed to make good predictions in real product development environments. We carried out a comprehensive simulation study to evaluate the accuracy of the estimation methods in a laboratory-style setting with complete control of confounding factors. The study is based on Monte Carlo experiments, which are a special class of computational algorithms that rely on repeated random sampling of the work processes to numerically evaluate their results. The repeated random sampling and the statistical analysis was carried out in a simulation environment that we developed in-house. For the Monte Carlo experiments, a PVAR model of cooperative task processing is formulated that connects the dynamics of module design and integration in a vehicle door development project with component development. This model forms a reference model that can then simulate the work processes and generate samples of different size. We try to use this data to reconstruct the reference model representation. Technically speaking, we investigate the identifiability of the reference model and the associated parameter uncertainty in repeated trials. The main question concerns how the accuracy of the estimation of the model matrices (A_0^{C}, A_0^{CS}, A_0^{SC}, A_0^{S}, A_0^{SH}, A_0^{HC}, C_1 and C_s) is influenced by the length T of the time series of simulated task processing that is used for numerical estimation, and whether the introduced heuristic model selection procedure can reliably select the correct regression orders of the reference PVAR model. Another important question is whether the obtained model representation can also accurately capture the underlying geometry of the subspaces and whether or not it just leads to a small root mean square deviation between the corresponding model matrices as a conventional measure of distance. To answer this question, we have to use subspace angles from the theory of linear dynamical systems. Section 2.9 will provide an extensive evaluation of this class of linear models with latent state variables. We follow the mathematical formulation given in de Cock

(2002) to compute these angles and to evaluate the squared cepstral distance as the last (Eq. 135) and probably most important independent variable in the Monte Carlo experiments (see below).

We start by defining the structure and parameters of the reference PVAR model. As previously stated, the idea is to model the dynamics of cooperative work in a vehicle door development project. To simplify the analysis and numerical computation, we focus on the door module of the vehicle door subsystem. A door module typically consists of a functional carrier plate and other components that are fitted onto it. The carrier plate is usually made from plastic or steel. It is rubber-sealed to separate the wet and dry sides of the door system. Various door components, such as the window lift mechanism, locks, wiring harness, switches, loud speakers, crash sensors and cables connecting the latch to the inner release handle have to be integrated. The major automotive manufacturers usually outsource the design, development and manufacturing of the door module to selected first-tier suppliers to save costs and weight. We focus on the periodically correlated work processes in the supplier's development organization. At the supplier, the corresponding development work is located on the first level of the work breakdown structure and is supported by additional functions, such as manufacturing, procurement, sales, controlling and quality management. To build our reference PVAR model, we focus on system design and the related component development and integration activities. In what follows, we have to greatly simplify the real industrial project organization so as to obtain simulation models of reasonable complexity that can be analyzed in Monte Carlo experiments conducted in our simulation environment. However, this does not limit the generality of the model-driven approach. The interested reader can easily upscale the reference model to build organization designs for projects that are much larger and have more complex hierarchical coordination structures.

We assume in the simulation study that the project work in the engineering design department of the first-tier supplier mirrors the system structure of the door module and breaks it down into development teams focusing on mechanical and electrical/electronic functions. A dedicated module design and integration testing team handles the systems engineering process and the integration of the components into a fully functional module. The team also coordinates the design of the interfaces to the complete door system and the car body. We focus on the cooperation between the module design and integration testing team and two subordinate teams dealing with the mechanical design of the functional carrier plate and the mechanical/kinematic design of the window lift mechanism (including safety and convenience electronics). As we said at the beginning of this chapter, the module design team inadvertently hided, for a certain amount of time, information on the integration and testing of geometric/kinematic entities and thus do not immediately pass it on to component-level teams. This kind of non-cooperative behavior is justified by the desire to make the module architecture more mature and to focus communication with customers and suppliers. To make it easier to evaluate the accuracy of the estimation methods, we define a PVAR model for the Monte Carlo experiments that includes only one module-level task and two component-level tasks. The three teams process the tasks simultaneously. Each team is assigned one task in the model. For the sake of simplicity, we do not consider individual task processing. The three vector components of the state variable X_{ns+v} that relate to the simultaneous processing of the module-level and component-level tasks represent the relative number of labor units required to complete the tasks. We assume that the first component-level team, which designs the functional carrier plate, works with autonomous task processing rate $a_{11}^{\mathrm{C}} = 0.91$. The second component-level team, which designs the window lift mechanisms, can use a standard mechanism and several internally standardized parts and therefore processes the task faster, with autonomous task processing rate $a_{22}^{\mathrm{C}} = 0.88$. The cooperative relationships within both component-level teams are very similar and therefore the tasks are coupled with symmetric strength $a_{12}^{\mathrm{C}} = a_{21}^{\mathrm{C}} = 0.03$. Due to a well-designed system architecture and experienced members, the team responsible for module design and integration testing works with autonomous task processing rate $a^{\mathrm{M}} = 0.80$. The module-level task generates 5% of finished work at each short iteration that is put in a hold state until it is released at time step $ns (n \in \mathbb{N})$. Hence, $a^{\mathrm{MH}} = 0.05$. Furthermore, both component-level teams generate 5% of finished work at each iteration for the module-level, and we have $a_{11}^{\mathrm{CM}} = a_{12}^{\mathrm{CM}} = 0.05$. Conversely, the module design team only feeds back 2% of the unresolved issues at each short iteration to the two component-level teams, and there is $a_{11}^{\mathrm{MC}} = a_{21}^{\mathrm{MC}} = 0.02$. The accumulated development issues of the module-level are released to the component-level team responsible for designing and developing the functional carrier plate at the end of the period ($a_{11}^{\mathrm{HC}} = 1$ and $a_{21}^{\mathrm{HC}} = 0$). This team transfers the accumulated unresolved issues to the second component-level team at the next time step and holds regular discussions to find appropriate solutions. Additional dynamical dependencies were not considered and therefore all other matrix entries were defined as zero. The release period s, in which component-level teams receive information on the overall module design and on integration testing of specific geometric/kinematic entities, is considered as an independent variable.

The complete PVAR representation for state Eq. 89 is as follows:

Combined dynamical operator $A_0^* = \left(\Phi_0^*\right)^{-1} \cdot \Phi_1^*$:

$$\left(\Phi_0^*\right)^{-1} \cdot \Phi_1^* = \begin{pmatrix} \Phi_1(s)(\Phi_1(1))^{s-1} & 0 & 0 & \cdots & 0 \\ (\Phi_1(1))^{s-1} & 0 & 0 & \cdots & 0 \\ (\Phi_1(1))^{s-2} & 0 & 0 & \cdots & 0 \\ \vdots & \vdots & 0 & \ddots & \vdots \\ \Phi_1(1) & 0 & 0 & \cdots & 0 \end{pmatrix}.$$

Work transformation sub-matrices, vectors and real-valued parameters:

$$A_0^{\mathrm{C}} = \begin{pmatrix} 0.91 & 0.03 \\ 0.03 & 0.88 \end{pmatrix} \tag{123}$$

$$A_0^{\mathrm{M}} = a^{\mathrm{M}} = 0.80 \tag{124}$$

$$A_0^{\mathrm{CM}} = \begin{pmatrix} 0.05 & 0.05 \end{pmatrix} \tag{125}$$

$$A_0^{\mathrm{MC}} = \begin{pmatrix} 0.02 \\ 0.02 \end{pmatrix} \tag{126}$$

$$A_0^{\mathrm{MH}} = a^{\mathrm{MH}} = 0.05 \tag{127}$$

$$A_0^{\mathrm{HC}} = \begin{pmatrix} 1 \\ 0 \end{pmatrix}. \tag{128}$$

Transformation matrices:

$$\begin{aligned}
\Phi_1(1) &= \begin{pmatrix} A_0^{\mathrm{C}} & A_0^{\mathrm{MC}} & 0 \\ A_0^{\mathrm{CM}} & A_0^{\mathrm{M}} & 0 \\ 0 & A_0^{\mathrm{MH}} & \{1-\varepsilon\} \end{pmatrix} \\
&= \begin{pmatrix} 0.91 & 0.03 & 0.02 & 0 \\ 0.03 & 0.88 & 0.02 & 0 \\ 0.05 & 0.05 & 0.8 & 0 \\ 0 & 0 & 0.05 & 0.9999 \end{pmatrix} \\
\Phi_1(s) &= \begin{pmatrix} A_0^{\mathrm{C}} & A_0^{\mathrm{MC}} & A_0^{\mathrm{HC}} \\ A_0^{\mathrm{CM}} & A_0^{\mathrm{M}} & 0 \\ 0 & 0 & \{\varepsilon\} \end{pmatrix} \\
&= \begin{pmatrix} 0.91 & 0.03 & 0.02 & 1 \\ 0.03 & 0.88 & 0.02 & 0 \\ 0.05 & 0.05 & 0.8 & 0 \\ 0 & 0 & 0 & 0.0001 \end{pmatrix}.
\end{aligned}$$

As explained in the previous chapter, the variable ε is necessary for an explicit complexity evaluation. We calculated with $\varepsilon = 10^{-4}$ to set the finished work, after release, back to a nonzero but negligible amount in terms of productivity.

The initial state x_0^* was defined using the assumption that all concurrent development tasks are to be fully completed initially and that no issues are left unresolved from the previous phase. Hence, for the minimum release period $s_{min} = 2$ we have the initial state vector:

$$x_0^* = \begin{pmatrix} 1 \\ 1 \\ 1 \\ 0 \\ 0 \\ 0 \\ 0 \\ 0 \end{pmatrix}.$$

For larger release periods, we appended additional zeros to the initial state vector.

Furthermore, we have to make reasonable assumptions about the variances and covariances of the unpredictable performance fluctuations of the development teams. We assumed that the standard deviation c_{ii} of performance fluctuations (Eq. 10) influencing task i in the module development project is proportional to the autonomous task processing rate: as the task processing rate increases (i.e. as the speed of task processing slows down), so the expected root square deviation from the mean will also increase. We chose a proportionality constant of $r = 0.02$. Additional correlations between vector components were not considered. We also assumed that the variance of the fluctuations related to the issues put in a hold state is reduced by the factor $\varepsilon' = 10^{-3}$ and that the same reduced variance holds for the fluctuations when releasing the hidden information. Through these variance reductions, the performance variability related to the augmented substate accounting for the periodic hold-and-release policy of module-level design information is extremely small and does not influence the basic mechanisms of cooperation between teams in the model. Hence, we have the covariance matrix $C^* = E\left[\varepsilon_n^* \varepsilon_n^{*\mathrm{T}}\right]$:

$$C^* = \begin{pmatrix} C_s & 0 & 0 & 0 \\ 0 & C_1 & 0 & 0 \\ 0 & 0 & \ddots & 0 \\ 0 & 0 & 0 & C_1 \end{pmatrix},$$

where the sub-matrices are given by

$$C_1 = \{r^2\} \cdot \begin{pmatrix} (a_{11}^{\mathrm{C}})^2 & 0 & 0 & 0 \\ 0 & (a_{22}^{\mathrm{C}})^2 & 0 & 0 \\ 0 & 0 & (a^{\mathrm{M}})^2 & 0 \\ 0 & 0 & 0 & \varepsilon'(1-\varepsilon)^2 \end{pmatrix}$$
$$= \{0.02^2\} \cdot \begin{pmatrix} 0.91^2 & 0 & 0 & 0 \\ 0 & 0.88^2 & 0 & 0 \\ 0 & 0 & 0.80^2 & 0 \\ 0 & 0 & 0 & 10^{-3}(1-10^{-4})^2 \end{pmatrix} \tag{129}$$

and

$$C_s = \{r^2\} \cdot \begin{pmatrix} (a_{11}^{\mathrm{C}})^2 & 0 & 0 & 0 \\ 0 & (a_{22}^{\mathrm{C}})^2 & 0 & 0 \\ 0 & 0 & (a^{\mathrm{M}})^2 & 0 \\ 0 & 0 & 0 & \varepsilon'(1-\varepsilon)^2 \end{pmatrix}$$
$$= \{0.02^2\} \cdot \begin{pmatrix} 0.91^2 & 0 & 0 & 0 \\ 0 & 0.88^2 & 0 & 0 \\ 0 & 0 & 0.80^2 & 0 \\ 0 & 0 & 0 & 10^{-3}(1-10^{-4})^2 \end{pmatrix}. \tag{130}$$

The covariance of the transformed error vector $\varepsilon_n^\star$ is given by $C^\star = (\Phi_0^*)^{-1} \cdot C^* \cdot (\Phi_0^*)^{-\mathrm{T}}$. Due to space limitations, we do not show the covariance matrix.

As mentioned earlier, our initial aim was to estimate as many independent parameters from data as possible. Therefore, we defined the matrix R from Eq. 103 in a way that only posed linear constraints on the cells in the transformation matrices $\Phi_1(1)$ and $\Phi_1(s)$ that are needed to incorporate the essential mechanisms of putting a certain amount of finished module work at each short iteration in a hold state and releasing it at time step ns. To model the hold part of the hold-and-release policy, the entries [[4,1]], [[4,2]], [[1,4]], [[2,4]] and [[3,4]] in $\Phi_1(1)$ must be zero and protected from least square estimation. Conversely, to model the release part of the policy, the entries [[4,1]], [[4,2]], [[4,3]], [[2,4]] and [[3,4]] of the transformation matrix $\Phi_1(s)$ must be null entries, and on entry [[1,4]] we have to impose the additional constraint that it must equal one. Through the positive integer constraint on entry [[1,4]], we express that all the accumulated finished work in a hold state is released to the component-level teams at the corresponding time step. Hence, for the regression coefficient $\beta(1)$ related to the transformation matrix $\Phi_1(1)$, we defined

$$R_1 = \begin{pmatrix} 1 & 0 & 0 & 0 & 0 & 0 & 0 & 0 & 0 & 0 & 0 \\ 0 & 1 & 0 & 0 & 0 & 0 & 0 & 0 & 0 & 0 & 0 \\ 0 & 0 & 1 & 0 & 0 & 0 & 0 & 0 & 0 & 0 & 0 \\ 0 & 0 & 0 & 0 & 0 & 0 & 0 & 0 & 0 & 0 & 0 \\ 0 & 0 & 0 & 1 & 0 & 0 & 0 & 0 & 0 & 0 & 0 \\ 0 & 0 & 0 & 0 & 1 & 0 & 0 & 0 & 0 & 0 & 0 \\ 0 & 0 & 0 & 0 & 0 & 1 & 0 & 0 & 0 & 0 & 0 \\ 0 & 0 & 0 & 0 & 0 & 0 & 0 & 0 & 0 & 0 & 0 \\ 0 & 0 & 0 & 0 & 0 & 0 & 1 & 0 & 0 & 0 & 0 \\ 0 & 0 & 0 & 0 & 0 & 0 & 0 & 1 & 0 & 0 & 0 \\ 0 & 0 & 0 & 0 & 0 & 0 & 0 & 0 & 1 & 0 & 0 \\ 0 & 0 & 0 & 0 & 0 & 0 & 0 & 0 & 0 & 1 & 0 \\ 0 & 0 & 0 & 0 & 0 & 0 & 0 & 0 & 0 & 0 & 0 \\ 0 & 0 & 0 & 0 & 0 & 0 & 0 & 0 & 0 & 0 & 0 \\ 0 & 0 & 0 & 0 & 0 & 0 & 0 & 0 & 0 & 0 & 0 \\ 0 & 0 & 0 & 0 & 0 & 0 & 0 & 0 & 0 & 0 & 1 \end{pmatrix}.$$

For the regression coefficient $\beta(s)$ related to the transformation matrix $\Phi_1(s)$ we have:

$$R_s = \begin{pmatrix} 1 & 0 & 0 & 0 & 0 & 0 & 0 & 0 & 0 & 0 \\ 0 & 1 & 0 & 0 & 0 & 0 & 0 & 0 & 0 & 0 \\ 0 & 0 & 1 & 0 & 0 & 0 & 0 & 0 & 0 & 0 \\ 0 & 0 & 0 & 0 & 0 & 0 & 0 & 0 & 0 & 0 \\ 0 & 0 & 0 & 1 & 0 & 0 & 0 & 0 & 0 & 0 \\ 0 & 0 & 0 & 0 & 1 & 0 & 0 & 0 & 0 & 0 \\ 0 & 0 & 0 & 0 & 0 & 1 & 0 & 0 & 0 & 0 \\ 0 & 0 & 0 & 0 & 0 & 0 & 0 & 0 & 0 & 0 \\ 0 & 0 & 0 & 0 & 0 & 0 & 1 & 0 & 0 & 0 \\ 0 & 0 & 0 & 0 & 0 & 0 & 0 & 1 & 0 & 0 \\ 0 & 0 & 0 & 0 & 0 & 0 & 0 & 0 & 1 & 0 \\ 0 & 0 & 0 & 0 & 0 & 0 & 0 & 0 & 0 & 0 \\ 0 & 0 & 0 & 0 & 0 & 0 & 0 & 0 & 0 & 0 \\ 0 & 0 & 0 & 0 & 0 & 0 & 0 & 0 & 0 & 0 \\ 0 & 0 & 0 & 0 & 0 & 0 & 0 & 0 & 0 & 0 \\ 0 & 0 & 0 & 0 & 0 & 0 & 0 & 0 & 0 & 1 \end{pmatrix}.$$

It holds that

$$R = \begin{bmatrix} R_1 \\ R_s \end{bmatrix}.$$

The intercept vector b_1 (cf. Eq. 103) for the regression coefficient $\beta(1)$ was defined as a null vector with $\mathrm{Dim}[vec[\Phi_1(1)]] = 16$ components, that is

$$b_1 = (0\ \ 0\ \ 0\ \ 0\ \ 0\ \ 0\ \ 0\ \ 0\ \ 0\ \ 0\ \ 0\ \ 0\ \ 0\ \ 0\ \ 0\ \ 0)^{\mathrm{T}},$$

because we only posed constraints on $\Phi_1(1)$ that said selected entries must be zero. On the other hand, the intercept vector b_s for the regression coefficient $\beta(s)$ has to indicate that the first component of the parameter vector A_0^{HC} must be one and therefore, on the corresponding entry [[1,4]] of $\Phi_1(s)$, a positive integer constraint is defined. We have

$$b_s = (0\ \ 0\ \ 0\ \ 0\ \ 0\ \ 0\ \ 0\ \ 0\ \ 0\ \ 0\ \ 0\ \ 0\ \ 1\ \ 0\ \ 0\ \ 0)^{\mathrm{T}}.$$

The aggregated intercept vector is, simply

$$b = \begin{bmatrix} b_1 \\ b_s \end{bmatrix}.$$

We used the Mathematica software package from Wolfram Research to carry out the Monte Carlo experiments. We developed the functions and procedures for least squares estimation with linear constraints and model selection based on Schwarz's adapted BIC criterion ourselves. Dr. Ursu kindly provided us with the original Matlab routines that Ursu and Duchesne (2009) used to fit a PVAR model to quarterly seasonally unadjusted West German income and consumption data for the years 1960–1987. This meant that we could verify and validate our own Mathematica code. To "reconstruct" the introduced reference model representation for different release periods s using data on periodically correlated task processing, we computed repeated Monte Carlo trials for each PVAR model instance. We systematically varied the release period between $s_{min} = 2$ and $s_{max} = 5$ time steps, increasing the value of this variable by one per step. We also systematically varied length T of the time series being used for parameter estimation. The range was $T_{min} = 100$ and $T_{max} = 1000$ time steps with increments of 100 time steps. The minimum length $T_{min} = 100$ was chosen so that, even for the longest release period s_{max} of 5 time steps, the mean work remaining for all three tasks was less than 0.05. To obtain a good statistic, a total of 1000 separate and independent Monte Carlo runs (which the literature also refers to as "replications") were calculated for each setting of the independent parameters. To increase external validity, we decided not to use a "warm-up interval" in the simulated task processing, and therefore the T data points were all the information available.

In order to answer to the main question of how the accuracy of the estimation of the independent parameters is influenced by the length T of the time series generated by the reference model with given release period s, we considered the root mean square deviation (RMSD) between the entries of the reference transformation and covariance matrices $\Phi_1(1)$, $\Phi_1(s)$, C_1 and C_s (embedding the work transformation sub-matrix A_0^{C}, the vectors A_0^{CM}, A_0^{MC}, and A_0^{HC}, as well as the real-valued parameters a^{M} and a^{MH}) and their corresponding least square estimates $\widehat{\Phi}_1(1)$,

$\widehat{\Phi}_1(s)$, $\hat{C}_1$ and $\hat{C}_s$. We used the following mathematical formulations to calculate these dependent variables in the Monte Carlo runs:

$$RMSD_{\Phi_1(1)} = \sqrt{\frac{\mathrm{Tr}\left[\left(\Phi_1(1) - \widehat{\Phi}_1(1)\right)\left(\Phi_1(1) - \widehat{\Phi}_1(1)\right)^{\mathrm{T}}\right]}{\mathrm{Dim}[vec[\Phi_1(1)]]}} \tag{131}$$

$$RMSD_{\Phi_1(s)} = \sqrt{\frac{\mathrm{Tr}\left[\left(\Phi_1(s) - \widehat{\Phi}_1(s)\right)\left(\Phi_1(s) - \widehat{\Phi}_1(s)\right)^{\mathrm{T}}\right]}{\mathrm{Dim}[vec[\Phi_1(s)]]}} \tag{132}$$

$$RMSD_{C_1} = \sqrt{\frac{\mathrm{Tr}\left[(C_1 - \hat{C}_1)(C_1 - \hat{C}_1)^{\mathrm{T}}\right]}{\mathrm{Dim}[vec[C_1]]}} \tag{133}$$

$$RMSD_{C_s} = \sqrt{\frac{\mathrm{Tr}\left[(C_s - \hat{C}_s)(C_s - \hat{C}_s)^{\mathrm{T}}\right]}{\mathrm{Dim}[vec[C_s]]}}. \tag{134}$$

We also computed the difference of the subspace angles between the reference model and its estimated representation to evaluate the accuracy of the introduced estimation methods. Following the mathematical analysis of de Cock and Moor (2002) and of de Cock (2002), the subspace angles are defined as the principal angles between the row spaces of the infinite observability matrix of the model and of the infinite observability matrix of the inverse model. In this sense, the principal angles between two subspaces are the angles between their principal directions and can be used to construct a cepstral distance for VAR and PVAR models (and other classes of dynamic models, see de Cock and Moor 2002). The subspace angles allow a holistic evaluation of the estimated model representation in view of the information dynamics of the true model, as the measurement shows whether the underlying geometry of the subspaces is accurately reflected and whether or not only apparent similarities between model matrices in Euclidian space exist. The cepstral distance between two autoregression models is zero if the underlying geometry of the subspaces is equivalent and therefore the canonical correlations (see Section 4.1.3) of the past and future output processes are one, given that both models are driven by the same zero-mean periodic white noise process $\{\varepsilon_n^*\}$. To evaluate the subspace angles in Monte Carlo experiments, we have to bring the PVAR reference model represented by parameter tuple $(A_0^*, C^\star)$ and its estimated representation $(\hat{A}_0^*, \hat{C}^\star)$ into the form of a linear dynamical system with additive zero-mean white noise (see e.g. Gharahmani 2001; Puri 2010). This class of systems will be introduced and extensively discussed in the next chapter. We therefore concentrate on the essential preliminaries to calculate the squared cepstral distance, and sketch only the main steps. To further simplify the analytical evaluation of the squared cepstral distance, we focus on estimated model representations whose autoregression order does not exceed an order of one for all iterations ν. Let

$$\begin{aligned} A^{(1)} &= A_0^* \in \mathbb{R}^{ds \times ds} \\ A^{(2)} &= \hat{A}_0^* \in \mathbb{R}^{ds \times ds} \end{aligned}$$

be the combined dynamical operators of the first-order autoregressive representations that are compared by the cepstral distance, and let

$$\begin{aligned} C^{(1)} &= C^{\star} \in \mathbb{R}^{ds \times ds} \\ C^{(2)} &= \hat{C}^{\star} \in \mathbb{R}^{ds \times ds} \end{aligned}$$

be the corresponding covariance matrices. Furthermore, let the output operators (see Section 2.9) be the identity matrices

$$\begin{aligned} H^{(1)} &= I_{ds} \\ H^{(2)} &= I_{ds} \end{aligned}$$

and the covariance matrices of the observation process (see Section 2.9) be the null matrices

$$\begin{aligned} V^{(1)} &= 0_{ds} \\ V^{(2)} &= 0_{ds}. \end{aligned}$$

Based on these parametric representations of the "forward form" of a linear dynamical system, we compute the associated forward innovation form. The forward innovation form is an equivalent form in the sense that the first-order and second-order statistics of the sequence of observations generated by the system in steady state are the same, but only a single error process is used to model performance fluctuations (see Section 2.9). In both forms the recursive state equations run forward in time. The forward innovation form can be obtained by solving the Lyapunov Eq. 27 and the algebraic Ricatti Eq. 296. Let

$$\begin{aligned} K^{(1)} &\in \mathbb{R}^{ds \times ds} \\ K^{(2)} &\in \mathbb{R}^{ds \times ds} \end{aligned}$$

be the Kalman gain matrices of the forward innovation representations of $(A^{(1)}, C^{(1)}, H^{(1)}, V^{(1)})$ and $(A^{(2)}, C^{(2)}, H^{(2)}, V^{(2)})$ according to the definition from Eq. 163. We assume that both parametric representations correspond to stable and minimum phase models. The state space matrices of the combined VAR model as a function of the state-space representations of the separate models is then given by (de Cock 2002):

$$A^{(12)} = \begin{pmatrix} A^{(1)} & 0_{ds} \\ K^{(2)} \cdot C^{(1)} & A^{(2)} - K^{(2)} \cdot C^{(2)} \end{pmatrix}$$
$$K^{(12)} = \begin{pmatrix} K^{(1)} \\ K^{(2)} \end{pmatrix}$$
$$C^{(12)} = \begin{pmatrix} C^{(1)} & -C^{(2)} \end{pmatrix}.$$

By combined VAR model, we mean a model whose transfer function is equal to the quotient of the individual transfer functions. The squared cepstral distance $d\left[\left(A_0^*, C^\star\right), \left(\hat{A}_0^*, \hat{C}^\star\right)\right]^2$ between the models parameterized by $\left(A_0^*, C^\star\right)$ and $\left(\hat{A}_0^*, \hat{C}^\star\right)$ is then, according to de Cock (2002), defined as

$$d\left[\left(A_0^*, C^\star\right), \left(\hat{A}_0^*, \hat{C}^\star\right)\right]^2 := \log_2 \mathrm{Det}\left[I_{2ds} - Q_z^{(12)} \cdot P^{(12)}\right], \tag{135}$$

where the controllability Gramian $P^{(12)}$ of the combined model is given by solving a Lyapunov equation (cf. Eq. 27)

$$P^{(12)} = A^{(12)} \cdot P^{(12)} \cdot A^{(12)\mathrm{T}} + K^{(12)} \cdot K^{(12)\mathrm{T}}$$

and the observability Gramian $Q_z^{(12)}$ of the inverse combined model is given by solving another Lyapunov equation

$$Q_z^{(12)} = \left(A^{(12)} - K^{(12)} \cdot C^{(12)}\right)^{\mathrm{T}} Q_z^{(12)} \left(A^{(12)} - K^{(12)} \cdot C^{(12)}\right).$$

The analytical considerations of de Cock (2002) show that a direct theoretical connection exists between the squared cepstral distance and the effective measure complexity. The effective measure complexity is an information-theoretic measure from basic research that can be used to evaluate the "informational structure" of stochastic processes and is therefore interesting as a way of evaluating self-generated (emergent) complexity of open organizational systems. It will be introduced and discussed in detail in Section 3.2.4, while in Chapter 4 it will be used to find different strengths of closed-form complexity solutions for different classes of models. The theoretical connection exists because the squared cepstral distance between two VAR models can also be expressed by the mutual information $I\left[X_{-\infty}^{-1}; X_0^{\infty}\right]$ (Eq. 237) that is communicated from the infinite past to the infinite future by the combined state process, and it holds that

$$d\left[\left(A_0^*, C^\star\right), \left(\hat{A}_0^*, \hat{C}^\star\right)\right]^2 = 2I\left[X_{-\infty}^{-1}; X_0^{\infty}\right].$$

The quantities $X_{-\infty}^{-1}$ and X_0^{∞} denote the infinite past and future histories of the output process $\{X_t\}$ of the combined model that are generated under the assumption that both embedded models are driven by the same zero-mean periodic white noise process $\{\varepsilon_n^*\}$. These and other quantities are defined and explained in Section 3.2.4.

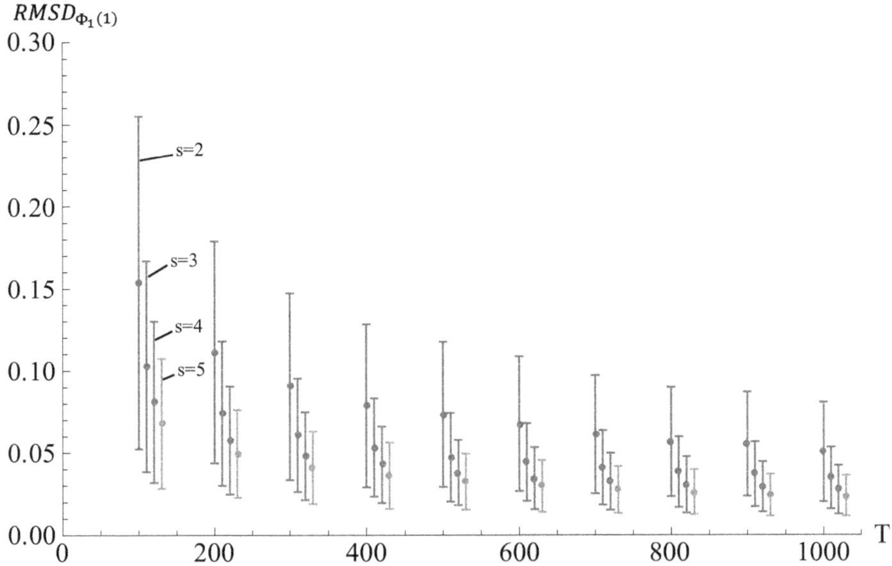

Fig. 2.12 List plot of root mean square deviation (RMSD) between reference transformation matrix $\Phi_1(1)$ and its least-squares estimate $\hat{\Phi}_1(1)$. The release period s of finished work that is put in a hold state at each short iteration and released by the module design and integration testing team at time step ns was varied systematically between $s_{min} = 2$ and $s_{max} = 5$. The length T of the time series that was used for least-squares estimation was also varied systematically within the range of $T_{min} = 100$ and $T_{max} = 1000$ time steps. The plot shows the mean values as note points and 95% confidence intervals as error bars. The note points have been slightly offset to make it easier to see the error bars. The Monte Carlo simulation was based on state equation 89 and the definition of the independent parameters from Eqs. 123 to 130. A total of 1000 separate and independent runs were calculated

In the following, we use the shorthand notation d^2_{cepstr} to denote the squared cepstral distance.

The aggregated results of the Monte Carlo experiments are shown in Figs. 2.12–2.16. Figures 2.12, 2.13, 2.14 and 2.15 show the list plots of the root mean square deviation between the reference model matrices and their corresponding least squares estimates according to the definitions from Eqs. 131 to 134. Each plot shows the obtained mean values of the independent variables as note points, and their 95% confidence intervals as error bars. The note points have been slightly offset to make it is easier to see the error bars. The 95% confidence intervals were calculated under the assumption of a normal distribution and therefore correspond to ±1.96 standard deviations.

The root mean square deviations between the transformation matrices $\Phi_1(1)$ and $\widehat{\Phi}_1(1)$ related to the short iterations of the development teams, and the covariance matrices C_1 and $\hat{C}_1$ of the corresponding performance fluctuations show a very similar pattern: as the number of data points T that are available in the time series being used for least-squares estimation increases, so the root mean square deviation

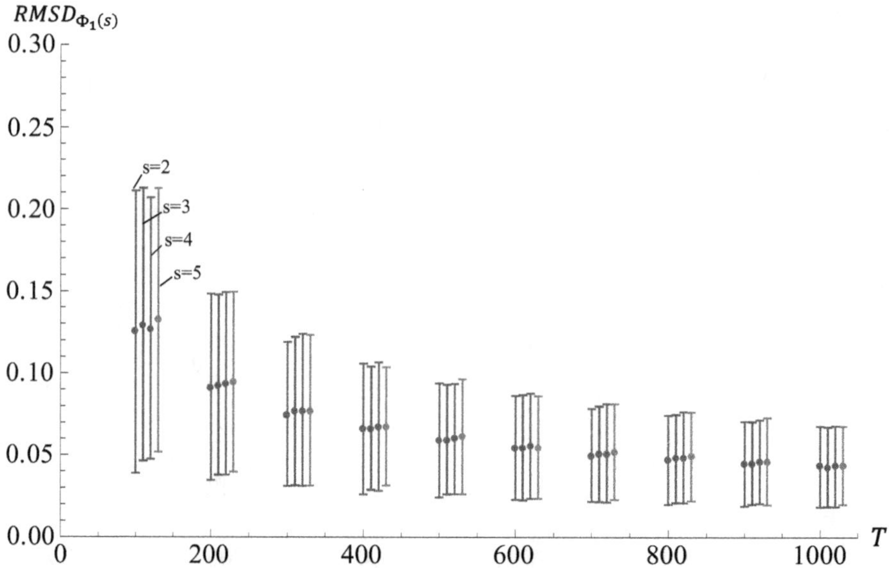

Fig. 2.13 List plot of root mean square deviation (RMSD) between reference transformation matrix $\Phi_1(s)$ and its corresponding least square estimate $\widehat{\Phi}_1(s)$. The conditions and parameters of the Monte Carlo experiments are the same as in Fig. 2.12

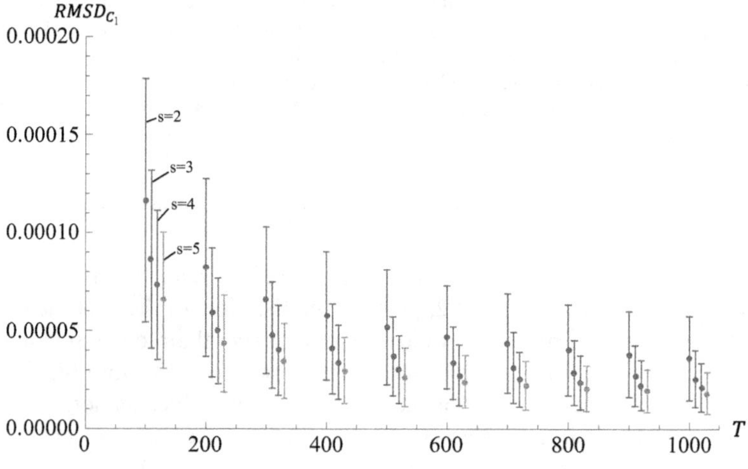

Fig. 2.14 List plot of root mean square deviation (RMSD) between reference covariance matrix C_1 and its corresponding least square estimate $\hat{C}_1$. The conditions and parameters of the Monte Carlo experiments are the same as in Fig. 2.12

shrinks, the release period s of finished work that was put in a hold state for a given time-series length increases, and the average deviation becomes smaller. Furthermore, the 95% confidence interval shows that the more data points available for

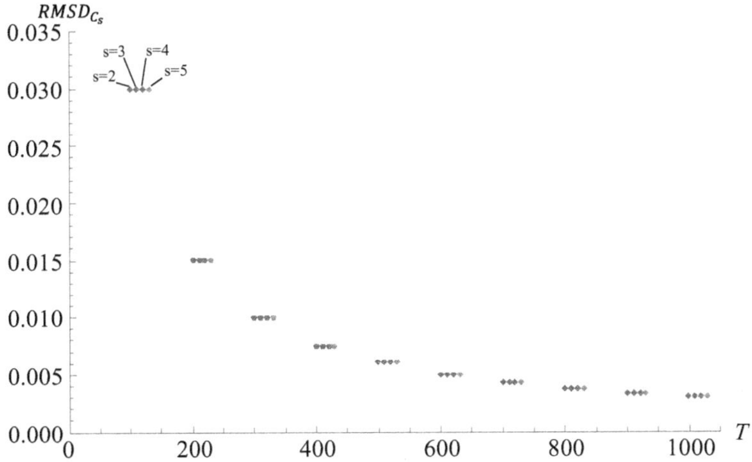

Fig. 2.15 List plot of root mean square deviation (RMSD) between reference covariance matrix C_s and its corresponding least square estimate $\hat{C}_s$. The conditions and parameters of the Monte Carlo experiments are the same as in Fig. 2.12

estimation and the longer the release period, the smaller the confidence interval and therefore the more certain we can be of the estimates. For more than $T = 300$ data points, we can expect a $RMSD_{\Phi_1(1)}$ that is smaller than 0.1 and a $RMSD_{C_1}$ that is smaller than 0.0007 (see Figs. 2.12 and 2.14). The overall deviations are low and the estimation results are consistent.

When we analyze the root mean square deviations between the transformation matrices $\Phi_1(s)$ and $\widehat{\Phi}_1(s)$ related to the release of finished work by the module design and integration testing team, and their corresponding covariance matrices C_s and $\hat{C}_s$, we find a different but internally consistent pattern: the more data available for least squares estimation, the smaller the mean deviations and the smaller the 95% confidence intervals. However, the means and 95% confidence intervals show very little sensitivity to variations in the release period s of finished work that was put in a hold state. The aggregated means of $RMSD_{\Phi_1(s)}$ are slightly lower than the means of $RMSD_{\Phi_1(1)}$ (see Figs. 2.12 and 2.13), and for more $T = 200$ data points, we can expect a $RMSD_{\Phi_1(s)}$ that is smaller than 0.1.

On the other hand, the means of $RMSD_{C_s}$ are significantly larger than the means of $RMSD_{C_1}$ (compare Figs. 2.14 and 2.15) and can be as large as 0.030 for the smallest sample size of $T = 100$ data points (Fig. 2.15). This result was unexpected because the release of hidden information can be observed for the smallest release period $s_{min} = 2$ at every second time step and therefore equally often as the task processing on the short time scale. A detailed numerical analysis showed that under all experimental conditions, the average finished work put in a hold state is quite small in relation to the cumulative variance of the performance fluctuations of the system integration and testing team, and that this low "signal-to-noise ratio" leads

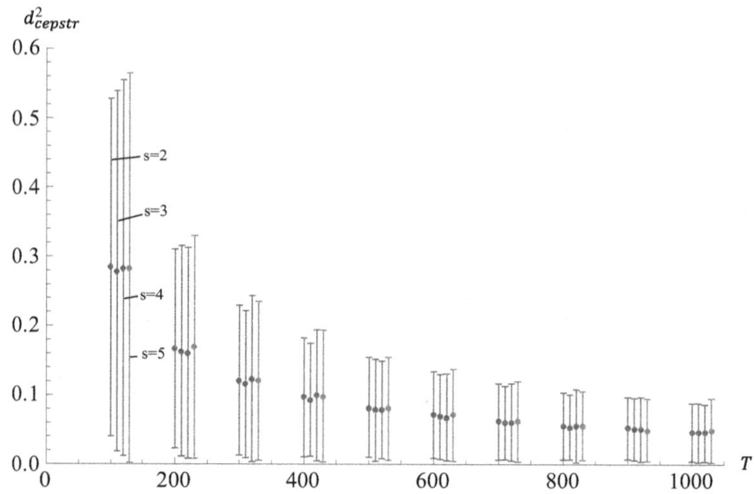

Fig. 2.16 List plot of squared cepstral distance d^2_{cepstr} between the reference PVAR models parameterized by (A_0^*, C^*) and their estimated representations $(\hat{A}_0^*, \hat{C}^*)$. The squared cepstral distance goes back to the work of de Cock and Moor (2002) and is defined in Eq. 135. The conditions and parameters of the Monte Carlo experiments are the same as in Fig. 2.12

to an disproportionate increase in the magnitude of the estimation error. However, in view of an industrial application of the estimation methods, the overall deviations between the reference covariance matrix C_s and its corresponding least square estimate $\hat{C_s}$ are acceptable, and the estimation error converges to zero fast.

The results of the analyses of the root mean square deviations show that the least squares estimation methods that were developed by Ursu and Duchesne (2009) and adapted to our own model formulation can identify the parametric representation of the reference model of cooperative work in a vehicle door module development project with low uncertainty. The estimation results are also highly consistent in the sample size and in the length of the release period of accumulated development issues from the module-level to the component-level. It is also important to note that the estimation methods showed a very good numerical stability in the Monte Carlo experiments. Re-estimations of the model matrices due to badly conditioned intermediate matrices were only necessary in one of 40,000 independent runs. These findings mean that we can turn our attention to the question of whether or not the estimated parameters only lead to small deviations between the corresponding model matrices, as they implicitly ignore the geometry of the sub-spaces but can also accurately capture the higher-order informational and statistical properties of the true model in non-Euclidian space. The calculated squared cepstral distances d^2_{cepstr} in Fig. 2.16 show that this seems to be the case for all investigated settings of the independent parameters, as the mean distances and the 95% confidence intervals smoothly and quickly converge to zero for a growing sample size T. The rate of convergence is not significantly influenced by the release period s (Fig. 2.16). Additional Monte Carlo experiments showed that the linear

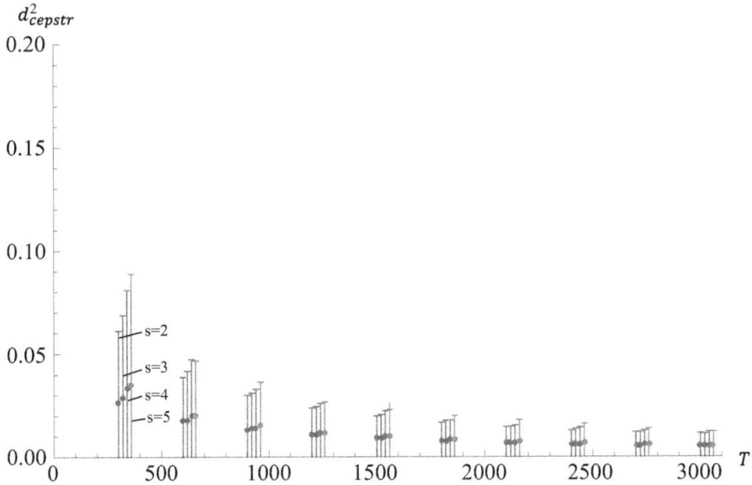

Fig. 2.17 List plot of squared cepstral distance d^2_{cepstr} between the reference PVAR models parameterized by (A_0^*, C^*) and their estimated representations $(\hat{A}_0^*, \hat{C}^*)$. The conditions and parameters of the Monte Carlo experiments are the same as in Fig. 2.12, except that we reduced the proportionally constant of the standard deviation of performance fluctuations from $r = 0.02$ to $r = 0.001$ to accelerate convergence to zero, and we considered larger sample sizes $T > 1000$ in the least squares estimation

constraints on the entries of the dynamic operators $\Phi_1(1)$ and $\Phi_1(s)$ play a very important role in ensuring small squared cepstral distances. For an unconstrained model formulation, the obtained cepstral distances are usually 20 (for $T = 100$) to 50 (for $T = 1000$) times larger. The distances are especially sensitive to the constraint necessary for modeling the release mechanism of the hold-and-release policy: if we do not subject entry [[1, 4]] of the transformation matrix $\Phi_1(s)$ to the constraint that it must equal one, i.e. $\Phi_1(s)_{[[1,4]]} = 1$, and instead treat it as a free parameter, the resulting squared cepstral distances are five times larger on average. Therefore, it is necessary to incorporate the complete hold-and-release policy, and not just the hold mechanism, in the formulation of the constrained estimation problem. The overall squared cepstral distance in the Monte Carlo experiments is low and shows that the basic estimation concept is valid and that the accuracy of the investigated estimation methods is good.

To verify that the squared cepstral distance d^2_{cepstr} between the reference PVAR models and their estimated representations from Fig. 2.16 does indeed converge to zero, and not to any other non-negative constant value for large sample sizes, we reduced the proportionality constant of the standard deviation c_{ii} of performance fluctuations (Eq. 10) from $r = 0.02$ to $r = 0.001$ and repeated the least squares estimation. The results are shown in Fig. 2.17 for sample sizes ranging from $T = 400$ to $T = 3000$ under the same regime of the four release periods as before. Note that the scale of the ordinate in Fig. 2.17 is reduced by a factor of three compared to Fig. 2.16. As one would expect, Fig. 2.17 shows that reducing the proportionality constant significantly reduces the means and standard deviation of the squared cepstral distance. It also shows that the means quickly converge to values less than 0.01. Additional Monte Carlo runs showed that the squared cepstral distance is negligible for sample sizes larger than $T = 10,000$. We observed similar convergent behavior for the root mean square deviation between the entries of the reference transformation and covariance matrices $\Phi_1(1)$, $\Phi_1(s)$, C_1 and C_s and the corresponding least squares estimates.

Finally, we investigated whether the heuristic model selection procedure, which was developed by Ursu and Duchesne (2009) and introduced in the previous chapter, leads to reliable decisions about the regression order of the reference PVAR model from data. In the Monte Carlo experiments, it turned out that the heuristic model selection procedure based on Schwarz's BIC criterion is extremely reliable: the first-order autoregression of the reference model was identified correctly for all parameter settings in all runs. This result was rather unexpected, as we knew from the comprehensive Monte Carlo study of Lütkepohl (1985) and our own studies (Schlick et al. 2013) that, for ordinary first and second-order VAR models, it is rare to find levels of accuracy of model selection that exceed 99%. The typical range is between 95% and 99%. We hypothesize that this maximum reliability is the result of the additive formulation of the aggregate model selection criterion from Eq. 117, the comparably low complexity of the four-dimensional reference model, and the overall small prediction errors. In conclusion, we can say that, in light of the standards of organizational modeling and simulation (see e.g. Rouse and Boff 2005), the identifiability of the reference model of periodically correlated, cooperative work in a vehicle door development project is high and that the uncertainty in parameter estimation is low. The results are highly consistent and replicable. The estimation and model selection methods are numerically efficient and showed a very good numerical stability. To provide additional insights into modeling and simulation of periodically correlated work processes, we present and discuss a six-dimensional project model in a similar application scenario in Section 5.2. That chapter also shows that simulated traces of outstanding work can help improve our understanding of the hold-and-release policy for selected components in new or updated engineering design information. We also discuss values of key performance indicators for optimizing problems with and without constraints on the expected total amount of work x_{tot} according to Eq. 91. In addition, we investigated the identifiability of this more complex reference model and the resulting parameter uncertainty in Monte Carlo simulations using the same experimental design and the same methods. The results show that even the six-dimensional reference model can be accurately identified from data if constraints are imposed on the auxiliary variable ε. In this case, the squared cepstral distances are, on average, no larger than 0.44 and follow a convergence pattern to zero that is similar to the pattern shown in Fig. 2.17. Furthermore, model selection is extremely reliable. However, if the auxiliary variable ε is also considered as a free parameter in least square estimation, the model identifiability is not good. Therefore, we recommend encoding all constraints of the developed PVAR model formulation in the description of the regression coefficient matrix R and the intercept vector b from linear relation 103 that are known in advance. If this is not possible, we recommend reducing the number of independent parameters by setting selected entries in the dynamical operators $\Phi_1(v), \ldots, \Phi_{n(v)}(v)$ to zero, as Ursu and Duchesne (2009) showed for the quarterly seasonally unadjusted West German income and consumption data (Section 2.7). Otherwise, very large sample sizes will be needed to accurately estimate the true representation from data.

2.9 Stochastic Formulation with Hidden State Variables

A theoretical extension of the previously introduced approaches to modeling cooperative work in PD projects through autoregressive processes with periodically correlated or non-correlated components is the formulation of a stochastic state-space model with "hidden" (latent) variables (Gharahmani 2001). In statistics, an independent variable is termed a latent variable (as opposed to an observable variable) if it cannot be directly observed but is rather inferred through a causal model from other variables that are directly measured. In our context, the hidden state variable $X_t \in \mathbb{R}^q$ represents the comprehensive internal state of the project at a specific time instant t in vector form. We assume the state vector x_t not only captures the essential dynamics of the work remaining from the q predefined component-level and system-level tasks (which can be measured, for instance, by the time left to finalize a specific design or the definable labor units required to complete a particular development activity or component of the work breakdown structure; see Yassine et al. 2003 and Sections 2.1 and 2.5) but also the efforts that must be made to communicate design decisions. The communication conveys the design information from one person or team to another and contributes to the common understanding of the design problem, product and processes. Communication is initiated more or less spontaneously and can also occur across the organizational hierarchy (Gibson and Hodgetts 1991). If communication occurs between hierarchically positioned persons, we speak of vertical communication. Horizontal communication occurs on the same hierarchical level. Diagonal communication refers to communication between managers and working persons located in different functional divisions. Due to these multiple channels, the internal state information is not completely known to the project manager but must be estimated through the mental model of the possible state evolution in conjunction with readings from dedicated performance measurement instruments (earned value analysis, etc.). It is assumed that the estimates are obtained periodically in the form of observation vectors $Y_t \in \mathbb{R}^p$. These vectors directly refer to the work remaining of the predefined work breakdown structure and can be associated to the corresponding internal state X_t without sequencing errors. Furthermore, it is assumed that the state process can be decomposed into q interacting subtasks. These subtasks either represent concurrent development activities on individual or team levels or vertical, horizontal, or diagonal communication processes based on speech acts. It is important to note that the dimensions of the state space can differ from the observation space. In most cases of practical interest, the internal state vectors have significantly more components than the observation vectors ($\mathrm{Dim}[X_t] > \mathrm{Dim}[Y_t]$). Because we are aiming at a predictive model of a complex sociotechnical system, the inherent performance fluctuations must also be taken into account for the representation of the hidden process. We represent the performance fluctuations by the random variable ε_t and assume that they have no systematic component, that is $E[\varepsilon_t] = 0_q$. Furthermore, we develop the model under the assumption that the reliability of the measurement instruments is limited and non-negligible fluctuations ν_t of the readings around the true means occur. However, the instruments are not biased and there is $E[\nu_t] = 0_p$.

Formally, we define the state process $\{X_t\}$ to be linear and influenced by Gaussian noise ε_t $(t = 0, \ldots, T)$. It is assumed that the observation process $\{Y_t\}$ directly depends on the state process in the sense that each vector Y_t being acquired through observation at time instant t is linearly dependent on the state vector X_t and not on other instances of the state process. The observation process itself is perturbed by another Gaussian variable ν_t. Hence, we have the simultaneous system of equations

$$X_{t+1} = A_0 \cdot X_t + \varepsilon_t \tag{136}$$

$$Y_t = H \cdot X_t + \nu_t. \tag{137}$$

The Gaussian vectors ε_t and ν_t have zero means and covariances C and V, respectively:

$$\varepsilon_t \sim \mathcal{N}(0_q, C)$$
$$\nu_t \sim \mathcal{N}(0_p, V).$$

In contrast to the vector autoregression models, we assume a Gaussian initial state density with location π_0 and covariance Π_0 in the above state-space model:

$$X_0 \sim \mathcal{N}(\pi_0, \Pi_0). \tag{138}$$

Like the stochastic model formulation without hidden variables (Section 2.2), we make the reasonable assumption that the performance fluctuations and measurements errors are uncorrelated from time step to time step and it holds for all time steps $\{\mu, \nu\} \in \mathbb{Z}$ that:

$$E\left[\begin{pmatrix} \varepsilon_\mu \\ \nu_\mu \end{pmatrix} \cdot \begin{pmatrix} \varepsilon_\mu \\ \nu_\mu \end{pmatrix}^T\right] = \begin{pmatrix} C & S_{\varepsilon\nu} \\ S_{\varepsilon\nu}^{\mathrm{T}} & V \end{pmatrix} \cdot \{\delta_{\mu\nu}\}.$$

$\delta_{\mu\nu}$ is the Kronecker delta which was defined in Eq. (14). To simplify the parameter estimation and complexity analysis, we focus in the following on processes in which the partial covariances $S_{\varepsilon\nu}$ and $S_{\varepsilon\nu}^{\mathrm{T}}$ are zero:

$$E\left[\begin{pmatrix}\varepsilon_\mu \\ \nu_v\end{pmatrix}\cdot\begin{pmatrix}\varepsilon_\mu \\ \nu_v\end{pmatrix}^T\right]=\begin{pmatrix}C & 0 \\ 0 & V\end{pmatrix}\cdot\{\delta_{\mu v}\}.$$

In the literature, the state-space model formulation introduced above is termed the "forward form" (e.g. van Overschee and de Moor 1996; de Cock 2002). Eq. 136 is termed the state equation and Eq. 137 the output equation. Additional input (predictor) variables are not considered in the following. The linear state and output processes correspond to matrix operations, which are denoted by the operators A_0 and H, respectively. The dynamical operator A_0 is a square matrix of size $q \times q$. The output operator H is a rectangular matrix of size $p \times q$. The literature often calls A_0 the state transition matrix, and H the measurement, observation or generative matrix. Both matrices can be considered as the system matrices. We assume that A_0 is of rank q, H of rank p and that C, V and Π_0 are always of full rank. The complete parameter vector is $\theta = [A_0 \quad H \quad C \quad V \quad \pi_0 \quad \Pi_0]$.

In the engineering literature, the complete state-space model of cooperative work in PD projects is termed a linear dynamical system (LDS) with additive Gaussian noise (de Cock 2002) or—using a more historical terminology—discrete time Kalman Filter (Puri 2010). In this model, only the vector Y_t can be observed in equidistant time steps, whilst the true state vector X_t and its past history $\{X_t\}_1^{t-1} = (X_1, \ldots, X_{t-1})$ must be inferred through the stochastic linear model from the observable variables. A graphical representation in the form of a dynamic Bayesian network is shown in Fig. 2.18 (cf. Fig. 2.4, Gharahmani 2001).

An LDS model is one of the most prominent models in statistical signal processing. The model has also proven very useful for sensor data fusion and target tracking (see e.g. Bar-Shalom et al. 2001; Koch 2010, 2014). Ghahramani and Hinton (1996), Yamaguchi et al. (2007) and others developed numerical methods to estimate the independent parameters based on multivariate time series of the observable variables.

In view of the theory of LDS it is important to point out that the generated stochastic process $\{Y_t\}$ can have a large memory depth in the sense that the past

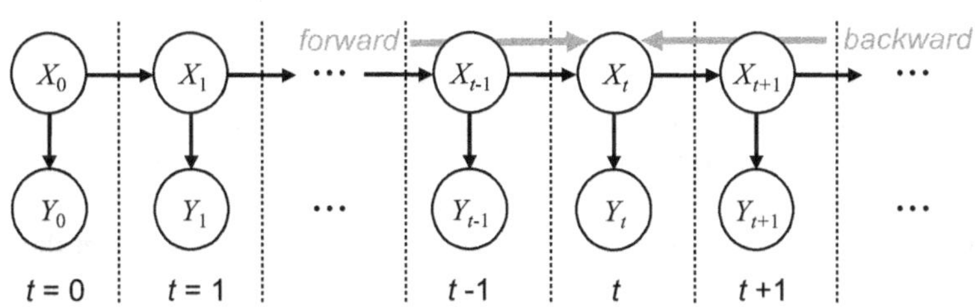

Fig. 2.18 Graphical representation of the linear dynamical system with additive Gaussian noise in the form of a dynamic Bayesian network. The nodes in the graph represent the random state and observation variables of the stochastic process. The directed arcs encode conditional dependencies between the variables. The grey arrows indicate the forward and backward messages passed during the re-estimation of the parameters using the expectation-maximization algorithm (see Section 2.10 for a detailed explanation of the algorithm)

must be observed across a long time interval in order to make good predictions of the future. It is evident that the Markov property (cf. Eq. 18) holds for the hidden state process $\{X_t\}$ and the conditional *pdf* can be expressed as

$$f_\theta[x_{t+1}|x_t, \ldots, x_0] = f_\theta[x_{t+1}|x_t] \quad \forall t \geq 0.$$

Therefore, the state evolution seems to be memorylessness in the sense that properties of random variables related to the future depend only on information about the present state and not on information from past instances of the process (see Section 2.2). However, for the sequence of observations $\{Y_t\}$ the Markov property does not necessarily need to be satisfied. This is due to the fact that the information which is communicated from the past to the future by the hidden process must not completely flow through the observation Y_t. Part of the predictive information can be "kept secret" from the external observer even over long time intervals. The predictive information is formally defined in Section 3.2.4 (Eq. 226). Due to the latent dependency structure, the stochastic process generated by an LDS can have a certain "crypticity" (Ellison et al. 2009) in that it may not reveal all internal correlations and structures during the observation time. The process' crypticity can be analyzed systematically on the basis of "elusive information" (Marzen and Crutchfield 2014), also defined in Section 3.2.4 (Eq. 231). Formally speaking, for the conditional *pdf* of the output process it usually holds that

$$f_\theta\left[y_{t+1}|y_t, \ldots, y_0\right] \neq f_\theta\left[y_{t+1}|y_t\right]$$

or equivalently expressed based on Bayes theorem as

$$\frac{f_\theta\left[y_0, \ldots, y_{t+1}\right]}{f_\theta[y_0, \ldots, y_t]} \neq \frac{f_\theta\left[y_t, y_{t+1}\right]}{f_\theta[y_t]}.$$

As a consequence, the LDS representation theoretically allows an infinite memory depth of the stochastic process and therefore properly subsumes the vector autoregression model of order n ($VAR(n)$ model) from Section 2.4. As the $VAR(n)$ model's memory only reaches back n time steps into the past because its state is the weighted value of the last n observations, the fact that the observations can covary with one another through the state process means that in an LDS an observation's distribution can be affected by an event occurring arbitrarily far in the past (Rudary 2009).

For the introduced state-space model we can factorize the joint *pdf* of the system over time steps $t = 0$ to $t = T$ as

$$f_\theta[x_0, \ldots, x_T, y_0, \ldots, y_T] = f_\theta[x_0] \prod_{t=0}^{T-1} f_\theta[x_{t+1}|x_t] \prod_{t=0}^{T} f_\theta[y_t|x_t] \tag{139}$$

and calculate the marginal *pdf* by integrating out the hidden states of the process:

$$\begin{aligned} f_\theta[y_0,\ldots,y_T] &= \int_{\mathbb{X}^q}\cdots\int_{\mathbb{X}^q} f_\theta[x_0,\ldots,x_T,y_0,\ldots,y_T]\; dx_0\ldots dx_T \\ &= \int_{\mathbb{X}^q}\cdots\int_{\mathbb{X}^q} f_\theta[x_0]\prod_{t=0}^{T-1} f_\theta[x_{t+1}|x_t]\prod_{t=0}^{T} f_\theta[y_t|x_t]\; dx_0\ldots dx_T. \end{aligned} \tag{140}$$

If all densities are Gaussian,

$$\begin{aligned} f_\theta[x_0] &= \mathcal{N}(x_0;\pi_0,\Pi_0) \\ f_\theta[x_t|x_{t-1}] &= \mathcal{N}(x_t;A_0x_{t-1},C) \\ f_\theta[y_t|x_t] &= \mathcal{N}(y_t;Hx_t,V), \end{aligned}$$

the integration with respect to the hidden state variables can be carried out analytically. Using the explicit form for Gaussian densities, we have for a number of $\Delta t = T+1$ time steps for the marginal *pdf* the expression

$$\begin{aligned} f_\theta[y_0,\ldots,y_T] &= c_y\int \mathrm{Exp}\Bigg[-\frac{1}{2}(x_0-\pi_0)^{\mathrm{T}}\Pi_0^{-1}(x_0-\pi_0) \\ &-\frac{1}{2}\sum_{t=0}^{T-1}(x_{t+1}-A_0x_t)^{\mathrm{T}}C^{-1}(x_{t+1}-A_0x_t) \\ &-\frac{1}{2}\sum_{t=0}^{T}(y_t-Hx_t)^{\mathrm{T}}V^{-1}(y_t-Hx_t)\Bigg]dx_0\cdots dx_T, \end{aligned} \tag{141}$$

with the normalization constant

$$c_y = (2\pi)^{-\frac{\Delta t}{2}(p+q)}(\mathrm{Det}\ \Pi_0)^{-1/2}(\mathrm{Det}\ C)^{-\frac{\Delta t-1}{2}}(\mathrm{Det}\ V)^{-\frac{\Delta t}{2}}. \tag{142}$$

It is quite easy to show that the joint *pdf* is normalized as

$$\int f_\theta[x_0,\ldots,x_T,y_0,\ldots,y_T]dx_0\cdots dx_T dy_0\cdots dy_T = 1.$$

First, we have to integrate over all y_t. As $f[y_t|x_t]$ is already a properly normalized Gaussian, all integrals over y_t equal one. We can then carry out the integrals over $x_0\cdots x_T$ by starting with the one over x_T and integrating down to x_0. Each integral is also a normalized Gaussian and finally we obtain unity for all integrals together. A proper normalization is essential to a correctly computing the effective measure complexity in Section 4.2.

Based on the notation from Section 2.2, we write the vectors in a more convenient form:

$$\begin{aligned} \boldsymbol{y}^{\mathrm{T}} &= \left(y_0^{\mathrm{T}} \cdots y_T^{\mathrm{T}}\right) \\ \boldsymbol{x}^{\mathrm{T}} &= \left(x_0^{\mathrm{T}} \cdots x_T^{\mathrm{T}}\right) \\ \boldsymbol{b}^{\mathrm{T}} &= \left(y_0^{\mathrm{T}} V^{-1} H + \pi_0 \Pi_0^{-1} | y_1^{\mathrm{T}} V^{-1} H | \cdots | y_T^{\mathrm{T}} V^{-1} H\right) \end{aligned}$$

and matrices $\mathcal{V} = I_{\Delta t} \otimes V^{-1}$ and $\mathcal{C} = \mathcal{C}_1 + \mathcal{C}_2$ with $\mathcal{C}_1 = I_{\Delta t} \otimes H^{\mathrm{T}} V^{-1} H$ and

$$\mathcal{C}_2 = \begin{pmatrix} \Pi_0^{-1} + A_0^{\mathrm{T}} C^{-1} A_0 & -A_0^{\mathrm{T}} C^{-1} & & & 0 \\ -C^{-1} A_0 & C^{-1} + A_0^{\mathrm{T}} C^{-1} A_0 & -A_0^{\mathrm{T}} C^{-1} & & \\ & \ddots & \ddots & \ddots & \\ & & -C^{-1} A_0 & C^{-1} + A_0^{\mathrm{T}} C^{-1} A_0 & -A_0^{\mathrm{T}} C^{-1} \\ 0 & & & -C^{-1} A_0 & C^{-1} \end{pmatrix}.$$

The above joint *pdf* can then be written as

$$f_\theta[y_0, \cdots, y_T] = c_y \mathrm{Exp}\left[-\frac{1}{2} \boldsymbol{y}^{\mathrm{T}} \mathcal{V} \boldsymbol{y} - \frac{1}{2} \pi_0^{\mathrm{T}} \Pi_0^{-1} \pi_0\right] \int \mathrm{Exp}\left[-\frac{1}{2} \boldsymbol{x}^{\mathrm{T}} \mathcal{C} \mathbf{x} + \boldsymbol{b}^{\mathrm{T}} \boldsymbol{x}\right] dx_0 \cdots dx_T$$

and computed as:

$$f_\theta[y_0, \cdots, y_T] = c_y \mathrm{Exp}\left[-\frac{1}{2} \boldsymbol{y}^{\mathrm{T}} \mathcal{V} \boldsymbol{y} - \frac{1}{2} \pi_0^{\mathrm{T}} \Pi_0^{-1} \pi_0\right] \frac{(2\pi)^{\Delta t q/2}}{\sqrt{\mathrm{Det}\ \mathcal{C}}} \mathrm{Exp}\left[\frac{1}{2} \boldsymbol{b}^{\mathrm{T}} \mathcal{C}^{-1} \boldsymbol{b}\right]. \quad (143)$$

As we can write

$$\boldsymbol{b} = \left(I \otimes H^{\mathrm{T}} V^{-1}\right) \boldsymbol{y},$$

it is convenient to introduce the matrix

$$\mathcal{B} = \left(I \otimes V^{-1} H\right) \mathcal{C}^{-1} \left(I \otimes H^{\mathrm{T}} V^{-1}\right).$$

With this definition, the joint *pdf* for the observations can be written as

$$f_\theta[\boldsymbol{y}] = \frac{c_y (2\pi)^{\Delta t q/2}}{\sqrt{\mathrm{Det}\ \mathcal{C}}} \mathrm{Exp}\left[-\frac{1}{2} \boldsymbol{y}^{\mathrm{T}} \mathcal{V} \boldsymbol{y} - \frac{1}{2} \pi_0^{\mathrm{T}} \Pi_0^{-1} \pi_0 + \frac{1}{2} \boldsymbol{y}^{\mathrm{T}} \mathcal{B} \boldsymbol{y}\right]. \quad (144)$$

The inverse $\mathcal{C}^{-1}$ of the block tridiagonal matrix $\mathcal{C}$ is also a block matrix containing $\Delta t \times \Delta t$ blocks of size $q \times q$. As $\mathcal{C}$ is symmetric, its inverse is also symmetric. To illustrate this, Fig. 2.19 shows the inverse $\mathcal{C}^{-1}$ of the matrix $\mathcal{C}$, and Fig. 2.20 shows the original matrix $\mathcal{C}$ for a state process with dimension $q = 4$ and randomly generated system matrices A_0 and H.

We can observe the following three properties of the inverse, which will be investigated analytically later on: (1) the matrix elements in the center of $\mathcal{C}^{-1}$ tend to a constant block-Töplitz matrix; (2) the magnitude of the off-diagonal elements

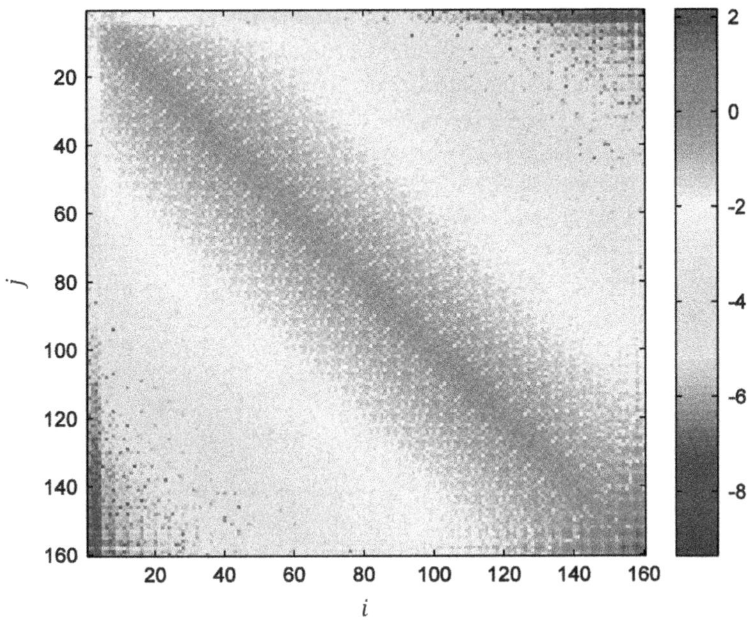

Fig. 2.19 Matrix elements of the covariance matrix C_{ij}^{-1} for a state process with dimension $q = 4$ and randomly generated system matrices A_0 and H. The color coding is based on a logarithmic scale

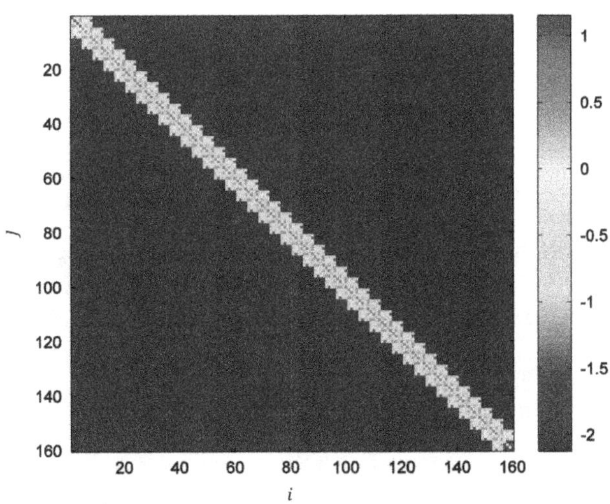

Fig. 2.20 Matrix elements of the original covariance matrix C_{ij} (cf. Fig. 2.19) for a state process with dimension $q = 4$ and randomly generated system matrices A_0 and H. The color coding is based on a logarithmic scale

decays exponentially; (3) in the upper and lower left-hand corners the inverse C^{-1} shows a transient behaviour that is caused by boundary conditions and depends on the particular covariance of the initial state Π_0^{-1}.

The ij-th block of C^{-1} which we denote as $\mathbf{X}_{ij}$, $j \geq i$, can be computed explicitely using the following recursions proven by Bowden (1989):

$$\mathbf{X}_{ij} = \mathbf{K}_i \mathbf{K}_{\Delta t+1}^{-1} \mathbf{N}_j.$$

The forward recursions for $\mathbf{K}_i$ are:

$$\begin{aligned} \mathbf{K}_1 &= I_q \\ \mathbf{K}_2 &= -\mathbf{A}^{-1}\mathbf{B}_1 \\ \mathbf{K}_{i+1} &= -\mathbf{A}^{-1}\left(\mathbf{B}\mathbf{K}_i + \mathbf{A}^{\mathrm{T}}\mathbf{K}_{i-1}\right), i = 2, \cdots, \Delta t - 1 \\ \mathbf{K}_{\Delta t+1} &= -\mathbf{B}_{\Delta t}\mathbf{K}_{\Delta t} - \mathbf{A}^{\mathrm{T}}\mathbf{K}_{\Delta t-1} \end{aligned} \tag{145}$$

and the backward recursions for $\mathbf{N}_i$ read:

$$\begin{aligned} \mathbf{N}_{\Delta t} &= I_q \\ \mathbf{N}_{\Delta t-1} &= -\mathbf{B}_{\Delta t}\mathbf{A}^{-1} \end{aligned} \tag{146}$$

$$\begin{aligned} \mathbf{N}_{j-1} &= -\left(\mathbf{N}_j\mathbf{B} + \mathbf{N}_{j+1}\mathbf{A}^{\mathrm{T}}\right)\mathbf{A}^{-1}, j = \Delta t - 1, \cdots, 2 \\ \mathbf{N}_0 &= -\mathbf{N}_1\mathbf{B}_1 - \mathbf{N}_2\mathbf{A}^{\mathrm{T}} \end{aligned} \tag{147}$$

with

$$\begin{aligned} \mathbf{B}_1 &= H^{\mathrm{T}}V^{-1}H + \Pi_0^{-1} + A_0^{\mathrm{T}}C^{-1}A_0 \\ \mathbf{B} &= H^{\mathrm{T}}V^{-1}H + C^{-1} + A_0^{\mathrm{T}}C^{-1}A_0 \\ \mathbf{B}_{\Delta t} &= H^{\mathrm{T}}V^{-1}H + C^{-1} \\ \mathbf{A} &= -A_0^{\mathrm{T}}C^{-1}. \end{aligned}$$

As we are interested in the likelihood of some infinitely long observations $\left(y_{-\infty}^{-1}, y_0^{\infty}, y_{-\infty}^{\infty}\right)$, we have to study the limiting value of the recursions for $\mathbf{K}_{i\to\infty}$ and N_0 for $\Delta t \to \infty$. This is accomplished as follows: we transform the recursions of second order in the following first-order recursions:

$$\begin{pmatrix} \mathbf{K}_i \\ \mathbf{K}_{i+1} \end{pmatrix} = \underbrace{\begin{pmatrix} 0 & I_p \\ -\mathbf{A}^{-1}\mathbf{A}^{\mathrm{T}} & \mathbf{A}^{-1}\mathbf{B} \end{pmatrix}}_{=:\mathbf{M}_f} \begin{pmatrix} \mathbf{K}_{i-1} \\ \mathbf{K}_i \end{pmatrix}.$$

Then the matrices at iteration step i can be expressed explicitly in terms of the initial matrices:

$$\begin{pmatrix} \mathbf{K}_i \\ \mathbf{K}_{i+1} \end{pmatrix} = \mathbf{M}_f^{i-1} \begin{pmatrix} \mathbf{K}_1 \\ \mathbf{K}_2 \end{pmatrix} = \mathbf{M}_f^{i-1} \mathbf{K}_f^{(0)}.$$

The initial matrix $\mathbf{K}_f^{(0)}$ depends explicitly on the covariance of the initial state and the system matrices:

$$\begin{aligned} \mathbf{K}_f^{(0)} &= \begin{pmatrix} \mathbf{K}_1 \\ \mathbf{K}_2 \end{pmatrix} \\ &= \begin{pmatrix} I_q \\ CA_0^{-T}\left(H^{\mathrm{T}}V^{-1}H + \Pi_0^{-1} + A_0^{\mathrm{T}}C^{-1}A_0\right) \end{pmatrix}. \end{aligned}$$

Expanding the initial matrix $\mathbf{K}_f^{(0)}$ in the eigenvectors $v_{k,f}$ of the forward transition matrix $\mathbf{M}_f$

$$\begin{aligned} \mathbf{M}_f \mathbf{V}_f &= \mathbf{V}_f \Lambda \\ \mathbf{V}_f &= \left(\left[v_{1,f} \cdots v_{2q,f}\right]\right) \\ \Lambda &= \begin{pmatrix} \lambda_{1,f} & & \\ & \ddots & \\ & & \lambda_{2q,f} \end{pmatrix}, \end{aligned}$$

where the eigenvalues are arranged in descending order $\left|\lambda_{1,f}\right| \geq \left|\lambda_{2,f}\right| \geq \cdots \geq \left|\lambda_{2q,f}\right|$, the coefficients T_f of the basis expansion

$$\mathbf{K}_f^{(0)} = \mathbf{V}_f \mathbf{T}_f$$

are easily obtained as

$$\mathbf{T}_f = \mathbf{V}_f^{-1} \mathbf{K}_f^{(0)}.$$

Note that the rate for reaching the asymptotic behavior depends on the ratio of the leading eigenvalue to the second largest one. Inserting the above expansion in the recursions, we find

$$\begin{aligned} \begin{pmatrix} \mathbf{K}_i \\ \mathbf{K}_{i+1} \end{pmatrix} &= \mathbf{M}_f^{i-1} \mathbf{V}_f \mathbf{T}_f \\ &= \left(\mathbf{M}_f^{i-1} v_{1,f} \cdots \mathbf{M}_f^{i-1} v_{2q,f}\right) \mathbf{T}_f \\ &= \left(\lambda_{1,f}^{i-1} v_{1,f} \cdots \lambda_{2q,f}^{i-1} v_{1,f}\right) \mathbf{T}_f \\ &\overset{i\to\infty}{\rightarrow} \left(\lambda_{1,f}^{i-1} v_{1,f} \cdots \lambda_{m,f}^{i-1} v_{m,f} 0 \cdots 0\right) \mathbf{T}_f. \end{aligned}$$

Here, m is the multiplicity of the largest eigenvalue. Let $\widehat{\mathbf{T}}_f$ be the m first rows of $\mathbf{T}_f$, and let $\widehat{\mathbf{V}}_f$ be the m most left columns of $\mathbf{V}_f$, then we have:

$$\begin{pmatrix} \mathbf{K}_i \\ \mathbf{K}_{i+1} \end{pmatrix} \overset{i\to\infty}{\to} \lambda_{1,f}^{i-1} \widehat{\mathbf{V}}_f \widehat{\mathbf{T}}_f.$$

It is easy to show that the eigenvectors can be partitioned as

$$v_{i,f} = \begin{pmatrix} \tilde{v}_{i,f} \\ \lambda_{i,f} \tilde{v}_{i,f} \end{pmatrix},$$

because the invariance condition for the forward recursions can be transformed into the quadratic eigenvalue problem

$$-\lambda_{i,f} \mathbf{A}^{-1} \mathbf{B} \tilde{v}_{i,f} - \mathbf{A}^{-1} \mathbf{A}^{\mathrm{T}} \tilde{v}_{i,f} = \lambda_{i,f}^2 \tilde{v}_{i,f}.$$

Finally, for large i the matrix $\mathbf{K}_i$ tends to

$$\begin{aligned} \mathbf{K}_{i\to\infty} &= \lambda_1^{i-1} (I_q 0) \begin{pmatrix} \tilde{v}_{1,f} & \cdots & \tilde{v}_{m,f} \\ \lambda_{1,f} \tilde{v}_{1,f} & \cdots & \lambda_{1,f} \tilde{v}_{m,f} \end{pmatrix} \widehat{\mathbf{T}}_f \\ &= \lambda_1^{i-1} \left(\tilde{v}_{1,f} \cdots \tilde{v}_{m,f} \right) \widehat{\mathbf{T}}_f \\ &= \lambda_1^{i-1} \tilde{\mathbf{V}}_f \widehat{\mathbf{T}}_f, \end{aligned}$$

i.e. it is the product of a constant matrix $\tilde{\mathbf{V}}_f \widehat{\mathbf{T}}_f$ and a pre-factor λ_1^{i-1}. A similar equation can be derived for the backward recursions. Consequently, the matrix block $\boldsymbol{X}_{ij}$ in the middle of $\mathcal{C}$ also tends to a constant matrix.

We can derive another expression for $\boldsymbol{X}_{ij}$ that avoids the numerical instability of the above recursions for large index $i \to \infty$: The block on the diagonal is given by

$$\boldsymbol{X}_{ii} = \lim_{N\to\infty} \frac{1}{N} \sum_{n=0}^{N} \left(\mathbf{B} + e^{\frac{2\pi i n}{N}} \mathbf{A} + e^{\frac{2\pi i n (N-1)}{N}} \mathbf{A}^T \right)^{-1}.$$

Moreover, we derive the joint *pdf* for the observations if the system is in steady state. Using the result for the joint *pdf* of the observed states, Eq. 144, we see that the total covariance of all observations in steady state ($\boldsymbol{\mu} = 0$, see Eq. 26) is given by

$$\mathcal{C}_y = (\mathcal{V} - \mathcal{B})^{-1}.$$

Using the Woodbury matrix identity (Higham 2002)

$$(A + UCV)^{-1} = A^{-1} - A^{-1} U \left(C^{-1} + V A^{-1} U \right)^{-1} V A^{-1}, \tag{148}$$

we can calculate

$$\begin{aligned}(\mathcal{V}-\mathcal{B})^{-1} &= \left((I_{\Delta t}\otimes V^{-1}) - (I_{\Delta t}\otimes V^{-1}H)\mathcal{C}^{-1}(I_{\Delta t}\otimes H^{\mathrm{T}}V^{-1})\right)^{-1}\\ &= \left(I_{\Delta t}\otimes V^{-1}\right)^{-1} - \left(I_{\Delta t}\otimes V^{-1}\right)^{-1}\left(I_{\Delta t}\otimes V^{-1}H\right)\\ &\quad\cdot\left(-\mathcal{C} + \left(I_{\Delta t}\otimes H^{\mathrm{T}}V^{-1}\right)\left(I_{\Delta t}\otimes V^{-1}\right)^{-1}\left(I_{\Delta t}\otimes V^{-1}H\right)\right)^{-1}\\ &\quad\cdot\left(I_{\Delta t}\otimes H^{\mathrm{T}}V^{-1}\right)\left(I_{\Delta t}\otimes V^{-1}\right)^{-1}.\end{aligned}$$

Using the identities $(A\otimes B)^{-1} = A^{-1}\otimes B^{-1}$ and $(A\otimes B)(C\otimes D) = AC\otimes BD$, we find

$$\begin{aligned}(\mathcal{V}-\mathcal{B})^{-1} &= I_{\Delta t}\otimes V + (I_{\Delta t}\otimes H)\left(\mathcal{C} - I_{\Delta t}\otimes H^{\mathrm{T}}V^{-1}H\right)^{-1}\left(I_{\Delta t}\otimes H^{\mathrm{T}}\right)\\ &= I_{\Delta t}\otimes V + (I_{\Delta t}\otimes H)(\mathcal{C}_2)^{-1}\left(I_{\Delta t}\otimes H^{\mathrm{T}}\right).\end{aligned} \quad (149)$$

The matrix $\mathcal{C}_2$ appeared in the explicit form for the joint *pdf* of the Markov process (see Eq. 33 in Section 2.2). Its inverse could be computed explicitly in Eq. 34 and was denoted as $\mathcal{C}_x$, which is the covariance of the hidden variables. Finally, the covariance of the observations is given as

$$\mathcal{C}_y = I_{\Delta t}\otimes V + (I_{\Delta t}\otimes H)\mathcal{C}_x\left(I_{\Delta t}\otimes H^{\mathrm{T}}\right). \quad (150)$$

This shows that the covariance for the observations in steady state is also block Toeplitz. The joint *pdf* for the observed states is given by

$$f[y] = \frac{1}{(2\pi)^{p\Delta t/2}\left(\mathrm{Det}\ \mathcal{C}_y\right)^{1/2}}\mathrm{Exp}\left[-\frac{1}{2}y^{\mathrm{T}}\mathcal{C}_y^{-1}y\right]. \quad (151)$$

It is evident that in the case of an LDS with an arbitrary structure of the system matrices A_0 and H, not only the evolution of observables between all consecutive time steps must be considered in order to make good predictions but also all possible transitions between hidden states of the process in the past that could give rise to the sequence of observations (see Eq. 140). For two consecutive time steps, we find the conditional density:

$$\begin{aligned}f_\theta[y_{t+1}|y_t] &= \frac{f_\theta[y_{t+1},y_t]}{f_\theta[y_t]}\\ &= \frac{1}{f_\theta[y_t]}\int_{\mathbb{X}^q}\int_{\mathbb{X}^q} f_\theta[y_{t+1},y_t,x_{t+1},x_t]\ dx_{t+1}dx_t\\ &= \frac{1}{f_\theta[y_t]}\int_{\mathbb{X}^q}\int_{\mathbb{X}^q} f_\theta[x_t]f_\theta[y_t|x_t]f_\theta[x_{t+1}|x_t]f_\theta[y_{t+1}|x_{t+1}]\ dx_t dx_{t+1}.\end{aligned}$$

For Gaussian noise vectors ε_t and v_t, the joint *pdf* of the system can be written as

$$f_\theta[x_0,\ldots,x_T,y_0,\ldots,y_T] = \mathcal{N}(x_0;\pi_0,\Pi_0)\prod_{t=0}^{T-1}\mathcal{N}(x_{t+1};A_0x_t,C)\prod_{t=0}^{T}\mathcal{N}(y_t;Hx_t,V), \tag{152}$$

where the Gaussian density $\mathcal{N}(.)$ with location μ_x and covariance Σ_x is defined as (cf. Eq. 13)

$$\mathcal{N}(x;\mu_x,\Sigma_x) = \frac{1}{(2\pi)^{p/2}(\mathrm{Det}[\Sigma_x])^{1/2}}\mathrm{Exp}\left[-\frac{1}{2}(x-\mu_x)^{\mathrm{T}}\Sigma_x^{-1}(x-\mu_x)\right]. \tag{153}$$

The density function $f_\theta[y_t]$ of state Y_t given the initial location π_0, and the system and covariance matrices can be written explicitly as (cf. Eq. 19)

$$f_\theta[y_t] = \frac{1}{(2\pi)^{p/2}\left(\mathrm{Det}\left[\Sigma_{y;t}\right]\right)^{1/2}}\mathrm{Exp}\left[-\frac{1}{2}(y_t - HA_0^t\pi_0)^{\mathrm{T}}\Sigma_{y;t}^{-1}(y_t - HA_0^t\pi_0)\right], \tag{154}$$

where

$$\Sigma_{y;t} = H\Sigma_{x;t}H^{\mathrm{T}} + V$$

and

$$\Sigma_{x;t} = A_0^t\Pi_0\left(A_0^{\mathrm{T}}\right)^t + \sum_{v=0}^{t-1}A_0^vC\left(A_0^{\mathrm{T}}\right)^v.$$

The conditional density of observation Y_{t+1} given observation $Y_t = y_t$ is (cf. Eq. 20):

$$f_\theta\left[y_{t+1}|y_t\right] = \frac{1}{f_\theta[y_t]}\int_{\mathbb{R}^q}\int_{\mathbb{R}^q}\mathcal{N}(x_t;A_0^t\pi_0,\Sigma_{x;t})\mathcal{N}(y_t;Hx_t,V)$$
$$\mathcal{N}(x_{t+1};A_0x_t,C)\mathcal{N}\left(y_{t+1};Hx_{t+1},V\right)\ dx_tdx_{t+1}.$$

We will derive the explicit form of this density for the steady-state process in Section 4.2.

Following the procedure set out in Sections 2.1 and 2.2, a closed-form solution to the total amount of work done in all tasks during the iteration process (modeled by hidden initial state x_0 and operators A_0 and H) until the stopping criterion δ is satisfied is given by

$$
\begin{aligned}
E\left[\sum_{t=0}^{T_\delta} Y_t\right] &= \sum_{t=0}^{T_\delta} E[Y_t] \\
&= \sum_{t=0}^{T_\delta} H \cdot \left(A_0^t \cdot x_0\right) \\
&= H \cdot \left(\sum_{t=0}^{T_\delta} A_0^t\right) \cdot x_0 \\
&= H \cdot \left(I_p - A_0\right)^{-1} \cdot \left(I_p - (A_0)^{T_\delta+1}\right) \cdot x_0 \,.
\end{aligned}
$$

If all subtasks of the state process are initially to be completed 100%, the initial state is simply

$$
x_0 = \begin{pmatrix} 1 \\ \vdots \\ 1 \end{pmatrix}.
$$

As with the vector autoregression models, the expected total amount of work y_{tot} across the time interval $t = 0, \ldots, T_\delta$ is estimated by:

$$
y_{tot} = \text{Total}\left[H \cdot \left(I_p - A_0\right)^{-1} \cdot \left(I_p - (A_0)^{T_\delta+1}\right) \cdot x_0\right]. \tag{155}
$$

In the limit $T_\delta \to \infty$ we have for an asymptotically stable project phase the total amount of work done during the iteration process:

$$
\lim_{T_\delta \to \infty} E\left[\sum_{t=0}^{T_\delta} Y_t\right] = H \cdot \left(I_p - A_0\right)^{-1} \cdot x_0 \,.
$$

Following the concept from Section 2.3, we can also transform the LDS into the spectral basis and therefore decompose the state process into a system with uncoupled processes with correlated noise. To transform the state-space coordinates, the state transitions matrix A_0 is diagonalized through an eigendecomposition as shown in Eq. 35. The eigenvectors $\vartheta_i(A_0) = S_{:i} (i = 1 \ldots q)$ of the state transition matrix are the column vectors of the linear transformation represented by the transformation matrix S. The stochastic processes $\{X_t'\}$ and $\{Y_t'\}$ transformed in the spectral basis are generated by the system of equations

$$
\begin{aligned}
X_{t+1}' &= \Lambda_S \cdot X_t' + \varepsilon_t' \\
Y_t' &= H' \cdot X_t' + v_t'
\end{aligned}
$$

with the additional representation

$$H' = S^{-1} \cdot H$$
$$\varepsilon_t' \sim (0_q, C')$$
$$C' = S^{-1} \cdot C \cdot \left(\left[S^{\mathrm{T}}\right]^*\right)^{-1}$$
$$\nu_t' \sim (0_p, V')$$
$$V' = S^{-1} \cdot V \cdot \left(\left[S^{\mathrm{T}}\right]^*\right)^{-1}$$
$$X_0' \sim (\pi_0', \Pi_0')$$
$$\pi_0' = S^{-1} \cdot \pi_0$$
$$\Pi_0' = S^{-1} \cdot \Pi_0 \cdot \left(\left[S^{\mathrm{T}}\right]^*\right)^{-1}.$$

The transformed covariance matrices C' and V' are also positive-semidefinite. The transformed LDS can be helpful in evaluating emergent complexity of the modeled PD project, because the steady-state covariance matrix Σ' of the state process can be expressed in a simple and expressive matrix form (see Eq. 258).

Moreover, in specific application contexts it can also be interesting to use the inherent "degeneracy" in the LDS model (see e.g. Roweis and Gharahmani 1999). Degeneracy means that the complete informational structure contained in the covariance matrix C of the performance fluctuations can be shifted into the state transition matrix A_0 and the observation matrix H. The informational structure can be shifted by decomposing C into independent covariance components through the same eigendecomposition that was used to transform the state-space coordinates:

$$C = U \cdot \Lambda_U \cdot U^{-1}$$

with

$$\Lambda_U = \mathrm{Diag}[\lambda_i(C)] \quad 1 \le i \le q.$$

Because C is symmetric by definition, the eigenvectors $\vartheta_i(C) = U_{:i}$ are mutually orthogonal and $U^{-1} = U^{\mathrm{T}}$ holds. Therefore, for any LDS that is not driven by performance fluctuations represented by a standard normal distribution with identity covariance matrix I_q, we can build an equivalent model with rescaling based on the transformation $T = U \cdot \Lambda_U^{1/2}$:

$$X' = T^{-1} \cdot X \tag{156}$$
$$A_0' = T^{-1} \cdot A_0 \cdot T \tag{157}$$
$$H' = H \cdot T\,. \tag{158}$$

The rescaling can be interpreted as a "whitening" of the hidden random vectors (DelSole and Tippett 2007, cf. Eqs. 253 and 254). A random vector in a sample space is said to be white if the vector components are statistically independent of each other. If the independent vector components are also identically distributed, as in our case, then the random vector is said to be an i.i.d. random vector. In practice, variables can be transformed to whitened space by projecting them onto their principal components and then normalizing the principal components to unit variance. It is evident that the informational structure of the covariance matrix V cannot be changed in the same way since the realizations Y_t are definitely observed and we are not free to rescale them.

To support the time-dependent statistical analysis of the work processes and to simplify the analytical complexity evaluation in Section 4.2, we introduce the autocovariance function

$$\begin{aligned} C_{YY}(t,s) &:= E\left[(Y_t - \mu_t)(Y_s - \mu_s)^{\mathrm{T}}\right] \\ &= E\left[Y_t Y_s^{\mathrm{T}}\right] - \mu_t \mu_s^{\mathrm{T}} \end{aligned}$$

of the observation process $\{Y_t\}$, based on the assumption that the state process $\{X_t\}$ is in steady state. If the modeled project is asymptotically stable and therefore the modulus of the largest eigenvalue of the dynamical operator A_0 is less than 1, the mean of the state process in steady state is equal to the zero vector, indicating that there is no remaining work and we have $\mu_X = 0$. A detailed analysis of steady-state process dynamics will be carried out in Section 4.1, and we only present some basic results from system theory. If $\{X_t\}$ is a stationary process, the autocovariance of the observation process becomes

$$\begin{aligned} C_{YY}(k,l) &= C_{YY}(l-k) \\ &= C_{YY}(\tau), \end{aligned}$$

where $\tau = l - k$ is the number of time steps by which the observation has been shifted. As a result, the autocovariance function can be expressed as a function with the lead time as the only argument:

$$\begin{aligned} C_{YY}(\tau) &= E\left[(Y_t - \mu_Y)(Y_{t+\tau} - \mu_Y)^{\mathrm{T}}\right] \\ &= E\left[Y_t Y_{t+\tau}^{\mathrm{T}}\right] - \mu_Y \mu_Y^{\mathrm{T}} \\ &= E\left[Y_t Y_{t+\tau}^{\mathrm{T}}\right] - (H \cdot \mu_X)(H \cdot \mu_X)^{\mathrm{T}} \\ &= E\left[Y_t Y_{t+\tau}^{\mathrm{T}}\right]. \end{aligned} \tag{159}$$

Hence, in a signal processing sense, the autocovariance $C_{YY}(\tau)$ and the autocorrelation

$$R_{YY}(\tau) := E\left[Y_t Y_{t+\tau}^{\mathrm{T}}\right]$$

are equal in steady state for every lead time and we have

$$C_{YY}(\tau) = R_{YY}(\tau). \tag{160}$$

According to the work of van Overschee and de Moor (1996) on subspace identification of purely stochastic systems the autocovariance function of the observation process is given by

$$C_{YY}(\tau) = \begin{cases} H \cdot \Sigma \cdot H^{\mathrm{T}} + V & \tau = 0 \\ H \cdot A_o^{\tau-1} \cdot G & \tau > 0, \end{cases} \tag{161}$$

where the coefficient G can be expressed as the expected value

$$\begin{aligned} G &= E\left[X_{t+1} Y_t^{\mathrm{T}}\right] \\ &= A_0 \cdot \Sigma \cdot H^{\mathrm{T}}. \end{aligned}$$

In this sense G describes the cross-covariance between hidden state X_{t+1} and observed state Y_t. The matrix Σ denotes the covariance of the states in steady-state of the process $\{X_t\}$. It satisfies the Lyapunov equation (see Eq. 27) and can be expressed in closed-form in the original state-space coordinates according to Eq. 245.

For the autocovariance function the following symmetry condition holds:

$$C_{YY}(-\tau) = C_{YY}(\tau)^{\mathrm{T}}.$$

As can be seen in the autocovariance function, the correlations between observations over $\tau > 0$ time steps can be significantly larger for an LDS in steady state than for the previously formulated VAR and PVAR models.

Following a similar line of thought as in the forcing matrix concept from Section 2.1 (see Eq. 23), the LDS model according to Eqs. 136 and 137 can be transformed into a more compact "forward innovation model" (e.g. van Overschee and de Moor 1996; de Cock 2002). In this context, "more compact" means that only a single noise source is used to model performance fluctuations. The forward innovation model is an equivalent representation in the sense that the first-order and second-order statistics of the sequence of observations generated by the model in steady state are the same, i.e., the autocovariances $E\left[Y_t Y_{t+\tau}^{\mathrm{T}}\right]$ and cross-covariances $E\left[X_{t+1} Y_t^{\mathrm{T}}\right]$ are identical. The same property holds for the "backward form" and the corresponding "backward innovation form" that will be derived later in this chapter.

The forward innovation representation results from applying a Kalman filter (Kalman 1960; for a comprehensive consideration of theoretical and practical aspects see e.g. Kailath 1981; Bar-Shalom et al. 2001; Honerkamp 2002 or Puri 2010) to the state-space model. In general, the Kalman filter operates recursively on time series of noisy input data from a PD project (or other dynamical systems) to calculate an estimate of the hidden system state that is statistically optimal. The filter is named after Rudolf E. Kálmán, who was one of the principal developers of its theoretical foundations. As will be shown in the analysis below, the algorithm follows a two-step procedure. In the first step—the so-called prediction step—the Kalman filter calculates an unbiased and linear estimate of the current state vector in conjunction with the covariances. Once the result of the next observation is obtained, the estimates are updated. The update of the state is done by weighting the previous state estimate and the measurement prediction error. These weights are determined in a way which assigns larger weights to state estimates with higher certainty.

Following the textbook of Bar-Shalom et al. (2001), we start the derivation of the forward innovation representation by defining the hidden-state variable as the conditional mean of the state X_{t+1} given all measurements up to time t, that is

$$\begin{aligned}\hat{x}_{t+1|t} &:= E\left[X_{t+1}|Y_0^t\right] \\ &:= \int_{\mathbb{X}^q} x_{t+1}\, f_\theta\left(x_{t+1}|y_0^t\right) dx_{t+1}.\end{aligned}$$

Following the preferred notation, the term Y_0^t represents the sequence of observations of task processing in the PD project that have been made across an interval of t and are used to compute the conditional mean.

Using the state-space Eqs. 136 and 137 and the fact that ε_t has zero-mean, the state prediction is

$$\begin{aligned}\hat{x}_{t+1|t} &= E\left[(A_0 \cdot X_t + \varepsilon_t)|Y_0^t\right] \\ &= A_0 \cdot E\left[X_t|Y_0^t\right] \\ &= A_0 \cdot \hat{x}_{t|t}.\end{aligned}$$

and the state prediction error is given by

$$\begin{aligned}\tilde{X}_{t+1|t} &:= X_{t+1} - \hat{x}_{t+1|t} \\ &= A_0 \cdot X_t + \varepsilon_t - A_0 \cdot \hat{x}_{t|t} \\ &= A_0 \cdot \tilde{X}_{t|t} + \varepsilon_t.\end{aligned}$$

The standard Kalman filter calculates $\hat{x}_{t|t}$, which is an unbiased and linear Minimum Mean Square Error (MMSE, see e.g. Honerkamp 2002) estimate of the state vector X_t, given the current sequence of observations Y_0^t.

The state prediction covariance $\Phi_{t+1|t}$ is computed as follows:

$$\begin{aligned}\Phi_{t+1|t} &:= E\left[\tilde{X}_{t+1|t}\tilde{X}_{t+1|t}^{\mathrm{T}}|Y_0^t\right] \\ &= A_0 \cdot \Phi_{t|t} \cdot A_0^{\mathrm{T}} + C.\end{aligned}$$

The predicted measurements, defined as

$$\hat{y}_{t+1|t} := E\left[Y_{t+1}|Y_0^t\right]$$

follow from the observation (measurement) equation and the fact that v_t has zero mean:

$$\hat{y}_{t+1|t} = H \cdot \hat{x}_{t+1|t}.$$

The measurement prediction error

$$\tilde{Y}_{t+1|t} := Y_{t+1} - \hat{y}_{t+1|t},$$

also simply called "innovation," is given by

$$\begin{aligned}\tilde{Y}_{t+1|t} &= H \cdot X_{t+1} + v_{t+1} - H \cdot \hat{x}_{t+1|t}\\ &= H \cdot \tilde{X}_{t+1|t} + v_{t+1}.\end{aligned}$$

Moreover, the innovation covariance is given by

$$\begin{aligned}S_{t+1|t} &:= E\left[\tilde{Y}_{t+1|t}\tilde{Y}^{\mathrm{T}}_{t+1|t}|Y_0^t\right]\\ &= H \cdot E\left[\tilde{X}_{t+1|t}\tilde{X}^{\mathrm{T}}_{t+1|t}|Y_0^t\right] \cdot H^{\mathrm{T}} + V\\ &= H \cdot \Phi_{t+1|t} \cdot H^T + V\end{aligned} \tag{162}$$

and the covariance between state and measurement is

$$\begin{aligned}Q_{t+1|t} &:= E\left[\tilde{X}_{t+1|t}\tilde{Y}^{\mathrm{T}}_{t+1|t}|Y_0^t\right]\\ &= \Phi_{t+1|t} \cdot H^{\mathrm{T}}.\end{aligned}$$

Defining the filter gain,

$$\begin{aligned}W &:= Q_{t+1|t} \cdot S^{-1}_{t+1|t}\\ &= \Phi_{t+1|t} \cdot H^{\mathrm{T}} \cdot S^{-1}_{t+1|t}\end{aligned}$$

the update of the state is the cited MMSE estimate, which is given for Gaussian random variables in closed form as

$$\hat{x}_{t+1|t+1} = \hat{x}_{t+1|t} + W \cdot \tilde{Y}_{t+1|t}.$$

Now the forward innovation representation can be written down: From

$$\begin{aligned}
\hat{x}_{t+1|t} &= A_0 \cdot \hat{x}_{t|t} \\
&= A_0\left(\hat{x}_{t|t-1} + W \cdot \tilde{Y}_{t|t-1}\right) \\
&= A_0 \cdot \hat{x}_{t|t-1} + A_0 \cdot W \cdot \tilde{Y}_{t|t-1}
\end{aligned}$$

with the definition of the Kalman gain

$$\begin{aligned}
K &:= A_0 \cdot W \\
&= A_0 \cdot \Phi_{t+1|t} \cdot H^{\mathrm{T}} \cdot S_{t+1|t}^{-1}
\end{aligned} \tag{163}$$

and the more convenient notation

$$\begin{aligned}
X_{t+1}^{f} &:= \hat{x}_{t+1|t} \\
\eta_t &:= \tilde{Y}_{t|t-1},
\end{aligned}$$

we obtain the simultaneous system of equations:

$$X_{t+1}^{f} = A_0 \cdot X_t^{f} + K \cdot \eta_t \tag{164}$$

$$Y_t = H \cdot X_t^{f} + \eta_t. \tag{165}$$

In this alternative representation form, the time-independent Kalman gain matrix K can be interpreted as forcing matrix of the state process noise η_t (cf. Eq. 23), which is driven by the single-source fluctuations. The time-dependent Kalman gain matrix will be calculated in the next chapter.

An explicit calculation of the state prediction covariance in steady-state, using the fact that $\hat{x}_{t+1|t} = E\left[X_{t+1}|Y_0^t\right]$ is already an expected value, leads to

$$\begin{aligned}
\Phi_{t+1|t} &= E\left[\tilde{X}_{t+1|t}\tilde{X}_{t+1|t}^{\mathrm{T}}|\{Y_i,\ i \le t\}\right] \\
&= E\left[\left(X_{t+1} - \hat{x}_{t+1|t}\right)\left(X_{t+1} - \hat{x}_{t+1|t}\right)^{\mathrm{T}}|\{Y_i,\ i \le t\}\right] \\
&= E\left[X_{t+1}X_{t+1}^{\mathrm{T}}|\{Y_i,\ i \le t\})\right] - E\left[\left(\hat{x}_{t+1|t}\hat{x}_{t+1|t}\right)^{\mathrm{T}}\right] \\
&= \Sigma - \Sigma^{f}.
\end{aligned} \tag{166}$$

Σ is the covariance of the original state variable. In steady state it satisfies the Lyapunov criterion

$$\Sigma = A_0 \cdot \Sigma \cdot A_0^{\mathrm{T}} + C$$

from Eq. 27. The Lyapunov equation is explained in great detail in Lancaster and Tismenetsky (1985) and will also be discussed in Sections 4.1.1 and 4.2.2. Σ^f is the covariance of the state variable in the forward innovation representation, that is the conditional mean of the state x_{t+1} given all measurements from the infinite past up to time t. Σ^f satisfies another Lyapunov equation (cf. Eq. 27):

$$\Sigma^f = A_0 \cdot \Sigma^f \cdot A_0^{\mathrm{T}} + K \cdot S \cdot K^T. \tag{167}$$

The entries of Σ^f can be determined by solving an algebraic Ricatti equation (see Eq. 296), which will be introduced in the implicit formulation of a complexity solution in Section 4.2.2.

In steady-state $\Phi_{t+1|t}$ converges to Φ and so $S_{t+1|t}$ approaches the constant covariance matrix S, which was used in the above Lyapunov equation. With the autocovariance of the observable variables in the innovation representation

$$\begin{aligned} C_{YY}(0) &= E\left[Y_t Y_t^{\mathrm{T}}\right] \\ &= H \cdot \Sigma^f \cdot H^{\mathrm{T}} + S \end{aligned} \tag{168}$$

we arrive at an expression for the Kalman gain that is equivalent to the solution in the work of de Cock (2002):

$$K = \left(A_0 \cdot \Sigma \cdot H^{\mathrm{T}} - A_0 \cdot \Sigma^f \cdot H^{\mathrm{T}}\right)\left(C_{YY}(0) - H \cdot \Sigma^f \cdot H^{\mathrm{T}}\right)^{-1}. \tag{169}$$

It is evident that the single-source performance fluctuations that drive the state process in the forward innovation form can be expressed as

$$\eta_t \sim \mathcal{N}(0_q, S)$$

with covariance

$$S = C_{YY}(0) - H \cdot \Sigma^f \cdot H^{\mathrm{T}}. \tag{170}$$

Finally, let us show that both representations have the same autocovariances:

$$\begin{aligned} C_{YY}(0) &= E\left[Y_t Y_t^{\mathrm{T}}\right] \\ &= H \cdot \Sigma \cdot H^{\mathrm{T}} + V \\ C_{YY}^f(0) &= E\left[Y_t Y_t^{\mathrm{T}}\right] \left(Y_t \text{ generated based on } X_t^f\right) \\ &= H \cdot \Sigma^f \cdot H^{\mathrm{T}} + S \end{aligned}$$

and same cross-covariance between hidden and observable states

$$\begin{aligned} G &= E\left[X_{t+1} Y_t^{\mathrm{T}}\right] \\ &= A_0 \cdot \Sigma \cdot H^{\mathrm{T}} \\ G^f &= E\left[X_{t+1}^f \left(Y_t^f\right)^{\mathrm{T}}\right] \\ &= A_0 \cdot \Sigma^f \cdot H^{\mathrm{T}} + K \cdot S. \end{aligned}$$

To show that $C_{YY}(0) = C_{YY}^f(0)$, we simply insert Eq. 162 into Eq. 166:

$$S = H\left(\Sigma - \Sigma^f\right)H^{\mathrm{T}} + V.$$

Rearranging the above equality proves the statement. Secondly, rearranging the definition of the Kalman gain (Eq. 163)

$$K = \left(A_0 \cdot \Sigma \cdot H^{\mathrm{T}} - A_0 \cdot \Sigma^f \cdot H^{\mathrm{T}}\right)S^{-1}$$

yields

$$K \cdot S = A_0 \cdot \Sigma \cdot H^{\mathrm{T}} - A_0 \cdot \Sigma^f \cdot H^{\mathrm{T}},$$

from which $G = G^f$ follows immediately.

In conclusion, the covariance matrices of the hidden states differ among both representations and it holds that $\Sigma \neq \Sigma^f$, in general. Furthermore, the covariances of the single-source performance fluctuations driving the state process in the forward innovation form differ from the covariances of the measurement error in the regular forward form and for arbitrary dynamics we have $V \neq S$. However, the autocovariances and covariances of the observed processes remain unchanged and the cross-covariance between hidden and observable states are the same. The parameters of the state-space model are (A_0, H, C, V) whereas in the innovations representation the parameters are (A_0, H, K, S). As we will show in Section 4.2.2, the differences do not necessarily lead to a conflict in terms of a different complexity evaluation according to de Cock (2002) and this work: The explicit computation of the complexity measure EMC in Section 4.2.1 leads to a result which depends only on the combined quantity $C_{YY}(0) = R_{YY}(0) = H \cdot \Sigma \cdot H^T + V$ and $G = A_0 \cdot \Sigma \cdot H$.

Concerning the analytical complexity evaluation that will be presented in Section 4.2, it is also helpful to formulate a complementary "backward model" in which the autocovariances of the observed process and the cross-covariance between hidden and observable states are also left unchanged in steady state (van Overschee and de Moor 1996; de Cock 2002). In this model the recursive state Eq. 136 runs not forward but backward in time. Due to the backward recursion the backward model is formulated by considering the MMSE estimate of X_t given X_{t+1}:

$$\begin{aligned}\hat{x}_{t|t+1} &:= E[X_t|X_{t+1}] \\ &= E\left[X_t X_{t+1}^{\mathrm{T}}\right]\left(E\left[X_{t+1} X_{t+1}^{\mathrm{T}}\right]\right)^{-1} X_{t+1},\end{aligned}$$

where the last equation holds true because all random variables are Gaussian. From the state-space representation according to Eq. 136 we compute

$$E\left[X_t X_{t+1}^{\mathrm{T}}\right] = E\left[X_t (A_0 \cdot X_t + \varepsilon_t)^{\mathrm{T}}\right]$$
$$= \Sigma \cdot A_0^{\mathrm{T}},$$

and due to the satisfied stationary condition for the state covariance $E\left[X_{t+1} X_{t+1}^{\mathrm{T}}\right] = \Sigma$ we can express the MMSE estimate of X_t given X_{t+1} as

$$\hat{x}_{t|t+1} = \Sigma \cdot A_0^{\mathrm{T}} \cdot \Sigma^{-1} \cdot X_{t+1}.$$

Now, we define the error

$$\varepsilon_{t|t+1} := X_t - \hat{x}_{t|t+1}$$

and the backward state as

$$\overline{X}_{t-1} := \Sigma^{-1} \cdot X_t. \tag{171}$$

(note that the hat symbol denotes the hidden state variable of the backward model and not the means). Transposing the second to last equation and inserting the above one, we obtain the recursion for the hidden state in the backward model

$$\begin{aligned} \overline{X}_{t-1} &= \Sigma^{-1}\left(\hat{x}_{t|t+1} + \varepsilon_{t|t+1}\right) \\ &= \Sigma^{-1}\left(\Sigma \cdot A_0^{\mathrm{T}} \cdot \Sigma^{-1} \cdot X_{t+1} + \varepsilon_{t|t+1}\right) \\ &= A_0^{\mathrm{T}} \cdot \overline{X}_t + \overline{\varepsilon}_t \\ &= \overline{A}_0 \cdot \overline{X}_t + \overline{\varepsilon}_t \end{aligned} \tag{172}$$

with the definition of the backward dynamical operator

$$\overline{A}_0 := A_0^{\mathrm{T}} \tag{173}$$

and the error term

$$\overline{\varepsilon}_t := \Sigma^{-1} \cdot \varepsilon_{t|t+1}. \tag{174}$$

The output equation in backward form is obtained by considering the MMSE estimate of Y_t given X_{t+1}:

$$\begin{aligned}\hat{y}_{t|t+1} &:= E[Y_t|X_{t+1}] \\ &= E\left[Y_t X_{t+1}^{\mathrm{T}}\right]\left(E\left[X_{t+1}X_{t+1}^{\mathrm{T}}\right]\right)^{-1} X_{t+1} \\ &= E\left[(H\cdot X_t + v_t)\left(A_0^{\mathrm{T}}\cdot X_t + \varepsilon_t\right)^{\mathrm{T}}\right]\Sigma^{-1}\cdot X_{t+1} \\ &= H\cdot\Sigma\cdot A_0^{\mathrm{T}}\cdot\Sigma^{-1}\cdot X_{t+1} \\ &= H\cdot\Sigma\cdot A_0^{\mathrm{T}}\cdot\overline{X}_t \\ &= H\cdot\Sigma\cdot\overline{A}_0\cdot\overline{X}_t.\end{aligned}$$

Re-arranging the error equation

$$\overline{v}_t = Y_t - \hat{y}_{t|t+1}$$

we obtain

$$\begin{aligned}Y_t &= \hat{y}_{t|t+1} + \overline{v}_t \\ &= H\cdot\Sigma\cdot\overline{A}_0\cdot\overline{X}_t + \overline{v}_t \\ &= \overline{H}\cdot\overline{X}_t + \overline{v}_t,\end{aligned} \tag{175}$$

where the backward output operator was defined as

$$\overline{H} = H\cdot\Sigma\cdot\overline{A}_0.$$

The joint covariance matrix of the zero-mean Gaussian processes $\{\overline{\varepsilon}_\mu\}$ and $\{\overline{v}_v\}$ is defined as

$$E\left[\begin{pmatrix}\overline{\varepsilon}_\mu \\ \overline{v}_v\end{pmatrix}\cdot\begin{pmatrix}\overline{\varepsilon}_\mu \\ \overline{v}_v\end{pmatrix}^T\right] = \begin{pmatrix}\overline{C} & \overline{S}_{\overline{\varepsilon}\overline{v}} \\ \overline{S}_{\overline{\varepsilon}\overline{v}}^{\mathrm{T}} & \overline{V}\end{pmatrix}\cdot\{\delta_{\mu v}\}.$$

Let $\overline{\Sigma}$ denote the covariance of the states in steady-state of the backward process $\{\overline{X}_t\}$. According to the definition from Eq. 171, $\overline{\Sigma}$ can be expressed as the inverse of the forward state covariance:

$$\overline{\Sigma} = \Sigma^{-1}.$$

Since in the backward form the variables $\overline{\varepsilon}_t$ and $\overline{v}_t$ and their past histories are also independent of state $\overline{X}_t$, the backward state covariance matrix $\overline{\Sigma}$ also satisfies the Lyapunov equation (cf. Eq. 101)

$$\overline{\Sigma} = A_0^{\mathrm{T}}\cdot\overline{\Sigma}\cdot A_0 + \overline{C} \Leftrightarrow \Sigma^{-1} = A_0^{\mathrm{T}}\cdot\Sigma^{-1}\cdot A_0 + \overline{C}$$

and can be expressed in closed-form similar to Eq. 245. Hence, we can express the individual covariances of the zero-mean Gaussian processes as:

$$\overline{C} = \Sigma^{-1} - A_0^{\mathrm{T}} \cdot \Sigma^{-1} \cdot A_0$$
$$\overline{V} = C_{YY}(0) - G^{\mathrm{T}} \cdot \Sigma^{-1} \cdot G.$$

The autocovariance function is

$$C_{YY}(\tau) = \begin{cases} \overline{H} \cdot \overline{\Sigma} \cdot \overline{H}^{\mathrm{T}} + \overline{V} = G^{\mathrm{T}} \cdot \Sigma^{-1} \cdot G + \overline{V} & \tau = 0 \\ \overline{H} \cdot \overline{A}_o^{\tau-1} \cdot \overline{G} = G^{\mathrm{T}} \cdot \overline{A}_o^{\tau-1} \cdot G & \tau > 0, \end{cases}$$

where the cross-covariance between hidden and observed state is given by

$$\begin{aligned} \overline{G} &= E\left[\overline{X}_{t-1} Y_t^{\mathrm{T}}\right] \\ &= \overline{A}_0 \cdot \overline{\Sigma} \cdot \overline{H}^{\mathrm{T}} + \overline{S}_{\overline{\varepsilon}\overline{v}}. \end{aligned}$$

Due to the definition of the backward state from Eq. 171 the cross-covariance can be simply written as

$$\overline{G} = H^{\mathrm{T}}.$$

We can also develop a corresponding backward innovation form. The derivation of the backward innovation form follows exactly the same procedure that gave the forward innovation representation and we therefore only present the essential steps. The backward oriented hidden-state variable $\hat{\overline{x}}_{t-1|t}$ is defined as the conditional mean of the state $\overline{x}_{t-1}$ given all measurements from the last time step T down to time step t, that is

$$\begin{aligned} \hat{\overline{x}}_{t-1|t} &:= E\left[\overline{X}_{t-1} | Y_t^T\right] \\ &:= \int_{\mathbb{X}^q} \overline{x}_{t-1} f_{\overline{\theta}}\left(\overline{x}_{t-1} | y_t^T\right) \, d\overline{x}_{t-1}. \end{aligned}$$

The state retrodiction (backward oriented state prediction) is

$$\begin{aligned} \hat{\overline{x}}_{t-1|t} &= E\left[\left(\overline{A}_0 \cdot \overline{X}_t + \overline{\varepsilon}_t\right) | Y_t^T\right] \\ &= \overline{A}_0 \cdot E\left[\overline{X}_t | Y_t^T\right] \\ &= \overline{A}_0 \cdot \hat{\overline{x}}_{t|t}. \end{aligned}$$

and the state retrodiction error is

$$\begin{aligned}\tilde{\overline{X}}_{t-1|t} &:= \overline{X}_{t-1} - \hat{\overline{x}}_{t-1|t} \\ &= \overline{A}_0 \cdot \overline{X}_t + \overline{\varepsilon}_t - \overline{A}_0 \cdot \hat{\overline{x}}_{t|t} \\ &= \overline{A}_0 \cdot \tilde{\overline{X}}_{t|t} + \overline{\varepsilon}_t.\end{aligned}$$

The state retrodiction covariance $\overline{\Phi}_{t+1|t}$ is

$$\begin{aligned}\overline{\Phi}_{t-1|t} &:= E\left[\tilde{\overline{X}}_{t-1|t}\tilde{\overline{X}}^{\mathrm{T}}_{t-1|t}|Y_t^T\right] \\ &= \overline{A}_0 \cdot \overline{\Phi}_{t|t} \cdot \overline{A}_0^{\mathrm{T}} + \overline{C}.\end{aligned}$$

The measurements retrodiction is

$$\begin{aligned}\hat{y}_{t-1|t} &:= E\left[Y_{t-1}|Y_t^T\right] \\ &= \overline{H} \cdot \hat{\overline{x}}_{t-1|t}.\end{aligned}$$

The measurement retrodiction error $\tilde{\overline{Y}}_{t-1|t} := Y_{t-1} - \hat{\overline{y}}_{t-1|t}$ is given by

$$\begin{aligned}\tilde{\overline{Y}}_{t-1|t} &= \overline{H} \cdot \overline{X}_{t-1} + \overline{v}_{t-1} - \overline{H} \cdot \hat{\overline{x}}_{t-1|t} \\ &= \overline{H} \cdot \tilde{\overline{X}}_{t-1|t} + \overline{v}_{t-1}.\end{aligned}$$

The innovation covariance is given by

$$\begin{aligned}\overline{S}_{t-1|t} &:= E\left[\tilde{\overline{Y}}_{t-1|t}\tilde{\overline{Y}}^{\mathrm{T}}_{t-1|t}|Y_t^T\right] \\ &= \overline{H} \cdot E\left[\tilde{\overline{X}}_{t-1|t}\tilde{\overline{X}}^{\mathrm{T}}_{t-1|t}|Y_t^T\right] \cdot \overline{H}^{\mathrm{T}} + \overline{V} \\ &= \overline{H} \cdot \overline{\Phi}_{t-1|t} \cdot \overline{H}^T + \overline{V}.\end{aligned} \tag{176}$$

The covariance between state and measurement is

$$\begin{aligned}\overline{Q}_{t-1|t} &:= E\left[\tilde{\overline{X}}_{t-1|t}\tilde{\overline{Y}}^{\mathrm{T}}_{t-1|t}|Y_t^T\right] \\ &= \overline{\Phi}_{t-1|t} \cdot \overline{H}^{\mathrm{T}}.\end{aligned}$$

Based on the filter gain

$$\begin{aligned}\overline{W} &:= \overline{Q}_{t-1|t} \cdot \overline{S}_{t-1|t}^{-1} \\ &= \overline{\Phi}_{t-1|t} \cdot \overline{H}^{\mathrm{T}} \cdot \overline{S}_{t-1|t}^{-1}\end{aligned}$$

the update of the state can be written as

$$\hat{\bar{x}}_{t-1|t-1} = \hat{\bar{x}}_{t-1|t} + \overline{W} \cdot \tilde{\bar{Y}}_{t-1|t}.$$

From the state retrodiction

$$\begin{aligned}\hat{\bar{x}}_{t-1|t} &= \overline{A}_0 \cdot \hat{\bar{x}}_{t|t} \\ &= \overline{A}_0 \left(\hat{\bar{x}}_{t|t+1} + \overline{W} \cdot \tilde{\bar{Y}}_{t|t+1}\right) \\ &= \overline{A}_0 \cdot \hat{\bar{x}}_{t|t+1} + \overline{A}_0 \cdot \overline{W} \cdot \tilde{\bar{Y}}_{t|t+1}\end{aligned}$$

using the Kalman gain

$$\begin{aligned}\overline{K} &= \overline{A}_0 \cdot \overline{W} \\ &= \overline{A}_0 \cdot \overline{\Phi}_{t-1|t} \cdot \overline{H}^{\mathrm{T}} \cdot \overline{S}_{t-1|t}^{-1},\end{aligned} \tag{177}$$

we finally arrive at the simultaneous system of equations:

$$\overline{X}_{t-1}^{b} = \overline{A}_0 \cdot \overline{X}_t^{b} + \overline{K} \cdot \overline{\eta}_t \tag{178}$$

$$Y_t = \overline{H} \cdot \overline{X}_t^{b} + \overline{\eta}_t, \tag{179}$$

where we have used the more convenient notation

$$X_{t-1}^{b} := \hat{\bar{x}}_{t-1|t}$$

$$\overline{\eta}_t := \tilde{\bar{Y}}_{t-1|t}$$

and the definitions $\overline{A}_0 = A_0^{\mathrm{T}}$ and $\overline{H} = G^{\mathrm{T}} = \left(G^f\right)^{\mathrm{T}}$ as before.

Because $\hat{\bar{x}}_{t-1|t} = E\left[\overline{X}_{t-1} | Y_t^T\right]$ is an expected value, the state retrodiction covariance in steady-state can be expressed as

$$\begin{aligned}\overline{\Phi}_{t-1|t} &= E\left[\tilde{\overline{X}}_{t-1|t} \tilde{\overline{X}}_{t-1|t}^{\mathrm{T}} | \{Y_i, i \geq t\}\right] \\ &= E\left[\left(\overline{X}_{t-1} - \hat{\bar{x}}_{t-1|t}\right)\left(\overline{X}_{t-1} - \hat{\bar{x}}_{t-1|t}\right)^{\mathrm{T}} \middle| \{Y_i, i \geq t\}\right] \\ &= E\left[\overline{X}_{t-1} \overline{X}_{t-1}^{\mathrm{T}} | \{Y_i, i \geq t\})\right] - E\left[\left(\hat{\bar{x}}_{t-1|t} \hat{\bar{x}}_{t-1|t}\right)^{\mathrm{T}}\right] \\ &= \overline{\Sigma} - \overline{\Sigma}^{b}.\end{aligned} \tag{180}$$

In steady state $\overline{\Sigma}$ and $\overline{\Sigma}^{b}$ satisfy the Lyapunov equations:

$$\Sigma = \overline{A}_0 \cdot \Sigma \cdot \overline{A}_0^{\mathrm{T}} + C$$
$$\overline{\Sigma}^b = \overline{A}_0 \cdot \overline{\Sigma}^b \cdot \overline{A}_0^{\mathrm{T}} + \overline{K} \cdot \overline{S} \cdot \overline{K}^{\mathrm{T}}.$$

The entries of $\overline{\Sigma}^b$ can also be calculated by solving an algebraic Ricatti equation (see Eq. 298 in Section 4.2.2.).

As before, in steady-state $\overline{\Phi}_{t-1|t}$ converges to $\overline{\Phi}$ and $\overline{S}_{t-1|t}$ to $\overline{S}$. Based on the autocovariance

$$\begin{aligned} C_{YY}(0) &= E\left[Y_t Y_t^{\mathrm{T}}\right] \\ &= \overline{H} \cdot \overline{\Sigma}^b \cdot \overline{H}^{\mathrm{T}} + \overline{S} \end{aligned}$$

we can express the backward Kalman gain (cf. de Cock 2002) as:

$$\begin{aligned} \overline{K} &= \left(\overline{A}_0 \cdot \Sigma \cdot \overline{H}^{\mathrm{T}} - \overline{A}_0 \cdot \overline{\Sigma}^b \cdot \overline{H}^{\mathrm{T}}\right)\left(C_{YY}(0) - \overline{H} \cdot \overline{\Sigma}^b \cdot \overline{H}^{\mathrm{T}}\right)^{-1} \\ &= \left(H^{\mathrm{T}} - A_0^{\mathrm{T}} \cdot \overline{\Sigma}^b \cdot G\right)\left(C_{YY}(0) - G^{\mathrm{T}} \cdot \overline{\Sigma}^b \cdot G\right)^{-1}, \end{aligned} \tag{181}$$

where we have used the previous definitions $\overline{A}_0 = A_0^{\mathrm{T}}$ and $\overline{H} = G^{\mathrm{T}}$. We define $G^b := \overline{H}$. It holds that $G^b = G^{\mathrm{T}} = \left(G^f\right)^{\mathrm{T}}$. The autocovariance function therefore can be expressed as

$$C_{YY}(\tau) = \begin{cases} \overline{H} \cdot \overline{\Sigma}^b \cdot \overline{H}^{\mathrm{T}} + \overline{S} = G^{\mathrm{T}} \cdot \overline{\Sigma}^b \cdot G + \overline{S} & \tau = 0 \\ \overline{H} \cdot \overline{A}_o^{\tau-1} \cdot \overline{G} = G^{\mathrm{T}} \cdot \left(A_o^{\tau-1}\right)^{\mathrm{T}} \cdot H^{\mathrm{T}} & \tau > 0. \end{cases}$$

The performance fluctuations in the backward innovation model can therefore be expressed as

$$\overline{\eta}_t = \mathcal{N}\left(\xi; 0_q, \overline{S}\right)$$

with covariance

$$\overline{S} = C_{YY}(0) - G^{\mathrm{T}} \cdot \overline{\Sigma}^b \cdot G. \tag{182}$$

After comprehensively analyzing different forward and backward representations of an LDS, we will direct our attention toward a robust technique for estimating the independent parameters from data. An iterative maximum likelihood estimation technique for minimizing the deviation of the data from the predictions of the forward model will be presented in the next chapter. It is not difficult to transform the parameter estimates into the alternative forms by using the above definitions.

2.10 Maximum Likelihood Parameter Estimation with Hidden State Variables

Minimizing the deviation of observations from the model's predictions is equivalent to maximizing the likelihood of the sequence of observations conditioned on the model structure and independent parameters. Unfortunately, to date the only known methods for carrying out a maximum likelihood estimation of the parameters of an LDS have been iterative, and therefore computationally demanding. The objective function to be maximized is the logarithmic probability ln $\mathcal{L}$ (.) of the fixed sequence of observations $\{y_t\}_0^T = (y_0, y_1, \ldots, y_T)$ given an underlying LDS model with parameter vector $\theta = [A_0 \quad H \quad C \quad V \quad \pi_0 \quad \Pi_0]$:

$$\ln \mathcal{L}\left(\theta | \{y_t\}_0^T\right) = \ln f_\theta[y_0, \ldots, y_T] \to \max. \tag{183}$$

The objective function is also known as the log-likelihood function. The log-likelihood function can be obtained from the joint *pdf* $f_\theta[x_0, \ldots, x_T, y_0, \ldots, y_T]$ of the LDS (Eq. 139) through marginalization. The marginal *pdf* $f_\theta[y_0 \ldots y_T]$ can be calculated by integrating out the hidden states of the process (Eq. 140) and can be simplified by factorization:

$$\begin{aligned} \ln \mathcal{L}\left(\theta | \{y_t\}_0^T\right) &= \ln f_\theta[y_0, \ldots, y_T] \\ &= \ln \int_{\mathbb{X}^q} \cdots \int_{\mathbb{X}^q} f_\theta[x_0, \ldots, x_T, y_0, \ldots y_T] \; dx_0 \ldots dx_T \\ &= \ln \int_{\mathbb{X}^q} \cdots \int_{\mathbb{X}^q} f_\theta[x_0] \prod_{t=0}^{T-1} f_\theta[x_{t+1}|x_t] \prod_{t=0}^{T} f_\theta[y_t|x_t] \; dx_0 \ldots dx_T. \end{aligned} \tag{184}$$

While the gradient and Hessian of the log-likelihood function are usually difficult to compute, it is relatively easy to calculate the logarithmic joint probability and its expected value for a particular setting of the independent parameters. The observability, i.e. whether the independent parameters can be identified uniquely for a particular sequence of observations, is discussed in more depth later on. In the following we will use a specific algorithm developed by Ghahramani and Hinton (1996) to indirectly optimize the log-likelihood of the observations by iteratively maximizing expectations. The general principle behind the Ghahramani-Hinton algorithm is the expectation-maximization (EM) principle. The principle behind the EM algorithm was first proposed by Hartley (1958). Hartley's iterative maximum likelihood procedure with latent variables was first termed "EM algorithm" in the classic paper of Dempster et al. (1977) in which they generalized Hartely's approach and proved local convergence. The searching direction of the algorithm has a positive projection on the gradient of the log-likelihood. Each iteration alternates between two steps, the estimation (E) and the maximization (M). The maximization step maximizes an expected log-likelihood function for a given estimate of the parameters that is recalculated in each iteration by the expectation step. As Roweis and Gharahmani (1999) show, we can use any parameterized probability distribution $g_{\breve{\theta}}[x_0, \ldots, x_T]$—not necessarily multivariate normal—over the hidden state variables to obtain a lower bound on the log-likelihood function $\ln \mathcal{L}(\theta|\{y_t\}_0^T)$:

$$
\begin{aligned}
\ln & \int_{\mathbb{X}^q} \cdots \int_{\mathbb{X}^q} f_\theta[x_0, \ldots, x_T, y_0, \ldots, y_T] dx_0 \ldots dx_T \\
&= \ln \int_{\mathbb{X}^q} \cdots \int_{\mathbb{X}^q} g_{\breve{\theta}}[x_0, \ldots, x_T] \frac{f_\theta[x_0, \ldots, x_T, y_0, \ldots, y_T]}{g_{\breve{\theta}}[x_0, \ldots, x_T]} dx_0 \ldots dx_T \\
&\geq \int_{\mathbb{X}^q} \cdots \int_{\mathbb{X}^q} g_{\breve{\theta}}[x_0, \ldots, x_T] \ln \frac{f_\theta[x_0, \ldots, x_T, y_0, \ldots, y_T]}{g_{\breve{\theta}}[x_0, \ldots, x_T]} dx_0 \ldots dx_T \\
&= \int_{\mathbb{X}^q} \cdots \int_{\mathbb{X}^q} g_{\breve{\theta}}[x_0, \ldots, x_T] \ln f_\theta[x_0, \ldots, x_T, y_0, \ldots, y_T] dx_0 \ldots dx_T \\
&\quad - \int_{\mathbb{X}^q} \cdots \int_{\mathbb{X}^q} g_{\breve{\theta}}[x_0, \ldots, x_T] \ln g_{\breve{\theta}}[x_0, \ldots, x_T] dx_0 \ldots dx_T \\
&= \mathcal{F}\left(g_{\breve{\theta}}, \theta\right).
\end{aligned}
$$

The inequality in the middle of the above expression is known as Jensen's inequality. For greater clarity, the parameter vector $\breve{\theta}$ of the auxiliary distribution $g_{\breve{\theta}}$ $[x_0, \ldots, x_T]$ is not explicitly declared in the following. In the EM literature, the auxiliary distribution $g_{\breve{\theta}}$ is referred to as the auxiliary function or simply the $\mathcal{Q}$-function (Dellaert 2002).

If the "energy" of the complete configuration $(X_0, X_1, \ldots, X_T, Y_0, Y_1, \ldots, Y_T)$ is defined as

$$
-\ln f_\theta[x_0, \ldots, x_T, y_0, \ldots, y_T],
$$

then the lower bound $\mathcal{F}(g, \theta) \leq \ln \mathcal{L}(\theta|\{y_t\}_0^T)$ is the negative of a quantity that is known in statistical physics as the "free energy" (Honerkamp 2002). The free energy is the expected energy under the distribution $g[x_0, \ldots, x_T]$ minus the differential entropy of that distribution (Neal and Hinton 1998; Roweis and Ghahramani 1999). The definition of the differential entropy is given in Eq. 232 and will be used in later chapters to evaluate emergent complexity of the modeled phase of the PD project.

The EM algorithm alternates between maximizing the function $\mathcal{F}(g, \theta)$ by the auxiliary distribution g and by the parameter vector θ, while holding the other fixed. The iteration number is denoted by k. Starting from an initial parameter setting θ_0 it holds that

$$\text{E}-\text{step}: \quad g_{k+1} = \arg\max_g \mathcal{F}(g, \theta_k) \tag{185}$$

$$\text{M}-\text{step}: \quad \theta_{k+1} = \arg\max_\theta \mathcal{F}\left(g_{k+1}, \theta\right). \tag{186}$$

It can be shown that the maximum of the E-step is obtained when g is exactly the conditional *pdf* f_θ of $(X_0, X_1, \ldots, X_T)$ given $(Y_0, Y_1, \ldots, Y_T)$ and it holds that:

$$g_{k+1}[x_0, \ldots, x_T] = f_{\theta_k}[x_0, \ldots, x_T | y_0, \ldots, y_T].$$

Hence, the maximum in the M-step results when the term

$$\int_{\mathbb{X}^q} \cdots \int_{\mathbb{X}^q} g[x_0, \ldots, x_T] \ln f_\theta[x_0, \ldots, x_T, y_0, \ldots, y_T] \; dx_0 \ldots dx_T$$

in the function $\mathcal{F}(g, \theta)$ is maximized, since the differential entropy does not depend on the parameters θ. Therefore, we can also express the EM algorithm in a single maximization step:

$$\text{M}-\text{step}: \quad \theta_{k+1}$$

$$= \arg\max_\theta \int_{\mathbb{X}^q} \cdots \int_{\mathbb{X}^q} f_{\theta_k}[x_0, \ldots, x_T | y_0, \ldots, y_T] \ln f_\theta[x_0, \ldots, x_T, y_0, \ldots, y_T] \; dx_0 \ldots dx_T. \tag{187}$$

In this sense the EM principle can be interpreted as coordinate ascent in $\mathcal{F}(g, \theta)$. At the beginning of each M-step it holds that $\mathcal{F}\left(g_{k+1}, \theta_k\right) = \ln \mathcal{L}\left(\theta_k \middle| \{y_t\}_0^T\right)$. Since the E-step does not change θ, the likelihood is guaranteed not to decrease after each combined EM-step (Neal and Hinton 1998; Roweis and Ghahramani 1998). Therefore, in the EM algorithm the solutions to the filtering and smoothing problem that are incorporated in the conditional distribution $f_{\theta_k}[x_0, \ldots, x_T | y_0, \ldots, y_T]$ are applied to estimate the hidden states given the observations and the re-estimated model parameters. These virtually complete data points are used to solve for new model parameters. In Fig. 2.21 a graphical illustration of the first three iteration steps of the EM procedure are shown. Each lower bound $\mathcal{F}(g, \theta)$ touches the objective function $\ln \mathcal{L}(\theta | \{y_t\}_0^T)$ at the current estimate θ_k and is in this sense optimal. However, it is clear from the figure that the approached maximum of the objective function is only locally optimal and therefore a proper initial setting θ_0 of the independent parameters is crucial for the effectiveness of the EM algorithm. We will return later in this chapter to this issue.

According to Eq. 184 the joint probability $f_\theta[x_0, \ldots, x_T, y_0, \ldots, y_T]$ of hidden states and observations is multivariate normal and therefore the conditional distribution $f_{\theta_k}[x_0, \ldots, x_T | y_0, \ldots, y_T]$ also follows a multivariate normal distribution. There are reasonably efficient algorithms—the introduced Kalman filter and the corresponding smoother algorithm (see e.g. Shumway and Stoffer 1982)—for inferring this distribution for a given setting of the parameters. Using the estimated

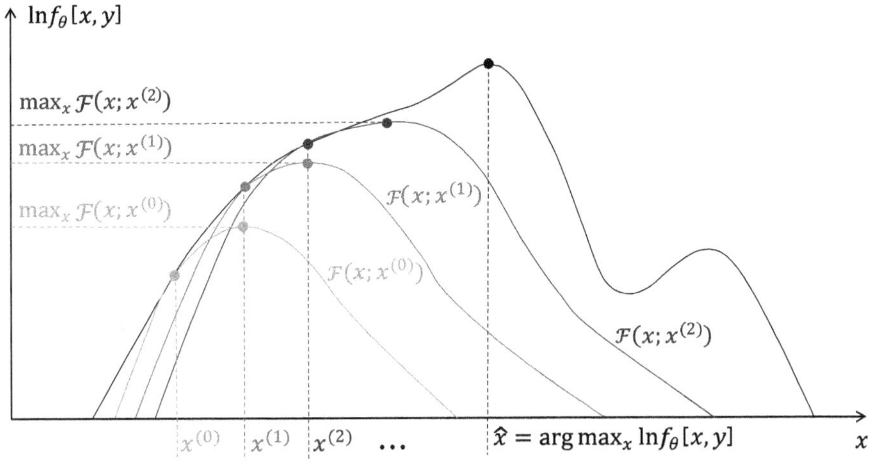

Fig. 2.21 Graphical illustration of the Expectation-Maximization (EM) algorithm (Dempster et al. 1977). The illustration was adapted from Streit (2006) and Wienecke (2013). The EM algorithm iterates between computing a lower bound $\mathcal{F}(x; x^{(i)})$ and maximization. The figure shows the first three iterations for parameter settings $x^{(0)}$, $x^{(1)}$ and $x^{(2)}$ in blue, red and green. In the illustrated case the EM algorithm converges to the global maximum of the objective function shown in the middle and not to the local maximum shown on the right. For greater clarity, the sequences of hidden state variables $\{x_t\}_0^T$ and observations $\{y_t\}_0^T$ are denoted by the single vectors x and y

states obtained from the filtering and smoothing algorithms, it is usually easy to calculate solutions for new parameter values. For LDS, solving for new parameter values typically involves minimizing quadratic expressions such as Eq. 188. This process is repeated using the new model parameters to infer the hidden states again, until the log-likelihood grows only very slowly.

The Ghahramani-Hinton algorithm makes the EM principle for LDS fully operational and can accurately estimate the complete set θ of independent parameters. The algorithm is guaranteed to converge to a local maximum of the log-likelihood function (Dempster et al. 1977, cf. Fig. 2.21). For some applications in project management, the local optima can be a significant problem because sometimes many symmetries exist in the LDS parameter space. The convergence rate is typically linear in T and polynomial in q and it therefore can also take many iterations to reach a satisfactory predictive accuracy (for simplicity, we assume that $q \geq p$, Martens 2009). This can be a real challenge for long series of observations and a large state space because each iteration involves recursive estimations for all hidden states over T time steps. An alternative approach is the combined deterministic-stochastic subspace identification algorithm of van Overschee and de Moor (1996). The algorithm is called a subspace algorithm because it retrieves the system related matrices as subspaces of projected data matrices. The combined deterministic-stochastic subspace identification algorithm reduces the estimation problem to that of solving a large singular value decomposition problem (cf. canonical correlation analysis in Section 4.1.3), subject to some heuristics in the form of user-defined weighting matrices. The user-defined weighting matrices define which part of the predictive information in the data is important to retain. The algorithm allows for computing the non-steady-state Kalman filter sequence (cf. Section 2.9) directly from the observation sequences, without knowledge of the system matrices. Because subspace identification is not iterative, it tends to be more efficient than EM when the state space is high dimensional. However, because it implicitly optimizes an objective function representing a sum of squared prediction errors instead of the log-likelihood function, and because implicitly uses a rather simple inference procedure for the hidden states, subspace identification is not statistically optimal. In particular, point estimates of hidden states lacking covariance information are obtained by conditioning on only the past l data points, where l is an independent parameter known as the "prediction horizon" (Martens 2009). In contrast, the EM algorithm estimates hidden states with mean and covariance information, conditioned on the entire observation sequence $\{y_t\}_0^T$ and the current estimate θ_k of the independent parameters in the kth iteration. Subspace-identification also tends to scale poorly to high-dimensional time series. When initialized properly the EM algorithm will find parameter settings with larger values of the log-likelihood function and lower prediction errors than those found by the deterministic-stochastic subspace identification algorithm. However, because of the speed of subspace identification, its ability to estimate the dimensionality of the state space, and the fact that it optimizes an objective that is similar to that of EM, subspace identification is often considered an excellent method for initializing iterative learning algorithms such as EM (Martens 2009). Smith and Robinson (2000) and Smith et al. (1999) have carried out comprehensive theoretical and empirical analyses of subspace identification and EM. Another standard approach for estimating LDS parameters are prediction error methods (Ljung 1999). In this approach, a one-step prediction-error objective is minimized. This is usually done via gradient-based optimization methods. In this sense these methods have a close kinship with the maximum likelihood method. However, prediction error methods are often based on a stationary predictor and for finite-length sequences of observation vectors therefore are not equivalent to the maximum likelihood principle, even if Gaussian noise is incorporated in the system equations. Prediction error methods are also embodied in a software package and therefore have become the dominant algorithms for system identification (Ljung 2000). But these methods do have some shortcomings. Typical implementations use either gradient descent and therefore require many iterations to converge, or use second-order optimization methods but then become impractical for large models (Martens 2010). Under certain circumstances the search for the parameters can be very laborious, involving search surfaces that may have many local minima. This parameter search is typically carried out using the damped Gauss-Newton method and therefore having good initial parameter values is of crucial importance for estimation accuracy. Because of the cited drawbacks of alternative methods and the theoretically very satisfactory formulation and solution of the objective function, we will direct our attention to the EM procedure in the following presentation. An additional approach for estimating parameters of linear Gaussian models of stochastic dynamical systems was introduced recently by Rudary (2009). He introduced a new version of the predictive linear-Gaussian model (Rudary et al. 2005, 2006) that models discrete-time dynamical systems with real-vector-valued states and observations and not just continuous scalar observations. He shows that the new predictive linear-Gaussian model subsumes LDSs of equal dimensions and introduces an efficient algorithm to obtain consistent estimates of the parameters from data. The only significant limitation on the systems that predictive linear-Gaussian models can be applied to is that the observation vectors cannot have a larger dimension than the model itself and therefore the observation vector space cannot be underranked in expectation (Rudary 2009).

According to Eq. 187 the EM procedure requires the evaluation of the expected log-likelihood function of observations and hidden states for the particular setting θ_k of the independent parameters in the kth iteration:

$$\wp_k(\theta) = E_{\theta_k}\left[\ln f_\theta[x_0, \ldots, x_T, y_0, \ldots, y_T] \middle| \{y_t\}_0^T, \theta_k\right].$$

$\{y_t\}_0^T$ denotes the fixed sequence of observations made in the PD project across an interval of $T+1$ time steps that used to compute the expectations. Similarly, $\{x_t\}_0^t$ in the following equations denotes the sequence of state vectors that starts at time instant 0 and ends at instant $t \le T$. We attached the index θ_k to the expectation operator to indicate that the expectation has to be calculated with respect to the *pdf* corresponding to the actual parameter estimate $f_{\theta_k}[x_0, \ldots, x_T|y_0, \ldots, y_T]$.

Considering the definition of the joint *pdf* of the dynamic system from Eqs. 152 and 153, the log-likelihood function over all hidden states and observations can be expressed as a sum of three simple quadratic terms:

$$\begin{aligned}
&\ln f_\theta[x_0,\ldots,x_T,y_0,\ldots,y_T] \\
&\quad = -(T+1)\frac{(p+q)}{2}\ln 2\pi - \frac{1}{2}\ln \mathrm{Det}[\Pi_0] - \frac{1}{2}T\ln \mathrm{Det}[C] - \frac{1}{2}(T+1)\ln \mathrm{Det}[V] \\
&\quad\quad - \frac{1}{2}(x_0-\pi_0)^{\mathrm{T}}\Pi_0^{-1}(x_0-\pi_0) - \frac{1}{2}\sum_{t=1}^{T}(x_t - A_0x_{t-1})^{\mathrm{T}}C^{-1}(x_t - A_0x_{t-1}) \\
&\quad\quad - \frac{1}{2}\sum_{t=0}^{T}(y_t - Hx_t)^{\mathrm{T}}V^{-1}(y_t - Hx_t).
\end{aligned} \tag{188}$$

It can be shown that the expected log-likelihood function depends on three expected values related to the hidden state variables given the observations, namely $E_{\theta_k}\left[X_t|Y_0^T\right]$, $E_{\theta_k}\left[X_tX_t^{\mathrm{T}}|Y_0^T\right]$ and $E_{\theta_k}\left[X_tX_{t-1}^{\mathrm{T}}|Y_0^T\right]$. Following the notation of the creators of the algorithm, we will use the variables

$$\hat{x}_t := E_{\theta_k}\left[X_t \middle| Y_0^T\right]$$
$$\hat{P}_t := E_{\theta_k}\left[X_t X_t^{\mathrm{T}} \middle| Y_0^T\right]$$
$$\hat{P}_{t,t-1} := E_{\theta_k}\left[X_t X_{t-1}^{\mathrm{T}} \middle| Y_0^T\right]$$

to encode these expectations. The variable $\hat{x}_t$ denotes the (re-)estimated state of the process $\{X_t\}$ at time instant t given the complete series of observations Y_0^T. Interestingly, the state estimate $\hat{x}_t$ is not only based on past observations $(Y_0, \ldots, Y_t)$ but also on the future history $(Y_t, \ldots, Y_T)$. This is in contrast to the classic Kalman filter in which only the estimates $E_{\theta_k}\left[X_t \middle| Y_0^t\right]$ are considered (Puri 2010). The above first- and second-order statistics over the hidden states allow one to easily evaluate and optimize $\mathcal{L}(\theta)$ with respect to θ in the M-step. These quantities can be considered as the "full smoother" estimates of the Kalman smoother (Roweis and Ghahramani 1999, see below).

As explained in the M-step each of the independent parameters is re-estimated (see intuitive graphical representation in Fig. 2.21). The new parameter estimate is obtained by maximizing the expected log-likelihood. To find a closed form of the maximum of the expected log-likelihood function the partial derivatives are calculated for each parameter from Eq. 188. The following equations summarize the results (Ghahramani and Hinton 1996). The re-estimated quantities are indicated by the prime symbol.

Initial state location π_0:

$$\frac{\partial \wp_k}{\partial \pi_0} = (\hat{x}_0 - \pi_0)\Pi_0^{-1} = 0$$
$$\Rightarrow \pi_0' = \hat{x}_0$$

Initial state covariance Π_0:

$$\frac{\partial \wp_k}{\partial \Pi_0^{-1}} = \frac{1}{2}\Pi_0 - \frac{1}{2}\left(\hat{P}_0 - \hat{x}_0 \pi_0^{\mathrm{T}} - \pi_0 \hat{x}_0^{\mathrm{T}} + \pi_0 \pi_0^{\mathrm{T}}\right)$$
$$\Pi_0' = \hat{P}_0 - \hat{x}_0 \hat{x}_0^{\mathrm{T}}.$$

Dynamical operator A_0:

$$\frac{\partial \wp_k}{\partial A_0} = -\sum_{t=1}^{T} C^{-1} \hat{P}_{t,t-1} + \sum_{t=1}^{T} C^{-1} A_0 \hat{P}_{t-1} = 0$$
$$\Rightarrow A_0' = \left(\sum_{t=1}^{T} \hat{P}_{t,t-1}\right)\left(\sum_{t=1}^{T} \hat{P}_{t-1}\right)^{-1}$$

State fluctuations covariance C:

$$
\begin{aligned}
\frac{\partial \wp_k}{\partial C^{-1}} &= \frac{T}{2}C - \frac{1}{2}\sum_{t=1}^{T}\left(\hat{P}_t - A_0\hat{P}_{t-1,t} - \hat{P}_{t,t-1}A_0^{\mathrm{T}} + A_0\hat{P}_{t-1}A_0^{\mathrm{T}}\right) \\
&= \frac{T}{2}C - \frac{1}{2}\left(\sum_{t=1}^{T}\hat{P}_t - A_0'\sum_{t=1}^{T}\hat{P}_{t-1,t}\right) = 0 \\
\Rightarrow C' &= \frac{1}{T}\left(\sum_{t=1}^{T}\hat{P}_t - \sum_{t=1}^{T}\hat{P}_{t,t-1}\left(\sum_{t=1}^{T}\hat{P}_t\right)^{-1}\sum_{t=1}^{T}\hat{P}_{t-1,t}\right)
\end{aligned}
$$

Output operator H:

$$
\begin{aligned}
\frac{\partial \wp_k}{\partial H} &= -\sum_{t=0}^{T}V^{-1}y_t\,\hat{x}_t^{\mathrm{T}} + \sum_{t=0}^{T}V^{-1}H\hat{P}_t = 0 \\
\Rightarrow H' &= \left(\sum_{t=0}^{T}y_t\,\hat{x}_t^{\mathrm{T}}\right)\left(\sum_{t=0}^{T}P_t\right)^{-1}
\end{aligned}
$$

Output fluctuations covariance V:

$$
\begin{aligned}
\frac{\partial \wp_k}{\partial V^{-1}} &= \frac{T+1}{2}V - \sum_{t=0}^{T}\left(\frac{1}{2}y_ty_t^{\mathrm{T}} - H\hat{x}_ty_t^{\mathrm{T}} + \frac{1}{2}H\hat{P}_tH^{\mathrm{T}}\right) = 0 \\
\Rightarrow V' &= \frac{1}{T+1}\sum_{t=0}^{T}\left(y_ty_t^{\mathrm{T}} - H'\hat{x}_ty_t^{\mathrm{T}}\right)
\end{aligned}
$$

Note that the covariances possess the correct symmetry property as each of the corresponding right-hand sides of the resulting formulas are already symmetric. This can be expressed simply as $\hat{P}_{t,t-1}^{\mathrm{T}} = \hat{P}_{t-1,t}$. To simplify the detailed description of the E-step, we will use three intermediate variables:

$$
\begin{aligned}
x_t^{\tau} &:= E_{\theta_k}\left[X_t \middle| Y_0^{\tau}\right] \\
\Sigma_t^{\tau} &:= \mathrm{Var}_{\theta_k}\left[X_t \middle| Y_0^{\tau}\right] \\
\Sigma_{t,t-1}^{\tau} &:= E_{\theta_k}\left[X_tX_{t-1}^{\mathrm{T}} \middle| Y_1^{\tau}\right] - x_t^{\tau}\left(x_t^{\tau}\right)^{\mathrm{T}}.
\end{aligned}
$$

The E-step consists of two sub-steps. The first sub-step is a forward recursion that uses the sequence of observations from y_0 to y_t for state estimation. This forward recursion is the well-known Kalman filter which was introduced in the previous chapter. The second sub-step carries out a backward recursion that uses the observations from y_T to y_{t+1} (Rauch 1963). The combined forward and backward recursions are known as the Kalman smoother (Shumway and Stoffer 1982). The following Kalman-filter forward recursions hold for $t = 0$ to T:

$$
\begin{aligned}
x_t^{t-1} &= A_0 x_{t-1}^{t-1} \\
\Sigma_t^{t-1} &= A_0 \Sigma_{t-1}^{t-1} A_0^{\mathrm{T}} + C \\
K_t &= \Sigma_t^{t-1} H^{\mathrm{T}} \left(H \Sigma_t^{t-1} H^{\mathrm{T}} + V \right)^{-1} \\
x_t^t &= x_t^{t-1} + K_t \left(y_t - H x_t^{t-1} \right) \\
\Sigma_t^t &= \Sigma_t^{t-1} - K_t H \Sigma_t^{t-1}.
\end{aligned}
$$

It holds that $x_0^{-1} = \pi_0$ and $\Sigma_0^{-1} = \Pi_0$. K_t is the time-dependent Kalman gain matrix (cf. Section 2.9).

To compute $\hat{x}_t = x_t^T$ and $\hat{P}_t = \Sigma_t^T + x_t^T \left(x_t^T \right)^{\mathrm{T}}$ the following backward recursions have to be carried out from $t = T$ to $t = 1$ (Shumway and Stoffer 1982):

$$
\begin{aligned}
J_{t-1} &= \Sigma_{t-1}^{t-1} A_0^{\mathrm{T}} \left(\Sigma_t^{t-1} \right)^{-1} \\
\hat{x}_{t-1} &= x_{t-1}^{t-1} + J_{t-1} \left(\hat{x}_t - A_0 x_{t-1}^{t-1} \right) \\
\Sigma_{t-1}^T &= \Sigma_{t-1}^{t-1} + J_{t-1} \left(\Sigma_t^T - \Sigma_t^{t-1} \right) J_{t-1}^{\mathrm{T}}.
\end{aligned}
$$

Moreover, if $t < T$ the conditional covariance $\hat{P}_{t,t-1} = \Sigma_{t,t-1}^T + x_t^T \left(x_{t-1}^T \right)^{\mathrm{T}}$ of the hidden states across two time steps can be obtained through the backward recursion:

$$
\Sigma_{t,t-1}^T = \Sigma_t^t J_{t-1}^{\mathrm{T}} + J_t \left(\Sigma_{t+1,t}^T - A_0 \Sigma_t^t \right) J_{t-1}^{\mathrm{T}}.
$$

The recursion is initialized with $\Sigma_{T,T-1}^T = \left(I_q - K_T H \right) A_0 \Sigma_{T-1}^{T-1}$.

If not only a single sequence of empirically acquired observation vectors y_t of work remaining is given but multiple realizations of the work processes had also been acquired in N independent measurement trials, then the above equations can be easily generalized. The basic procedure involves calculating the expected values in the E-step for each sequence separately and summing up the individual quantities to accumulated expectations. The only difficulty here is estimating the initial state covariance. According to Ghahramani and Hinton (1996) we can define $\hat{x}_t^{[i]}$ as the state estimate of sequence $[i]$ at time instant t and $\hat{x}_{N,t}$ as the mean estimate at the same time instant:

$$
\hat{x}_{N,t} = \frac{1}{N} \sum_{i=1}^{N} \hat{x}_t^{[i]}.
$$

Based on these estimates, we can then calculate the initial covariance as follows:

$$\Pi_0' = \hat{P}_0 - \hat{x}_{N,t}\hat{x}_{N,t}^{\mathrm{T}} + \frac{1}{N}\sum_{i=1}^{N}\left(\hat{x}_t^{[i]} - \hat{x}_{N,t}\right)\left(\hat{x}_t^{[i]} - \hat{x}_{N,t}\right)^{\mathrm{T}}.$$

In the M-step, the accumulated expectations are used to re-estimate the independent parameters.

An interesting way of significantly simplifying the EM algorithm is to replace the time-dependent matrices in the E- and M-steps with their steady-state values. There are efficient methods for finding these matrices (e.g. the doubling algorithm, Anderson and Moore 1979). Martens (2009) improved the efficiency of the EM algorithm of Ghahramani and Hinton (1996) by using a steady-state approximation, which simplifies inference of the hidden state and by using the fact that the M-step requires only a small set of expected second-order statistics that can be approximated without doing complete inference for each x_t as shown above. Martens' experiments show that the resulting approximate EM algorithm performs nearly as well as the EM algorithm given the same number of iterations (Martens 2009). Since the calculations required per iteration of the approximate EM do not depend on T, it can be much more computationally efficient when T is large.

In application areas with no or only shallow prior knowledge about the dynamic dependency structure, the (true or approximate) EM iterations are usually initialized with a setting θ_0 of the independent parameters in which Gaussian random numbers are assigned to the initial state, the dynamical operator and the output operator. In a similar manner the diagonal and off-diagonal entries of the corresponding covariance matrices are set to Gaussian random numbers at the start of the iterations. Clearly, randomizing the initial covariance matrices has to be done under the constraint of matrix symmetry. Let κ denote the number of iterations that were calculated using the EM algorithm in a specific modeling and simulation environment. The re-estimated setting of the independent parameters in the (last) κth iteration is denoted by $\hat{\theta}_\kappa$. We assume that the independent parameters were re-estimated sufficiently often. Sufficiently often means that the log-likelihood grew only very slowly in the final iterations and a stable local optimum was found. In many practical cases 20 to 30 iterations are sufficient to reach a stable optimum.

It is important to note that the LDS system identification based on the maximum likelihood principle lacks identifiability: Owing to the fact that the likelihood function is invariant under an arbitrary invertible transform $\Psi \in \mathbb{R}^{q \times q}$ with $\mathrm{Det}[\Psi] = 1$, which transforms the set of parameters as

$$\begin{aligned}
x'_t &= \Psi \cdot x_t \\
A'_0 &= \Psi \cdot A_0 \cdot \Psi^{-1} \\
H' &= H \cdot \Psi^{-1} \\
C' &= \Psi \cdot C \cdot \Psi^{\mathrm{T}} \\
V' &= V \\
\pi'_0 &= \Psi \cdot \pi_0 \\
\Pi'_0 &= \Psi \cdot \Pi_0 \cdot \Psi^{\mathrm{T}}
\end{aligned}$$

the parameters identified with any ML procedure are not uniquely determined. This statement can easily be proved simply by computing the likelihood of the new parameter set $\mathcal{L}(\{x'_t\}_0^T, A'_0, H', C', V', \pi'_0, \Pi'_0 | \{y_t\}_0^T)$ using the above expressions. The result then equals $\mathcal{L}(\{x_t\}_0^T, A_0, H, C, V, \pi_0, \Pi_0 | \{y_t\}_0^T)$.

Therefore, the estimated initial state, initial covariance, dynamical operator, state covariance, output operator and output covariance are all dependent on the particular initial setting θ_0 of the independent parameters that is used in the first iteration and there are many different groups of system matrices (A_0 and H) which can generate a given set of observations. To ensure that the method does not lack identifiability, constraints are imposed on the parameters to make the transform Ψ unique. There are several different options for constraining the parameters, e.g., the one option given in Yamaguchi et al. (2007) in cases where the dimension of the observed states is not smaller than the dimension of the hidden states. Another constraint, which also holds in the case $p < q$, corresponds to the invariants of the dynamical operator A_0 under the transform Ψ, namely the eigenvalues $\lambda_i = \lambda_i(A_0)$ of the dynamical operator (see eigendecomposition in Section 2.3). Furthermore, it is assumed that the initial state x_0 is known (cf. Eq. 2). Hence, we have the constraints:

- The dynamical operator is a diagonal matrix with unknown distinct diagonal elements:

$$A_0 = \Lambda_A = \mathrm{Diag}\left[\lambda_1, \ldots, \lambda_q\right], \lambda_i \neq \lambda_j.$$

- The initial state is known, e.g.

$$x_0 = \begin{pmatrix} 1 \\ \vdots \\ 1 \end{pmatrix}.$$

To prove that imposing the above constraints is sufficient to make the transform Ψ unique, we start from the invariance condition for A_0. Given that $A_0 = \Lambda_A = A'_0 = \Lambda'_A$, we must have $\Lambda_A = \Psi \cdot \Lambda_A \cdot \Psi^{-1}$, or equivalently $\Lambda_A \cdot \Psi = \Psi \cdot \Lambda_A$ as Ψ is invertible. Considering this equation in elements, it yields $\{\lambda_i\} \cdot \Psi_{i,j} = \{\lambda_j\} \cdot \Psi_{i,j}$, which can only be solved with $\Psi_{i,j} = 0$ for $i \neq j$ as the diagonal elements $\lambda_i \neq \lambda_j$ are distinct. To fix the diagonal elements of Ψ we consider the invariance condition for x_0, $x'_0 = x_0 = \Psi \cdot x_0$, which leads to a unique solution $\Psi = I_q$.

In general, the eigenvalues of the dynamical operator may be complex-valued. This would, in turn, lead to complex-valued unobserved state vectors. Note, that, in principle, the transform Ψ can be complex-valued. To circumvent unobserved complex valued dynamics, it is preferable to use a constraint which diagonalizes the covariance C of the unobserved process fluctuations. As C is real-valued and symmetric the eigenvalues $\lambda_i = \lambda_j(C)$ are real and an arbitrary covariance can be transformed by an orthogonal matrix into a diagonal matrix. Similar to the proof given above, it is easy to show that the following constraints lead to a unique solution for the transform Ψ:

- The covariance C is a diagonal matrix with unknown distinct diagonal elements:

$$C = \Lambda_C = \mathrm{Diag}\left[\lambda_1, \ldots, \lambda_q\right], \lambda_i \neq \lambda_j.$$

- The signs of the real-valued components of the initial state are known.

However, where emergent complexity of PD projects with hidden state variables has to be evaluated on the basis of the effective measure complexity (EMC, Grassberger 1986) the lack of identifiability is not a critical issue, because the measure is invariant under the above transform of model parameters. This is proved in Section 4.2.1. The EMC is an information-theoretic quantity that measures the mutual information between the infinite past and future histories of a stationary stochastic process. Section 3.2.4 formally introduces the EMC and gives a detailed explanation of its interrelationships to other key invariants of stochastic processes. In the sense of an information-theoretic learning curve, EMC measures the amount of apparent randomness at small observation windows during the stochastic task processing that can be resolved by considering correlations between blocks with increasing length. For a completely randomized work process with independent and identically distributed state variables, the apparent randomness cannot be reduced by any means, and it therefore holds that EMC $= 0$. For all other processes that have a persistent internal organization, EMC is strictly positive. PD projects with more states and larger correlation length are assigned higher complexity values. If optimal predictions are influenced by events in the arbitrarily distant past, EMC can also diverge. However, this is not relevant for the complexity analysis of the investigated linear systems.

If not only the coefficients and covariance matrices of an LDS have to be estimated from data but also the dimensionality of the hidden state process $\{X_t\}$, a good trade-off between the predictive accuracy gained by increasing the dimension of independent parameters and the danger of overfitting the model to random fluctuations and not to rules that generalize to other datasets has to be found. In an analogous manner to the model selection procedure that was introduced in Section 2.4, information-theoretic or Bayesian criteria can be used to evaluate approximating LDS models. An alternative method is to use the previously cited combined deterministic-stochastic subspace identification algorithm of van Overschee and de Moor (1996) to estimate the dimensionality of the state space and to initialize the EM algorithm accordingly. However, in the following we focus on the Schwarz-Bayes information criterion (BIC, cf. Eq. 71) because of its close theoretical connection to the minimum description length principle. This principle aims to select the model with the briefest recording of all relevant attribute information and builds on the intuitive notion that model fitting is equivalent to finding an efficient encoding of the data. However, in searching for an efficient code, it is important to not only consider the number of bits required to describe the deviations of the data from the model's predictions, but also the number of bits required to specify the independent parameters of the model (Bialek et al. 2001). The minimum description length principle will be elaborated in Section 3.2.2.

According to the work of Yamaguchi et al. (2007), the BIC can be defined for an LDS with dimension q of the hidden states as:

$$\begin{aligned} BIC(q) &= -\frac{2}{T}\ln f_{\hat{\theta}_\kappa}[y_0, \ldots, y_T] + \frac{\ln T}{T}k \\ &= -\frac{2}{T}\ln \prod_{t=0}^{T} f_{\hat{\theta}_\kappa}[y_t | y_{t-1}, \ldots, y_0] + \frac{\ln T}{T}k \\ &= -\frac{2}{T}\sum_{t=0}^{T} \ln f_{\hat{\theta}_\kappa}[y_t | y_{t-1}, \ldots, y_0] + \frac{\ln T}{T}k. \end{aligned} \tag{189}$$

It holds that $f_{\hat{\theta}_\kappa}[y_0|y_{-1}] = f_{\hat{\theta}_\kappa}[y_0]$. The term $\ln f_{\hat{\theta}_\kappa}[y_0, \ldots, y_T]$ denotes the best estimate of the local maximum of the log-likelihood function. For a converging estimation process, the best estimate is obtained in the κ-th iteration of the EM algorithm, and the particular setting of the parameters is $\hat{\theta}_\kappa = \left[\hat{A}_0^{(\kappa)} \quad \hat{H}^{(\kappa)} \quad \hat{C}^{(\kappa)} \quad \hat{V}^{(\kappa)} \quad \hat{\pi}_0^{(\kappa)} \quad \hat{\Pi}_0^{(\kappa)}\right]$. According to the analysis in Section 2.4, the corresponding quantity

$$-\frac{2}{T}\sum_{t=0}^{T} \ln f_{\hat{\theta}_\kappa}[y_t | y_{t-1}, \ldots, y_0] = \ln \mathrm{Det}\left[\widehat{\Sigma}_{(q,p)}\right]$$

is used to evaluate the one-step prediction error $\widehat{\Sigma}_{(q,p)}$ (cf. Eq. 66) of the parameterized LDS and therefore indicates the goodness-of-fit of data and model. The second term $(\ln T/T)k$ penalizes model complexity. Equivalent to the definition of the BIC for a vector autoregression model of order n in Eq. 71, the factor

$$k = q + qp + \frac{q(q+1)}{2} + \frac{p(p+1)}{2} \tag{190}$$

denotes the effective number of parameters of the approximating model. To calculate the effective number of parameters, we have only considered the eigenvalues $\lambda_i(A_0)$ of the dynamical operator A_0 as they are invariant under the invertible transform Ψ and therefore well-determined parameters. This is in contrast to Wang et al. (2011), who count all q^2 entries of A_0 and ignore that system identification based on any ML procedure lacks identifiability. We prefer the reduced number of effective parameters, because it is not only an estimation theoretical necessity but can also be traced back to our information-theoretic approach to complexity evaluation (Chapter 3.2.4).

The log-likelihood $f_{\hat{\theta}_\kappa}[y_0, \ldots, y_T]$ can be written in a simple parametric form (Martens 2009). To derive this form, we define the conditional covariance of the observations at time step t as

$$\begin{aligned}\hat{S}_t &= \mathrm{Var}_{\hat{\theta}_\kappa}[Y_t|Y_{t-1}, \ldots, Y_0] \\ &= \hat{H}^{(\kappa)}\mathrm{Var}_{\hat{\theta}_\kappa}\left[X_t\middle|\{Y_t\}_0^{t-1}\right]\left(\hat{H}^{(\kappa)}\right)^{\mathrm{T}} + \hat{V}^{(\kappa)} \\ &= \hat{H}^{(\kappa)}\Sigma_t^{t-1}\left(\hat{H}^{(\kappa)}\right)^{\mathrm{T}} + \hat{V}^{(\kappa)} \\ &= \hat{H}^{(\kappa)}\hat{A}_0^{(\kappa)}\Sigma_{t-1}^{t-1}\left(\hat{A}_0^{(\kappa)\mathrm{T}}\right)\left(\hat{H}^{(\kappa)}\right)^{\mathrm{T}} + \hat{V}^{(\kappa)}.\end{aligned}$$

It holds that $\Sigma_0^{-1} = \hat{\Pi}_0^{(\kappa)}$ for $t = 0$.

It is evident that

$$f_{\hat{\theta}_\kappa}[y_t|y_{t-1}, \ldots, y_0] = \mathcal{N}\left(y_t; \overline{y}_t, \hat{S}_t\right),$$

where

$$\overline{y}_t = E_{\hat{\theta}_\kappa}\left[Y_t\middle|\{Y_t\}_0^{t-1}\right] = \hat{H}^{(\kappa)}x_t^{t-1}.$$

It holds that $x_0^{-1} = \hat{\pi}_0^{(\kappa)}$ for $t = 0$.

Hence, the log-likelihood of the sequence of observations can be expressed by the equation:

$$\begin{aligned}\ln f_{\hat{\theta}_\kappa}[y_0, \ldots, y_T] &= \sum_{t=0}^{T}\ln f_{\hat{\theta}_\kappa}[y_t|y_{t-1}, \ldots, y_0] \\ &= -\frac{1}{2}\sum_{t=0}^{T}\left(p\ln 2\pi + \ln\mathrm{Det}\left[\hat{S}_t\right] + \left(y_t - \hat{H}^{(\kappa)}x_t^{t-1}\right)^{\mathrm{T}}\left(\hat{S}_t\right)^{-1}\left(y_t - \hat{H}^{(\kappa)}x_t^{t-1}\right)\right) \\ &= -\frac{1}{2}\left((T+1)p\ln 2\pi + \ln\mathrm{Det}\left[\hat{\Pi}_0^{(\kappa)}\right] + \left(y_0 - \hat{H}^{(\kappa)}\hat{\pi}_0^{(\kappa)}\right)^{\mathrm{T}}\left(\hat{\Pi}_0^{(\kappa)}\right)^{-1}\left(y_0 - \hat{H}^{(\kappa)}\hat{\pi}_0^{(\kappa)}\right)\right) \\ &\quad - \frac{1}{2}\sum_{t=1}^{T}\left(\ln\mathrm{Det}\left[\hat{S}_t\right] + \left(y_t - \hat{H}^{(\kappa)}x_t^{t-1}\right)^{\mathrm{T}}\left(\hat{S}_t\right)^{-1}\left(y_t - \hat{H}^{(\kappa)}x_t^{t-1}\right)\right).\end{aligned}$$

In the second summand in the last row of the above equation we have $x_t^{t-1} = E_{\hat{\theta}_\kappa}\left[x_t\middle|y_0^{t-1}\right] = \hat{A}_0^{(\kappa)}x_{t-1}^{t-1}$ and $\hat{S}_t = \mathrm{Var}_{\hat{\theta}_\kappa}\left[Y_t\middle|y_0^{t-1}\right] = \hat{H}^{(\kappa)}\hat{A}_0^{(\kappa)}\Sigma_{t-1}^{t-1}\left(\hat{A}_0^{(\kappa)}\right)^{\mathrm{T}}\left(\hat{H}^{(\kappa)}\right)^{\mathrm{T}}$ $+\hat{V}^{(\kappa)}$ for $t > 0$.

The number of dimensions q_{opt} of the hidden states of the LDS is considered as the optimal one if it is assigned minimum scores, that is

$$q_{opt} = \arg\min_q BIC(q). \tag{191}$$

2.11 Product Development Project Management Example Revisited

To demonstrate the more complex concept of a state process with hidden variables and validate the developed project model with field data, the above recursions were implemented into the Mathematica® modeling and simulation environment. The estimation routines were also verified through the Bayesian Net Toolbox for Matlab (BNT), which was developed by Murphy (2001). The field data came from the previous case study of product development in the small German industrial company (see Section 2.5, Schlick et al. 2012). For simplicity, we also focus in the following on the first two overlapping development tasks of project A, "conceptual sensor design" (task 1) and "design of circuit diagram" (task 2), and model only their overlapping range. Due to the barcode-based labor time system in the company, only the series of the time-on-task could be used as an objective observation to estimate the evolution of the hidden state process. According to Section 2.5, the conceptual sensor design had reached a completion level of 39.84% when the design of the circuit diagram began. The observed work remaining at the initial time step is therefore $y_0 = (0.6016 \quad 1.0000)^{\mathrm{T}}$.

Moreover, we assumed that each development task is accompanied by two latent subtasks representing horizontal and lateral communication. Due to the small size of the company, diagonal communication was not relevant. Under these conditions a six dimensional state process seems to be adequate to capture the essential dynamics of the development project and we have $\mathrm{Dim}[x_t] = 6$. We term the corresponding LDS model an LDS(6, 2), because the hidden states are six dimensional and the observations two dimensional. We used the implemented estimation routines to re-estimate the independent parameters in 20 iterations of the EM algorithm. We chose an initial setting θ_0 of the independent parameters, in which Gaussian random numbers were assigned to the initial state, the dynamical operator and the output operator. The covariance matrices were set to a multiple of the identity matrix.

As shown in the previous chapter, the parameters can only be identified up to an unknown transformation. As the eigenvalues of the covariance of the hidden state process and the eigenvalues of the dynamical operator are invariant under the unknown transform, they are the only quantities which are estimated unambiguously. We found the following values:

$$\mathrm{Eig}\left[\hat{A}_0^{(20)}\right] = \{0.9075 \pm 0.0273i, 0.8312 \pm 0.2945i, -0.2638 \pm 0.2188i\} \quad (192)$$

$$\mathrm{Eig}\left[\hat{\Pi}_0^{(20)}\right] = 10^{-4} \cdot \{12.346, 1.198, 0.453, 0.0194, 0.0067, 0.0021\} \quad (193)$$

$$\mathrm{Eig}\left[\hat{C}^{(20)}\right] = 10^{-4} \cdot \{6.764, 3.952, 1.769, 0.928, 0.137, 0.085\}. \quad (194)$$

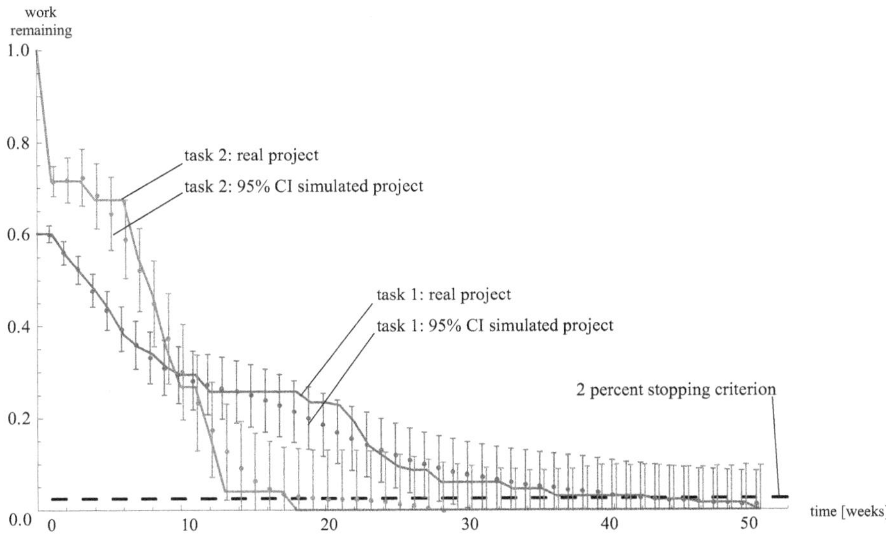

Fig. 2.22 List plot of work remaining in the real and simulated product development projects. In an analogous way to Figs. 2.9 and 2.10, only the overlapping range of the first two tasks is shown in conjunction with the means of simulated traces of task processing as note points and 95% confidence intervals as error bars. The Monte Carlo simulation was based on the system of equations 136 and 137. A six dimensional state vector was used to represent the state of the project (*LDS*(6, 2) model). The recursion formula for maximum likelihood parameter estimation based on Expectation-Maximization are given in the text. A total of 1000 separate and independent runs were calculated. Note points have been offset to distinguish the error bars. The stopping criterion of 2% is marked by a *dashed line* at the *bottom* of the plot

The estimated covariance matrix $\hat{V}$ of the normally distributed random variable η_t related to the observation process is given by

$$\hat{V}^{(20)} = \left\{10^{-5}\right\} \cdot \begin{pmatrix} 1.53 & 0.25 \\ 0.25 & 7.13 \end{pmatrix}. \tag{195}$$

Based on the parameterized LDS model, an additional Monte Carlo simulation was carried out within the Mathematica® software environment. One thousand separate and independent simulation runs were calculated.

In an analogous way to Figs. 2.9 and 2.10, Fig. 2.22 shows the list plots of the empirically acquired work remaining for both tasks as well as the means and 95% confidence intervals of simulated time series of task processing, which were calculated for the overlapping range over 50 weeks. The stopping criterion of $\delta = 0.02$ was left unchanged and is plotted as a dashed line at the bottom of the chart. Interestingly, according to Fig. 2.22 all 95% confidence intervals of the simulated work remaining of both tasks include the empirical data points from the real project before the stopping criterion is met. Furthermore, the confidence intervals are small. Therefore, the LDS model can be fitted much better to the data than the rather simple VAR(1) and VAR(2) models from Section 2.5 (compare Figs. 2.9 and 2.10 to Fig. 2.22). For instance, in the parameterized VAR(1) model only 49 out of 50 confidence intervals (95% level) for task 1 and 28 out of 31 intervals for task 2 covered the empirical data points before the stopping criterion is met. It is important to point out that this high goodness-of-fit only holds under the assumption of a six dimensional state space.

If the parameter space is collapsed into three dimensions, for instance, the goodness-of-fit is usually not better than with the vector autoregression model of first order. The six dimensional state process is also able to accurately predict the quite abrupt completion of task 2 in week 18, because the mean work remaining at this point in time is equal to the stopping criterion. The root-mean-square deviation between the predicted work remaining in task 1 and the field data is $\mathrm{RMSD}_{\mathrm{task1}} = 0.037$ and therefore 20% lower than the deviation for the VAR(1) model. For task 2, the deviation is $\mathrm{RMSD}_{\mathrm{task2}} = 0.055$. This value is 48% lower than the deviation obtained for the autoregression model of first order (see Section 2.5, Schlick et al. 2012).

If the dimension of the hidden states cannot be specified by the model developer based on knowledge about the task and communication structure in the project as shown before, the Schwarz-Bayes criterion (BIC, see Eq. 189) is a theoretically convincing alternative, because it favors the approximating model which is a posteriori most probable. The drawback is that the criterion is only valid in a setting with large sample size and therefore must be carefully applied in our case. We systematically varied the dimensions of the hidden states between two and six ($\mathrm{Dim}[x_t] = \{2,3,4,5,6\}$) and calculated the corresponding $BIC(q)$ values based on Eq. 189. For each approximating model, 20 iterations of the EM algorithm were computed to estimate the independent parameters. As before, we used an initial setting θ_0 of the independent parameters for each model in which Gaussian random numbers were assigned to the initial state, the dynamical operator and the output operator. The covariance matrices were set to a multiple of the identity matrix.

The model selection procedure showed that BIC is minimal for an LDS with a four dimensional state process and we have $BIC\left(q_{opt} = 4\right) = -19.311$. For this model the invariant parameters are:

$$\mathrm{Eig}\left[\hat{A}_0^{(20)}\right] = \{0.9342, 0.8956 \pm 0.1836i, 0.0443\} \tag{196}$$

$$\mathrm{Eig}\left[\hat{\Pi}_0^{(20)}\right] = 10^{-4} \cdot \{6.9595, 0.1380,\ 0.0609, 0.0044\} \tag{197}$$

$$\mathrm{Eig}\left[\hat{C}^{(20)}\right] = 10^{-4} \cdot \{9.2319, 2.4869, 0.8116,\ 0.2221\} \tag{198}$$

and the observation noise covariance is:

$$\hat{V}^{(20)} = \{10^{-5}\} \cdot \begin{pmatrix} 1.17 & 0.12 \\ 0.12 & 8.76 \end{pmatrix}. \tag{199}$$

We investigated the sensitivity of the optimal model order that is selected on the basis of the Schwarz-Bayes criterion. 200 independent trials of repeated parameter estimation and model order selection were carried out. As before, the dimensionality of the state process was varied between two and six. In each trial, the independent parameters for the given model order were estimated through 20 iterations of the EM algorithm. The estimates were based on an initial parameter setting θ_0, in which Gaussian random numbers were assigned to the initial state, the dynamical operator and the output operator. The covariance matrices were set to a multiple of the identity matrix. The results show that on average a LDS with a four dimensional state process is selected.

In addition, 1000 separate and independent simulation runs were calculated to determine the root mean square deviation between the predicted work remaining in task 1 and the field data for the parameterized LDS(4, 2) model according to Eqs. 196 to 199. The result is $\text{RMSD}_{\text{task1}} = 0.034$. This empirical deviation is approximately 9% lower than the value that was computed for the LDS(6, 2) model and more than 25% lower than the corresponding values for the VAR(1) and VAR (2) models (Section 2.5). Interestingly, for task 1 the predictive accuracy of the LDS (4, 2) model is higher than the LDS(6, 2) model, even though the state space was reduced from six to four dimensions. For task 2, the empirical deviation is $\text{RMSD}_{\text{task2}} = 0.070$ and therefore approximately 35% and 38% lower than the values obtained for the first- and second-order regression models respectively (Section 2.5). Compared with the previously analyzed LDS, the value is approximately 27% larger. We conclude that the predictive accuracy of an LDS with a four dimensional state process whose independent parameters were estimated by the introduced EM algorithm can be not Pareto-inferior to an LDS(6, 2) model, and the predictions of the work remaining can be almost as accurate as for the more complex model. For both tasks the total root mean square deviation is only 13% higher. The predictive accuracy of the LDS(4, 2) model is significantly higher than the accuracy of the VAR(1) and VAR(2) models for both tasks and shows the Pareto-superiority of the approach with hidden state variables. It is important to point out that these conclusions only hold if the independent parameters are estimated using a favorable initial setting. We will return to this issue in the sensitivity analyses.

Figure 2.23 shows the list plots of the time series from the real project as well as the means and 95% confidence intervals of simulated task processing for the LDS (4, 2) model. As before, the stopping criterion of $\delta = 0.02$ is plotted as a dashed line at the bottom of the chart. For both tasks, the means and 95% confidence intervals follow a pattern that is very similar to the LDS(6, 2) model (Fig. 2.22). Therefore, all confidence intervals of the simulated work remaining include the data points from the real project before the stopping criterion is met. The only important difference between both models with hidden state variables is that for task 2 the

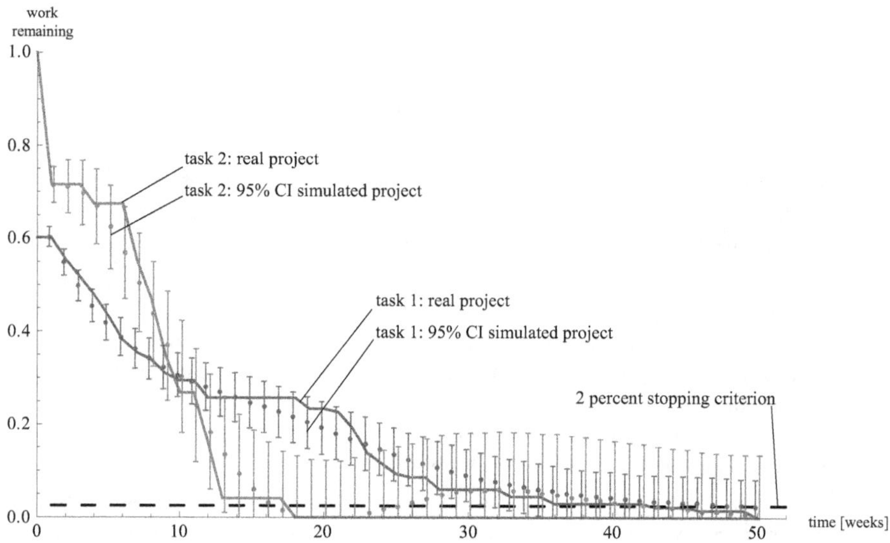

Fig. 2.23 List plot of work remaining in the real and simulated product development projects. In an analogous way to Figs. 2.9, 2.10 and 2.22, only the overlapping range of the first two tasks is shown in conjunction with the means of simulated traces of task processing as note points and 95% confidence intervals as error bars. The Monte Carlo simulation was based on the simultaneous system of equations 136 and 137. In contrast to Fig. 2.22, only a four dimensional state vector was used to represent the work remaining of the project (known as the *LDS*(4, 2) model). The recursion formula for maximum likelihood parameter estimation based on Expectation-Maximization are given in the text. A total of 1000 separate and independent runs were calculated. Note points have been offset to distinguish the error bars. The stopping criterion of 2% is marked by a *dashed line* at the *bottom* of the plot

lower dimensional model shows an oscillatory behavior of the means of work remaining after the real completion of the task in week 18. The means first undershoot the abscissa from week 18 to 21, then overshoot it from week 22 to week 32 and finally follow a smoothly decaying geometric series until week 50 (see bottom of Fig. 2.23). Interestingly, these oscillations lead to an average system behavior that makes it possible to predict almost perfectly without using a stopping criterion the abrupt completion of task 2 in week 18.

Finally, we carried out different sensitivity analyses for the LDS models. The importance of these analyses should not be underestimated, given that the previously introduced EM procedure lacks identifiability and therefore an infinite number of parameterizations $\hat{\theta}_{\kappa}$ theoretically exist that yield the same maximum log-likelihood. In the first sensitivity analysis, we investigated the sensitivity of the predictive accuracy of the LDS models in terms of the root-mean-square deviation between the predicted work remaining and the field data for both development tasks. For each model, 200 independent trials of repeated project execution were simulated and analyzed. In each trial, the independent parameters were estimated through 20 iterations of the EM algorithm with an initial setting θ_0 of the independent parameters, in which Gaussian random numbers were assigned to the initial state, the dynamical operator and the output operator. The covariance matrices were set to a multiple of the identity matrix. Based on the estimated independent parameters, 1000 separate and independent simulation runs of project dynamics were computed and the root-mean-square deviations for each case were calculated. The complete dataset was used to calculate the mean $\mathrm{RMSD}_{\mathrm{task1}}^{\mathrm{all}}$ and $\mathrm{RMSD}_{\mathrm{task2}}^{\mathrm{all}}$ over all trials. These overall means were compared to the corresponding $\mathrm{RMSD}_{\mathrm{task1}}$ and $\mathrm{RMSD}_{\mathrm{task2}}$ values that were obtained in the previous Monte Carlo simulation. We were thus able to assess whether the parameterized LDS(6, 2) model according to Eqs. 192–195 and the LDS(4, 2) model according to Eqs. 196–199 have an above-average predictive accuracy compared to alternative models of the same class. If badly conditioned covariance matrices occurred during the simulation and therefore the EM results possibly contained significant numerical errors, the complete trial was recalculated.

For the LDS(6, 2) model, an overall $\mathrm{RMSD}_{\mathrm{task1}}^{\mathrm{all}} = 0.0397$ was computed for task 1 and $\mathrm{RMSD}_{\mathrm{task2}}^{\mathrm{all}} = 0.0750$ for task 2. Both overall deviations are larger than the $\mathrm{RMSD}_{\mathrm{task1}}$ and $\mathrm{RMSD}_{\mathrm{task2}}$ values from the previous simulation. Therefore, the predictive accuracy of the LDS(6, 2) model shown is higher than the accuracy of an average model. An additional analysis of the empirical cumulative distribution function showed for task 1 that the $\mathrm{RMSD}_{\mathrm{task1}}$ value is equivalent to the 40th percentile of the distribution. In other words, the $\mathrm{RMSD}_{\mathrm{task1}}$ value is exceeded by 60% of all 200 simulated trials. For task 2 the $\mathrm{RMSD}_{\mathrm{task2}}$ value is equivalent to the 30th percentile of the distribution. The relative accuracy deviation is approximately 7% for task 1 and 26% for task 2 and therefore seems to be negligible for almost every application in project management. The LDS(4, 2) model leads to an overall $\mathrm{RMSD}_{\mathrm{task1}}^{\mathrm{all}} = 0.0455$ for task 1 and an overall $\mathrm{RMSD}_{\mathrm{task2}}^{\mathrm{all}} = 0.1065$ for task 2. Similar to the LDS(6, 2) model, both values are larger than the previously calculated $\mathrm{RMSD}_{\mathrm{task1}}$ and $\mathrm{RMSD}_{\mathrm{task2}}$ values. However, the predictive accuracy of the LDS(4, 2) model is much higher than the accuracy of average models and demonstrates unexpected predictive power for such a low-dimensional model. The empirical cumulative distribution function showed for task 1 that the $\mathrm{RMSD}_{\mathrm{task1}}$ value represents approximately a fourth percentile and for task 2 the $\mathrm{RMSD}_{\mathrm{task2}}$ value a 13th percentile. For task 1 the relative accuracy deviation is approximately 28% and for task 2 approximately 36% for the benefit of the given model. These values can be relevant for project controlling. For application in project management, it therefore makes sense to estimate the independent parameters in independent runs of the EM iterations with a randomized initial setting θ_0 of the independent parameters and to select the model with the most favorable properties for the given application. Desired properties can include the highest total predictive accuracy, the highest predictive accuracy in certain tasks, etc.

The second sensitivity analysis evaluated the sensitivity of emergent complexity associated with the parameterized LDS models as constructive representations of a complex PD project. The dependent variable to evaluate emergent complexity was the cited effective measure complexity (EMC) according to Grassberger (1986). As in the first sensitivity analysis, 200 independent trials of repeated project execution were considered for each LDS model and the corresponding EMC values were calculated. In each trial, the independent parameters were re-estimated in 20 iterations of the EM algorithm. We used an initial setting θ_0 of the independent parameters, in which Gaussian random numbers were assigned to the initial state, the dynamical operator and the output operator. The covariance matrices were set to a multiple of the identity matrix. The implicit solution of de Cock (2002) from Eq. 299 was used to calculate the EMC values based on estimated independent parameters $\hat{\theta}_{20}$. The infinitive sum was approximated by a partial sum of 100 terms. Under this boundary condition the other closed-form solutions from Eqs. 291 and 308 produce results that are correct to three decimal places and can therefore be considered numerically equivalent. For the LDS(6, 2) model the complexity value is $\text{EMC}_{\text{LDS}(6,2)} = 3.1363$. The 200 independent trials lead to a mean complexity of $\text{EMC}^{\text{all}}_{\text{LDS}(6,2)} = 3.0946$. The relative difference is less than 2% and therefore negligible for almost every application in project management. An analysis of the empirical cumulative distribution function showed that the $\text{EMC}_{\text{LDS}(6,2)}$ value is equivalent to the 54th percentile of the empirical distribution and therefore close to the median. Regarding the parameterized LDS(4, 2) model a smaller complexity value of $\text{EMC}_{\text{LDS}(4,2)} = 2.9312$ was obtained as one would expect for a model with lower dimensional (hidden) state space. The mean complexity over 200 trials was $\text{EMC}^{\text{all}}_{\text{LDS}(4,2)} = 3.3041$. The relative difference is approximately 13% and therefore much larger than for LDS(6, 2) model. This difference could be relevant for practical complexity evaluations. The $\text{EMC}_{\text{LDS}(4,2)}$ value is equivalent to the 14th percentile of the empirical distribution. This means that 86% of the models in the sample exceed the emergent complexity that is associated with the introduced dynamical operators and covariance matrices. Interestingly, the mean complexity related to the LDS(4, 2) models is slightly larger than the mean related to the higher-dimensional LDS(6, 2) models. This counterintuitive result of the sensitivity analysis shows that a stochastic process generated by an LDS with a comparatively low dimensionality of the state space must not necessarily be less complex than higher dimensional representations, because strong interactions between internal states might exist that lead to strong correlations between observations of task processing and therefore to increasing time-dependent complexity.

In the third and final sensitivity analysis, we carried out an additional evaluation of the sensitivity of emergent complexity of cooperative task processing represented by the LDS models. However, we did not use the introduced independent parameter vector $\hat{\theta}_{20}$ to directly calculate the EMC values based on the closed-form solution from Eq. 299. Instead, we used a parameterized LDS model to simulate task processing, generate time series and (re-)estimate the independent parameters purely from data. The data consisted of ten independent cases of the work remaining for both development tasks over $T = 100$ time steps. In line with the procedure used for the previous sensitivity analyses, 200 independent trials of repeated data generation were considered for each LDS model. In each trial, the ten independent cases of work remaining were input into the EM algorithm, and 20 iterations were calculated based on an initial setting θ_0, in which Gaussian random numbers were assigned to the initial state, the dynamical operator and the output operator. The covariance matrices were set to a multiple of the identity matrix. After the 20th iteration, the re-estimated independent parameters $\hat{\theta}'_{20}$ were used to calculate the EMC values based on Eq. 130. We hypothesized that although the introduced EM procedure lacks identifiability, the emergent complexity can be accurately estimated fusing data from the "true" LDS(6, 2) and LDS(4, 2) models by averaging over EMC values which were obtained in repeated Monte Carlo trials. This is a pivotal hypothesis, because if it proved unverifiable, the introduced maximum likelihood estimation procedure could lead to inaccurate and unreliable complexity evaluations in the application domain. Concerning the parameterized LDS(6, 2) model, the mean complexity value that was estimated from the data is $\mathrm{EMC}_{\mathrm{LDS}(6,2)}^{\mathrm{all,\,data}} = 3.1254$. The relative difference between this value and the reference $\mathrm{EMC}_{\mathrm{LDS}(6,2)}$ value is less than 0.5%. This difference is significantly smaller than the difference obtained in the purely analytical evaluation, which means it is also negligible for almost every application in project management. An analysis of the empirical cumulative distribution function showed that the $\mathrm{EMC}_{\mathrm{LDS}(6,2)}$ value is equivalent to the 54th percentile of the empirical distribution of the $\mathrm{EMC}_{\mathrm{LDS}(6,2)}^{\mathrm{all,data}}$ values and therefore also very close to the median. For the parameterized LDS(4, 2) model the corresponding mean complexity is $\mathrm{EMC}_{\mathrm{LDS}(4,2)}^{\mathrm{all,\,data}} = 3.0466$. The relative difference is approximately 4%. This value is also significantly lower than the one obtained in the purely analytical evaluation and seems to be negligible for most applications. The $\mathrm{EMC}_{\mathrm{LDS}(4,2)}$ value is equivalent to the 15th percentile of the empirical distribution. Interestingly, the indirect, data-driven sensitivity analysis leads, for both LDS models, to mean complexity values that are significantly closer to the true values than the direct analytical approach. For use in project management, it can therefore be beneficial to evaluate emergent complexity using the introduced indirect estimation methods rather than doing so directly by using the closed-form solution from Section 4.2.1 or 4.2.2 after completed EM runs. Finally—and most importantly—the results show that although the EM procedure lacks identifiability, emergent complexity can be accurately estimated from data simply by averaging over EMC values which were obtained in repeated Monte Carlo trials. Additional methods of dealing with divergent and inconsistent findings by introducing auxiliary concepts seem to be unnecessary.

In conclusion, in terms of emergent complexity the introduced LDS(6, 2) model seems to be a very good dynamical representation of all models within this class that are parameterized based on the EM algorithm. The emergent complexity associated with the alternative LDS(4, 2) model shown is within the first quartile and therefore too low to be representative. However, we preferred to present that model with reduced state space dimensionality because the root-mean-square deviation between the predicted work remaining and the field data is within the first quantiles for both tasks, and for task 1 it is also lower than the deviation obtained of the LDS(6, 2) model. This finding holds even though lower complexity values are assigned to the LDS(4, 2) model. This is a good example of how emergent complexity in the sense of Grassberger's theory cannot simply be evaluated by classic predictability measures. A good model not only has to be able to make accurate predictions; it must also employ an internal state representation that is not unnecessarily complex to capture the set of essential regularities in a process. We will return to this important issue in the next chapter.

References

Akaike, H.: Autoregressive model fitting for control. Ann. Inst. Stat. Math. **23**, 163–180 (1971)

Akaike, H.: Information theory and an extension of the maximum likelihood principle. In: Petrov, B.N., Csaki, F. (eds.) Second International Symposium of Information Theory, pp. 267–281. Akademia Kiado, Budapest (1973)

Akaike, H.: A new look at the statistical model identification. IEEE Trans. Automat. Contr. **19**, 716–723 (1974)

Anderson, B.D.O., Moore, J.B.: Optimal Filtering. Prentice-Hall, New York (1979)

Baker, K.R., Trietsch, D.: Principles of Sequencing and Scheduling. Wiley, New York (2009)

Bar-Shalom, Y., Li, X.R., Kirubarajan, T.: Estimation with Application to Tracking and Navigation: Theory, Algorithms and Software. Wiley Interscience, New York (2001)

Beal, D.J.: Information criteria methods in SAS® for multiple linear regression models. In: Proceedings of the South East SAS User Group Meeting SESUG 2007, pp. 1–10 (2007)

Bialek, W., Nemenman, I., Tishby, N.: Predictability, complexity and learning. Neural Comput. **13** (1), 2409–2463 (2001)

Boots, B.: Learning stable linear dynamical systems. M.S. Thesis in Machine Learning, Carnegie Mellon University (2009)

Bowden, K.: A direct solution to the block tridiagonal matrix inversion problem. Int. J. Gen. Syst. **15**(3), 185–198 (1989)

Box, G., Jenkins, G.: Time Series Analysis, Forecasting and Control. Holden-Day, San Francisco, CA (1976)

Braha, D., Bar-Yam, Y.: The statistical mechanics of complex product development: empirical and analytical results. Manage. Sci. **53**(7), 1127–1145 (2007)

Brockwell, P.J., Davis, R.A.: Time Series: Theory and Methods, 2nd edn. Springer, New York (1991)

Bronstein, I.N., Semendjajew, K.A., Musiol, G., Mühlig, H.: Taschenbuch der Mathematik (in German). Wissenschaftlicher Verlag Harri Deutsch, Frankfurt am Main (2000)

Burnham, K.P., Anderson, D.R.: Model Selection and Multimodel Inference: A Practical Information-Theoretic Approach. Springer, New York (2002)

Cavanaugh, J.E., Neath, A.A.: Generalizing the derivation of the Schwarz information criterion. Commun. Stat. Theory Methods **28**(1), 49–66 (1999)

Cooper, R.G.: Winning at New Products: Creating Value Through Innovation, 4th edn. Basic Books, New York (2011)

Cover, T.M., Thomas, J.A.: Elements of Information Theory. Wiley, New York (1991)

de Cock, K.D.: Principal angles in system theory, information theory and signal processing. Ph.D. Thesis, Katholieke Universiteit Leuven (2002)

de Cock, K.D., Moor, D.B.: Subspace angles and distances between ARMA models. Syst. Contr. Lett. **64**, 265–270 (2002)

Dellaert, F.: The expectation maximization algorithm. Technical Report GIT-GVU-02-20, College of Computing, Georgia Institute of Technology (2002)

DelSole, T., Tippett, M.K.: Predictability: recent insights from information theory. Rev. Geophys. **45** (2007)

Dempster, A.P., Laird, N.M., Rubin, D.B.: Maximum likelihood from incomplete data via the EM algorithm. J. R. Stat. Soc. Ser. B **39**(1), 1–38 (1977)

Ellison, C.J., Mahoney, J.R., Crutchfield, J.P.: Prediction, retrodiction, and the amount of information stored in the present. Santa Fe Institute Working Paper 2009-05-017 (2009)

Eppinger, S.D., Browning, T.: Design Structure Matrix Methods and Applications. MIT Press, Cambridge, MA (2012)

Fazar, W.: Program evaluation and review technique. Am. Stat. **13**(2), 10–21 (1959)

Franses, P.H., Paap, R.: Periodic Time Series Models. Oxford University Press, Oxford (2004)

Gentle, J.E.: Matrix Algebra. Theory, Computations, and Applications in Statistics. Springer, New York (2007)

Gharahmani, Z.: An introduction to hidden Markov models and Bayesian networks. Int. J. Pattern Recognit. Artif. Intell. **15**(1), 9–42 (2001)

Gharahmani, Z., Hinton, G.E.: Parameter estimation for linear dynamical systems. University of Toronto Technical Report CRG-TR-96-2 (1996)

Gibson, J.W., Hodgetts, R.M.: Organizational Communication—A Managerial Perspective, 2nd edn. HarperCollins Publishers, New York (1991)

Grassberger, P.: Toward a quantitative theory of self-generated complexity. Int. J. Theor. Phys. **25** (9), 907–938 (1986)

Grünwald, P.: The Minimum Description Length Principle. MIT Press, Cambridge, MA (2007)

Halanay, A., Rasvan, V.: Stability and Stable Oscillations in Discrete Time Systems. CRC Press, Boca Raton (2000)

Hansen, M.H., Yu, B.: Model selection and the principle of minimum description length. J. Am. Stat. Soc. **96**(454), 746–774 (2001)

Hartley, H.O.: Maximum likelihood estimation from incomplete data. Biometrics **14**(2), 174–194 (1958)

Hauser, J.R.: Note on Product Development, p. 3. MIT Sloan Courseware, Cambridge, MA (2008)

Higham, N.J.: Accuracy and Stability of Numerical Algorithms, 2nd edn. Society for Industrial and Applied Mathematics (SIAM), Philadelphia, PA (2002)

Hinrichsen, D., Pritchard, A.J.: Mathematical Systems Theory I. Modelling, State Space Analysis, Stability and Robustness. Springer, Berlin (2005)

Honerkamp, J.: Statistical Physics, 2nd edn. Springer, Berlin (2002)

Huberman, B.A., Wilkinson, D.M.: Performance variability and project dynamics. Comput. Math. Organ. Theory **11**(4), 307–332 (2005)

Hurvich, C.M., Tsai, C.L.: A corrected Akaike information criterion for vector autoregressive model selection. J. Time Ser. Anal. **14**(3), 271–279 (1993)

Kailath, T.: Lectures Notes on Wiener and Kalman Filtering. Springer, Berlin (1981)

Kalman, R.E.: A new approach to linear filtering and prediction problems. Trans. Am. Soc. Mech. Eng. Ser. D J. Basic Eng. **82**, 35–45 (1960)

Kass, R.E., Wasserman, L.: A reference Bayesian test for nested hypotheses and its relationship to the Schwarz criterion. J. Am. Stat. Assoc. **90**(2), 928–934 (1995)

Katz, G.: Rethinking the product development funnel. Visions (J. Prod. Dev. Manage. Assoc.) **35** (2), 24–31 (2011)

Kay, S.: Modern Spectral Estimation: Theory and Application. Prentice Hall, Upper Saddle River, NJ (1988)

Kelley, J., Walker, M.: Critical-path planning and scheduling. In: 1959 Proceedings of the Eastern Joint Computer Conference, pp. 160–173 (1959)

Klein, M., Braha, D., Sayama, H., Bar-Yam, Y.: Editorial for special issue on a complex system perspective on concurrent engineering. Concurr. Eng. Res. Appl. **11**(3), 163 (2003)

Koch, W.: On Bayesian tracking and data fusion: a tutorial introduction with examples. IEEE Aerosp. Electron. Syst. Mag. **25**(7), 29–51 (2010)

Koch, W.: Tracking and Sensor Data Fusion: Methodological Framework and Selected Applications. Springer, Berlin (2014)

Lancaster, P., Tismenetsky, M.: The Theory of Matrices, 2nd edn. Academic, Orlando, FL (1985)

Ljung, L.: System Identification—Theory for the User, 2nd edn. Prentice-Hall, Upper Saddle River, NJ (1999)

Ljung, L.: System Identification Toolbox for use with Matlab. Version 5, 5th edn. The MathWorks, Natick, MA (2000)

Luenberger, D.G.: Introduction to dynamic systems. Wiley, New York (1979)

Lütkepohl, H.: Comparison of criteria for estimating the order of a vector autoregressive process. J. Time Ser. Anal. **6**, 35–52 (1985)

Lütkepohl, H.: New Introduction to Multiple Time Series Analysis. Springer, Berlin (2005)

Maier, M.: Architecting principles for systems-of-systems. Syst. Eng. **1**(4), 267–284 (1998)

Marple, S.: Digital Spectral Analysis with Applications. Prentice-Hall, Upper Saddle River, NJ (1987)

Martens, J.: A new algorithm for rapid parameter learning in linear dynamical systems. Master's Thesis (Computer Science), University of Toronto (2009)

Martens, J.: Learning the linear dynamical system with ASOS. In: Proceedings of the 27th International Conference on Machine Learning, pp. 743–750, Haifa, Israel (2010)

Marzen, S., Crutchfield, J. P.: Circumventing the curse of dimensionality in prediction: causal rate-distortion for infinite-order Markov processes. Santa Fe Institute Working Paper 2014-12-047 (2014)

McGrath, M.E.: Setting the PACE® in Product Development. Butterworth-Heinemann, Boston MA (1996)

McLeod, A.I.: Diagnostic checking periodic autoregression models with applications. J. Time Ser. Anal. **15**(1), 221–231 (1994)

Mihm, J., Loch, C.: Spiraling out of control: Problem-solving dynamics in complex distributed engineering projects. In: Braha, D., Minai, A.A., Bar-Yam, Y. (eds.) Complex Engineered Systems: Science Meets Technology, pp. 141–158. Springer, Berlin (2006)

Mihm, J., Loch, C., Huchzermeier, A.: Problem-solving oscillations in complex engineering. Manage. Sci. **46**(6), 733–750 (2003)

Murphy, K.P.: The Bayes Net Toolbox for Matlab. Comput. Sci. Stat. **33**(1), 1–20 (2001)

Murthy, S., Ramachandran, P., Uppal, T., Kumar, K.: A generalized deterministic analysis of product development process using DSM. In: Proceedings of the 14th International Dependency and Structure Modeling Conference, DSM 2012 (the paper is only available on the proceedings CD-ROM) (2012)

Neal, R.M., Hinton, G.E.: A view of the EM algorithm that justifies incremental, sparse and other variants. In: Jordan, M.I. (ed.) Learning in Graphical Models, pp. 355–368. Kluwer, Dordrecht, MA (1998)

Neumaier, A., Schneider, T.: Estimation of parameters and eigenmodes of multivariate autoregressive models. ACM Trans. Math. Softw. **27**, 27–57 (2001)

Oispuu, M.: Passive emitter localization by direct position determination with moving array sensors. Ph.D. Thesis (electrical engineering), Siegen University, Siegen, Germany (2014)

Papoulis, A., Pillai, S.U.: Probability, Random Variables and Stochastic Processes. McGraw-Hill, Boston, MA (2002)

Puranam, P., Goetting, M., Knudsen, T.: Organization design: the epistemic interdependence perspective. Working paper, London Business School, London, UK (2011)

Puri, N.N.: Fundamentals of Linear Systems for Physical Scientists and Engineers. CRC Press, Boca Raton, FL (2010)

Raftery, A.E.: Approximate Bayes factors and accounting for model uncertainty in generalized linear models. Biometrika **83**(2), 251–266 (1996)

Rauch, H.E.: Solutions to the linear smoothing problem. IEEE Trans. Automat. Contr. **8**(1), 371–372 (1963)

Rissanen, J.: Stochastic Complexity in Statistical Inquiry. World Scientific, Singapore (1989)

Rissanen, J.: Information and Complexity in Statistical Modeling. Springer, Berlin (2007)

Rouse, B., Boff, K. (eds.): Organizational Simulation. Wiley, New York (2005)

Roweis, S., Ghahramani, Z.: A unifying review of linear Gaussian models. Neural Comput. **11**(2), 305–345 (1999)

Roweis, S., Ghahramani, Z.: Learning nonlinear dynamical systems using an EM algorithm. In: Kearns, M.S., Solla, S.A., Cohn, D.A. (eds.) Advances in Neural Information Processing Systems 11, pp. 599–605. MIT Press, Cambridge, MA (1998)

Rudary, M.: On predictive linear models. Ph.D. Thesis (computer science and engineering), The University of Michigan, USA (2009)

Rudary, M., Singh, S.: Predictive linear-Gaussian models of controlled stochastic dynamical systems. In: Proceedings of the 23rd International Conference on Machine Learning, pp. 777–784 (2006)

Rudary, M., Singh, S., Wingate, D.: Predictive linear-Gaussian models of stochastic dynamical systems. In: Bacchus, F., Jaakkola, T. (eds.) Uncertainty in Artificial Intelligence 21, pp. 501–508 (2005)

Salmon, R.: Critical path scheduling by the linear programming method. Report ORNL-TM-394 of the Oak Ridge National Laboratory (1962)

Sarkar, S., Dong, A.: A spectral analysis software to detect modules in a DSM. In: Proceedings of the 16th International Dependency and Structure Modeling Conference, DSM 2014, pp. 55–64 (2014)

Sarkar, S., Dong, A., Henderson, J.A., Robinson, P.A.: Spectral characterization of hierarchical modularity in product architectures. J. Mech. Des. **136**(1), 011006 (2013)

Sawa, T.: Information criteria for discriminating among alternative regression models. Econometrica **46**(1), 1273–1282 (1978)

Schlick, C.M., Beutner, E., Duckwitz, S., Licht, T.: A complexity measure for new product development projects. In: Proceedings of the 19th International Engineering Management Conference, pp. 143–150 (2007)

Schlick, C.M., Duckwitz, S., Gärtner, T., Schmidt, T.: A complexity measure for concurrent engineering projects based on the DSM. In: Proceedings of the 10th International DSM Conference, pp. 219–230 (2008)

Schlick, C.M., Schneider, S., Duckwitz, S.: Modeling of periodically correlated work processes in large-scale concurrent engineering projects based on the DSM. In: Proceedings of the 13th International Dependency and Structure Modeling Conference, DSM 2011, pp. 273–290 (2011)

Schlick, C.M., Schneider, S., Duckwitz, S.: Modeling of cooperative work in concurrent engineering projects based on extended work transformation matrices with hidden state variables. In: Proceedings of the 14th International Dependency and Structure Modeling Conference, DSM 2012, pp. 411–422 (2012)

Schlick, C.M., Schneider, S., Duckwitz, S.: A universal complexity criterion for model selection in dynamic models of cooperative work based on the DSM. In: Proceedings of the 15th International Dependency and Structure Modeling Conference, DSM 2013, pp. 99–105 (2013)

Schlick, C.M., Schneider, S., Duckwitz, S.: Estimation of work transformation matrices for large-scale concurrent engineering projects. In: Proceedings of the 16th International Dependency and Structure Modeling Conference, DSM 2014, pp. 211–221 (2014)

Schneider, T., Neumaier, A.: Algorithm 808: ARfit—a Matlab package for the estimation of parameters and eigenmodes of multivariate autoregressive models. ACM Trans. Math. Softw. **27**(1), 58–65 (2001)

Schwarz, G.: Estimating the dimension of a model. Ann. Stat. **6**, 61–64 (1978)

Shtub, A., Bard, J.F., Globerson, S.: Project Management—Processes, Methodologies, and Economics, 2nd edn. Prentice Hall, Upper Saddle River, NJ (2004)

Shumway, R.H., Stoffer, D.S.: An approach to time series smoothing and forecasting using the EM algorithm. J. Time Ser. Anal. **3**(4), 253–264 (1982)

Siddiqi, W.M.: Learning latent variable and predictive models of dynamical systems. Doctoral dissertation, Technical Report CMU-RI-TR-09-39, Robotics Institute, Carnegie Mellon University, January, 2010

Smith, G., De Freitas, J.F., Niranjan, M., Robinson, T.: Speech modelling using subspace and EM techniques. Adv. Neural Inf. Process. Syst. **12**, 796–802 (1999)

Smith, G.A., Robinson, T.: A comparison between the EM and subspace identification algorithms for time-invariant linear dynamical systems. Technical Report, Cambridge University (2000)

Smith, R.P., Eppinger, S.D.: Identifying controlling features of engineering design iteration. Manage. Sci. **43**(3), 276–293 (1997)

Smith, R.P., Eppinger, S.D.: Deciding between sequential and concurrent tasks in engineering design. Concurr. Eng. Res. Appl. **6**(1), 15–25 (1998)

Soderstrom, T., Stoica, P.: System Identification. Prentice Hall, Upper Saddle River, NJ (1989)

Songsiri, J., Dahl, J., Vanderberghe, L.: Graphical models of autoregressive processes. In: Eldar, Y., Palomar, D. (eds.) Convex Optimization in Signal Processing and Communications, pp. 89–116. Cambridge University Press, Cambridge (2010)

Sterman, J.D.: Business Dynamics: Systems Thinking and Modeling for a Complex World. McGraw-Hill Higher Education, Burr Ridge, IL (2000)

Steward, D.V.: The design structure system: a method for managing the design of complex systems. IEEE Trans. Eng. Manage. **28**(3), 71–74 (1981)

Stoica, P., Moses, R.: Introduction to Spectral Analysis. Prentice Hall, Upper Saddle River, NJ (1997)

Stoica, P., Selen, Y.: Model-order selection. A review of information criterion rules. IEEE Signal Process. Mag. **21**(4), 36–47 (2004)

Streit, R.L.: PMHT and related applications of mixture densities (plenary talk). In: Proceedings of the 9th International Conference on Information Fusion 2006 (FUSION), Florence, Italy (2006)

Trietsch, D., Mazmanyan, L., Gevorgyan, L., Baker, K.R.: Modeling activity times by the Parkinson distribution with a lognormal core: theory and validation. Eur. J. Oper. Res. **216** (2), 445–452 (2012)

Urban, G.L., Hauser, J.R.: Design and Marketing of New Products, 2nd edn. Prentice-Hall, Englewood Cliffs, NJ (1993)

Ursu, E., Duchesne, P.: On modelling and diagnostic checking of vector periodic autoregressive time series models. J. Time Ser. Anal. **30**(1), 70–96 (2009)

Van Overschee, P., de Moor, B.: Subspace Identification for Linear Systems: Theory, Implementations, Applications. Kluwer Academic Publishers, Boston, MA (1996)

von Storch, H., Bürger, G., Schnur, R., von Storch, J.S.: Principal oscillation patterns: a review. J. Climate **8**(1), 377–400 (1995)

Wang, P., Shi, L., Du, L., Liu, H., Xu, L., Bao, Z.: Radar HRRP statistical recognition with temporal factor analysis by automatic Bayesian Ying-Yang harmony learning. Front. Electr. Electron. Eng. Chin. **6**(2), 300–317 (2011)

Wheelwright, S.C., Clark, K.B.: Revolutionizing Product Development. The Free Press, New York (1992)

Wienecke, M.: Probabilistic framework for person tracking and classification in security assistance systems. Ph.D. thesis (computer science), Rheinische Friedrich-Wilhelms Universität Bonn, Bonn, Germany (2013)

Williams, T.: Modeling Complex Projects. Wiley, New York (2002)

Yamaguchi, R., Yoshida, R., Imoto, S., Higuchi, T., Miyano, S.: Finding module-based gene networks with state-space models. IEEE Signal Process. Mag. **24**(1), 37–46 (2007)

Yassine, A., Joglekar, N., Eppinger, S.D., Whitney, D.: Information hiding in product development: the design churn effect. Res. Eng. Des. **14**(3), 145–161 (2003)

Chapter 3
Evaluation of Complexity in Product Development

The term "complexity" stems from the Latin word "complexitas," which means comprehensive or inclusive. In current usage, it is the opposite of simplicity, though this interpretation does not appear to be underpinned by any explicit concept that could be directly used for the development of scientifically rigorous models or metrics. Various disciplines have studied the concepts and principles of complexity in basic and applied scientific research. Several frameworks, theories and measures have been developed, reflecting the differing views of complexity between disciplines. An objective evaluation of structural and dynamic complexity in PD would benefit project managers, developers and customers alike, because it would enable them to compare and optimize different systems in analytical and experimental studies. To obtain a comprehensive view of organizational, process and product elements and their interactions in the product development environment, a thorough review of the notion of complexity has to start from organizational theory (Section 3.1). The literature on organizational theory shows that the complexity of PD projects results from different "sources" and the consideration of the underlying organizational factors and their interrelationships is essential to successful project management (Kim and Wilemon 2009). However, our analyses have shown that static factor-based approaches are not sufficient to evaluate emergent complexity in open organizational systems and therefore the complexity theories and measures of basic scientific research must also be taken into account to capture the inherently complex nature of the product development flow (cf. Amaral and Uzzi 2007). These theories and measures can provide deeper insights into emergent phenomena of complex sociotechnical systems and dynamic mechanisms of cooperation (Section 3.2). Selected measures can also be used to optimize the project organization (Schlick et al. 2009, see Sections 5.2 and 5.3). The measures build upon our intuitive assessment that a system is complex if it is difficult to describe. The description can focus on structure, processes or both. In the description not only the length and the format are relevant but also the expressive power of the "description language." Furthermore, in a process-centered view, for many nontrivial systems the difficulty of prediction and retrodiction have to be

© Springer International Publishing Switzerland 2016

C. Schlick, B. Demissie, *Product Development Projects*, Understanding Complex Systems, DOI 10.1007/978-3-319-21717-8_3

simultaneously taken into account to obtain valid results. Comprehensive overviews of this and related concepts including detailed mathematical analyses and illustrations can be found in Shalizi (2006), Prokopenko et al. (2009) and Nicolis and Nicolis (2007). We will describe the main concepts and methods of basic scientific research in Section 3.2 based on the material from Shalizi (2006). For effective complexity management in PD, the product-oriented measures from theories of systematic engineering design are also relevant (Section 3.3). Seminal work in this field has been done by Suh (2005) on the basis of information-theoretic quantities. These quantities are also the foundation of statistical complexity measures from basic scientific research, which means that Suh's complexity theory and recent extensions of it (see Summers and Shah 2010) must be discussed in the light of the latest theoretical developments. Moreover, the literature that has been published concerning the design structure matrix (Steward 1981) as a universal dependency modeling technique has to be considered (see e.g. Lindemann et al. 2009; Eppinger and Browning 2012). This literature also provides a firm foundation for quantitative modeling of cooperative work in PD projects by means of either time- or task-based design structure matrices (see e.g. Gebala and Eppinger 1991; Smith and Eppinger 1997; Schlick et al. 2007). In general, we have sought to restrict our analyses to mature scientific theories because of their universality, objectivity and validity.

3.1 Approaches from Organizational Theory

According to Murmann (1994) and Griffin (1997), complexity in the product development environment is determined by the number of (different) parts in the product and the number of embodied product functions. This basic approach can be used to assess complexity in different types of PD projects, for instance the five classic types defined by Wheelwright and Clark (1992): research and development, breakthrough, platform, derivative, and alliances and partnership projects. To make this approach fully operational, Kim and Wilemon (2003) developed a complexity assessment template covering these and other important "sources." The first source in their assessment template is "technological complexity," which can be divided into "component integration" and "technological newness." The second source is the "market (environmental) complexity" that results from the sensitivity of the project's attributes to market changes. "Development complexity" is the third source and is generated when different design decisions and components have to be integrated, qualified suppliers have to be found and supply chain relationships have to be managed. The fourth source is "marketing complexity," which results from the challenges of bringing the product to market. "Organizational complexity" is the fifth source, because projects usually require intensive cooperation and involve many areas of the firm. Their coordination leads to "intraorganizational complexity," the sixth source. When in large-scale engineering projects many other companies such as highly specialized engineering service providers are involved and must be coordinated in a continuous integration rhythm, this source should be extended and cover both inter- and intraorganizational complexity. In order to validate and prioritize sources of complexity, Kim and Wilemon (2009) conducted an extensive empirical investigation. An analysis of exploratory field interviews with 32 project leaders and team members showed that technological challenges, product concept/customer requirement ambiguities and organizational complexity are major issues that generate complexity in PD. The perceived dominant source was technological challenges, since roughly half of the respondents noted technological difficulties encountered in attempting to develop a product using an unproven technique or process. With regard to complexity in the management of projects of different types—not necessarily focusing on (new) product development projects—Mulenburg (2008) distinguishes between the following six sources: (1) Details: number of variables and interfaces, (2) Ambiguity: lack of awareness of events and causality, (3) Uncertainty: inability to pre-evaluate actions, (4) Unpredictability: inability to know what will happen, (5) Dynamics: rapid rate of change, and (6) Social structure: number and types of interactions between actors.

Hölttä-Otto and Magee (2006) developed a project complexity framework based on the seminal work of Summers and Shah (2003). They identified three dimensions: the product itself (artifact), the project mission (design problem), and the tasks required to develop the product (process). The key indicators for each of these dimensions are size, interactions and stretch (solvability). Hölttä-Otto and Magee conducted interviews in five divisions of large corporations competing in different industries on the North American market. Their findings show that the effort estimation is primarily based on the scale and the stretch of the project. Surprisingly, they found no utilization of the level of either component or task interactions in estimating project complexity. Further, they found no empirical evidence for interactions being a determinant of project difficulty (Hölttä-Otto and Magee 2006). Tatikonda and Rosenthal (2000) focus on the task dimension and relate project complexity to the nature, quantity and magnitude of the organizational subtasks and subtask interactions required by a project.

A recent work combining a literature review and their own empirical work on the elements that contribute to complexity in large engineering projects was published by Bosch-Rekveldt et al. (2011). The analysis of the literature sources and 18 semi-structured interviews in which six completed projects were studied in depth led to the development of the TOE framework. The framework covers 50 different elements, which are grouped into three main categories: "technical complexity" (T), "organizational complexity" (O) and "environmental complexity" (E). Additional subcategories of TOE are defined on a lower level: "goals," "scope," tasks," "experience," "size," "resources," "project team," "trust," "stakeholders," "location," "market conditions," and "risks," showing that organizational and environmental complexity are more often linked with softer, qualitative aspects. Interestingly, Bosch-Rekveldt et al. (2011) distinguish between project complexity and project management (or managerial) complexity. Project management complexity is seen as a subset of project complexity. Various normative organizing principles for coping with managerial complexity can be found in the standard literature on project management (e.g. Shtub et al. 2004; Kerzner 2009). If, for instance, the level of managerial complexity is low, project management within the classic functional organizational units of the company is usually most efficient and cross-functional types of project organization can create unnecessary overhead. However, if coordination needs between functional, spatial and temporal boundaries are high, a matrix organization is often the better choice, as it allows development projects to be staffed with specialists from across the organization (Shtub et al. 2004). The preferred organizational structure for large-scale, long-term engineering projects is pure project organization. The inherent advantage of this type of structure is that responsibilities for the project lie with one team, which works full-time on the project tasks throughout the entire project life cycle. Specific sources of managerial complexity and their impact on performance were also examined in the literature, e.g. communication across functional boundaries (Carlile 2002), cross-boundary coordination (Kellogg et al. 2006), spatial and temporal boundaries in globally distributed projects (Cummings et al. 2009), and the effects of a misalignment in the geographic configuration of globally distributed teams (O'Leary and Mortensen 2010). Maylor et al. (2008) developed an integrative model of perceived managerial complexity in project-based operations. Based on a multistage empirical study elements of complexity were identified and classified under the dimensions of "mission," "organization," "delivery," "stakeholder," and "team."

The literature review shows that there are a large variety of nomenclatures and definitions for the sources of complexity in PD projects. However, the underlying factors have not yet been integrated into a single objective and valid framework. According to Lebcir (2011) there is an urgent need for a new, non-confusing, and comprehensive framework that is derived from the extensive body of available knowledge. He suggests a framework in which "project complexity" is decomposed into "product complexity" and "innovation." Product complexity refers to structural complexity (see Section 3.3) and is determined by "product size" in terms of the number of elements (components, parts, subsystems, functions) in the product and by "product interconnectivity," which represents the level of linkages between elements. On the other hand, innovation refers to "product newness" and "project uncertainty." Product newness represents the degree of redesign of the product compared to previous generations of the same or similar products. Project uncertainty represents the fact that methods and capabilities are often not clearly defined at the start of a project. The results of a dynamic simulation indicate that an increase in uncertainty has a significant impact on the development time. The other factors also tend to increase development time as they increase, but their impact is not significantly different in projects involving medium or high levels of these factors.

In reviews of the more practice-oriented project management literature, two complexity models have received considerable attention, especially at large-scale development organizations: (1) the UCP (uncertainty, complexity and pace) model and the (2) NTCP (novelty, technology, complexity and pace) model. Both models were developed by Shenhar and colleagues (Shenhar and Dvir 1996, 2007; Shenhar

1998). In principle, these models can be applied to all types of projects. The UCP model is based on a conceptual two-dimensional taxonomy which classifies a project according to four levels of technological uncertainty and three levels of system scope. The four levels of technological uncertainty (low-tech, medium-tech, high-tech and super-high-tech) mainly refer to the uncertainty as perceived by the organization at the time of the project's initiation and thereby indicate how soon the product functions can be concretized. Moreover, they characterize the extent of new and therefore possibly premature technologies that are needed to reach the project goals. In the UCP model, the second dimension of system scope is based on the complexity of the system as expressed by the different hierarchies inside the product (assembly, system and array). Since systems are composed of subsystems and subsystems of components, hierarchies usually involve many levels. Hierarchies apply to systems as well as to tasks, which together determine the overall complexity of the project. The element of time in terms of "pace" was added to the model to account for the urgency and criticality of reaching milestones, as milestones with different time constraints call for different managerial strategies (Dvir et al. 2006). When complexity, uncertainty or pace increase, project planning becomes more difficult and the risk of project failure increases. Consequently, the formality of project management must also increase. The UCP model is based on quantitative and qualitative analyses of more than 250 projects within the US and Israeli defense and industry sectors. However, since it is usually used in retrospect rather than at the outset of new projects, the UCP model has a descriptive character. In contrast, the NTCP model, also called the "Diamond Framework," was developed as a prescriptive model in order to analyze projects and provide a better understanding of what needs to be done in order to ensure their success. The Diamond Framework is based on four pillars. The first pillar, "novelty," refers to the degree of newness (derivate vs. platform vs. breakthrough) of project results or crucial aspects of the project. With varying degrees of novelty, different requirements must be satisfied and corresponding action plans have to be developed. The second pillar represents the level of "technological uncertainty" in a project. Technological uncertainty is primarily determined by the level of new and mature technology required (low vs. medium vs. high vs. super-high technology). As the level of technological uncertainty rises, the risk of failure and efficiency loss increases. "Complexity," the third pillar, describes the types of arrangement between elements within the system, especially their hierarchical structure (assembly vs. system vs. array). Higher degrees of complexity entail more interaction between elements, which in turn demands higher project management formality. The fourth and final pillar of the NTCP model is "pace." As within the UCP model, pace refers to the urgency of reaching time goals and milestones. It chiefly depends on the available time for project completion and is divided into four types: regular, fast/competitive, time-critical and blitz. By considering all four pillars of the NTCP model at the beginning of the project and revisiting them as it progresses, project managers are provided with a methodology for assessing the uniqueness of their project and selecting appropriate management methods and techniques for coping with complexity.

Like the UCP and NCTP models, the Project Complexity Model developed by Hass (2009) has received a great deal of attention in the practice-oriented project management community. This model offers a broad framework for identifying and diagnosing the aspects of complexity within a project so that the project team can make appropriate management decisions. The model captures a number of sources of project complexity, including project duration and value; team size and composition; urgency; schedule, cost, and scope flexibility; clarity of the problem and solution; stability of requirements; strategic importance; stakeholder influence; level of organizational and commercial change; external constraints and dependencies; political sensitivity; and unproven technology (Hass 2009). The detailed complexity dimensions are shown in Table 3.1. The Project Complexity Model can also be used to evaluate the complexity of a particular project in an enterprise. To carry out the evaluation, Hass (2009) developed a corresponding "Project Complexity Formula," which is summarized in Table 3.2.

The complexity templates and frameworks that have been developed in organization theory and neighboring disciplines are especially beneficial for the management of product development projects because they help to focus managerial intervention on empirically validated performance-shaping factors and key elements of complexity. It must be criticized, though, that without a quantitative theory of emergent complexity it is almost impossible to identify the essential variables and their interrelationships. Furthermore, it is very difficult to consolidate them into one consistent complexity metric. In the literature very few authors, such as Mihm et al. (2003, 2010), Rivkin and Siggelkow (2003, 2007), and Braha and Bar-Yam (2007) build upon quantitative scientific concepts for the analysis of complex sociotechnical systems. Mihm et al. (2003) present analytical results from random matrix theory predicting that the larger the project, as measured by components or interdependencies, the more likely are problem-solving oscillations are and the more severe they become—failure rates grow exponentially. In the work of Rivkin and Siggelkow (2003, 2007), Kaufman's the famous biological evolution theory and the NK model are used to study organizations as systems of interacting decisions. Different interaction patterns such as block diagonal, hierarchical, scale-free, and so on are integrated into a simulation model to identify local optima. The results show that, by keeping the total number of interactions between decisions fixed, a shift in the pattern can alter the number of local optima by more than one order of magnitude. In a similar fashion Mihm et al. (2010) use a statistical model and Monte Carlo experiments to explore the effect of an organizational hierarchy on search solution stability, quality and speed. Their results show that assigning a lead function to "anchor" a solution speeds up problem-solving, that the choice of local solutions should be delegated to the lowest hierarchical level, and that organizational structure is comparatively unimportant at the middle management level, but does indeed matter at the "front line," where groups should be kept small. Braha and Bar-Yam (2007) examine the statistical properties of networks of people engaged in distributed development and discuss their significance. The autoregression models of cooperative work that were introduced in Chapter 2 (Eq. 8 and 39) are quite closely related to their dynamical model. However, there

Table 3.1 Complexity dimensions and project complexity profiles of the Project Complexity Model developed by Hass (2009)

Complexity dimensions	Project complexity profile		
	Independent	Moderately complex	Highly complex
Time/Cost	<3 months < $250K	3–6 months $250K–$750K	>6 months > $750K
Team Size	3–4 team members	5–10 team members	>10 team members
Team Composition and Performance	• Strong project leadership • Team staffed internally, has worked together in the past, and has a track record of reliable estimates • Formal, proven PM, BA and SE methodology with QA and QC processes defined and operational	• Competent project leadership • Team staffed with internal and external resources; internal staff has worked together in the past and has track record of reliable estimates • Contract for external resources is straightforward; contractor performance is known • Semi-formal methodology with QA/QC processes defined	• Project manager inexperienced in leading complex projects • Complex team structure of varying competencies (e.g., contractor, virtual, culturally diverse, outsourced) • Complex contracts; contractor performance unknown • Diverse methodologies
Urgency and Flexibility of Cost, Time and Scope	• Minimized scope • Small milestones • Flexible schedule, budget and scope	• Schedule, budget and scope can undergo minor variations, but deadlines are firm • Achievable scope and milestones	• Over-ambitious schedule and scope • Deadline is aggressive, fixed, and cannot be changed • Budget, scope and quality leave no room for flexibility
Clarity of Problem, Opportunity and Solution	• Clear business objectives • Easily understood problem, opportunity or solution	• Defined business objectives • Problem or opportunity is partially defined • Solution is partially defined	• Unclear business objectives • Problem or opportunity is ambiguous and undefined • Solution is difficult to define
Requirements Volatility and Risk	• Strong customer/user support • Basic requirements are understood, straightforward and stable	• Adequate customer/user support • Basic requirements are understood but are expected to change • Moderately complex functionality	• Inadequate customer/user support • Requirements are poorly understood, volatile and largely undefined • Highly complex functionality
Strategic Importance, Political Implications, Multiple Stakeholders	• Strong executive support • No political implications • Straightforward communications	• Adequate executive support • Some direct impact on mission • Minor political implications	• Mixed/inadequate executive support • Impact on core mission • Major political implications

(continued)

Table 3.1 (continued)

Complexity dimensions	Project complexity profile		
	Independent	Moderately complex	Highly complex
		• 2–3 stakeholder groups • Challenging communication and coordination effort	• Visible at highest levels of the organization • Multiple stakeholder groups with conflicting expectations
Level of Organizational Change	• Impacts a single business unit, one familiar business process and one IT system	• Impacts 2–3 somewhat familiar business units, processes and IT systems	• Large-scale organizational change that impacts the enterprise • Spans functional groups or agencies • Shifts or transforms the organization • Impacts many business processes and IT systems
Level of Commercial Change	• Minor changes to existing commercial practices	• Enhancements to existing commercial practices	• Groundbreaking commercial practices
Risks, Dependencies, and External Constraints	• Considered low risk • Some external influences • No challenging integration issues • No new or unfamiliar regulatory requirements • No punitive exposure	• Considered moderate risk • Some project objectives are dependent on external factors • Challenging integration effort • Some new regulatory requirements • Acceptable exposure	• Considered high risk • Overall project success largely depends on external factors • Significant integration required • Highly regulated or novel sector • Significant exposure
Level of IT Complexity	• Solution is readily achievable using existing, well-understood technologies • IT complexity is low	• Solution is difficult to achieve or technology is proven but new to the organization • IT complexity and legacy integration are moderate	• Solution requires groundbreaking innovation • Solution is likely to use immature, unproven or complex technologies provided by outside vendors • IT complexity and legacy integration are high

Table 3.2 Decision table of the Project Complexity Formula developed by Hass (2009)

Highly Complex	Moderately Complex	Independent
Level of change = large-scale enterprise impacts *or* Both the problem and the solution are difficult to define or understand, and the solution is difficult to achieve. The solution is likely to use unproven technologies. *or* Four or more categories in the "highly complex" column	Two or more categories in the "moderately complex" column *or* One category in the "highly complex" column and three or more in the "moderately complex" column	No more than one category in the "moderately complex" column *and* No categories in the "highly complex" column

To evaluate the complexity of a particular project, the boxes in the Project Complexity Model from Table *3.1* that best describe the project must be shaded out. Then, the complexity formula can be applied by following the decision rules above

are important differences: the VAR(1) models are defined over a continuous range of state values and can therefore represent different kinds of cooperative relationships as well as precedence relations (e.g. overlapping); each task is unequally influenced by other tasks; and finally, correlations ρ_{ij} between performance fluctuations among tasks i and j can be captured.

3.2 Approaches from Basic Scientific Research

3.2.1 Algorithmic Complexity

Historically, the most important measure from basic scientific research is algorithmic complexity, which dates back to the great mathematicians Kolmogorov, Solomonoff and Chaitin. They independently developed a measure known today as the "Kolmogorov–Chaitin complexity" (Chaitin 1987; Li and Vitányi 1997). In terms of information processing, the complexity of the intricate mechanisms of a nontrivial system can be evaluated using output signals, signs and symbols that are communicated to an intelligent observer. In this sense, complexity is manifested to an observer through the complicated way in which events unfold in time and are organized in state space. According to Nicolis and Nicolis (2007), the characteristic hallmarks of such spatiotemporal complexity are nonrepetitiveness, a pronounced variability extending over many scales of place and time, and sensitivity to initial conditions and to the other parameters. Furthermore, a given system can generate a variety of dependencies of this kind associated with the different states simultaneously available. If the transmitted output of a complex system is symbolic, it can be concatenated in the form of a data string x and may be sequentially stored in a computer file for post-hoc analysis. The symbols are typically chosen from a predefined alphabet $\mathcal{X}$. If the output is a time- or space-continuous signal, it can be effectively encoded with methods of symbolic dynamics (Lind and Marcus 1995; Nicolis and Nicolis 2007). The central idea put forward by Kolmogorov, Solomonoff and Chaitin is that a generated string is "complex" if it is difficult for the observer to describe. The observer can describe the string by writing a computer program that reproduces it. The difficulty of description is measured by the length of the computer program on a Universal Turing Machine U. If x is transformed into binary form, the algorithmic complexity of x, termed $K_U(x)$, is the length of the shortest program with respect to U that will print x and then halt. According to Chaitin (1987), an additional requirement is that the string x has to be encoded by a prefix code $d(x)$. A prefix code is a type of code system that has no valid code word that is a prefix (start substring) of any other valid code word in the set. The corresponding universal prefix computer U has the property that if it is defined for a string s, then $U(st)$ is undefined for every string t that is not the empty string ε (Li and Vitányi 1997). The complete definition of the Kolmogorov–Chaitin complexity is:

$$K_U(x) = \min\{|d(p)| : U(p) = x\}. \tag{200}$$

In this sense, $K_U(x)$ is a measure of the computational resources needed to specify the data string x in the language of U. We can directly apply this algorithmic complexity concept to project management by breaking down the total amount of work involved in the project into fine-grained activities a_i and labeling the activities unambiguously by using discrete events e_i from a predefined set $\mathcal{X}$ $(i = 1, \ldots, |\mathcal{X}|)$. During project execution it is recorded when activity a_i is successfully completed and this is indicated by scheduling the corresponding event e_i. The sequence of scheduled events $x = \left(e_{j(o)}, e_{j(1)} \ldots\right)$ $\left(e_{j(i)} \in \mathcal{X}, j(\tau) \in \{1, \ldots, |\mathcal{X}|\}, \tau = 0, 1, \ldots\right)$ encodes how the events unfold in time and are organized in a goal-directed workflow. The index $j(\tau)$ can be interpreted as a pointer to the event e that occurred at position τ in the data sequence x. It is evident that a simple periodic work process whose activities are processed in strict cycles, like in an assembly line, is not complex because we can store a sample of the period and write a program that repeatedly outputs it. At the opposite end of the complexity range in the algorithmic sense, a completely unpredictable work process without purposeful internal organization cannot be described in any meaningful way except by storing every feature of task processing, because we cannot identify any persisting structure that could offer a shorter description. This example quite clearly shows that the algorithmic complexity is not a good measure for emergent complexity in PD projects, because it is maximal in the case of purely random task processing. Intuitively, such a state of "amnesia," in which no piece of information from the project history is valuable for improving the forecasts of the project manager and the team members, is not truly complex. Nor can the algorithmic complexity reveal the important long-range interactions between tasks or evaluate multilayer interactions in the hierarchy of an organization either. An additional conceptual weakness of the algorithmic complexity measure and its later refinements is that it aims for an exact description of patterns. Many of the details of any configuration are simply random fluctuations from different sources such as human performance variability. Clearly, it is impossible to identify regularities from random fluctuations that generalize to other datasets from the same complex system; to assess complexity, the focus must be on the underlying regularities and rules shaping system dynamics. These regularities and rules must be distinguished from noise by employing specific selection principles. Therefore, a statistical representation is necessary that refers not to individual patterns but to a joint ensemble generated by a complex system in terms of an information source. In complex systems, the deterministic and probabilistic dimensions become two facets of the same reality: the limited predictability of complex systems (in the sense of the traditional description of phenomena) necessitates adopting an alternative view, and the probabilistic description allows us to sort out regularities of a new kind. On the other hand, far from being applied in a heuristic manner, in which observations have to fit certain preexisting laws imported from classical statistics, the probabilistic description we are dealing with here is "intrinsic" (Nicolis and Nicolis 2007), meaning that it is self-generated by the underlying system dynamics. Depending on the scale of the phenomenon, a complex system may have to develop mechanisms for controlling randomness to sustain a global behavioral pattern or, in contrast, to thrive on randomness and to acquire in a transient manner the variability and flexibility needed for its evolution between two such configurations. In addition to these significant conceptual weaknesses, a fundamental computational problem is that $K_U(x)$ cannot be calculated exactly. We can only approximate it "from above," which is the subject of the famous Chaitin theorem (Chaitin 1987). Later extensions of the classic concept of algorithmic complexity focus on complementary computational resources. In Bennett's (1988) logical depth the number of computing steps is counted that the minimum length program on a Universal Turing Machine U requires to generate the data string x. In Koppel and Atlan's (1991) theory of "sophistication" only the length of the part of the program on U is evaluated that captures all regularities of the data string. This means that, as with effective complexity (Gell-Mann 1995; Gell-Mann and Lloyd 1996; Gell-Mann and Lloyd 2004, see Section 3.2.3), irreducible random fluctuations that do not generalize to other datasets are sorted out. As with the Kolmogorov–Chaitin complexity, logical depth and sophistication are not computable, even with a generative model (Crutchfield and Marzen 2015).

3.2.2 *Stochastic Complexity*

The most prominent statistical complexity measure is Rissanen's (1989, 2007) stochastic complexity. It is rooted in the construction of complexity penalties for model selection (see procedure for VAR(n) model in Section 2.4), where a good trade-off between the prediction accuracy gained by increasing the number of free parameters and the danger of overfitting the model to random fluctuations and not regularities that generalize to other datasets has to be found. In an early paper,

Wallace and Boulton (1968) hypothesized that this trade-off could best be achieved by selecting the model with "the briefest recording of all attribute information." Akaike (1973, 1974) developed an important quantitative step along this line of thought by formulating a simple relationship between the expected Kullback-Leibler information and Fisher's maximized log-likelihood function (see deLeeuw 1992). He created his model selection criterion—which is today known as the Akaike Information Criterion (AIC, see Section 2.4)—without explicit links to complexity theory. Yet even from a complexity-theoretical perspective the AIC is not arbitrary, as it represents the asymptotic bias correction term of the maximized log-likelihood from each approximating model to full reality and can therefore be interpreted as a "complexity penalty" for increasing the number of free parameters beyond a point that is justified by the data (Burnham and Anderson 2002). Mathematically speaking, the AIC is defined as (Burnham and Anderson 2002)

$$AIC = -2 \ln \mathcal{L}\left(\widehat{\theta}|x\right) + 2k, \tag{201}$$

where the expression $\ln \mathcal{L}\left(\widehat{\theta}|x\right)$ denotes the numerical value of the log-likelihood at its maximum point, and k denotes the effective number of parameters (see Section 2.4). The maximum point of the log-likelihood function corresponds to the values of the maximum likelihood estimates $\widehat{\theta}$ of the free parameters of the approximating model given data x. In terms of a heuristic complexity-theoretic interpretation, the first term in AIC, $-2 \ln \mathcal{L}\left(\widehat{\theta}|x\right)$ can be considered as a measure of lack of model fit, while the second term $2k$ represents the cited complexity penalty for increasing the freely estimated parameters beyond a point that is compatible with the data-generating mechanisms. In the above definition, the dependency of the criterion on the number of data points is only implicit through the likelihood function. According to Section 2.4, for $VAR(n)$ models assuming normally distributed errors with a constant covariance, the dependency can easily be made explicit from least square regression statistics (Eq. 67) as

$$AIC(n) = \ln \mathrm{Det}\left[\widehat{\Sigma}_{(n)}\right] + \frac{2}{T}k,$$

where

$$\widehat{\Sigma}_{(n)} = \frac{\widehat{\Delta}_{(n)}}{T} \tag{202}$$

is the maximum likelihood estimate of the one-step prediction error of order n and

$$k = np^2 + \frac{p(p+1)}{2}$$

denotes according to Eq. 68 the effective number of parameters related to the coefficient matrices $A_0, \ldots, A_{n-1}$ and the covariance matrix C of the inherent one-step prediction error (sensu Akaike 1973).

Akaike's fundamental ideas were systematically developed by Rissanen in a series of papers and books starting from 1978. Rissanen (1989, 2007) emphasizes that fitting a statistical model to data is equivalent to finding an efficient encoding of that data, and that in searching for an efficient code we need to measure not only the number of bits required to describe the deviations of the data from the model's predictions, but also the number of bits required to specify the independent parameters of the model (Bialek et al. 2001). This specification has to be made with a level of precision that is supported by the data.

To clarify this theoretically convincing concept, it is assumed that we carried out a work sampling study in a complex PD project involving many and intensive cooperative relationships between the development teams. Based on a large number of observations the proportion of time spent by the developers in predefined categories of activity $\mathcal{X} = \{x_1, \ldots, x_m\}$ (e.g. sketching, drawing, calculating, communicating etc.) was estimated with high statistical accuracy. In addition to the observations made at random times, a comprehensive longitudinal observation of the workflows of different development teams was carried out in a specific project phase at regular intervals. The observations were made in R independent trials and encoded by the same categories of activity $\mathcal{X}$. We define the r-th workflow in the specific project phase in formal terms as a data string $x_r^{\mathcal{T}} = \left(x_{j_r(0)}, \ldots, x_{j_r(\mathcal{T})}\right)$ of length $(\mathcal{T}+1)$ $\left(x_{j_r(\tau)} \in \mathcal{X},\ j_r(\tau) \in \{1, \ldots, |\mathcal{X}|\},\ \tau = 0, 1, \ldots, \mathcal{T}, r = 1, \ldots, R\right)$. In a similar manner as in the previous section the index $j_r(\tau)$ can be interpreted as a pointer to activity $x_{j_r(\tau)} \in \mathcal{X}$ observed at time instant τ in the r-th workflow encoded by $x_r^{\mathcal{T}}$. All empirically acquired data strings are stored in a database of ordered sequences $DB = \left\{x_1^{\mathcal{T}}, \ldots, x_R^{\mathcal{T}}\right\}$. We aim at developing an integrative workflow model that can be used for the prediction and evaluation of development activities in the project phase based on the theory of discrete random processes. Therefore, we start by defining a finite one-dimensional random process $(X_0, \ldots, X_{\mathcal{T}})$ of discrete state variables. In terms of information theory the process communicates to an observer how the development activities unfold and are organized in time. In formal terms, $(X_0, \ldots, X_{\mathcal{T}})$ is a joint ensemble $\mathbb{E}$, in which each outcome is an ordered sequence $\left(x_{j(0)}, \ldots, x_{j(\mathcal{T})}\right)$ with $x_{j(0)} \in \mathcal{X} = \{x_1, \ldots, x_m\}$, $x_{j(1)} \in \mathcal{X}, \ldots,$ $x_{j(\tau)} \in \mathcal{X}, \ldots, x_{j(\mathcal{T})} \in \mathcal{X}$ (see, e.g. MacKay 2003). Each component X_τ of the joint ensemble $\mathbb{E} = X_0, \ldots, X_\tau, \ldots, X_{\mathcal{T}}$ is an ensemble. An ensemble X_τ is a triple $\left(x_{j(\tau)}, A_{X_\tau}, P_{X_\tau}\right)$, where the outcome $x_{j(\tau)}$ is the value of a random variable that can take on one of a set of possible values $A_X = \left(a_1, a_2, \ldots, a_{|\mathcal{X}|}\right)$, having probabilities $P_X = \left(p_1, p_2, \ldots, a_{|\mathcal{X}|}\right)$, with $P(X_\tau = a_i) = p_i$ (MacKay 2003). It holds that $p_i \geq 0$ and $\Sigma_{a_i \in A_X} P(X = a_i) = 1$. We call the term

$$P\left(X_0 = x_{j(0)}, \ldots, X_{\mathcal{T}} = x_{j(\mathcal{T})}\right) \quad j(\tau) \in \{1, \ldots, |\mathcal{X}|\}$$

the joint probability of $\left(x_{j(0)}, \ldots, x_{j(\mathcal{T})}\right)$. The joint probability describes the statistical properties of the joint ensemble in the sense that, when evaluated at a given data point $\left(x_{j(o)}, \ldots, x_{j(\mathcal{T})}\right)$, we get the probability that the realization of the random sequence will be equal to that data point. A joint ensemble is therefore a probability distribution on $\{x_1, \ldots, x_m\}^{\mathcal{T}}$. Similar to the definition of the probability density function of a continuous-type random variable from Section 2.2, we can make the functional relationship between the values and their joint probability explicit and use a joint probability mass function $p_{(X_0,\ldots,X_\mathcal{T})}$:

$$p_{(X_0,\ldots,X_\mathcal{T})}\left(x_{j(0)}, \ldots, x_{j(\mathcal{T})}\right) = P\left(X_0 = x_{j(0)}, \ldots, X_\mathcal{T} = x_{j(\mathcal{T})}\right) \quad j(\tau) \in \{1, \ldots, |\mathcal{X}|\}.$$

The joint probability mass function completely characterizes the probability distribution of a joint ensemble. Without limiting the generality of the approach, the joint probability $P\left(X_0 = x_{j(0)}, \ldots, X_\mathcal{T} = x_{j(\mathcal{T})}\right)$ of $\left(x_{j(o)}, \ldots, x_{j(\mathcal{T})}\right)$ as an integrative workflow model of the specific phase of the PD project can be factorized over all $\mathcal{T}$ time steps using, iteratively, the definition for the conditional probability $P(X|Y) = P(X,Y)/P(Y)$ as:

$$\begin{aligned} &P\left(X_0 = x_{j(0)}, \ldots, X_\mathcal{T} = x_{j(\mathcal{T})}\right) \\ &\quad = P\left(X_0 = x_{j(0)}\right) \prod_{\tau=1}^{\mathcal{T}} P\left(X_\tau = x_{j(\tau)} \middle| X_{\tau-1} = x_{j(\tau-1)}, \ldots, X_0 = x_{j(0)}\right). \end{aligned}$$

The above decomposition of the joint probability into conditional distributions $P\left(X_\tau = x_{j(\tau)} \middle| X_{\tau-1} = x_{j(\tau-1)}, \ldots, X_0 = x_{j(0)}\right)$ with correlations of increasing length τ can theoretically capture interactions between activities of long range and therefore holds true under any circumstances of cooperative relationships in the given phase. It is assumed that there are persistent workflow patterns in the project phase and we can express them by means of a reduced dependency structure capturing only short correlations, e.g. by using a Markov chain of order $n \ll \mathcal{T}$ or an equivalent dynamic Bayesian network (see Gharahmani 2001). As such, the reduced dependency structure reflects only the essential signature of spatiotemporal coordination in the project phase on a specific time scale. In the simplest case, only transitions between two consecutive development activities must be taken into account and a Markov chain of first order is an adequate candidate model for capturing these direct dynamic dependencies. In this model the conditional probability distribution of development activities at the next time step—and in fact all future steps—depends only on the current activity and not on past instances of the process when the current activity is known. Accordingly, the current activity shields the future from past histories, and the joint probability can be expressed as:

$$P\left(X_0 = x_{j(0)}, \ldots, X_\mathcal{T} = x_{j(\mathcal{T})}\right) = P\left(X_0 = x_{j(0)}\right) \prod_{\tau=1}^{\mathcal{T}} P\left(X_\tau = x_{j(\tau)} \middle| X_{\tau-1} = x_{j(\tau-1)}\right).$$

After the model structure of the Markov chain of first order has been defined by the above factorization of the joint probability, we have to specify the free parameters.

Continuing the notation of the previous chapters we denote the parameter vector by $\theta \in \mathbb{R}^k$. Due to the intrinsic "memorylessness" of the chain, only the initial distribution

$$\pi_0 = \left(P(X_0 = x_1) \quad \ldots \quad P\left(X_0 = x_{|\mathcal{X}|}\right) \right) \in [0;1]^{|\mathcal{X}|}$$

of the probability mass over the state space $\mathcal{X}$ and the transition probabilities

$$P = \left(p_{ij}\right) = \begin{pmatrix} P\left(X_\tau = x_1 \middle| X_{\tau-1} = x_1\right) & P\left(X_\tau = x_2 \middle| X_{\tau-1} = x_1\right) & \ldots \\ P\left(X_\tau = x_1 \middle| X_{\tau-1} = x_2\right) & P\left(X_\tau = x_2 \middle| X_{\tau-1} = x_2\right) & \\ \vdots & & \ddots \end{pmatrix} \in [0;1]^{|\mathcal{X}|^2}$$

between consecutive activities are relevant for making good predictions. Hence, we have the ordered pair of parameters:

$$\theta_1 = \begin{bmatrix} \pi_0 & P \end{bmatrix}.$$

Note that only $(|\mathcal{X}| - 1)$ parameters of the initial distribution π_0 and $|\mathcal{X}|(|\mathcal{X}| - 1)$ of the transition matrix P are freely estimated parameters, because a legitimate probability distribution has to be formed and the constraints

$$\sum_{i=1}^{|\mathcal{X}|} \pi_0^{(i)} = 1 \quad \text{and} \quad \forall i : \sum_{j=1}^{|\mathcal{X}|} p_{ij} = 1$$

have to be satisfied.

We can use Maximum Likelihood Estimation (MLE, see Section 2.4) to minimize the deviations of the empirically acquired data sequences from the model's predictions (see e.g. Papoulis and Pillai 2002; Shalizi 2006). In other words, the goodness of fit is maximized. The maximum likelihood estimate of the parameter pair θ_1 is denoted by $\widehat{\theta}_{1,\mathcal{T}}$. MLE was pioneered by R. A. Fisher (cf. Edwards 1972) under a repeated-sampling paradigm and is the most prominent estimation technique. As an estimation principle, maximum likelihood is supported by $\widehat{\theta}_{1,\mathcal{T}}$'s asymptotic efficiency in a repeated sampling setting under mild regularity conditions and its attainment of the Cramér-Rao lower bound in many exponential family examples in the finite-sample case (Hansen and Yu 2001). For a first-order Markov chain, the estimate $\widehat{\theta}_{1,\mathcal{T}}$ can be determined by solving the objective function:

$$\begin{aligned} \widehat{\theta}_{1,\mathcal{T}} &= \arg\max_{\theta_1} \prod_{r=1}^{R} P\left(X_0 = x_{j_r(0)} \middle| \theta_1\right) \prod_{\tau=1}^{\mathcal{T}} P\left(X_\tau = x_{j_r(\tau)} \middle| X_{\tau-1} = x_{j_r(\tau-1)}, \theta_1\right) \\ &= \arg\max_{(\pi_0, P)} \prod_{r=1}^{R} P\left(X_0 = x_{j_r(0)} \middle| \pi_0\right) \prod_{\tau=1}^{\mathcal{T}} P\left(X_\tau = x_{j_r(\tau)} \middle| X_{\tau-1} = x_{j_r(\tau-1)}, P\right). \end{aligned}$$

Note that the objective function is only valid if all R data sequences had been acquired in independent trials.

Due to the inherent memorylessness of the first-order Markov chain, this model is usually not expressive enough to capture the complicated dynamic dependencies between activities in a project phase. Consequently, a second-order Markov chain is considered as a second approximating model with extended memory capacity. For this model, the joint probability can be expressed as:

$$\begin{aligned} P\big(X_0 = x_{j(0)}, \ldots, X_T = x_{j(T)}\big) = {} & P\big(X_0 = x_{j(0)}\big) P\big(X_1 = x_{j(1)} \big| X_0 = x_{j(0)}\big) \\ & \cdot \prod_{\tau=2}^{T} P\big(X_\tau = x_{j(\tau)} \big| X_{\tau-1} = x_{j(\tau-1)}, X_{\tau-2} = x_{j(\tau-2)}\big). \end{aligned}$$

It is evident that the conditional distribution $P\big(X_\tau = x_{j(\tau)} \big| X_{\tau-1} = x_{j(\tau-1)}, X_{\tau-2} = x_{j(\tau-2)}\big)$ cannot only be used to predict direct transitions between current and future activities but can also model transitions between activities of the process that are conditioned on two time steps in the past. To parameterize this extended chain, three quantities are required: The initial distribution

$$\pi_0 = \big(P(X_0 = x_1) \quad \ldots \quad P\big(X_0 = x_{|\mathcal{X}|}\big) \big) \in [0;1]^{|\mathcal{X}|},$$

the transition probabilities between consecutive activities at the first two time steps

$$P_0 = \big(p_{0,ij}\big) = \begin{pmatrix} P\big(X_1 = x_1 \big| X_0 = x_1\big) & P\big(X_1 = x_2 \big| X_0 = x_1\big) & \cdots \\ P\big(X_1 = x_1 \big| X_0 = x_2\big) & P\big(X_1 = x_2 \big| X_0 = x_2\big) & \\ \vdots & & \ddots \end{pmatrix} \in [0;1]^{|\mathcal{X}|^2}$$

and the transition probabilities for the next activity given both preceding activities at arbitrary time steps

$$P = \big(p_{ij}\big) = \begin{pmatrix} p\big(x_1 \big| x_1, x_1\big) & p\big(x_1 \big| x_1, x_2\big) & \cdots & p\big(x_1 \big| x_1, x_{|\mathcal{X}|}\big) \\ p\big(x_1 \big| x_2, x_1\big) & p\big(x_1 \big| x_2, x_2\big) & \cdots & p\big(x_1 \big| x_2, x_{|\mathcal{X}|}\big) \\ \vdots & \vdots & & \vdots \\ p\big(x_1 \big| x_{|\mathcal{X}|}, x_1\big) & p\big(x_1 \big| x_{|\mathcal{X}|}, x_2\big) & & p\big(x_1 \big| x_{|\mathcal{X}|}, x_{|\mathcal{X}|}\big) \\ p\big(x_2 \big| x_1, x_1\big) & p\big(x_2 \big| x_1, x_2\big) & & p\big(x_2 \big| x_1, x_{|\mathcal{X}|}\big) \\ p\big(x_2 \big| x_2, x_1\big) & p\big(x_2 \big| x_2, x_2\big) & & p\big(x_2 \big| x_2, x_{|\mathcal{X}|}\big) \\ \vdots & \vdots & & \vdots \\ \vdots & \vdots & & \vdots \\ p\big(x_{|\mathcal{X}|} \big| x_{|\mathcal{X}|}, x_1\big) & p\big(x_{|\mathcal{X}|} \big| x_{|\mathcal{X}|}, x_2\big) & & p\big(x_{|\mathcal{X}|} \big| x_{|\mathcal{X}|}, x_{|\mathcal{X}|}\big) \end{pmatrix} \in [0;1]^{|\mathcal{X}|^3}.$$

In the above matrix the shorthand notation $p\big(x_i \big| x_j, x_k\big) = P\big(X_\tau = x_i \big| X_{\tau-1} = x_j, X_{\tau-2} = x_k\big)$ was used. Hence, we have the parameter triple

$$\theta_2 = [\pi_0 \quad P_0 \quad P].$$

In this triple $(|\mathcal{X}| - 1)$ parameters of the initial distribution π_0, $|\mathcal{X}|(|\mathcal{X}| - 1)$, parameters of the initial transition matrix P_0 and $|\mathcal{X}|^2(|\mathcal{X}| - 1)$ of the general transition matrix P are freely estimated parameters, because a legitimate probability distribution has to be formed. The ordered pair $\begin{bmatrix}\pi_0 & P_0\end{bmatrix}$ can be regarded as the initial state of the chain. We denote the maximum likelihood estimate for the parameterized model by $\widehat{\theta}_{2,\mathcal{T}}$. The corresponding objective function is:

$$\begin{aligned}\widehat{\theta}_{2,\mathcal{T}} &= \arg\max_{\theta_2}\prod_{r=1}^{R} P\left(X_0 = x_{j_r(0)}\middle|\theta_2\right)P\left(X_1 = x_{j_r(1)}\middle|X_0 = x_{j_r(0)},\theta_2\right)\\ &\cdot\prod_{\tau=2}^{\mathcal{T}} P\left(X_\tau = x_{j_r(\tau)}\middle|X_{\tau-1} = x_{j_r(\tau-1)}, X_{\tau-2} = x_{j_r(\tau-2)},\theta_2\right)\\ &= \arg\max_{(\pi_0,P_0,P)}\prod_{r=1}^{R} P\left(X_0 = x_{j_r(0)}\middle|\pi_0\right)P\left(X_1 = x_{j_r(1)}\middle|X_0 = x_{j_r(0)},P_0\right)\\ &\cdot\prod_{\tau=2}^{\mathcal{T}} P\left(X_\tau = x_{j_r(\tau)}\middle|X_{\tau-1} = x_{j_r(\tau-1)}, X_{\tau-2} = x_{j_r(\tau-2)},P\right).\end{aligned}$$

It is not difficult to prove that the solutions of the objective functions for Markov chains of first and second order (as well as all higher orders) are equivalent to the relative frequencies of observed subsequences of activity in the database *DB* (Papoulis and Pillai 2002). In other words, the MLE results can be obtained by simple frequency counting of data substrings of interest. Let the #-operator be a unary counting operator that counts the number of times the data string $(x_{j(o)}x_{j(1)}\ldots)$ in the argument occurred in $DB = \{x_1^{\mathcal{T}},\ldots,x_R^{\mathcal{T}}\}$. Then the MLE yields

$$\widehat{\pi}_0 = \left\{\frac{1}{R}\right\}\cdot\begin{pmatrix}\#(x_1)_{\tau=0} & \ldots & \#(x_{|\mathcal{X}|})_{\tau=0}\end{pmatrix}$$

$$\widehat{P} = \left\{\frac{1}{R\mathcal{T}}\right\}\cdot\begin{pmatrix}\#(x_1x_1) & \#(x_1x_2) & \ldots\\ \#(x_2x_1) & \#(x_2x_2) & \\ \vdots & & \ddots\end{pmatrix}$$

for the first-order Markov chain and

$$\widehat{\pi}_0 = \left\{\frac{1}{R}\right\}\cdot\begin{pmatrix}\#(x_1)_{\tau=0} & \ldots & \#(x_{|\mathcal{X}|})_{\tau=0}\end{pmatrix}$$

$$\widehat{P}_0 = \left\{\frac{1}{R}\right\}\cdot\begin{pmatrix}\#(x_1x_1)_{\tau=0} & \#(x_1x_2)_{\tau=0} & \ldots\\ \#(x_2x_1)_{\tau=0} & \#(x_2x_2)_{\tau=0} & \\ \vdots & & \ddots\end{pmatrix}$$

$$\widehat{P} = \left\{\frac{1}{R(\mathcal{T}-1)}\right\}\cdot\begin{pmatrix}\#(x_1x_1x_1) & \#(x_2x_1x_1) & \cdots & \#\left(x_{|\mathcal{X}|}x_1x_1\right)\\ \#(x_1x_2x_1) & \#(x_2x_2x_1) & \cdots & \#\left(x_{|\mathcal{X}|}x_2x_1\right)\\ \vdots & \vdots & & \vdots\\ \#\left(x_1x_{|\mathcal{X}|}x_1\right) & \#\left(x_2x_{|\mathcal{X}|}x_1\right) & & \#\left(x_{|\mathcal{X}|}x_{|\mathcal{X}|}x_1\right)\\ \#(x_1x_1x_2) & \#(x_2x_1x_2) & & \#\left(x_{|\mathcal{X}|}x_1x_2\right)\\ \#(x_1x_2x_2) & \#(x_2x_2x_2) & & \#\left(x_{|\mathcal{X}|}x_2x_2\right)\\ \vdots & \vdots & & \vdots\\ \vdots & \vdots & & \vdots\\ \#\left(x_1x_{|\mathcal{X}|}x_{|\mathcal{X}|}\right) & \#\left(x_2x_{|\mathcal{X}|}x_{|\mathcal{X}|}\right) & & \#\left(x_{|\mathcal{X}|}x_{|\mathcal{X}|}x_{|\mathcal{X}|}\right)\end{pmatrix}$$

for the second-order chain. To estimate the initial state probabilities $\hat{\pi}_0$ only the observations $\left(\#(x_1)_{\tau=0} \quad \ldots \quad \#(x_{|\mathcal{X}|})_{\tau=0}\right)$ in the first time step $\tau = 0$ must be counted. To calculate the initial transition matrix P_0 of the Markov chain of second order, only the data points in the first two time steps have to be considered, and we therefore use $\#(x_. x_.)_{\tau=0}$ to indicate the number of all leading substrings of length two. The estimate of the initial state distribution can be refined by using the data from the cited work sampling study that was carried out prior to the longitudinal observation of workflows.

The above solutions show that in a complex PD project that already manifests its intrinsic complexity in a single project phase by a rich body of data sequences with higher-order correlations, the data can usually be predicted much better with a second-order Markov chain than with a first-order model. This is due to the simple fact that the second-order chain has additional $|\mathcal{X}|^2(|\mathcal{X}| - 1)$ free parameters for encoding specific activity patterns and therefore a larger memory capacity. By inductive reasoning we can proceed with nesting Markov models of increasing order n

$$\begin{aligned}
&P\left(X_0 = x_{j(0)}, \ldots, X_T = x_{j(T)}\right) = \\
&P\left(X_0 = x_{j(0)}\right) P\left(X_1 = x_{j(1)} \middle| X_0 = x_{j(0)}\right) \ldots \\
&P\left(X_{n-1} = x_{j(n-1)} \middle| X_{n-2} = x_{j(n-2)}, \ldots, X_0 = x_{j(0)}\right) \\
&\cdot \prod_{\tau=n}^{T} P\left(X_\tau = x_{j(\tau)} \middle| X_{\tau-1} = x_{j(\tau-1)}, \ldots, X_{\tau-n} = x_{j(\tau-n)}\right)
\end{aligned} \tag{203}$$

and capture more and more details of the workflows. Formally speaking, the n-th order Markov model is the set of all n-th order Markov chains, i.e. all statistical representations that are equipped with a starting state and satisfy the above factorization of the joint probability. Given the order n of the chain, the probability distribution of X_τ depends only on the n observations preceding τ. However, beyond an order that is supported by the data, we begin to encounter the problem of "not seeing the forest for the trees" and incrementally fitting the model to random fluctuations that do not generalize to other datasets from the same project phase.

In order to avoid this kind of overfitting, the maximum likelihood paradigm has to be extended, because for an approximating model of interest, the likelihood function only reflects the conformity of the model to the data. As the complexity of the model is increased and more freely estimated parameters are included, the model usually becomes more capable of adapting to specific characteristics of the data. Therefore, selecting the parameterized model that maximizes the likelihood often leads to choosing the most complex model in the approximating set. Rissanen's minimum description length (MDL) principle (1989) provides a natural safeguard against overfitting by using the briefest encoding of not only the attribute information related to the data sequences but also to the parameters of the approximating models. In general, let θ be a parameter vector of model

$$\mathcal{M}^{(n)} = \left\{P\left(X_0 = x_{j(0)}, \ldots, X_T = x_{j(T)} \middle| \theta\right) : \theta \in \Theta \subset \mathbb{X}^n\right\}$$

whose support is a set $\mathbb{X}$ of adequate dimensionality and consider the class

$$\mathcal{M} = \bigcup\nolimits_{n=1}^{N} \mathcal{M}^{(n)}$$

consisting of all models represented by parametric probability distributions $P\left(X_0 = x_{j(0)}, \ldots, X_T = x_{j(T)} \middle| \theta\right)$ from the first order up to order N (Rissanen 2012). Note that Rissanen (2012) also calls $\mathcal{M}^{(n)}$ a model class that is defined by the independent parameters. For the sake of simplicity and to remain consistent with the previously used notation, we simply speak of an approximating model. The sequence of discrete state variables $\left(X_0, \ldots, X_T \middle| \theta\right)$ forms a one-dimensional random process encoding a joint ensemble of histories that can be explained by the structure and independent parameters of an approximating model within the class $\mathcal{M}$. By using a model with a specific structure and parameters, the joint probability can usually be decomposed into predictive distributions whose conditional part does not scale with the length of the sequence and therefore does not need an exponentially growing number of freely estimated parameters. In the following the number of parameters incorporated in the vector θ is the only variable of interest that is related to a specific model representation within class $\mathcal{M}$.

As previously shown, a model from class $\mathcal{M}$ with parameter vector θ assigns a certain probability

$$p_\theta\left(x^T\right) = P\left(X_0 = x_{j(0)}, \ldots, X_T = x_{j(T)} \middle| \theta\right) \tag{204}$$

to a data sequence $\left(x_{j(0)}, \ldots, x_{j(T)}\right)$ of interest. If we take the definition of the Shannon information content of an ensemble X

$$I[x] := \log_2 \frac{1}{P(X = x)}, \tag{205}$$

then the likelihood function $p_\theta\left(x^T\right)$ can be transformed into an information-theory loss function L

$$\begin{aligned} L\left[\theta, x^T\right] &= I\left[p_\theta\left(x^T\right)\right] \\ &= \log_2 \frac{1}{P\left(X_0 = x_{j(0)}, \ldots, X_T = x_{j(T)} \middle| \theta\right)} \\ &= -\log_2 P\left(X_0 = x_{j(0)}, \ldots, X_T = x_{j(T)} \middle| \theta\right). \end{aligned} \tag{206}$$

According to Eq. 203 we can interpret $L\left[\theta, x^T\right]$ in a predictive view as the loss incurred when forecasting X_τ sequentially based on the conditional distributions $P\left(X_\tau = x_{j(\tau)} \middle| X_{\tau-1} = x_{j(\tau-1)}, \ldots, X_{\tau-n} = x_{j(\tau-n)}, \theta\right)$. The loss is measured using a logarithmic scale. In the predictive view MLE aims at minimizing the accumulated logarithmic loss. We denote the maximum likelihood estimate by the member $\widehat{\theta}_T$. In the sense of information theory, minimizing the loss can also be thought of as minimizing the encoded length of the data based on an adequate prefix code $d(x)$. Shannon's famous source coding theorem (see e.g. Cover and Thomas 1991) tells us that for an ensemble X there exists a prefix code $d(x)$ with expected length $L[d(x), X]$ satisfying

$$\begin{aligned} -\sum_{x \in \mathcal{X}} P(X=x) \log_2 P(X=x) &\leq L[d(x), X] \\ &< -\sum_{x \in \mathcal{X}} P(X=x) \log_2 P(X=x) + 1. \end{aligned} \tag{207}$$

The term on the left of the inequality is the "information entropy" (see Eq. 210). It measures in [bits] the amount of freedom of choice in the coding process. This fundamental quantity will be explained in detail in the next chapter. A beautifully simple algorithm for finding a prefix code with minimal expected length is the Huffman coding algorithm (see e.g. Cover and Thomas 1991). In this algorithm the two least probable data points in $\mathcal{X}$ are taken and assigned the longest codewords. The longest codewords are of equal length and differ only in the last digit. In the next step, these two symbols are combined into a new single symbol and the procedure is repeated. Since each recursion reduces the size of the alphabet by one, the algorithm will have assigned strings to all symbols after $|\mathcal{X}| - 1$ steps. Following the predictive view, we can obtain an intuitive interpretation of the logarithmic loss in terms of coding: the code length needed to encode the data points $\left(x_{j(0)}, \ldots, x_{j(T)}\right)$ with prefix code $d(x)$ based on the joint distribution $P(.)$ is simply the accumulated logarithmic loss incurred when the corresponding conditional distributions $P(.\,|.)$ are used to sequentially predict the τ-th outcome on the basis of the previous $(\tau - 1)$ observations (Grünwald 2007).

It is evident that this interpretation is incomplete; we have an encoded version of the data, but we have still not said what the encoding scheme for the member $\widehat{\theta}_T$ is. Thus, the total description length DL must be divided into two parts,

$$DL\left[x^T, \theta, \Theta\right] = L\left[\theta, x^T\right] + D[\theta, \Theta],$$

where $D[\theta, \Theta]$ denotes the code length in terms of the number of bits needed to specify the member within class $\mathcal{M}$. The two parts of description length are usually obtained in a sequential two-stage encoding process (see Hansen and Yu 2001). In the first stage, the description length $D\left[\widehat{\theta}_T, \Theta\right]$ for the best-fitting member $\widehat{\theta}_T$ is calculated. The $\widehat{\theta}_T$'s maximizing the goodness-of-fit can be obtained both by MLE and Bayesian estimation. In the second stage, the description length of data $L\left[\widehat{\theta}_T, x^T\right]$ is determined on the basis of the parameterized probability mass function $p_{\hat{\theta}_T}\left(x^T\right)$.

Clearly, the model related to $D[\theta, \Theta]$ represents the part of the description that can be generalized, while $L\left[\theta, x^{\mathcal{T}}\right]$ includes the noisy part that does not generalize to other datasets. If $D[\theta, \Theta]$ assigns short code words to simple models, we have the desired tradeoff: we can reduce the part of the data that looks like noise only by using a more elaborate approximating model. Such an assignment provides an effective safeguard against overfitting. The minimum description length (MDL) principle supplied by Rissanen (1989, 2007) allow us to select the model that minimizes the total description length:

$$\theta_{MDL} := \arg\min_{\theta} DL\left[x^{\mathcal{T}}, \theta, \Theta\right].$$

The only requirement for the code length of the optimizing parameters $D\left[\widehat{\theta}_{\mathcal{T}}, \Theta\right]$ of this general MDL principle is that they be decodable (Rissanen 2012). The definition of a prior probability as in Bayesian estimation is therefore not required. Minimizing the total description length is apparently a consistent principle in connection with maximum likelihood estimation, because if we want to maximize the joint probability $DL\left[x^{\mathcal{T}}, \theta, \Theta\right]$ we need to calculate the probability of the coincidence of the observed data and the different approximating models and choose the maximizing model. It is important to point out that in MDL, one is never concerned with actual encodings but only with code length functions, e.g. $L[d(x), X]$ for an ensemble X encoded by a prefix code $d(x)$ (Grünwald 2007). The stochastic complexity C_{SC} of the joint ensemble $X^{\mathcal{T}}$ with reference to the model class $\mathcal{M}$ is simply the MDL:

$$C_{SC}\left[x^{\mathcal{T}}, \Theta\right] := \min_{\theta} DL\left[x^{\mathcal{T}}, \theta, \Theta\right]. \tag{208}$$

Under mild conditions for the underlying data-generating process in the model class, as we provide more data, θ_{MDL} will converge to the model that minimizes the generalization error.

Returning to our previous example of workflow modeling with Markov chains, we can follow the considerations of Hansen and Yu (2001) and, for didactic purposes, construct a simple but reasonable two-part code for the n-th order Markov chain $\mathcal{M}^{(n)}$ within the class $\mathcal{M}$ of finite-order Markov chains up to order N. The parameter vector of the n-th order Markov chain is denoted by $\theta_n \in \Theta_n$. Firstly, the order has to be described. We can start with a straightforward, explicit description for n that is based on a binary prefix code with $\lceil \log_2 n \rceil$ zeros followed by a one. The encoding of n can be done by using a simple uniform code for $\left\{1, \ldots, 2^{\lceil \log_2 n \rceil}\right\}$. Therefore, we need approximately $2\lceil \log_2 n \rceil + 1$ bits to describe the model order. By applying Huffman's algorithm here, we can also obtain a more efficient uniform code with a length function that is not greater than $\lfloor \log_2 n \rfloor$ for all values of $\{1, 2, \ldots, n\}$ but is equal to $\lfloor \log_2 n \rfloor$ for at least two values in this set. The function $\lfloor . \rfloor$ provides the integer part of the argument. Whereas we know from Shannon's source coding theorem (Eq. 207) that an expected length of such a code is optimal only for a true uniform distribution of the order of the model, this code is a reasonable choice when little is known about how the data was generated. Secondly, the $\Sigma_{i=0}^{n}|\mathcal{X}|^{i}(|\mathcal{X}|-1)=|\mathcal{X}|^{n+1}$ best-fitting free parameters $\widehat{\theta}_{n,\mathcal{T}}$ have to be described. We start by discretizing the range [0; 1] of a single ensemble into equal cells of size δ and then apply Huffman's algorithm. If we discretize the Cartesian product $\Theta_n=[0;1]^{|\mathcal{X}|^{n+1}}$ associated with the joint ensemble $X^{\mathcal{T}}$ in the same fashion, the quantity $-\log_2\left(p\left([0;1]^{|\mathcal{X}|^{n+1}}\right)\cdot\delta^{|\mathcal{X}|^{n+1}}\right)=-\log_2 p\left([0;1]^{|\mathcal{X}|^{n+1}}\right)-|\mathcal{X}|^{n+1}\log_2\delta$ can be viewed as the code length of a prefix code for $\widehat{\theta}_{n,\mathcal{T}}$ (Hansen and Yu 2001). Here, the probability density p can be regarded as an auxiliary density. It is used instead of the unknown true parameter-generating density f. Assuming a continuous uniform distribution with density $p(x)=1$ for $x\in[0;1]^{|\mathcal{X}|^{n+1}}$ (and $q(x)=0$ otherwise), an additional $|\mathcal{X}|^{n+1}\log_2\delta$ bits are needed to describe the free parameters. In a compact parameter space, we can refine the description and choose for the precision $\delta=\sqrt{1/(\mathcal{T}+1)}$ for each effective dimension. Rissanen (1989) showed that this choice of precision is optimal in regular parametric families. The intuitive explanation is that $\sqrt{1/(\mathcal{T}+1)}$ represents the magnitude of the estimation error in $\widehat{\theta}_{n,\mathcal{T}}$ and therefore there is no need to encode the estimator with greater precision (Hansen and Yu 2001). When the uniform encoder is used, one needs a total of $\left(|\mathcal{X}|^{n+1}/2\right)\log_2(\mathcal{T}+1)$ bits to communicate an estimated parameter $\widehat{\theta}_{n,\mathcal{T}}$ of dimension $|\mathcal{X}|^{n+1}$. Putting both partial descriptions together leads to

$$D[\theta_n,\Theta_n]=\log_2 n+\frac{|\mathcal{X}|^{n+1}}{2}\log_2(\mathcal{T}+1).$$

Interestingly, the formalized total description length of the n-th order Markov chain is similar to the Schwarz-Bayes Criterion (BIC) for the *VAR*(n) (Eq. 71) and LDS (Eq. 189) models of cooperative work in the sense that model complexity is penalized with a factor that increases linearly in the number of free parameters and logarithmically in the number of observations in the joint ensemble. This is a clear and unambiguous indication that there are deep theoretical connections between different approaches to model selection. The predictive view of Markovian models provides us with a refined interpretation of model selection based on the MDL principle: given two approximating models $\mathcal{M}^{(1)}$ and $\mathcal{M}^{(2)}$, we should prefer the model that minimizes the accumulated prediction error resulting from a sequential prediction of future outcomes given all past histories (Grünwald 2007).

Regarded as a principle of model selection, MDL has proven very successful in many areas of application (see e.g. Grünwald 2007; Rissanen 2007). Nevertheless, a part of this success comes from carefully tuning the model-coding term $D[\theta,\Theta]$ in such a manner that those models that do not generalize well turn out to have long encodings. Though not illegitimate, this approach relies on the intuition and knowledge of the human model builder. Motivated in part by this kind of theoretical incompleteness, Rissanen (2012) refined the above general MDL principle in his latest textbook on optimal estimation of parameters, formulating a "complete MDL principle." The complete MDL principle differs from the previously formulated principle in the requirement that the code length for the parameters defining the model $\mathcal{M}^{(k)}$ is the negative logarithm of the probability defined by the joint distribution

$$\hat{p}_k\left(x^T\right) = \frac{p_{\hat{\theta}\left(x^T\right)}\left(x^T\right)}{\hat{C}_k},$$

where $\hat{C}_k$ is a normalizing coefficient. $\widehat{\theta}\left(x^T\right)$ represents the ML estimator and k denotes the number of parameters incorporated in the parameter vector θ (Rissanen 2012). The requirement for the code length can also be generalized to the case where even the number of parameters is estimated, see Rissanen (2012). Since $\hat{p}_k\left(x^T\right)$ is determined by the model $\mathcal{M}^{(k)}$, its code length is common for all data sequences. The code of $\hat{p}_k\left(x^T\right)$ for fixed k is complete. The logarithm of the normalizing coefficient is given by the maximum capacity for the model $\mathcal{M}^{(k)}$ within class $\mathcal{M}$:

$$\log_2 \hat{C}_k = \log_2 \int_{\Theta} \sum\nolimits_{x^T : \hat{\theta}\left(x^T\right) = \theta} p_{\hat{\theta}\left(x^T\right)}\left(x^T\right) d\theta > 0.$$

The range Θ of the integration is selected to make the integral finite. Rissanen (2012) also calls the term $\log_2 \hat{C}_k$, representing the maximum capacity for model $\mathcal{M}^{(k)}$, the maximum estimation information, and interprets it as a measure of the maximum amount of information an estimator can obtain about the corresponding distribution. The estimator maximizing the estimation information agrees with the standard ML estimator. The model related to $\hat{p}_k\left(x^T\right)$ was introduced earlier by Shtarkov (1987) as a universal information-theoretic model for data compression.

In spite of these recent refinements, the complete MDL principle has limitations in terms of selecting an adequate family of model classes. An additional shortcoming is non-optimality if the model class cannot be well defined (Rissanen 2007, 2012). Whatever its merits as a model selection method, stochastic complexity is not a good metric of emergent complexity in open organizational systems for three reasons (sensu Shalizi 2006). (1) The dependence on the model-encoding scheme is very difficult to formulate in a valid form for project-based organizations. (2) The log-likelihood term, $L\left[\theta, x^T\right]$, can be decomposed into additional parts, one of which is related to the entropy rate of the information-generating work processes (h_μ, Eq. 223) and which therefore reflects their intrinsic unpredictability, not their complexity. Other parts indicate the degree to which even the most accurate model in $\mathcal{M}$ is misspecified, for instance, through an improper choice of the coordinate system. Thus, it largely reflects our unconscious incompetence as modelers, rather than a fundamental characteristic of the process. (3) The stochastic complexity reflects the need to specify some particular organizational model and to formally represent this specification. This is necessarily part of the process of model development but seems to have no significance from a theoretical point of view. For instance, a sociotechnical system being studied does not need to represent its organization; it simply has it (Shalizi 2006).

3.2.3 *Effective Complexity*

Effective complexity (EC) was developed by Seth Lloyd and the Nobel laureate Murray Gell-Mann. The fact that random strings without any purposeful informational structure display maximal Kolmogorov–Chaitin complexity (see Section 3.2.1) was one of the main reasons for Gell-Mann and Lloyd's criticism of the algorithmic complexity concept from Section 3.2.1 and for their attempt to define effective complexity as a more intuitive measure for scientific discourse. The concept of EC and its mathematical treatment were the subject of a series of papers that gained a great deal of attention in the scientific community (Gell-Mann 1995; Gell-Mann and Lloyd 1996, 2004). As with previous approaches for evaluating the complexity of an entity with inherent regularities in terms of its structure and behavior, it is assumed that its complexity is manifested to an observer in the form of a data string x, typically encoded in binary form. However, Gell-Mann and Lloyd do not consider the minimum description length of the string itself, which is what Wallace and Boulton (1968) and Rissanen (1989, 2007) did to evaluate stochastic complexity. Instead, they consider the joint ensemble $\mathbb{E}$ in which the string is embedded as a typical member (Ladyman et al. 2013). "Typicality" is defined using the theory of types (see e.g. Cover and Thomas 1991), which means that the negative binary logarithm of the joint probability distribution of $\mathbb{E}[x]$ on $\{x_1, \ldots, x_m\}^T$ is approximately equal to the information entropy $H[\mathbb{E}]$ (see below and next chapter). To evaluate the minimum description length of the ensemble $\mathbb{E}$, the (prefix) Kolmogorov–Chaitin complexity from Eq. 200 is used. This approach assumes that one can find a meaningful way to estimate what the ensemble is. The resulting informal definition of the EC[x] of an entity is the Kolmogorov–Chaitin complexity of the ensemble $\mathbb{E}$, in which the string x manifesting the object's complexity to an observer is embedded as a δ-typical member. Instead of Kolmogorov–Chaitin complexity, Gell-Mann and Lloyd use the equivalent term "algorithmic information content" (Gell-Mann and Lloyd 1996, 2004). The main idea of EC is therefore to split the algorithmic information content of the string x into two parts, where the first contains all regularities and the second contains all random features. The EC of x is defined as the algorithmic information content of the regularities alone (Ay et al. 2010). In contrast to previous approaches, the EC is therefore not a metric for evaluating the difficulty of describing all the attribute information of an entity, but rather the degree of organization (Ladyman et al. 2013). By degree of organization, we mean the internal structural and behavioral regularities that can be identified by using ensembles as models of the string. Following this concept of ensemble-based complexity measurement, in order to compute the ensemble $\mathbb{E}$ a computer program on a universal computer U takes as input the target string x and a precision parameter n and simply outputs $\mathbb{E}[x]$ to precision n. This approach can resolve the paradox from Section 3.2.1, whereby random strings without any internal structure display high Kolmogorov–Chaitin complexity because no underlying regularities or rules exist that could allow a shorter description. The ideal ensemble for modeling a random string is a joint ensemble with a uniform distribution of the probability mass that assigns equal probability to every string x' of length $|x|$, and it holds that (Foley and Oliver 2011):

$$\mathbb{E}_x^U[x'] = 2^{-|x|}.$$

The Kolmogorov–Chaitin complexity of this ensemble is apparently very low, because the computer program used to calculate it on U simply calculates $2^{-|x|}$ to precision n when confronted with input x'. The EC of a random string is thus low, although it is incompressible and the Kolmogorov–Chaitin complexity is maximal for its length $|x|$ (Foley and Oliver 2011).

Ay et al. (2010) introduced a more formal approach to defining EC and proving some of its basic properties. In the following, we summarize their main definitions and interpretations. First, we have to define the Kolmogorov–Chaitin complexity $K_U[\mathbb{E}]$ of a joint ensemble $\mathbb{E}$. As previously stated, a program to compute the ensemble $\mathbb{E}$ on a universal prefix computer U expects two inputs: the target string x and a precision parameter $n \epsilon \mathbb{N}$. It outputs the binary digits of the approximation $\mathbb{E}_x^U$ of $\mathbb{E}[x]$ with an accuracy of at least 2^{-n}. The Kolmogorov–Chaitin complexity $K_U[\mathbb{E}]$ of $\mathbb{E}$ is then the length of the shortest program for the universal prefix computer U that computes $\mathbb{E}$ on the basis of the approximation $\mathbb{E}_x^U$ (Ay et al. 2010). Unfortunately, not every ensemble is computable, as there is a continuum of string ensembles but only a finite number of algorithms computing ensembles. Another subtlety is that the information entropy $H[\mathbb{E}] = \Sigma_{x \epsilon \mathcal{X}^T} \mathbb{E}(x) \log_2 \mathbb{E}(x)$ (cf. Eq. 210) as a measure of the "ignorance" of the probability distribution of a computable ensemble $\mathbb{E}(x)$ for string x does not necessarily need to be computable. All that is known is that it can be enumerated from "below." Thus, it must be assumed in the following that all ensembles are computable and have computable and finite entropy. Even when we restrict the analysis to the set of ensembles that are computable and have computable and finite entropy, the map $\mathbb{E} \mapsto H[\mathbb{E}]$ is not necessarily a computable function. Hence, the approximate equality $K_U[\mathbb{E}, H[\mathbb{E}]] \stackrel{+}{=} K_U[\mathbb{E}]$ is not necessarily uniformly true in $\mathbb{E}$ (the operator $\stackrel{+}{=}$ denotes an equality to within a constant). Therefore, the definition of $K_U(\mathbb{E})$ has to be refined (Ay et al. 2010):

$$K_U[\mathbb{E}] := K_U[\mathbb{E}, H[\mathbb{E}]].$$

We therefore assume that the programs on the universal prefix computer U computing $\mathbb{E}$ when confronted with input x carry an additional subroutine to compute the information entropy $H[\mathbb{E}]$. The Kolmogorov–Chaitin complexity $K_U[\mathbb{E}]$ is integer-valued. Second, we have to define the "total information" $\Sigma[\mathbb{E}]$ of an ensemble $\mathbb{E}$ (Gell-Mann and Lloyd 1996, 2004). To explain the role of the total information within the theory, Gell-Mann and Lloyd (2004) consider a typical situation in which a theoretical scientist is trying to construct a theory to explain a large body of data. The theory is represented by a probability distribution over a set of bodies of data. One body consists of the real data, while the rest of the bodies are imagined. In this setting, the Kolmogorov–Chaitin complexity $K_U[\mathbb{E}]$ corresponds to the complexity of the theory, and the information entropy $H[\mathbb{E}]$ measures the extent to which the predictions of the theory are distributed widely over different possible bodies of data. Ideally, the theorist would like both quantities to be small: the Kolmogorov–Chaitin complexity $K_U[\mathbb{E}]$ so as to make the theory simple, and the information entropy $H[\mathbb{E}]$ so as to make it focus narrowly on the real data points. However, there can be a trade-off. By adding more details to the theory and more arbitrary parameters, the theoretical scientist might be able to focus on the real data, but only at the expense of complicating the theory. Similarly, by allowing appreciable probabilities for many possible bodies of data, the scientist might be able to develop a simple theory. In any case, it makes good sense to minimize the sum of the two quantities that is defined as the total information $\Sigma[\mathbb{E}]$:

$$\Sigma[\mathbb{E}] := K_U[\mathbb{E}] + H[\mathbb{E}].$$

This allows the scientist to deal with the possible trade-off: a good estimate of the ensemble that generated the string x should not only have a small Kolmogorov–Chaitin complexity and therefore provide a simple explanation in the language of U; it should also have a small information entropy, as the explanation should have a low level of arbitrariness and prefer outcomes that include the string x. The total information is a real number larger than or equal to one. Third, we have to explain what is meant by an ensemble $\mathbb{E}$ in which the string is embedded as a typical member. As previously stated, typicality is defined according to the theory of types (see, e.g. Cover and Thomas 1991). To briefly explain the concept of typicality, suppose that we toss a biased coin with probability p that it lands on heads and $q = 1 - p$ that it lands on tails n times. We call the resulting probability distribution the ensemble $\mathbb{E}'$. It is well known from theoretical and empirical considerations that typical outcomes x have a probability $\mathbb{E}'[x]$ that is close to 2^{-nH} (Cover and Thomas 1991). In this case the information entropy is defined as $H := -p \log_2 p - q \log_2 q$. We can prove that the probability that $\mathbb{E}'[x]$ lies between $2^{-n(H+\varepsilon)}$ and $2^{-n(H-\varepsilon)}$ for $\varepsilon > 0$ tends to one as the number of tosses n grows. This property is a simple consequence of the weak law of large numbers and is the subject of the "asymptotic equipartition theorem" (Cover and Thomas 1991). Generalizing this property, we consider a string x as typical for a joint ensemble $\mathbb{E}$ if its probability is not much smaller than $2^{-nH[\mathbb{E}]}$. We say x is δ–typical for $\mathbb{E}$ for some small constant $\delta \geq 0$ if

$$\mathbb{E}[x] \geq 2^{-H[\mathbb{E}](1+\delta)}.$$

Fourth, we have to define how small the total information $\Sigma[\mathbb{E}]$ should be for an ensemble $\mathbb{E}$ that explains the string x well but is not unnecessarily complex in the language of U. This lemma by Ay et al. (2010) shows that the total information should not be too small: it uniformly holds for $x \epsilon \mathcal{X}^*$ and $\delta \geq 0$ that

$$\frac{K_U(x)}{1+\delta} <^+ \mathrm{Inf}\{\Sigma[\mathbb{E}] : x \text{ is typical for } \mathbb{E}\} <^+ K_U(x).$$

The symbol $<^+$ denotes an inequality to within a constant. $K_U(x)$ is the (algorithmic) Kolmogorov–Chaitin complexity of x according to Eq. 200. The function Inf{.} denotes the infimum of the generated set of total information values. Put simply, the lemma tells us that the total information $\Sigma[\mathbb{E}]$ should not be much larger than the Kolmogorov–Chaitin complexity of the string of interest. Fifth, the ultimate question is, of all the "good" ensembles according to the previously defined criteria, which ensemble $\mathbb{E}$ is the best for evaluating an entity's degree of organization? In their simple yet convincing answer, Gell-Mann and Lloyd (1996, 2004) claim it is the ensemble with minimum Kolmogorov–Chaitin complexity. The exact definition (Ay et al. 2010) is that, given small constants $\delta \geq 0$ and $\Delta \geq 0$, the effective complexity EC[x] of any string $x \epsilon \mathcal{X}^*$ is defined as:

$$\mathrm{EC}[x] := \mathrm{Inf}\{\mathrm{K}[\mathbb{E}] : x \text{ is typical for } \mathbb{E} \text{ and } \Sigma[\mathbb{E}] \leq K(x) + \Delta\}, \tag{209}$$

or as ∞ if this set is the empty set. The right-hand side of the above definition defines the minimization domain of the string x for effective complexity. Ensembles $\mathbb{E}$ of the minimization domain of $x \epsilon \mathcal{X}^*$ satisfy

$$\frac{K_U(x)}{1+\delta} <^+ \Sigma[\mathbb{E}] \leq K(x) + \Delta.$$

As Gell-Mann and Lloyd (2004) point out, it is often necessary to extend this definition of effective complexity by imposing additional constraints on the ensembles allowed in the minimization domain. These additional constraints can refer to certain properties of the string x that are judged important from the standpoint of a general scientific theory, or they can involve additional information about the processes that generated x (Ay et al. 2010). Ay et al. (2010) prove several properties of EC[x], such as its finiteness, and they show that incompressible strings are effectively simple, which is desirable given the criticism of the algorithmic complexity concept from Section 3.2.1. They also show that strings exist that have effective complexity close to their length $|x|$. Finally, one can show that EC[x] is related to Bennett's logical depth (1988, see Section 3.2.1). If the effective complexity of a string x exceeds a certain threshold, then the string must have an extremely large depth (Ay et al. 2010).

Moreover, Duncan Foley recently presented an interesting re-phrased formalism based on Bayesian inference. The Bayesian formulation allows us to interpret effective complexity in terms of the minimum description length principle of Wallace and Boulton (1968) and Rissanen (1989, 2007) as a two-part code (see notes on facticity and effective complexity by Foley and Oliver, 2011). To apply the method of Bayesian inference, Foley regards the problem of assigning probabilities to joint ensembles $\mathbb{E}$ as hypotheses, and the target string x as data. In this case, Bayes' theorem can be written as

$$P(\mathbb{E}|x) = P(\mathbb{E})\frac{P(x|\mathbb{E})}{P(x)},$$

where $P(\mathbb{E})$ is the prior probability assigned to the joint ensemble $\mathbb{E}$, $P(x|\mathbb{E})$ is the probability of the data given the ensemble ("likelihood"), and $P(x)$ is a normalizing constant. $P(\mathbb{E}|x)$ is the posterior probability of the joint ensemble given the data string x. Given the prior probability distribution $P(\mathbb{E}) = 2^{-K_U[\mathbb{E}]}$, the posterior distribution will be

$$\begin{aligned} P(\mathbb{E}|x) &\propto 2^{-K_U[\mathbb{E}]}P(x|\mathbb{E}) \\ &\propto 2^{-K_U[\mathbb{E}]}E[x]. \end{aligned}$$

The term $E[x]$ denotes the expected value of the corresponding discrete sequence. When we take the logarithm to base 2 to express information content in bits, we have

$$\log_2 P(\mathbb{E}|x) \propto -K_U[\mathbb{E}] + \log_2 E[x].$$

From Shannon's source coding theorem (Eq. 207), we know that the quantity $-\log_2 E[x]$ is the prefix code $d(x)$ with expected length $L[d(x), X]$ assigned to the data string x as a message to minimize average code length when the probabilities of messages are given by the joint ensemble $\mathbb{E}$. The negative logarithm of the posterior probability of a joint ensemble can therefore be regarded as the sum of the number of bits required to encode the ensemble as a program on U and as the length of code required to identify the string x given the distribution corresponding to $\mathbb{E}$. The logarithm of the posterior probability can also be interpreted in terms of the minimum description length principle from Sect 3.2.2 as the negative of the length of the two-part code transmitting the string x given the joint ensemble $\mathbb{E}$ as a generative model. Hence, we have the intuitive definition (Foley and Oliver 2011):

$$\mathrm{EC}[x] := K_U\left[\widehat{\mathbb{E}}_x = \arg\min_{\mathbb{E}}\{K_U[\mathbb{E}] - \log_2 E[x]\}\right].$$

It is important to note that this direct definition is a limited concept of effective complexity, as the information entropy $H[\mathbb{E}]$ of the ensemble is not evaluated.

In spite of its convincing concept and its conformity with the aforementioned expectations for a consistent complexity measure, in the following we will not consider the effective complexity in evaluating the emergent complexity of PD projects, as it is not computable. As Gell-Mann says: "There can exist no procedure for finding the set of all regularities of an entity" (Gell-Mann 1995, p. 2). This severe practical limitation leaves us with information-theoretic quantities based on dynamic entropies of joint ensembles that possess many (though not all) of the theoretically desired properties and can be efficiently and robustly estimated from data in a product development environment. These quantities will be discussed in the next chapter.

3.2.4 Effective Measure Complexity and Forecasting Complexity

Motivated in part by the theoretical weaknesses of the concept of stochastic complexity that were cited in Section 3.2.2 and by the uncomputability of algorithmic measures, the German physicist Peter Grassberger (1986) developed a simple but highly satisfactory complexity theory. He posits that complexity is the amount of information required for optimal prediction. We will begin by analyzing why this concept is plausible, and then go on to look at how to develop measuring concepts and make them fully operational. In general, there is a limit to the accuracy of any prediction of a given sociotechnical system set by the characteristics of the system itself, e.g. the free will of the decision makers, spontaneous human error, limited precision of measurement, sensitive dependence on initial conditions, etc. Suppose we have a model that is maximally predictive, i.e. its predictions are at the theoretical limit of accuracy. Prediction is always a matter of mapping inputs to outputs. In our application context, the inputs are the encoded observations of single instances of task processing (encoding, for instance, the labor units required to finalize a specific component, open design issues that need to be addressed before design release, etc.) and the outputs are the expectations about the work remaining, as well as macroscopic key performance indicators such as the finishing time of the project phase. However, usually not all aspects of the entire past are relevant for making good predictions. In fact, if the task processing is strictly periodic with a predefined cycle time, one only needs to know which of the φ phases the work process is in. For a completely randomized work process with independent and identically distributed (iid) state variables, the past is completely irrelevant for predicting the future. Because of this "memorylessness," the clever, evidence-based estimates of an experienced project manager on average do not outperform naïve guesses of the outcome based on means. If we ask how much information about the past is relevant in these two extreme cases, the correct answers are $\log_2(\varphi)$ and 0, respectively. It is intuitive that these cases are of low complexity, and more informative dynamics "somewhere in between" must be assigned high complexity values. In terms of Shannon's famous information entropy $H[.]$ the "randomness" of the output either is simply a constant (low-period deterministic process with small algorithmic complexity) or grows precisely linearly with the length (completely randomized process with large algorithmic complexity). Hence, it can be concluded that both cases share the feature that corrections to the asymptotic behavior do not grow with the size of the dataset (Prokopenko et al. 2009). Grassberger considered the slow approach of the entropy to its extensive limit as an indicator of complexity. In other words, the subextensive components growing less rapidly with time than a linear function are of special interest for complexity evaluation.

When dealing with a Markovian model, such as the VAR model of cooperative task processing formulated in Section 2.2, only the present state of work remaining is relevant for predicting the future (see Eq. 8), so the amount of information needed for optimal prediction is simply equal to the amount of information needed to specify the current state. More formally, any predictor g will translate the one-dimensional infinite past $X_{-\infty}^{-1} = (X_{-\infty}, X_{-\infty+1}, \ldots, X_{-1})$ into an effective state $S = g\left[X_{-\infty}^{-1}\right]$ and then make its prediction on the basis of S. This is true whether or not $g[\cdot]$ is formally a state-space model as we have formulated. The amount of information required to specify the effective state in the case of discrete-type random variables (or discretized continuous-type random variables) can be expressed by Shannon's information entropy $H[S]$ (Cover and Thomas 1991). We will return to this point later in the chapter and take $H[S]$ to be the statistical complexity C_{GCY} of $g[\cdot]$ under the assumption of a minimal maximally predictive model of the stationary stochastic process $\{X_t\}$ ($t \in \mathbb{Z}$, see Eq. 228).

Shannon's information entropy represents the average information content of an outcome. Formally, it is defined for a discrete-type random variable X with values in the alphabet $\mathcal{X}$ and probability distribution $P(.)$ as

$$H[X] := -\sum_{x \in \mathcal{X}} P(X = x) \log_2 P(X = x). \tag{210}$$

The information entropy $H[.]$ is non-negative and measures in [bits] the amount of freedom of choice in the associated decision process or, in other words, the degree of randomness. If we focus on the set $\mathcal{M}$ of maximally predictive models, we can define what Grassberger called "the true measure complexity C_μ of the process" as the minimal amount of information needed for optimal prediction:

$$C_\mu := \min_{g \in \mathcal{M}} H\left[g\left[X_{-\infty}^{-1}\right]\right]. \tag{211}$$

The true measure complexity is also termed "forecasting complexity" (Zambella and Grassberger 1988), because it is defined on the basis of maximally predictive models requiring the least average information content of the memory variable. We will use the term "forecasting complexity" in the following, as it is well-established and more intuitive. Unfortunately, Grassberger provided no procedure for finding the maximally predictive models or for minimizing the information content. However, he did draw the following conclusion. A basic result of information theory, called "the data-processing inequality" (Cover and Thomas 1991), states that for any pair of random variables X and Y (or pair of sequences of random variables) the mutual information $I[.;.]$ follows the rule

$$I[X;Y] \geq I[g[X];Y].$$

It is therefore impossible to extract more information from observations by processing than was in the sample to begin with. Since the state S of the predictor is a function of the past, it follows that

$$I\left[X_{-\infty}^{-1};X_0^{\infty}\right] \geq I\left[g\left[X_{-\infty}^{-1}\right];X_0^{\infty}\right],$$

where $X_0^{\infty} = (X_0, X_1, \ldots, X_{\infty})$ represents the infinite future of the stochastic process including the "present" that is encoded in the observation X_0.

The mutual information $I[.;.]$ is another key quantity of information theory (Cover and Thomas 1991). It can be equivalently expressed on the basis of the joint $P(.,.)$ and marginal probability mass functions $P(.)$ as

$$I[X;Y] := \sum_{x \in \mathcal{X}} \sum_{y \in \mathcal{Y}} P(X = x, Y = y) \log_2 \frac{P(X = x, Y = y)}{P(X = x)P(Y = y)} \tag{212}$$

or in terms of the information entropy $H[.]$ as

$$\begin{aligned} I[X;Y] &= H[X] - H\left[X \middle| Y\right] \\ &= H[Y] - H\left[Y \middle| X\right] \\ &= H[X] + H[Y] - H[X,Y] \\ &= H[X,Y] - H\left[X \middle| Y\right] - H\left[Y \middle| X\right]. \end{aligned}$$

In the above equations, with the conditional entropy (also called equivocation, Cover and Thomas 1991) we have used another important information-theoretic quantity which measures the amount of information for the random variable X given the value of another random variable Y. It can be explicitly written as

$$H\left[X \middle| Y\right] = H[X,Y] - H[Y]. \tag{213}$$

The mutual information $I[.;.]$ is non-negative and measures the amount of information that can be obtained about one random variable by observing another. It is symmetric in terms of these variables. System designers often maximize the amount of information $I[A;B]$ shared by transmitted and received signals by choosing the best transmission technique. Channel coding guarantees that reliable communication is possible over noisy communication channels, if the rate of information transmission is below a certain threshold that is termed "the channel capacity," defined as the maximum mutual information for the channel over all possible probability distributions of the signal (see Cover and Thomas 1991). According to Polani et al. (2006) mutual information should not be regarded as something that is transported from a transmitter to a receiver as a "bulk" quantity. Instead, the mutual information makes it possible to evaluate the intrinsic dynamics that can provide deeper insights into the inner structure of information; maximization of information transfer through selected channels appears to be one of the main evolutionary processes (Bialek et al. 2001; Polani et al. 2006).

In a similar manner, the conditional mutual information $I[X; Y|Z]$ (Cover and Thomas 1991) can be defined on the basis of the joint $P(.,.,.)$, marginal $P(.)$ and conditional $P(.|.)$ probability mass functions as

$$I[X;Y|Z]:=\sum_{z\in\mathcal{Z}}P(Z=z)\sum_{x\in\mathcal{X}}\sum_{y\in\mathcal{Y}}P(X=x,Y=y|Z=z)\log_2\frac{P(X=x,Y=y|Z=z)}{P(X=x|Z=z)P(Y=y|Z=z)}. \tag{214}$$

The conditional mutual information can be interpreted in its most basic form as the expected value of the mutual information of two random variables given the value of a third one. Alternatively, we can write

$$I[X;Y|Z]=H[X|Z]+H[Y|Z]-H[X,Y|Z]. \tag{215}$$

Presumably, for optimal predictors, the amounts of information $I[X_{-\infty}^{-1};X_0^{\infty}]$ and $I[g[X_{-\infty}^{-1}];X_0^{\infty}]$ are equal and the predictor's state is just as informative as the original data. This is the case for so-called "ε-machines," which are analyzed below. Otherwise, the model would be missing potential predictive power. Another basic inequality is that $H[X]\geq I[X;Y]$, i.e. no variable contains more information about another than it does about itself (Cover and Thomas 1991). Even for the maximally predictive models it therefore holds that $H[X_{-\infty}^{-1}]\geq I[X_{-\infty}^{-1};X_0^{\infty}]$. Grassberger called the latter quantity $I[X_{-\infty}^{-1},X_0^{\infty}]$— the mutual information between the infinite past and future histories of a stochastic process—the effective measure complexity (EMC):

$$\mathrm{EMC}:=I[X_{-\infty}^{-1};X_0^{\infty}]. \tag{216}$$

Recall that EMC is defined with reference to infinite sequences of random variables and is therefore only valid for stationary stochastic processes. The same is true for the forecasting complexity. For the sequence $(\ldots,X_{-1},X_0,X_1,\ldots)$ stationarity implies that the joint probability distribution $P(.,\ldots,.)$ associated with any finite block of n variables $X^n:=X_{t+1}^{t+n}=(X_{t+1},\ldots,X_{t+n})$ is independent of t and only depends on the block length n. The independency of the joint probability distribution of t can limit the evaluation of PD projects in industry, as the dynamical dependencies between process and product can significantly change over time. In this case an alternative complexity measure—known as the "binding information"—developed by Abdallah and Plumbley (2012) should be taken into consideration, as it can be used to evaluate non-stationary processes of different kinds.

If optimal predictions of the stationary stochastic process are influenced by events in the arbitrarily distant past, the mutual information diverges and the measure EMC tends to infinity (see discussion of predictive information I_{pred} below).

Shalizi and Crutchfield (2001) proved that the forecasting complexity gives an upper bound of the EMC:

$$\text{EMC} \leq C_{\mu}. \tag{217}$$

In terms of a communication channel, EMC is the effective information transmission rate of the process. The units are bits. C_{μ} is the memory stored in that channel. Hence, the inequality above means that the memory needed to carry out an optimal prediction of the future cannot be less than the information that is transmitted from the past $X_{-\infty}^{-1}$ to the future X_0^{∞} (by storing it in the present). However, the specification of how the memory has to be designed and managed cannot be derived on the basis of information-theory considerations. Instead, a constructive and more structural approach based on a theory of computation must be developed. A highly satisfactory theory based on "causal states" was developed by Crutchfield and Feldman (2003). These causal states lead to the cited ε-machines, as well as the Grassberger–Crutchfield–Young statistical complexity C_{GCY}, which will be presented later in this chapter.

EMC can be estimated purely from historical data, without use of a generative stochastic model of cooperative work. If the data is generated by a model in a specific class but with unknown parameter values, we can derive closed-form solutions for EMC, as will be shown in Sections 4.1.1, 4.1.2 and 4.1.3 for a VAR (1) model (cf. Eq. 262). The mutual information between the infinite past and future histories of a stochastic process has been considered in many contexts. It is termed, for example, excess entropy **E** (Crutchfield and Feldman 2003; Ellison et al. 2009; Crutchfield et al. 2010), predictive information $I_{pred}(n \to \infty)$ (Bialek et al. 2001, see below), stored information (Shaw 1984), past-future information I_{p-f} (Li and Xie 1996, see Section 5.1) or simply complexity (Arnold 1996; Li 1991). Rissanen (1996, 2007) also refers to the part of stochastic complexity required for coding model parameters as model complexity. Hence, there should be a close connection between Rissanen's ideas of encoding a data stream based on generative models and Grassberger's ideas of extracting the amount of information required for optimal prediction. In fact, if the data allows a description by a model with a finite number of independent parameters, then mutual information between the data and the parameters is of interest, and this is also the predictive information about all of the future (Bialek et al. 2001). Rissanen's approach was further strengthened by a result put forward by Vitányi and Li (2000) showing that an estimation of parameters using the MDL principle is equivalent to Bayesian parameter estimations with a "universal" prior (Li and Vitányi 1997). Since the mutual information between the infinite past and future histories can quantify the statistical dependency structures of cooperative work processes, it will be used in the following to evaluate the emergent complexity in PD projects.

In addition to C_μ and EMC, another key invariant of stochastic processes that was discovered much earlier is Shannon's source entropy rate (Cover and Thomas 1991):

$$h_\mu := \lim_{\eta \to \infty} \frac{H[X^{n=\eta}]}{\eta}. \tag{218}$$

This limit exists for all stationary processes. The source entropy rate is the intrinsic randomness that cannot be reduced, even after considering statistics over longer and longer blocks of generating variables. The unit of h_μ is bits/symbol. It is also known as per-symbol entropy, thermodynamic entropy density, Kolmogorov–Sinai entropy or metric entropy. The source entropy rate is zero for periodic processes. Surprisingly, it is also zero for deterministic processes with infinite memory. The source entropy rate is larger than zero for irreducibly unpredictable processes like the cited iid process or Markov processes. The capacity of a communication channel must be larger than h_μ for error-free data transmission (Cover and Thomas 1991). Interestingly, the source entropy rate is related to the algorithmic complexity (Section 3.1): h_μ is equal to the average length (per variable) of the minimal program with respect to U that, when run, will cause the Universal Turing Machine to produce a typical configuration and then halt (Cover and Thomas 1991). In the above definition the variable $H[X^n]$ is the joint information entropy of length-n blocks $(X_{t+1}, \ldots, X_{t+n})$. This entropy is not the entropy of a finite string x^n with length n; rather, it is the entropy of sequences with length n drawn from mainly much longer or infinite output generated by the process in the steady state. The variable n is the nonnegative order parameter and can be interpreted as an expanding observation window of length n over the output. In the following, we will use the shorthand notation $H(n)$ to represent this kind of entropy, which is also termed Shannon block entropy (Grassberger 1986; Bialek et al. 2001). For discrete-type random variables the block entropy is defined as

$$\begin{aligned} H(n) &:= H[X^n] \\ &= H[X_{t+1}, \ldots, X_{t+n}] \\ &= -\sum_{\mathcal{X}} \ldots \sum_{\mathcal{X}} P\left(X_{t+1} = x_{j(t+1)}, \ldots, X_{t+n} = x_{j(t+n)}\right) \\ &\quad \cdot \log_2 P\left(X_{t+1} = x_{j(t+1)}, \ldots, X_{t+n} = x_{j(t+n)}\right) \end{aligned} \tag{219}$$

with

$$H(0) := 0. \tag{220}$$

The sums in Eq. 219 run over all possible blocks of length n. The corresponding definition for continuous-type variables will be given in Eq. 233. Interestingly, the length-n approximation $h_\mu(n)$ of the entropy rate h_μ can be defined as the two-point slope of the block entropy $H(n)$:

$$h_\mu(n) := H(n) - H(n-1), \tag{221}$$

with

$$h_\mu(0) := \log_2|\mathcal{X}|. \tag{222}$$

Vice versa, $h_\mu(n)$ is the discrete derivative of the block entropy with respect to the block length n. In this sense, the length-n approximation is a dynamic entropy representing the entropy gain (Crutchfield and Feldman 2003). It can be seen that the entropy gain can also be expressed as conditional entropy

$$h_\mu(n) := H\left[X_n|X^{n-1}\right].$$

In the limit of infinitely long blocks, it is equal to the source entropy rate

$$h_\mu = \lim_{\eta\to\infty} h_\mu(n=\eta). \tag{223}$$

In general $h_\mu(n)$ differs from the estimate $H(n)/n$ for any given n but converges to the same limit, namely the source entropy rate h_μ. According to Crutchfield and Feldman (2003), $h_\mu(n)$ typically overestimates h_μ at finite n, and each difference $h_n - h_\mu$ represents the difference between the entropy rate conditioned on n measurements and the entropy rate conditioned on an infinite number of measurements. As such, it estimates the information-carrying capacity in blocks in which the difference is not actually random but arises from correlations. The difference $h_n - h_\mu$ can therefore be interpreted as the local predictability. These local "overestimates" can be used to define a universal learning curve $\Lambda(n)$ (Bialek et al. 2001) as

$$\Lambda(n) := h_\mu(n) - h_\mu, \qquad n \geq 1. \tag{224}$$

EMC is simply the discrete integral of $\Lambda(n)$ with respect to the block length n, which controls the speed of convergence of the dynamic entropy to its limit (Crutchfield et al. 2010):

$$\text{EMC} := \sum_{n=1}^{\infty} \Lambda(n). \tag{225}$$

In the sense of a learning curve, EMC measures the amount of apparent randomness at small block length n that can be "explained away" by considering correlations between blocks with increasing lengths $n+1, n+2, \ldots$. Grassberger (1986) analyzed the manner in which $h_\mu(n)$ approaches its limit h_μ, noting that for certain classes of stochastic processes with long-range correlations, the convergence can be very slow and that this is an indicator of complexity. He also found that the approach of the limit can be so slow that $h_\mu(n)$ decays slower than $1/n$ and therefore EMC is infinite. These processes are termed infinitary processes (Travers and Crutchfield 2014). When EMC is infinite, then the manner of its divergence can provide additional information of how a system's internal state space is coarse grained (see e.g. Bialek et al. 2001 and Crutchfield and Feldman 2003). This phenomenon has been analyzed in greatest detail by Bialek et al. (2001). To carry out their analysis, they defined the predictive information $I_{pred}(n)$ $(n \geq 1)$ as the mutual information between a block of length n and the infinite future following the block:

$$\begin{aligned} I_{pred}(n) &:= \lim_{\eta \to \infty} I\left[X_{-n}^{-1}; X_0^{\eta}\right] \\ &= \lim_{\eta \to \infty} H(n) + H(\eta) - H(n+\eta). \end{aligned} \tag{226}$$

Bialek et al. (2001) showed that even if $I_{pred}(n)$ diverges as n tends to infinity, the way in which it grows is an indicator of a process's complexity in its own right. They also emphasized that the predictive information is the subextensive component of the entropy:

$$H(n) = nh_\mu + I_{pred}(n). \tag{227}$$

From the above equation, it can be seen that the sum of the first n terms of the discrete integral of the universal learning curve $\Lambda(n)$, that is, $H(n) - nh_\mu$, is equal to $I_{pred}(n)$ (Abdallah and Plumbley 2012):

$$I_{pred}(n) = \sum_{i=1}^{n} \Lambda(i).$$

As expected, $I_{pred}(n)$ (as well as EMC) is zero for an iid process. According to Bialek et al. (2001), it is positive in all other cases and grows less rapidly than a linear function (subextensive). $I_{pred}(n)$ may either stay finite or grow infinitely. If it stays finite, no matter how long we observe the past of a process, we gain only a finite amount of information about the future. This holds true, for instance, for the cited periodic processes after the period φ has been identified. A longer period results in larger complexity values and $I_{pred}(n \to \infty) = \text{EMC} = \log_2(\varphi)$. For some irregular processes, the best predictions may depend only on the immediate past, e.g. in our Markovian model of task processing or generally when evaluating a system far away from phase transitions or symmetry breaking. In these cases, I_{pred} $(n \to \infty) = \text{EMC}$ is also small and is bound by the logarithm of the number of accessible states. Systems with more accessible states and larger memories are assigned larger complexity values. On the other hand, if $I_{pred}(n)$ diverges and optimal predictions are influenced by events in the arbitrarily distant past, then the rate of growth may be slow (logarithmic) or fast (sublinear power). If the acquired data allows us to infer a model with a finite number of independent parameters, or to identify a set of generative rules that can be described by a finite number of parameters, then $I_{pred}(n)$ grows logarithmically with the size of the sample. The coefficient of this divergence counts the dimensionality of the model space (i.e. the effective number of independent parameters). Sublinear power-law growth can be associated with infinite parameter models or with nonparametric models, such as continuous functions with smoothness constraints. Typically these cases occur where predictability over long time scales is governed by a progressively more detailed description as more data points are observed.

To make the previously introduced key invariant C_μ (forecasting complexity, Eq. 211) of a stochastic process operational in terms of a theory of computation and to clarify its relationship to the other key invariant EMC (effective measure complexity, Eq. 225) by using a structurally rich model and not simply a purely mathematical representation of a communication channel, in the following we refer to the seminal work of Crutchfield and Young (1989, 1990) on computational mechanics. They provided a procedure for finding the minimal maximally predictive model and its causal states by means of an ε-machine (Ellison et al. 2009; Crutchfield et al. 2010). The general goal of building an ε-machine is to find a constructive representation of a nontrivial process that not only allows good predictions on the basis of the stored predictive information, but also reveals the essential mechanisms that produce a system's behavior. To build a minimal maximally predictive model of a stationary stochastic process, we can formally define an equivalence relation $x_{-\infty}^{-1} \sim \hat{x}_{-\infty}^{-1}$ that groups all process histories that give rise to the same prediction:

$$x_{-\infty}^{-1} \sim \hat{x}_{-\infty}^{-1} :\Leftrightarrow \left\{ P\left(X_0^{\infty} \middle| X_{-\infty}^{-1} = x_{-\infty}^{-1}\right) = P\left(X_0^{\infty} \middle| X_{-\infty}^{-1} = \hat{x}_{-\infty}^{-1}\right) \right\}.$$

Hence, for the purpose of forecasting, two different sequences of past observations are considered equivalent if they result in the same predictive distribution. The above equivalence relation determines the process's causal state, which partitions the space $X_{-\infty}^{-1}$ of pasts into sets that are predictively equivalent. The causal state $\varepsilon\left(x_{-\infty}^{-1}\right)$ of $x_{-\infty}^{-1}$ is its equivalence class

$$\varepsilon\left(x_{-\infty}^{-1}\right) := \left\{ \hat{x}_{-\infty}^{-1} : x_{-\infty}^{-1} \sim \hat{x}_{-\infty}^{-1} \right\},$$

and the causal state function $\varepsilon(.)$ defines a deterministic sufficient memory $\mathcal{M}_\varepsilon$ (see Shalizi and Crutchfield 2001; Löhr 2012). The set of memory states of the ε-machine is simply the set of causal states

$$\mathcal{M}_\varepsilon := \left\{ \varepsilon\left(x_{-\infty}^{-1}\right) : x_{-\infty}^{-1} \in \mathcal{X}^{\mathbb{N}} \right\}.$$

The set $\mathcal{X}$ represents the finite alphabet on which the stationary stochastic process is defined. The set of causal states $\mathcal{M}_\varepsilon$ does not need to be countable and can therefore represent either discrete or continuous state spaces. Shalizi and Crutchfield (2001) showed that the equivalence relation $x_{-\infty}^{-1} \sim \hat{x}_{-\infty}^{-1}$ is minimally sufficient and unique. Hence, it allows the highest compression of the data, while containing all the relevant information on local dynamics. For practical purposes, longer and longer histories are analyzed, from x_{-L}^{-1} up to a predefined maximum length $L = L_{\max}$, and the partition into classes for a fixed future horizon X_0^t is obtained. In principle, we start at the most coarse-grained level, grouping together those histories that have the same predictive distribution for the next observable X_0, and then refine the partition. The refinement is recursively carried out by further subdividing the classes using the predictive distributions of the next two observables X_0^1, the next three observables X_0^2, etc.

After all causal states have been identified, an ε-machine can be constructed. To simplify the definition of the forecasting complexity C_μ, we start by using an informal representation in the form of a stochastic output automaton that is expressed by the causal state function ε, a set of transition matrices $\mathcal{T}$ for the states defined by ε, and the start state s_0. The start state is unique. Given the current state $s \in \mathcal{M}_\varepsilon$ of the automaton, a transition to the next state $s' \in \mathcal{M}_\varepsilon$ is determined by the output symbol (or measurement) $x \in \mathcal{X}$. State-to-state transitions are probabilistic and must therefore be represented for each output symbol x by a separate transition matrix $T^{(x)} \in \mathcal{T}$. Each row and column of the transition matrices in the set $\mathcal{T}$ stands for an individual causal state. A stochastic output automaton can also be transformed into an equivalent edge-emitting hidden Markov model (Löhr 2012). A hidden Markov model is a universal machine that is defined over a set of non-observable internal states $\mathcal{M}_\varepsilon$. It therefore does not directly reveal its internal mechanisms to external observers; it only expresses them indirectly through emitted symbols. The emitted symbols are edge-labels of the hidden states. The model can be formally represented by the tuple $\left(\mathcal{M}_\varepsilon, \mathcal{X}, \pi, \left\{T^{(x)}\right\}\right)$. The start state of the hidden Markov model is not unique but determined by an initial probability distribution π. Depending on the current internal state s_t, at each time step t a transition to the new internal state s_{t+1} is made and an output symbol x_{t+1} from the alphabet $\mathcal{X}$ is emitted. The corresponding entry $T_{ij}^{(x)}$ of the transition matrix $T^{(x)}$ gives the probability $P\left(S_{t+1} = s_{t+1}, X_{t+1} = x_{t+1} \middle| S_t = s_t\right)$ of transitioning from current state s_t indexed by i to the next s_{t+1} indexed by j on "seeing" measurement x. This operation may also be thought of as a weighted random walk on the associated graphical model (Travers and Crutchfield 2011): from the current state s_t, the next state s_{t+1} is determined by selecting an outgoing edge from current state s_t according to their probabilities. After a transition has been selected, the model moves to the new state and outputs the symbol of the current state x labeling the edge. The transition matrices are usually non-symmetric. From the theory of Markov processes (see e.g. Puri 2010) it is well known that in a steady state the probability distribution over the hidden states is independent of the initial-state distribution. Edge-emitting hidden Markov models can also be expressed by an initial probability distribution π, by a state process $\{S_t\}$ and by an output process $\{X_t\}$, which means that they are theoretically similar to the continuous-type linear dynamical systems that were analyzed in Section 2.9. However, continuous-type linear dynamical systems usually do not possess the property of "unifilarity" (see below) and therefore cannot be used to directly calculate the entropy rate of the process.

To obtain the transition matrices $T^{(x)}$, one can parse the data sequence of interest in a sequential manner, identify all causal state transitions defined by ε over histories x_0^t and x_0^{t+1}, and estimate the transition probabilities $P(S',X=x_{t+1}|S)$ using frequency counting (MLE, see Section 2.4) or Bayesian methods. The transition probabilities allow calculation of an invariant probability distribution $P(S)$ over the causal states. This probability is obtained as the normalized principal eigenvector of the transition matrix $T=\Sigma_{x\in\mathcal{X}}T^{(x)}$ (Ellison et al. 2009). The matrix T is stochastic and $\Sigma_{j=1}^{|\mathcal{M}_\varepsilon|}T_{ij}=1$ holds for each i.

Interestingly, causal states have a Markovian property in that they render the past and future statistically independent. In other words, they shield the future from the past:

$$P(X_{-\infty}^{-1},X_0^{\infty}|S)=P(X_{-\infty}^{-1}|S)P(X_0^{\infty}|S).$$

Moreover, they are optimally predictive in the sense that knowing what causal state a process is in is as good as having the entire past: $P(X_0^{\infty}|S)=P(X_0^{\infty}|X_{-\infty}^{-1})$. Causal shielding is therefore equivalent to the fact that the causal states capture all of the information shared between past and future. Hence, $I[S,X_0^{\infty}]=\text{EMC}$. Out of all maximally predictive models $\mathcal{M}$ for which $I[\mathcal{M},X_0^{\infty}]=\text{EMC}$, the ε-machine captures the minimal amount of information that a stationary stochastic process must store in order to communicate all excess entropy from the past to the future. Accordingly, the ε-machine is as close to perfect determinism as any rival that has the same predictive power (Jänicke and Scheuermann 2009). The minimal amount of information that must be stored on a stationary stochastic process $X_{-\infty}^{\infty}=(\ldots,X_{-1},X_0,X_1,\ldots)$ for optimal prediction is the Shannon information entropy over the stationary distribution of its ε-machine's causal states—the forecasting complexity—and it holds that

$$C_\mu(X_{-\infty}^{\infty})=H[S].$$

Because of its significance in complex systems science, the forecasting complexity is also termed Grassberger–Crutchfield–Young statistical complexity C_{GCY} (Shalizi 2006). It should not be confused with Rissanen's stochastic complexity C_{SC} from Eq. 208, because the underlying concepts are based on a theory of computation. We have (Ellison et al. 2009)

$$\begin{aligned} C_{GCY}&=H[S]\leq H[\mathcal{M}]\\ C_{GCY}&=-\sum_{s\in\mathcal{M}_s}P(S)\log_2 P(S)\leq H[\mathcal{M}]. \end{aligned} \tag{228}$$

As we have argued, the causal states are an objective property of the stochastic process under consideration and therefore the associated statistical complexity C_{GCY} cannot be influenced by our ineptness as modelers or our (possibly poor) means of description. It is equal to the length of the shortest description of the past that is relevant to the actual dynamics of the system. As was shown above, for iid sequences it is exactly 0, and for periodic sequences it is $\log_2(\varphi)$. A detailed description of an algorithm providing an ε-machine reconstruction and calculation of C_{GCY} for one-dimensional and two-dimensional time series can be found in Shalizi and Shalizi (2004, 2003).

Moreover, the entropy rate h_μ can be directly calculated on the basis of a process's ε-machine (Ellison et al. 2009) because of unifilarity:

$$\begin{aligned} h_\mu &= H[X|S] \\ &= -\sum_{s\epsilon\mathcal{M}_s} P(S) \sum_{xs'\epsilon\mathcal{X}\mathcal{M}_s} T_{ss'}^{(x)} \log_2 \sum_{s'\epsilon\mathcal{M}_s} T_{ss'}^{(x)} . \end{aligned}$$

$\mathcal{X}\mathcal{M}_s$ denotes the set whose elements are generated by concatenating all elements of the sets $\mathcal{X}$ and $\mathcal{M}_s$. Unifilarity means that from the start state s_0 of the process, each generated sequence of observations corresponds to exactly one sequence of causal states. In a hidden Markov model representation of an ε-machine this property can easily be verified. For each hidden state, each emitted symbol appears on at most one edge. In the above equation, we used the shorthand notation $T_{ss'}^{(x)}$ to denote the matrix entry $T_{ij}^{(x)}$ corresponding to causal state s in row i and causal state s' in column j of the transition matrix associated with output symbol x. The probability $P(S)$ denotes the asymptotic probability of the causal states.

In a recent paper, Gu et al. (2012) extended the framework of ε-machines by allowing the casual states to have quantum mechanical properties. This extension also makes it possible to define the quantum complexity of a stochastic process. Interestingly, the quantum complexity of a process is bounded below by EMC and above by C_{GCY} (Wiesner 2015).

An especially interesting variant of Grassberger's classic definition of the effective measure complexity has recently been developed by Ball et al. (2010). These authors also quantify strong emergence within an ensemble of histories of a complex system in terms of mutual information between past and future history, but focus on the part of the information that persists across an interval of time $\tau > 0$. As such, we can specify the "persistent mutual information" as a complexity measure in its own right that evaluates the deficit in the information entropy in the joint history compared with that of past and future taken independently. Formally, the persistent mutual information can be defined on the basis of the EMC (Eq. 216) extended by the lead time τ to evaluate the persistent part as

$$\text{EMC}(\tau) := I\left[X_{-\infty}^{-1}; X_{\tau}^{\infty}\right], \tag{229}$$

where $X_{-\infty}^{-1}$ designates the history of the stochastic process from an infinite past to the present, and X_{τ}^{∞} is the corresponding future of the system from the later time

τ onwards. The key distinguishing feature of the definition above is that it ignores the information captured in block $X_0^{\tau-1}$, that is, the intervening interval of observations of length τ. For continuous state variables, EMC(τ) has the merit of being independent of continuous changes of the variable, as long as they preserve time labeling (Ball et al. 2010). EMC(τ) is known to be a Lyapunov function for the process, so that it decays with increasing lead time (Ay et al. 2012). For positive lead times the persistent mutual information is nonzero if a process has a memory mechanism to store the predictive information persistently and is therefore sensitive to how a system's state space is observed (Marzen and Crutchfield 2014). Li (2006) defines an information-regular process as a process whose persistent mutual information converges to zero as the lead time grows over all given limits and it holds that $\mathrm{EMC}(\tau) \to 0$ as $\tau \to \infty$. Otherwise, the process is information-irregular. The differences between the effective measure complexity and the persistent mutual information for continuous-state processes are presented in more detail in Section 4.1.6.

It is evident that the persistent mutual information enables the specification of an intuitive lower bound on EMC:

$$\mathrm{EMC}(\tau) \leq \mathrm{EMC}. \tag{230}$$

In fact, for zero lead time we have

$$\mathrm{EMC}(0) = \mathrm{EMC}.$$

The recent work of James et al. (2011), Marzen and Crutchfield (2014) and others has shown that a fine decomposition of the persistent mutual information can be carried out, essentially breaking it down into two pieces. With respect to emergent complexity, the most interesting piece is the so-called "elusive information" $\sigma_\mu(\tau)$, which is the mutual information between the past $X_{-\infty}^{-1}$ and the future X_τ^∞ conditioned on the length-τ present $X_0^{\tau-1}$ (cf. Eq. 214):

$$\sigma_\mu(\tau) := I\left[X_{-\infty}^{-1}; X_\tau^\infty \middle| X_0^{\tau-1}\right]. \tag{231}$$

According to the analysis by James et al. (2011) the elusive information has an especially interesting interpretation: it represents the Shannon information that is communicated from the past to the future, but does not flow through the currently observed length-τ sequence $X_0^{\tau-1}$. The key distinguishing feature of the persistent mutual information is that it is nonzero for positive length τ if a process necessarily has hidden states. In this case, all the information from the past that is relevant for generating future behavior has to be stored by an internal configuration to arrive at a complete description of the process. The internal configuration is necessary to keep track of the state information, because the present sequence of observations $X_0^{\tau-1}$ can only capture features of shorter term correlation and therefore does not have enough capacity to capture all the features that are relevant for forecasting. In the words of James et al. (2011): "This is why we build models and cannot rely on only collecting observation sequences." For instance, for the n-th order Markov chains that were introduced in Section 3.2.2, we have $\sigma_\mu(\tau) = 0$ for lead times τ that are larger than or equal to the model order n. In this case with only fully observable state variables, the length-n memory of the chain model serves as the effective state, rendering the process's past and future independent (Marzen and Crutchfield 2014). For infinite-order Markov chains EMC(τ) only vanishes asymptotically. Therefore, the elusive information is sensitive to which extent a system's internal state space is coarse grained (Marzen and Crutchfield 2014).

3.3 Complexity Measures from Theories of Systematic Engineering Design

The most prominent complexity theory in the field of systematic engineering design has been developed by Suh (2005). His theory aims at providing a systematic way of designing products and large-scale systems, as well as of determining the best designs from those proposed. Suh's complexity theory is based on his famous axiomatic design theory (Suh 2001). He defines complexity in the functional domain rather than in the physical domain of the design world. In the functional domain, uncertainty is measured through information-theoretic quantities like the information content that was already introduced and defined in Section 3.2.2. Alternative approaches to characterizing complexity in engineering design that are not based on information-theory and statistical models (see e.g. Lindemann et al. 2009; Kreimeyer and Lindemann 2011) are only very briefly addressed in the following, as they tend to be valid only for evaluating structural and not time-dependent complexity.

In Suh's axiomatic design theory, the product to be developed and the problem of solving the design issues are coupled through functional requirements (FRs) and design parameters (DPs). He proposes two axioms for design: the independence axiom and the information axiom. The independence axiom states that the FRs should be maintained by the designer or design team independent of each other. When there are two or more FRs, the design solution must be such that each of the FRs can be satisfied without affecting any of the other FRs. This means that a correct set of DPs is to be chosen so as to satisfy the FRs and maintain their independence. If the independence can be maintained for all FRs, the design is said to be "uncoupled." An uncoupled design is an optimal solution in the sense of the theory. Once the FRs are established, the next step in the design process is the conceptualization process, which occurs during the mapping process from the functional to the physical domain.

The conceptualization process may produce several designs, all of which may be satisfactory in terms of the independence axiom. Even for the same task defined by a given set of FRs, it is likely that different engineers will come up with different designs, because there are many solutions that satisfy a given set of m FRs ($FR_1,\ldots,FR_m$). The information axiom provides a quantitative measure of the merits of a given design, and is thus useful in selecting the best design from among those that are acceptable. The information axiom is formulated within an information-theory framework and states that the best design is that with the highest probability of success. Following the definition of the Shannon information content in Eq. 205 the information content I_i for a given functional requirement $FR_i(1 \leq i \leq m)$ is expressed as the logarithmic probability p_i of satisfying this specific FR:

$$\begin{aligned} I_i &= \log_2 \frac{1}{p_i} \\ &= -\log_2 p_i. \end{aligned}$$

In the general case of m specified FRs, the information content I_{sys} for the entire system under study is

$$I_{sys} = -\log_2 P(X^m),$$

where $P(X^m)$ denotes the joint probability that all m FRs are satisfied. When all FRs are statistically independent, as in an uncoupled design, the information content I_{sys} can be decomposed into independent summands and expressed as

$$\begin{aligned} I_{sys} &= \sum_{i=1}^{m} I_i \\ &= -\sum_{i=1}^{m} \log_2 p_i. \end{aligned}$$

When not all FRs are statistically independent (in the so called "decoupled design"), there holds

$$I_{sys} = -\sum_{i=1}^{m} \log_2 p_{i|\{j\}} \quad \text{for}\{j\} = \{1,\ldots,i-1\}$$

In the above equation $p_{i|\{j\}}$ is the conditional probability of satisfying FR_i given that all other correlated $\{FR_j\}_{j=1,\ldots,i-1}$ are also satisfied. It is assumed that the FRs are ordered according to their number of correlations. The information axiom states that the best design is that with the smallest I_{sys}, because the least amount of information in the sense of Shannon's theory is required to achieve the design goals. When all probabilities are one, the information content is zero and the design is optimal in the sense of the axiom. Conversely, when one or more probabilities are zero, the information required is infinite and the system has to be redesigned to satisfy the information axiom.

The probability of success p_i can be determined by the intersection of the design range defined by the designers to satisfy the FRs and the ability of the system to produce the part within the specified range. This probability can be computed by specifying the design range (r) for the FR and by determining the system range (sr) that the proposed design can provide to satisfy the FR. The lower bound of the specified design range for functional requirement FR_i is denoted by $r^l[FR_i]$, and the upper bound by $r^u[FR_i]$. The system range can be modeled in statistical terms on the basis of a probability density function (*pdf*, see Section 2.1). The *pdf* is specified over the theoretically feasible state space. The system *pdf* is denoted by $f_{sys}[FR_i]$. The overlap between the design and system ranges is called "the common range" (cr), and this is the only range where the FR is satisfied. Consequently, the area A_{cr} under the system *pdf* within the common range is the design's probability of achieving the specified goal. Hence, the information content I_i can be expressed as

$$\begin{aligned} I_i &= -\log_2 A_{cr} \\ &= -\log_2 \int_{r^l[FR_i]}^{r^u[FR_i]} f_{sys}[FR_i] \; dFR_i . \end{aligned}$$

Suh (2005) considers a design to be complex when its probability of success is low and hence the information content I_{sys} required to satisfy the FRs is high. Complex designs often arise when there are many components, because as their number increases through functional decomposition, the probability that some of them do not meet the specified requirements also increases, such as when the interfaces between components introduce additional errors. In order to steer the design process toward more effective, efficient and robust large-scale systems, a dedicated complexity axiom is defined that simply states "reduce the complexity of a system" (Suh 2005). The quantitative measure for complexity in the sense of this axiom is the information content, which was defined in the above equations. The rationale behind the axiom is that complex systems may require more information to make the system function. Therefore, Suh (2005) ties the notion of complexity to the design range for the FRs—the tighter the design range, the more difficult it becomes to satisfy the FRs. An uncoupled design is likely to be least complex. However, the complexity of a decoupled design can be high because of so-called "imaginary complexity" if we do not understand the system. It is not truly complex, but it appears to be so because of our lack of understanding of generalized or physical functions.

According to Suh (2005) complexity can also be a function of time if the system range changes as a function of time. In this case, we must differentiate between two types of time-dependent complexity: time-dependent combinatorial complexity and time-dependent periodic complexity. Time-dependent combinatorial complexity is defined as the complexity that increases as a function of time because of a continued expansion in the number of possible combinations of FRs and DPs in time, which may lead to chaotic behavior or system failure. It occurs because future events occur randomly in time and only have a limited predictability, even though they depend on the current state. Conversely, periodic complexity is defined as the complexity that only exists in a finite time period, resulting in a finite and limited number of probable configurations. Concerning a system subjected to combinatorial complexity, Suh (2005) concludes that the uncertainty of future outcomes continues to grow over time, and as a result, the system cannot have long-term stability and reliability. In the case of systems with periodic complexity, it is assumed that the system is deterministic and can renew itself over each period. Therefore, he concludes that a stable and reliable system must be periodic. It is readily apparent that a system with time-dependent combinatorial complexity can be changed to one with time-dependent periodic complexity by defining a set of functions that repeat periodically. This can be achieved temporally, geometrically, thermally, electrically and by other constructive means. In conclusion, engineered systems in PD should have small time-independent real and imaginary complexities and no time-dependent combinatorial complexity. If the system range must change as a function of time, the developer should be able to introduce time-dependent periodic complexity. These criteria need to be satisfied regardless of the size of the system or the number of FRs and DPs specified for the system.

Although Suh's complexity theory is grounded in axiomatic design theory and has been successfully applied in different domains, our criticism is that product and design problems are evaluated irrespective of the work processes, which are needed to decompose the FRs and DPs. The decomposition is a highly cooperative process that must be taken into account to satisfy all specified FRs on time and to avoid cycles of continual revision. Furthermore, the fact that Suh uses the information content I_{sys} directly as a complexity measure can be a point of criticism. I_{sys} is a simple additive measure that only represents the encoded length of the design in terms of binary design decisions; it does not take into account the encoding scheme. However, both parts of the description of a design are important because the description can always be simplified by formulating more complicated design rules, more complex standard components or interfaces (cf. Section 3.2.2). Lastly, Suh (2005) does not define specific measures for time-dependent complexity.

El-Haik and Yang (1999) have extended Suh's theory by representing the imaginary part of complexity through the differential entropy (Chapter 4) associated with the joint *pdf* of FRs with three components of variability, vulnerability and correlation. These components evaluate the product design according to the vector of DPs (see Summers and Shah 2010). Although this approach can be used to assess the mapping from the FRs to the DPs through an analysis of the topological structure of the design structure matrix (Browning 2001, see discussion below) and the variability of the design parameters (measured by the differential entropy of the joint pdf of DPs), the dynamics of the development processes in terms of a work transformation matrix (WTM, Section 2.2) are not taken into account. An alternative view introduced by Braha and Maimon (1998) suggests that complexity is a fundamental characteristic of the information content within either the product or the process. They introduce two measures that quantify either the structural representation of the information or the functional probability of achieving the specified requirements. The measures can be applied to compare products and processes at different levels of abstraction. The process is nominally defined as mapping between the product and problem, where the coupling determines process complexity. The size of the process is defined as the summation over the number of instances of operators (relationships) and operands (entities). A process instance is a sequence of the instances of operands and operators. The average information content of sequences can be evaluated on the basis of the block entropy (Eq. 219). As the design takes on different types of representations through the development stages, the average information contained changes. Braha and Maimon (1998) suggest that the ratio of the amount of average information content between the initial and current states is a measure of the current abstraction level. The effort required to move between abstraction levels is inversely proportional to this ratio. The proportionality constant is the information content of the current state. Summers and Shah (2010) follow these lines of reasoning and propose a process size complexity measure that includes the vocabulary of the specific representation for the problem, the product, the development process and the four operators available for sequencing the states of the design evolution. The measure is defined as

$$Cx_{size\_process} := \left(M^o + C^o + P_{op}\right)\ln\left(idv + ddv + dr + mg + a_{op} + e_{op} + s_{op} + r_{op}\right).$$

In the above definition the size of the vocabulary is represented by the total number of possible primitive modules (M^o), possible relations between these modules (C^o) and possible operators and operands (P_{op}). The additional parameters denote the variables whose values are controlled by the designer (*idv*), are derived from the independent design parameters, other dependent variables and design relations (*ddv*), are constraints that dictate the association between the other design variables (*dr*), or are used to determine how well the current design configuration meets the goals (*mg*), plus the four operators available for sequencing the states. Although the extended concepts based on information content within either the product or the process are appealing, the fact that the development process is only analyzed on stage-dependent hierarchical description levels, not on the basis of an explicit state-space model of cooperative work, opens it to criticism. Moreover, dynamic entropies in the sense of Grassberger's theory are not taken into account to evaluate time-dependent combinatorial complexity in an open organizational system. Last but not least, in real design problems, it is difficult to identify all operators and operands in advance and to specify valid sequences leading from one level of abstraction to the next.

In addition to methods for measuring characteristics of the design based on information-theoretic quantities, a large body of literature has been published on the design structure matrix (Steward 1981) as a dependency modeling technique supporting complexity management by focusing attention on the elements of a system and the dependencies through which they are related. Recent surveys can be found in the textbooks of Lindemann et al. (2009) or Eppinger and Browning (2012). Browning (2001) distinguishes between two basic types of DSMs: static and time-based. Static DSMs represent either product components or teams in an organization that exist simultaneously. Time-based DSMs either represent dynamic activities indicating precedence relationships or design parameters that change as a function of time. Generated static DSMs are usually analyzed for structural characteristics or by clustering algorithms (e.g. Rogers et al. 2006), whereas time-based DSMs are typically used to optimize workflows based on sequencing, tearing and banding algorithms (e.g. Gebala and Eppinger 1991; Maurer 2007). Kreimeyer and Lindemann (2011) review and discuss a comprehensive set of metrics that can be applied to assess the structure of engineering design processes encoded by DSMs (and other forms). According to Browning's taxonomy, the WTM as dynamical operator of state equation 8 is a static task-based DSM, because the development tasks are processed concurrently and persistent feedback/feed forward loops are modeled through the off-diagonal elements. The majority of work on complexity management with static DSMs focuses on the concept of modularity in identifying cluster structures (see Baldwin and Clark 2000). This work has been very influential in academia and industry. An important limitation, however, is its purely static view of the product structure and, consequently, of the task structure and the interactions between them. A task processing on different time scales corresponding to different autonomous task processing rates cannot be represented. Recent publications indicate that technical dependencies in product families tend to be volatile and therefore coordination needs among development tasks can evolve over time (e.g. Cataldo et al. 2006, 2008; Sosa 2008). When these evolving coordination needs are not adequately managed, significant misalignments of organizational structure and product architecture can occur that have a negative effect on product quality (Gokpinar et al. 2010). An effective method for dealing with volatility of dependencies is to use different WTMs for different phases of the project in which no task is theoretically processed independently of the others. Furthermore, additional task-mapping matrices can be specified at the transition points between phases. By doing so, the number of tasks as well as the kind and intensity of coordination needs can be adapted. It is also possible to specify phase-dependent covariances of performance fluctuations. In many PD projects the performance fluctuations tend to be larger for late development stages that are close to the desired start of production. Another limitation of the concept of product modularity is that the organizational patterns of a development project (e.g. communication links, team co-membership) do not necessarily mirror the technical dependency structures (Sosa et al. 2004). The literature review by Colfer and Baldwin (2010) shows that the "mirroring hypothesis" was supported in only 69% of the cases. Support for the hypothesis was strongest in the within-firm sample, less strong in the across-firm sample, and relatively weak in the open collaborative sample. As such, WTMs and covariance matrices represent dynamic dependency structures in their own right. They must be related to product components or organizational elements through additional multiple domain mapping matrices (Danilovic and Browning 2007) and cannot be substituted by the traditional modeling elements.

An approach to measuring structural complexity based on static component-based DSMs that is formally similar to our own analysis in the spectral basis (see Sections 2.3 and 4.2) has recently been developed by Sinha and de Weck

(2011; 2012). The three terms of their metric C_{SW} are related to the complexities of each of the n components in the system (local effect, represented by the α_i's), the number and complexity of each pairwise interaction (local effect, represented by the β_{ij}'s and a_{ij}'s) and the arrangement of the m interfaces (global, system level effect, represented by $E(A)$). Moreover, a normalization factor γ is introduced. The definition is (Denman et al. 2011; Sinha and de Weck 2012; Sinha 2014):

$$C_{SW} := \sum_{i=1}^{n} \alpha_i + \left(\sum_{i=1}^{n} \sum_{j=1}^{n} \beta_{ij} a_{ij} \right) \gamma E(A).$$

The normalization factor γ is taken as $1/n$ and used to map the n different components in the system onto a comparable scale. The matrix A is an adjacency matrix that corresponds to the component-based DSM of the product as follows:

$$A = \left(a_{ij} \right) = \begin{cases} 1 & \forall (i,j) : (i \neq j) \wedge (i,j) \in \Upsilon \\ 0 & otherwise. \end{cases}$$

The exogenous variable Υ represents the set of connected nodes in the system. Accordingly, the adjacency matrix is simply a binary form of the component-based DSM, in which ones are placed in the cells with marks and zeros elsewhere. The diagonal elements of A are zero. The underlying concept of the metric C_{SW} is that in order to develop the individual components, a non-zero complexity is involved. This complexity can vary across components and is represented by the α_i's, the so-called component complexity estimate (Sinha and de Weck 2012; Sinha 2014). Similar arguments hold true for the complexity β_{ij} of each interface, the so-called final interface complexity (Sinha and de Weck 2012; Sinha 2014). If there are multiple types of interface between two components (energy flow, material flow, control action flow etc.), large beta coefficients are assigned, since it would require more effort to implement them compared to a simpler (univariate) connection. An important aspect is that the correlation between the component complexity estimate and the final interface complexity can vary depending on the kind of product. For large-scale mechanical systems, the β_{ij}'s are often much smaller than the α_i's and α_j's. However, in micro or nanoscale systems it can be the opposite, because it is often much more difficult to develop the interfaces (Sinha 2014). The different interface complexities can be captured using a multiplicative model

$$\beta_{ij} = f_{ij} \alpha_i \alpha_j,$$

where f_{ij} stands for the interface complexity factor (Eppinger and Browning 2012; Sinha and de Weck 2012; Sinha 2014). Finally, the term $E(A)$ represents the graph energy of the adjacency matrix A. The graph energy is defined as the sum of the singular values σ_i of the orthogonal vectors:

$$E(A) := \sum_{i=1}^{n} \sigma_i,$$

where the singular values are computed by the decomposition

$$A = U \cdot \Sigma_A \cdot V^{\mathrm{T}}$$
$$\Sigma_A = \mathrm{Diag}[\sigma_i].$$

The graph energy is invariant under isomorphic transformations (Weyuker 1988) and therefore highly objective.

Ameri and Summers (Ameri et al. 2008; Summers and Ameri 2008) developed a complementary connectedness measure and an algorithm for assessing design connectivity complexity based on graphical models. In the graphical models, the development tasks are nodes of a graph and connected through variable dependency. The algorithm manipulates the graph in terms of connectivity. This manipulation starts by eliminating all unary relations, as they do not contribute to the connectivity complexity of the graph. Once the unary relations have been removed, the score keeping variables are initialized. From this point forward, the graph connectivity algorithm is a recursive algorithm that is applied against all subgraphs that are generated in the process. A cumulative score is maintained to quantify the connectedness of the whole structure (see Summers and Shah 2010). This approach also seems to have certain limitations for assessing emergent complexity in PD projects. The graph of development tasks is recursively decomposed into subgraphs, which tears apart potentially important indirect connections that can lead to higher-order interactions between activities. Furthermore, due to the deterministic approach to modeling the work processes it is impossible to analyze or evaluate the "problem-solving oscillations" (Mihm et al. 2003; Mihm and Loch 2006) emerging from cooperative task processing in conjunction with performance variability. Consequently, we will not consider the design connectivity complexity in the following.

The interested reader can find additional approaches to measuring and evaluating complexity in engineering design with a specific focus on structural characteristics in the excellent textbook by Kreimeyer and Lindemann (2011). The authors present a total of 52 complexity metrics from different disciplines and show in three case studies from process management in the automotive industry how different facets of complexity can materialize in real design processes. They also introduce the Structural Goal Question Metric framework for selecting metrics in a goal-oriented manner and guiding their application.

The information-theory and dependency-structure-based complexity metrics from theories of systematic engineering design are undoubtedly beneficial in facilitating studies that require the use of equivalent but different design problems and in comparing computer-aided design automation tools. Nevertheless, in the following analytical Chapter 4 we will shift our focus to the EMC metric first put forward in Grassberger's seminal theoretical work (1986), as it can both effectively measure self-generated complexity and provide a foundation for deriving closed-form solutions of different strengths from first principles. Furthermore, EMC stresses the dynamic nature of cooperative work in PD projects and can be calculated efficiently from generative models or from historical data.

Also very interesting for applications in project management is the later-formulated persistent mutual information EMC(τ) (Section 3.2.4). This is partly because of its intimate relationship with the famous Lyapunov function (Nicolis and Nicolis 2007) of a process, and partly because the generated complexity "landscape" often becomes more and more informative as the lead time increases. However, this phenomenon goes beyond the scope of this book and will be analyzed in detail in future work. To lay the analytical foundations for future studies of emergent complexity we will present closed-form solutions of the persistent mutual information for the developed vector autoregression models in the corresponding chapters. These solutions are generalized from the expressions for $\text{EMC} = \text{EMC}(\tau = 0)$, which will be presented in the beginning of Sections 4.1.1, 4.1.2 and 4.1.3 (see Eqs. 247, 253, 262 and 265). Due to the limited space in this book, the closed-form solutions of the persistent mutual information that is generated by a linear dynamical system (Section 2.9) will not be presented. The interested reader can develop them by applying the solution principles that will be introduced in Section 4.2.

The purely information-theoretic view on emergent complexity also opens EMC and the corresponding persistent mutual information EMC(τ) to criticism. In their latest paper on effective complexity (see also Section 3.2.4) Gell-Mann and Lloyd (2004) point out that, without modification, EMC assigns two identical and very long bit strings consisting entirely of 1's with high complexity values because the mutual information between them is very large, yet each process representation is obviously very simple. This is in stark contrast to the fundamental ideas of their EC metric (Eq. 209), which evaluates the algorithmic information content of the strings. The ideal ensemble for modeling an identical very long bit string x is the Dirac measure δ_x, i.e. the ensemble with $\delta_x(x) = 1$ and $\delta_x(x') = 0$ for $x \neq x'$. This ensemble has Kolmogorov–Chaitin complexity $K_U[\delta_x(x)] = K_U(x)$ and information entropy $H[\mathbb{E}] = 0$ (Ay et al. 2010). Its total information $\Sigma[\mathbb{E}]$ is therefore minimal. The algorithmic complexity $K_U(x)$ is apparently very low because the computer program used to calculate x on U simply outputs $|x|$ 1's in a simple pre- or post-test loop. Shiner et al. (2000) also criticize the fact that EMC is not uniquely defined for higher dimensional systems, e.g. spins in two dimensions. In spite of these apparent conceptual weaknesses, the ability of both measures to quantify the degree of informational structure between past and future histories of cooperative task processing and the value of that information in helping to make predictions mean that they are especially interesting and valuable for analyzing, evaluating and optimizing PD projects.

More details on complexity measures from statistical physics, information theory and computer science are presented in Shalizi (2006), Prokopenko et al. (2009), Nicolis and Nicolis (2007), Ellison et al. (2009) and Crutchfield et al. (2010). A focused review of complexity measures for the evaluation of human–computer interaction including two empirical validation studies can be found in Schlick et al. (2006, 2010).

References

Abdallah, S.A., Plumbley, M.D.: A measure of statistical complexity based on predictive information with application to finite spin systems. Phys. Lett. A **376**, 275–281 (2012)

Akaike, H.: Information theory and an extension of the maximum likelihood principle. In: Petrov, B.N., Csaki, F. (eds.) Second International Symposium of Information Theory, pp. 267–281. Akademia Kiado, Budapest, Budapest (1973)

Akaike, H.: A new look at the statistical model identification. IEEE Trans. Automatic Cont **19**, 716–723 (1974)

Amaral, L.A.N., Uzzi, B.: Complex systems: A new paradigm for the integrative study of management, physical, and technological systems. Manag. Sci **53**(7), 1033–1035 (2007)

Ameri, F., Summers, J.D., Mocko, G.M., Porter, M.: Engineering design complexity: An experimental study of methods and measures. Res. Eng. Des. **19**(2-3), 161–179 (2008)

Arnold, D.: Information-theoretic analysis of phase transitions. Complex Syst. **10**(2), 143–155 (1996)

Ay, N., Müller, M., Szkola, A.: Effective complexity and its relation to logical depth. IEEE Trans. Inform. Theor. **56**(9), 4593–4607 (2010)

Ay, N., Bernigau, H., Der, R., Prokopenko, M.: Information driven self-organization: The dynamic system approach to autonomous robot behavior. Theor. Biosci. **131**, 161–179 (2012)

Baldwin, C.Y., Clark, K.B.: Design Rules: The Power of Modularity. MIT Press, Cambridge, MA (2000)

Ball, R.C., Diakonova, M., MacKay, R.S.: Quantifying emergence in terms of persistent mutual information. (2010) arXiv:1003.3028v2 [nlin.AO]

Bennett, C.: Logical depth and physical complexity. In: Herken, R. (ed.) The Universal Turing Machine—a Half-Century Survey, pp. 227–257. Oxford University Press, Oxford (1988)

Bialek, W., Nemenman, I., Tishby, N.: Predictability, complexity and learning. Neural Comput. **13** (1), 2409–2463 (2001)

Bosch-Rekveldt, M., Jongkind, Y., Mooi, H., Bakker, H., Verbraeck, A.: Grasping project complexity in large engineering projects: The TOE (Technical, organizational and environmental) framework. Int. J. Project Manag. **29**(6), 728–739 (2011)

Braha, D., Bar-Yam, Y.: The statistical mechanics of complex product development: Empirical and analytical results. Manag. Sci. **53**(7), 1127–1145 (2007)

Braha, D., Maimon, O.: The measurement of a design structural and functional complexity. IEEE Trans. Syst. Man Cybernet. Part A: Syst. Hum **28**(4), 527–535 (1998)

Browning, T.: Applying the design structure matrix to system decomposition and integration problems: A review and new directions. IEEE Trans. Eng. Manag. **48**(3), 292–306 (2001)

Burnham, K.P., Anderson, D.R.: Model Selection and Multimodel Inference: A Practical Information-Theoretic Approach. Springer, New York, NY (2002)

Carlile, P.R.: A pragmatic view of knowledge and boundaries: Boundary objects in new product development. Organ. Sci. **13**(4), 442–455 (2002)

Cataldo, M., Wagstrom, P.A., Herbsleb, J.D., Carley, K.M.: Identification of Coordination requirements: Implications for the design of collaboration and awareness tools. In: Proceedings of the 2006 ACM Conference on Computer Supported Cooperative Work, CSCW 2006, Banff, Alberta, Canada, pp. 353–362, (2006)

Cataldo, M., Herbsleb, J.D., Carley, K.M.: Socio-technical congruence: A framework for assessing the impact of technical and work dependencies on software development productivity. In:

Proceedings of the 2nd International Symposium on Empirical Software Engineering and Measurement (ESEM'08), Kaiserslautern, Germany. pp. 2–11, (2008)
Chaitin, G.J.: Algorithmic Information Theory. Cambridge University Press, Cambridge (1987)
Colfer, L.J., Baldwin, C.Y.: The mirroring hypothesis: Theory, evidence and exceptions. Harvard Business School Working Paper 10-058, (2010)
Cover, T.M., Thomas, J.A.: Elements of Information Theory. John Wiley and Sons, New York (1991)
Crutchfield, J.P., Feldman, D.P.: Regularities unseen, randomness observed: Levels of entropy convergence. Chaos **13**(25), 25–54 (2003)
Crutchfield, J.P., Young, K.: Inferring statistical complexity. Phys. Rev. Lett. **63**, 105–108 (1989)
Crutchfield, J.P., Young, K.: Computation at the onset of Chaos. In: Zurek, W.H. (ed.) Complexity, entropy, and the physics of information, pp. 223–269. Addison-Wesley, Reading, MA (1990)
Crutchfield, J.P., Ellison, C.J., James, R.G., Mahoney, J.R.: Synchronization and control in intrinsic and designed computation: an information-theoretic analysis of competing models of stochastic computation. Santa Fe Institute Working Paper 2010-08-015. (2010)
Crutchfield, J.P., Marzen, S.: Signatures of infinity: Nonergodicity and resource scaling in prediction, complexity, and learning. Santa Fe Institute Working Paper 2015-04-010. (2015)
Cummings, J.N., Espinosa, J.A., Pickering, C.K.: Crossing spatial and temporal boundaries in globally distributed projects: A relational model of coordination delay. Inform. Syst. Res. **20** (3), 420–439 (2009)
Danilovic, M., Browning, T.R.: Managing complex product development projects with design structure matrices and domain mapping matrices. Int. J. Project Manag. **25**(3), 300–314 (2007)
deLeeuw, J.: Introduction to Akaike (1973) information theory and an extension of the maximum likelihood principle. In: Kotz, S., Johnson, N.L. (eds.) Breakthroughs in Statistics, vol. 1, pp. 599–609. Springer, London (1992)
Denman, J., Kaushik, S., de Weck, O.: Technology insertion in Turbofan Engine and assessment of architectural complexity. In: Proceedings of the 13th International Dependency and Structure Modeling Conference, DSM 2011, pp. 407–420, (2011).
Dvir, D., Sadeh, A., Malach-Pines, A.: Projects and Project Managers: The relationship between projects managers' personality, project types, and project success. Project Manag. J. **37**(5), 36–48 (2006)
Edwards, A.W.F.: Likelihood. Cambridge University Press, Cambridge U.K. (1972)
El-Haik, B., Yang, K.: The components of complexity in engineering design. IIE Trans. **31**(10), 925–934 (1999)
Ellison, C.J., Mahoney, J.R., Crutchfield, J.P.: Prediction, retrodiction, and the amount of information stored in the present. Santa Fe Institute Working Paper 2009-05-017. (2009)
Eppinger, S.D., Browning, T.: Design Structure Matrix Methods and Applications. MIT Press, Cambridge, MA (2012)
Foley, D.K., Oliver, D.: Notes on Bayesian inference and effective complexity. Unpublished manuscript. Available at http://www.american.edu/cas/economics/info-metrics/pdf/upload/Oct-2011-Workshop-Paper-Foley-and-Oliver.pdf (2011) (retrieved September 2013).
Gebala, D.A., Eppinger, S.D.: Methods for analyzing design procedures. In: Proceedings of the ASME Conference on Design Theory and Methodology, Miami, FL, pp. 227–233, (1991)
Gell-Mann, M.: What is complexity. Complexity **1**(1), 16–19 (1995)
Gell-Mann, M., Lloyd, S.: Information measures, effective complexity, and total information. Complexity **2**(1), 44–52 (1996)
Gell-Mann, M., Lloyd, S.: Effective complexity. In: Gell-Mann, M., Tsallis, C. (eds.) Nonextensive Entropy—Interdisciplinary Applications, pp. 387–398. Oxford University Press, Oxford (2004)
Gharahmani, Z.: An introduction to hidden Markov Models and Bayesian Networks. Int. J. Pattern Recogn. Artif. Intell. **15**(1), 9–42 (2001)

Gokpinar, B., Hopp, W.J., Iravani, S.M.R.: The impact of misalignment of organizational structure and product architecture on quality in complex product development. Manag. Sci. **56**(3), 468–484 (2010)

Grassberger, P.: Toward a quantitative theory of self-generated complexity. Int. J. Theor. Phys. **25** (9), 907–938 (1986)

Griffin, A.: The effect of project and process characteristics on product development cycle time. J. Market. Res. **34**(1), 24–35 (1997)

Grünwald, P.: The minimum description length principle. MIT Press, Cambridge, MA (2007)

Gu, M., Wiesner, K., Rieper, E., Vedral, V.: Quantum mechanics can reduce the complexity of classical models. Nat. Commun. **3**, Article number: 762. (2012)

Hansen, M.H., Yu, B.: Model selection and the principle of minimum description length. J. Am Stat. Soc. **96**(454), 746–774 (2001)

Hass, K.B.: Managing Complex Projects. A New Model. Management Concepts. Leesburg Pike, PA (2009)

Hölttä-Otto, K., Magee, C.L.: Estimating factors affecting project task size in product development—An empirical study. IEEE Trans. Eng. Manag. **53**(1), 86–94 (2006)

James, R., Ellison, C.J., Crutchfield, J.P.: Anatomy of a bit: Information in a time series observation. Santa Fe Institute Working Paper 2011-05-019. (2011)

Jänicke, H., Scheuermann, G.: Steady visualization of the dynamics in fluids using ε-machines. Comput. Graph. **33**(1), 597–606 (2009)

Kellogg, K.C., Orlikowski, W.J., Yates, J.: Life in the trading zone: Structuring coordination across boundaries in postbureaucratic organizations. Organ. Sci. **17**(1), 22–44 (2006)

Kerzner, H.: Project Management: A Systems Approach to Planning, Scheduling, and Controlling. John Wiley & Sons, Hoboken, NJ (2009)

Kim, J., Wilemon, D.: Sources and assessment of complexity in NPD projects. R&D Manag. **33** (1), 15–30 (2003)

Kim, J., Wilemon, D.: An empirical investigation of complexity and its management in new product development. Technol. Analysis Strat. Manag. **21**(4), 547–564 (2009)

Koppel, M., Atlan, H.: An almost machine-independent theory of program-length complexity, sophistication, and induction. Inform. Sci. **56**(1-3), 23–33 (1991)

Kreimeyer, M., Lindemann, U.: Complexity Metrics in Engineering Design—Managing the Structure of Design Processes. Springer, Berlin (2011)

Ladyman, J., Lambert, J., Wiesner, K.: What is a complex system? Eur. J. Philos. Sci. **3**(1), 33–67 (2013)

Lebcir, M.R.: Impact of Project Complexity Factors on New Product Development Cycle Time. University of Hertfordshire Business School Working Paper. https://uhra.herts.ac.uk/dspace/handle/2299/5549. (2011)

Li, W.: On the relationship between complexity and entropy for Markov chains and regular languages. Complex Syst. **5**(4), 381–399 (1991)

Li, L.: Some notes on mutual information between past and future. J. Time Ser. Anal. **27**(2), 309–322 (2006)

Li, M., Vitányi, P.: An introduction to Kolmogorov complexity and its applications, 2nd edn. Springer, New York, NY (1997)

Li, L., Xie, Z.: Model selection and order determination for time series by information between the past and the future. J Time Ser. Anal **17**(1), 65–84 (1996)

Lind, M., Marcus, B.: An Introduction to Symbolic Dynamics and Coding. Cambridge University Press, Cambridge (1995)

Lindemann, U., Maurer, M., Braun, T.: Structural Complexity Management. An Approach for the Field of Product Design. Springer, Berlin (2009)

Löhr, W.: Predictive models and generative complexity. J. Syst. Sci. Complex. **25**(1), 30–45 (2012)

MacKay, D.J.C.: Information Theory, Inference and Learning Algorithms. Cambridge University Press, Cambridge U.K. (2003)

Marzen, S., Crutchfield, J.P.: Circumventing the curse of dimensionality in prediction: Causal rate-distortion for infinite-order markov processes. Santa Fe Institute Working Paper 2014-12-047, (2014)

Maurer, M.: Structural awareness in complex product design. Doctoral dissertation, Technische Universität München. Dr. Hut Verlag, Munich, Germany, (2007)

Maylor, H., Vidgen, R., Carver, S.: Managerial complexity in project-based operations: A grounded model and its implications for practice. Project Manag. J. **39**(1), 15–26 (2008)

Mihm, J., Loch, C., Huchzermeier, A.: Problem-solving oscillations in complex engineering. Manag. Sci. **46**(6), 733–750 (2003)

Mihm, J., Loch, C.: Spiraling out of control: Problem-solving dynamics in complex distributed engineering projects. In: Braha, D., Minai, A.A., Bar-Yam, Y. (eds.) Complex Engineered Systems: Science Meets Technology, pp. 141–158. Springer, Berlin (2006)

Mihm, J., Loch, C., Wilkinson, D., Huberman, B.: Hierarchical structure and search in complex organisations. Manag. Sci. **56**(5), 831–848 (2010)

Mulenburg, J.: What does complexity have to do with it? Complexity and the management of projects. In: Proceedings of the 2008 NASA Project Management Challenge Conference. (2008)

Murmann, P.A.: Expected development time reductions in the german mechanical engineering industry. J. Product Innov. Manag. **11**(3), 236–252 (1994)

Nicolis, G., Nicolis, C.: Foundations of Complex Systems: Nonlinear Dynamics, Statistical Physics, Information and Prediction. World Scientific, Singapore (2007)

O'Leary, M.B., Mortensen, M.: Go (Con)figure: Subgroups, imbalance, and isolates in Geographically dispersed teams. Organ. Sci. **21**(1), 115–131 (2010)

Papoulis, A., Pillai, S.U.: Probability, random variables and stochastic processes. McGraw-Hill, Boston, MA (2002)

Polani, D., Nehaniv, C., Martinetz, T., Kim, J.T.: Relevant information in optimized persistence vs. progeny strategies. In: Proceedings of The 10th International Conference on the Simulation and Synthesis of Living Systems, Artificial Life X, pp. 337–343, (2006)

Prokopenko, M., Boschetti, F., Ryan, A.J.: An information-theoretic primer on complexity, self-organization and emergence. Complexity **15**(1), 11–28 (2009)

Puri, N.N.: Fundamentals of linear systems for physical scientists and engineers. CRC Press, Boca Raton, FL (2010)

Rissanen, J.: Stochastic complexity in statistical inquiry. World Scientific, Singapore (1989)

Rissanen, J.: Fisher information and stochastic complexity. IEEE Trans. Inform. Theor. **42**(1), 40–47 (1996)

Rissanen, J.: Information and Complexity in Statistical Modeling. Springer, Berlin (2007)

Rissanen, J.: Optimal Estimation of Parameters. Cambridge University Press, Cambridge (2012)

Rivkin, J.W., Siggelkow, N.: Balancing search and stability: Interdependencies among elements of organizational design. Manag. Sci. **49**(3), 290–311 (2003)

Rivkin, J.W., Siggelkow, N.: Patterned interactions in complex systems: Implications for exploration. Manag. Sci. **53**(7), 1068–1085 (2007)

Rogers, J.L., Korte, J.J., Bilardo, V.J. Development of a genetic algorithm to automate clustering of a dependency structure matrix. National Aeronautics and Space Administration, Langley Research Center, Technical Memorandum NASA/TM-2006-214279, (2006)

Schlick, C.M., Winkelholz, C., Motz, F., Luczak, H.: Self-generated complexity and human–Machine interaction. IEEE Trans. Syst. Man Cybernet, Part A: Syst. Hum **36**(1), 220–232 (2006)

Schlick, C.M., Beutner, E., Duckwitz, S., Licht, T.: A complexity measure for new product development projects. In: Proceedings of the 19th International Engineering Management Conference, pp. 143−150, (2007)

Schlick, C.M., Duckwitz, S., Gärtner, T., Tackenberg, S.: Optimization of concurrent engineering projects using an information-theoretic complexity metric. In: Proceedings of the 11th International DSM Conference, pp. 53–64, (2009)

Schlick, C.M., Winkelholz, C., Motz, F., Duckwitz, S., Grandt, M.: Complexity assessment of human–Computer interaction. Theor. Iss. Ergon. Sci. **11**(3), 151–173 (2010)

Shalizi, C.R.: Methods and techniques of complex systems science: An overview. In: Deisboeck, T.S., Kresh, J.Y. (eds.) Complex systems science in biomedicine, pp. 33–114. Springer, New York (2006)

Shalizi, C.R., Crutchfield, J.P.: Computational mechanics: Pattern and prediction, structure and simplicity. J. Stat. Phys. **104**, 817–879 (2001)

Shalizi, C.R., Shalizi, K.L.: Optimal nonlinear prediction of random fields on networks. In: Discrete Mathematics and Theoretical Computer Science, AB(DMCS), pp. 11–30, (2003)

Shalizi, C.R., Shalizi, K.L.: Blind construction of optimal nonlinear recursive predictors for discrete sequences. In: Proceedings of the Twentieth Conference on Uncertainty in Artificial Intelligence, pp. 504–511, (2004)

Shaw, R.: The Dripping Faucet as a Model Chaotic System. Aerial Press, Santa Cruz, CA (1984)

Shenhar, A.J.: From theory to practice: Toward a typology of project management styles. IEEE Trans. Eng. Manag. **45**(1), 33–48 (1998)

Shenhar, A.J., Dvir, D.: Toward a typological theory of project management. Res. Pol. **25**(4), 607–632 (1996)

Shenhar, A.J., Dvir, D.: Reinventing Project Management: The Diamond Approach to Successful Growth and Innovation. Harvard Business School Press, Boston, MA (2007)

Shiner, J.S., Davison, M., Landsberg, P.T.: Reply to comments on "Simple measure for complexity". Phys. Rev. E **62**(2), 3000–3003 (2000)

Shtarkov, Y.M.: Universal sequential coding of single messages. Prob. Inform. Transmis. **23**(3), 3–17 (1987) (translated from Russian)

Shtub, A., Bard, J.F., Globerson, S.: Project Management—Processes, Methodologies, and Economics, 2nd edn. Prentice Hall, Upper Saddle River, NJ (2004)

Sinha, K., de Weck, O.: Spectral and topological features of "Real-World" Product Structures. In: Proceedings of the 11th International Dependency and Structure Modeling Conference, DSM 2011, pp. 65–77, (2011).

Sinha, K., de Weck, O.: Structural complexity metric for engineered complex systems and its application. In: Proceedings of the 12th International Dependency and Structure Modeling Conference, DSM 2012, pp. 181–192, (2012)

Sinha, K.: Structural Complexity and its Implications for Design of Cyber-Physical Systems. Ph. D. Thesis, Massachusetts Institute of Technology, (2014)

Smith, R.P., Eppinger, S.D.: Identifying controlling features of engineering design iteration. Manag. Sci. **43**(3), 276–293 (1997)

Sosa, M.E.: A structured approach to predicting and managing technical interactions in software development. Res. Eng. Des. **19**, 47–70 (2008)

Sosa, M.E., Eppinger, S.D., Rowles, C.M.: The misalignment of product architecture and organizational structure in complex product development. Manag. Sci. **50**(12), 1674–1689 (2004)

Steward, D.V.: The design structure system: A method for managing the design of complex systems. IEEE Trans. Eng. Manag. **28**(3), 71–74 (1981)

Suh, N.P.: Axiomatic Design: Advances and Applications. Oxford University Press, Oxford (2001)

Suh, N.P.: Complexity—Theory and Applications. Oxford University Press, Oxford (2005)

Summers, J.D., Shah, J.J.: Mechanical engineering design complexity metrics: Size, coupling, and solvability. J. Mech Des. **132**(2), 1–11 (2010)

Summers, J.D., Shah, J.J.: Developing measures of complexity for engineering design. In: Proc. ASME DETC, Chicago, IL, Paper DTM-48633. pp. 381–392, (2003)

Summers, J.D., Ameri, F.: An algorithm for assessing design connectivity complexity. In: Tools and Methods for Competitive Engineering Conference, Izmir, Turkey, (2008)

Tatikonda, M.V., Rosenthal, S.R.: Technology novelty, project complexity and product development project execution success. IEEE Trans. Eng. Manag. **47**, 74–87 (2000)

Travers, N.F., Crutchfield, J.P.: Infinite excess entropy processes with countable-state generators. Entropy **16**(3), 1396–1413 (2014)

Travers, N.F., Crutchfield, J.P.: Equivalence of History and generator epsilon-machines. Santa Fe Institute Working Paper 2011-11-015, (2011).

Vitányi, P., Li, M.: Minimum description length induction, Bayesianism, and Kolmogorov Complexity. IEEE Trans. Inform. Theor **46**(2), 446–464 (2000)

Wallace, C.S., Boulton, D.M.: An information measure for classification. Comput J. **11**(2), 185–195 (1968)

Weyuker, E.: Evaluating software complexity measures. IEEE Trans. Softw. Eng. **14**(9), 1357–1365 (1988)

Wheelwright, S.C., Clark, K.B.: Creating project plans to focus product development. Harv. Bus. Rev. **70**(2), 70–82 (1992)

Wiesner, K.: Complexity measures and physical principles. In: Sanayei, A. et al. (eds). ISCS 2014: Interdisciplinary Symposium on Complex Systems, Emergence, Complexity and Computation, vol. 14, pp 15–20, (2015)

Zambella, D., Grassberger, P.: Complexity of forecasting in a class of simple models. Complex Syst. **2**(1), 269–303 (1988)

Chapter 4
Model-Driven Evaluation of the Emergent Complexity of Cooperative Work Based on Effective Measure Complexity

In this book, the main improvement on Grassberger's original definition of the effective measure complexity EMC, which is based on classic information-theoretic quantities like Shannon's information entropy that were developed to evaluate stochastic processes with discrete states, is the generalization of the theory and measures to continuous-state processes like that generated by the previously introduced VAR(1) model of cooperative work according to state Eq. 8. However, Li and Xie (1996), Bialek et al. (2001), de Cock (2002), Bialek (2003), Ellison et al. (2009) and others have already pioneered the generalization of Grassberger's concepts toward continuous systems in their works, and we can build upon their results. Their analyses show that we must primarily consider the so-called "differential block entropy" (Eq. 233) and the corresponding continuous-type mutual information (Eq. 234) as basic information-theoretic quantities.

In general, the differential entropy extends the basic idea of Shannon's information entropy as a universal measure of uncertainty about a discrete-type random variable with known probability mass function over the finite alphabet $\mathcal{X}$ to a p-dimensional continuous-type variable X with a probability density function $f[x]$ (*pdf*, see previous chapters) whose support is a set $\mathbb{X}^p$. The differential entropy is defined as:

$$H[X] := -\int_{\mathbb{X}^p} f[x] \log_2 f[x] \, dx. \tag{232}$$

The differential block entropy (cf. Eq. 219) is defined in an analogous manner as:

$$H(n) := H[X^n] = -\int_{\mathbb{X}^p} \cdots \int_{\mathbb{X}^p} f[x_1, \ldots, x_n] \log_2 f[x_1, \ldots, x_n] \, dx_1 \ldots \, dx_n. \tag{233}$$

In the above equation $f[x_1, \ldots, x_n]$ denotes the joint *pdf* of the vectors $(X_1, \ldots, X_n)$ with support $\mathbb{X}^{np}$.

© Springer International Publishing Switzerland 2016
C. Schlick, B. Demissie, *Product Development Projects*, Understanding Complex Systems, DOI 10.1007/978-3-319-21717-8_4

The information entropy of a discrete-type random variable is non-negative and can be used as a measure of average surprisal. This is slightly different for a continuous-type random variable, whose differential entropy can take any value from $-\infty$ to ∞ and is only used to measure changes in uncertainty (Cover and Thomas 1991; Papoulis and Pillai 2002). For instance, the differential entropy of a continuous random variable X that is uniformly distributed from 0 to a (and whose *pdf* is therefore $f[x] = 1/a$ from 0 to a, and 0 elsewhere) is $\log_2 a$. For $a < 1$ the differential entropy is negative and can become arbitrarily small as a approaches 0. The differential entropy measures the entropy of a continuous distribution relative to the uniformly distributed one. For a Gaussian distribution with a variance of σ^2 the differential entropy is $H[X] = 1/2\log_2\sigma^2 + const.$ Thus the differential entropy can be regarded as a generalization of the familiar notion of variance. With a normal distribution, the differential entropy is maximized for a given variance. An additional subtlety is that the differential entropy can be negative or positive depending on the coordinate system used for encoding the vectors. This also holds true for the differential block entropy. However, it can be proven that the complexity measure EMC calculated on the basis of dynamic entropies (cf. Eqs. 224 and 225) is always positive and may exist even in cases where the block entropies diverge. Under the assumption of an underlying VAR model, for instance, a closed-form solution for the EMC can be derived that is simply a logarithmic ratio of determinants of covariance matrices (cf. Eqs. 246 and 258), which in most industrial case studies is a real number that is much larger than zero. In this case, the generalized complexity measure can be interpreted similarly to discrete-state processes. Furthermore, it can be proven that for finite complexity values EMC is independent of the basis in which the state vectors of work remaining are represented, and is invariant under linear transformations of the state-space coordinates for any regular transformation matrix (Schneider and Griffies 1999). This invariance is due to the fact that the measure can be expressed as the continuous-type mutual information $I\left[X_{-\infty}^{-1}; X_0^{\infty}\right]$ between the infinite past and future histories of a stochastic process, where the base-independent mutual information $I[.;.]$ between the sequences $X_1^n = (X_1, \ldots, X_n)$ and $Y_1^m = (Y_1, \ldots, Y_m)$ of random vectors with support $\mathbb{X}^{nq}$ and $\mathbb{Y}^{mp}$ is defined as

$$I\left[X_1^n; Y_1^m\right] := \int_{\mathbb{X}^q} \cdots \int_{\mathbb{Y}^p} f[x_1, \ldots, x_n, y_1, \ldots, y_m]$$
$$\log_2 \frac{f[x_1, \ldots, x_n, y_1, \ldots, y_m]}{f[x_1, \ldots, x_n] f[y_1, \ldots, y_m]} dx_1 \ldots dx_n dy_1 \ldots dy_m . \tag{234}$$

For two random variables X and Y that are jointly normal with a correlation coefficient of ρ there is $I[X; Y] = 1/2\log_2(1 - \rho^2)$. As such, the mutual information can be viewed as a generalized covariance. Kraskov et al. (2004) published a simple proof that the mutual information as defined in Eq. 234 is not only invariant under linear transformations but also with respect to arbitrary reparameterizations based on smooth and uniquely invertible maps $x_1' = x_1'(x_1), \ldots, x_n' = x_n'(x_n), y_1' = y_1'(y_1), \ldots, y_m' = y_m'(y_m)$. Therefore, $I[.;.]$ provides a measure of statistical dependency structures between variables that is independent of the subjective choice of the measurement instrument. The analyses of Bialek et al. (2001) and other researchers show that this measure is a valid, expressive and consistent quantity for evaluating emergent complexity in open systems.

In the following chapters the generalization of the EMC to project organizations that are modeled by continuous state variables will be carried out step-by-step. Though some of the calculations are quite involved, the interested reader will find that they lay important groundwork for the complexity analysis of cooperative work in various kinds of open organizational systems, not only product development organizations. In Section 4.1, we start by calculating closed-form solutions with different strength for the vector autoregression models that were introduced in Sections 2.1, 2.2 and 2.4. These models do not have "hidden" state variables and therefore are quite easy to analyze in information-theoretic terms. To simplify the analysis a generalized solution for a VAR(1) process that does not refer to a specific family of *pdf*s of the unpredictable performance fluctuations is calculated in Section 4.1. We will use this generalized solution to derive closed-form solutions for the original state space (Section 4.1.1) and the spectral basis (Section 4.1.2) under the assumption of Gaussian behavior. Furthermore, a very compact closed-form solution will be obtained through a canonical correlation analysis (Section 4.1.3). For these three different approaches, we will also present the corresponding closed-form solutions of the persistent mutual information EMC(τ) (Eq. 229) according to Ball et al. (2010). Moreover, to clarify the concept of emergent complexity, polynomial-based solutions for simple processes with two and three tasks are presented in Section 4.1.4. This chapter also includes a short analytical study of minimizing emergent complexity subject to the constraint that the expected total amount of work done over all tasks is constant. Moreover, lower bounds are put on the EMC in Section 4.1.5. In Section 4.2.1, an additional explicit closed-form solution for a Markov process with hidden variables (a linear dynamical system, LDS, see Section 2.9) is calculated. This solution is, admittedly, complicated and difficult to interpret because the state variables of cooperative work that are not directly accessible can generate a significant number of long-range correlations between observations, and a great deal of linear algebra is needed to evaluate the associated infinite-dimensional integrals. Therefore, Section 4.2.2 will introduce two additional implicit formulations for the EMC. The first implicit solution is based on the seminal work of de Cock (2002) and allows analogical reasoning between the forward and backward innovation forms developed in Section 2.9, and the generated past-future mutual information. The second implicit solution is directly derived from the infinite-dimensional integrals and makes it possible for the interested reader to gain additional insights into the information-generating mechanisms by following the calculation step by step. Although the closed-form solutions for LDS are significantly more complicated, their derivations show that Grassberger's theory can, in principle, be applied in a straightforward manner to a larger model class that, thanks to its informational richness and predictive power, is especially attractive for applications in project management.

4.1 Closed-Form Solutions of Effective Measure Complexity for Vector Autoregression Models of Cooperative Work

To obtain analytical results, it is assumed that the parameterized VAR(1) process $\{X_t\}$ is strict-sense stationary (Puri 2010) and therefore all its statistical properties (especially the first and second moments) are invariant to a shift in the chosen time origin. Let $f_\theta[x_{t+1}, \ldots, x_{t+n}]$ $(t \in \mathbb{Z}, n \in \mathbb{N})$ be the joint *pdf* of the block of vectors $(X_{t+1}, \ldots, X_{t+n})$ generating the stochastic process, and let $f_\theta\left[x_{t+n}\middle|x_{t+1}, \ldots, x_{t+n-1}\right]$ be the conditional density of vector X_{t+n} given vectors $X_{t+1}, \ldots, X_{t+n-1}$. We use the shorthand notation $f[.]$ and $f\left[.\middle|.\right]$ in the following to denote these density functions. Due to strict sense stationarity the joint distribution of any sequence of samples does not depend on the sample's placement:

$$f[x_{t+1}, \ldots, x_{t+n}] = f[x_{t+1+\tau}, \ldots, x_{t+n+\tau}] \quad (t \in \mathbb{Z}, n \in \mathbb{N}, \tau \geq 0)\,.$$

We can use the index υ instead of t to express the shift-invariance. Therefore, $f[x_{\nu+1}, \ldots, x_{\upsilon+n}]$ denotes the joint *pdf* and $f\left[x_{\upsilon+n}\middle|x_{\upsilon+1}, \ldots, x_{\upsilon+n-1}\right]$ denotes the conditional density of the process in the steady state. The conditional density is given by (cf. Billingsley 1995):

$$f\left[x_{\upsilon+n}\middle|x_{\upsilon+1}, \ldots, x_{\upsilon+n-1}\right] = \frac{f[x_{\upsilon+1}, \ldots, x_{\upsilon+n}]}{f[x_{\upsilon+1}, \ldots, x_{\upsilon+n-1}]}\,.$$

Since the considered VAR(1) process is a Markov process (Eq. 18), the conditional density simplifies to

$$f\left[x_{\upsilon+n}\middle|x_{\upsilon+1}, \ldots, x_{\upsilon+n-1}\right] = f\left[x_{\upsilon+n}\middle|x_{\upsilon+n-1}\right] = \frac{f[x_{\upsilon+n-1}, x_{\upsilon+n}]}{f[x_{\upsilon+n-1}]}, \tag{235}$$

and the strict stationarity condition implies (Brockwell and Davis 1991)

$$\begin{gathered} f\left[x_{\upsilon+n}\middle|x_{\upsilon+n-1}\right] = f\left[x_{\upsilon}\middle|x_{\upsilon-1}\right] = f\left[x_2\middle|x_1\right] \\ \text{and } f[x_{\upsilon+n-1}] = f[x_{\upsilon}] = f[x_1] \ \forall \upsilon \geq 2\,. \end{gathered} \tag{236}$$

Furthermore, we assume that ergodicity holds, and the complexity measure can be conveniently derived using stochastic calculus based on an ensemble average or an infinite number of realizations of the unpredictable performance fluctuations (see

Puri 2010). To compute the EMC for the introduced VAR(1) process in the steady state, please recall from Eq. 216 that

$$\mathrm{EMC} = I\left[X_{-\infty}^{-1}; X_0^{\infty}\right].$$

According to the definition of the mutual information $I[.;.]$ from Eq. 234, we can write the information that is communicated from the past to the future as

$$I\left[X_{-\infty}^{-1}; X_0^{\infty}\right] = \int_{\mathbb{X}^p} \cdots \int_{\mathbb{X}^p} f\left[x_{-\infty}^{-1}, x_0^{\infty}\right] \log_2 \frac{f\left[x_{-\infty}^{-1}, x_0^{\infty}\right]}{f\left[x_{-\infty}^{-1}\right] f\left[x_0^{\infty}\right]}\, dx_{-\infty}^{-1}\, dx_0^{\infty}. \tag{237}$$

In the above equation the shorthand notation $f\left[x_{-\infty}^{-1}, x_0^{\infty}\right] = f[x_{-\infty}, x_{-\infty+1}, \ldots, x_{-1}, x_0, x_1, \ldots, x_{\infty-1}, x_{\infty}]$, $f\left[x_{-\infty}^{-1}\right] = f[x_{-\infty}, x_{-\infty+1}, \ldots, x_{-1}]$, $f\left[x_0^{\infty}\right] = f[x_0, x_1, \ldots, x_{\infty-1}, x_{\infty}]$, $dx_{-\infty}^{-1} = dx_{-\infty} dx_{-\infty+1} \ldots dx_{-1}$ and $dx_0^{\infty} = dx_0 dx_1 \ldots dx_{\infty}$ was used. Due to the Markov property (Eqs. 235 and 236) the joint *pdfs* can be factorized:

$$\begin{aligned}
f\left[x_{-\infty}^{-1}, x_0^{\infty}\right] &= f[x_{-\infty}] f\left[x_{-\infty+1} \middle| x_{-\infty}\right] \ldots f\left[x_{-1} \middle| x_{-2}\right] f\left[x_0 \middle| x_{-1}\right] f\left[x_1 \middle| x_0\right] \ldots f\left[x_{\infty} \middle| x_{\infty-1}\right] \\
f\left[x_{-\infty}^{-1}\right] &= f[x_{-\infty}] f\left[x_{-\infty+1} \middle| x_{-\infty}\right] \ldots f\left[x_{-1} \middle| x_{-2}\right] \\
f\left[x_0^{\infty}\right] &= f[x_0] f\left[x_1 \middle| x_0\right] \ldots f\left[x_{\infty} \middle| x_{\infty-1}\right].
\end{aligned}$$

Hence, we can simplify the mutual information:

$$\begin{aligned}
I\left[X_{-\infty}^{-1}; X_0^{\infty}\right] &= \int_{\mathbb{X}^p} \cdots \int_{\mathbb{X}^p} f\left[x_{-\infty}^{-1}, x_0^{\infty}\right] \log_2 \frac{f\left[x_0 \middle| x_{-1}\right]}{f[x_0]}\, dx_{-\infty}^{-1}\, dx_0^{\infty} \\
&= \int_{\mathbb{X}^p} \cdots \int_{\mathbb{X}^p} f\left[x_{-\infty}^{-1}, x_0^{\infty}\right] \log_2 f\left[x_0 \middle| x_{-1}\right] dx_{-\infty}^{-1}\, dx_0^{\infty} \\
&\quad - \int_{\mathbb{X}^p} \cdots \int_{\mathbb{X}^p} f\left[x_{-\infty}^{-1}, x_0^{\infty}\right] \log_2 f[x_0]\, dx_{-\infty}^{-1}\, dx_0^{\infty} \\
&= \int_{\mathbb{X}^p} \int_{\mathbb{X}^p} \log_2 f\left[x_0 \middle| x_{-1}\right] dx_0 dx_{-1} \int_{\mathbb{X}^p} \cdots \int_{\mathbb{X}^p} f\left[x_{-\infty}^{-1}, x_0^{\infty}\right] dx_{-\infty} \ldots dx_{-2} dx_1 \ldots dx_{\infty} \\
&\quad - \int_{\mathbb{X}^p} \log_2 f[x_0]\, dx_0 \int_{\mathbb{X}^p} \cdots \int_{\mathbb{X}^p} f\left[x_{-\infty}^{-1}, x_0^{\infty}\right] dx_{-\infty} \ldots dx_{-1} dx_1 \ldots dx_{\infty}.
\end{aligned} \tag{238}$$

On the basis of the definitions of the marginal density functions

$$f[x_0] = \int_{\mathbb{X}^p} \cdots \int_{\mathbb{X}^p} f\left[x_{-\infty}^{-1}, x_0^{\infty}\right] dx_{-\infty} \ldots dx_{-1} dx_1 \ldots dx_{\infty}$$

$$f[x_{-1}, x_0] = \int_{\mathbb{X}^p} \cdots \int_{\mathbb{X}^p} f\left[x_{-\infty}^{-1}, x_0^{\infty}\right] dx_{-\infty} \ldots dx_{-2} dx_1 \ldots dx_{\infty}$$

we can conclude that

$$\begin{aligned} I\left[X_{-\infty}^{-1}; X_0^{\infty}\right] &= \int_{\mathbb{X}^p}\int_{\mathbb{X}^p} f[x_{-1}, x_0] \log_2 f\left[x_0 | x_{-1}\right] dx_{-1} dx_0 - \int_{\mathbb{X}^p} f[x_0] \log_2 f[x_0] dx_0 \\ &= \int_{\mathbb{X}^p}\int_{\mathbb{X}^p} f\left[x_0 | x_{-1}\right] f[x_{-1}] \log_2 f\left[x_0 | x_{-1}\right] dx_{-1} dx_0 - \int_{\mathbb{X}^p} f[x_0] \log_2 f[x_0] dx_0, \end{aligned} \tag{239}$$

or equivalently

$$I\left[X_{-\infty}^{-1}; X_0^{\infty}\right] = \int_{\mathbb{X}^p}\int_{\mathbb{X}^p} f\left[x_1 | x_0\right] f[x_0] \log_2 f\left[x_1 | x_0\right] dx_0 dx_1 - \int_{\mathbb{X}^p} f[x_0] \log_2 f[x_0] dx_0.$$

It is evident that the second summand is the differential entropy of the random variable X_0 with probability density function $f[x_0]$. The first summand represents the entropy of the random variable X_1 conditioned on the variable X_0 taking a value in the support $\mathbb{X}^p$. The first summand therefore represents a conditional entropy that is obtained by averaging over all possible values for X_0.

Before we proceed with calculating the EMC on the basis of the generalized solution from Eq. 239 in the coordinates of the original state space $\mathbb{R}^p$, we summarize five essential properties that hold completely independent of the stochastic model generating a strict-sense stationary Gaussian process $\{X_t\}$. A Gaussian process is a stochastic process whose realizations consist of random values associated with every time step such that each random variable in the sequence has a normal distribution. In addition, every finite ensemble of random variables generating the process has a multivariate normal distribution (Puri 2010).

The five essential properties are as follows (cf. Boets et al. 2007):

1) The EMC of a strict-sense stationary Gaussian process equals zero if and only if the process is temporally uncorrelated:

$$\begin{aligned} \text{EMC} = I\left[X_{-\infty}^{-1}; X_0^{\infty}\right] = 0 \Leftrightarrow X_t = \nu_t \ \text{ with} \\ \nu_t = \mathcal{N}(\eta; \mu, V) \ \text{ and } \ E\left[\nu_t \nu_s^{\mathrm{T}}\right] = V\delta_{ts} \ . \end{aligned} \tag{240}$$

μ denotes the mean of the process and $s \in \mathbb{Z}$ an arbitrary time step. δ_{ts} is the Kronecker delta according to Eq. 14. The implication EMC $= 0$ can be easily deduced as Gaussian random variables being uncorrelated is equivalent to statistical independence, i.e. $f(X_{-\infty}^{-1};X_0^{\infty}) = f(X_{-\infty}^{-1}) \cdot f(X_0^{\infty})$. A proof of the implication that the process is temporally uncorrelated involves Jensen's inequality and can be found in elementary textbooks like Cover and Thomas (1991). Concerning the state and output equations of a LDS with additive Gaussian noise (Eqs. 136 and 137), this may be realized either by setting $H = 0$ or with $A_0 = 0$.

2) The range of values of the Effective Measure Complexity is

$$\mathrm{EMC} \in [0, +\infty). \tag{241}$$

This property follows directly from the canonical correlation analysis of the past $(X_{-\infty}^{-1})$ and future (X_0^{∞}) histories of the Gaussian process (see Eq. 265 in Section 4.1.3)

$$I[X_{-\infty}^{-1};X_0^{\infty}] = -\frac{1}{2}\log_2\prod_{i=1}^{q}(1-\rho_i^2)$$

and the fact that the canonical correlations ρ_i are confined to $\rho_i \in [0,1)$ (see e.g. de Cock 2002). The variable $q > p$ denotes the effective dimensionality of the process (see Section 4.2.1). The canonical correlation analysis was introduced by Hotelling (1935) and is often used for state-space identification. The goal is to find a suitable basis for cross-correlation between two random variables—in our case the infinite, one-dimensional sequences of random variables representing the past and future histories of the process. Based on the material of Creutzig (2008) we use a common variant of the canonical correlation analysis to provide a so-called balanced state-space representation (cf. Section 4.2). Given the ordered concatenation of the variables representing the past history

$$X_{past} = \begin{pmatrix} X_{-\infty}^{\mathrm{T}} & \cdots & X_{-2}^{\mathrm{T}} & X_{-1}^{\mathrm{T}} \end{pmatrix}^{\mathrm{T}}$$

and the future history

$$X_{fut} = \begin{pmatrix} X_0^{\mathrm{T}} & X_1^{\mathrm{T}} & \cdots & X_{\infty}^{\mathrm{T}} \end{pmatrix}^{\mathrm{T}}$$

of the Gaussian process we seek an orthonormal base $U = (U^{(1)}, \ldots, U^{(m)})$ for X_{past} and another orthonormal base $V = (V^{(1)}, \ldots, V^{(n)})$ for X_{fut} that have maximal correlations but are internally uncorrelated. Therefore, it must hold that $E[U^{(i)}V^{(j)}] = \rho_i\delta_{ij}$, for $i, j \le min(m,n)$. $U^{(i)}$ and $V^{(j)}$ are two zero-mean random variables of dimensions m and n, respectively. The resulting basis variables $(U^{(1)}, \ldots, U^{(m)})$ and $(V^{(1)}, \ldots, V^{(n)})$ are called canonical variates, and the correlation coefficients ρ_i between the canonical variates are called canonical correlations. The cardinalities of the bases must be chosen in a way that is compatible with the persistent informational structure of the process. The ρ_i's are not to be confused with the introduced ordinary correlations ρ_{ij} and ρ'_{ij} from Chapter 2.

To find the orthonormal bases, we normalize with Cholesky factors. The factors are given by

$$L_{past} \cdot L_{past}^{\mathrm{T}} := \left\{\frac{1}{N}\right\} \cdot X_{past} \cdot X_{past}^{\mathrm{T}}$$

$$L_{fut} \cdot L_{fut}^{\mathrm{T}} := \left\{\frac{1}{N}\right\} \cdot X_{fut} \cdot X_{fut}^{\mathrm{T}}.$$

N denotes the number of samples that are taken from the stochastic process. The sample size must be sufficiently large to uncover all canonical correlations. The normalized variables $\widehat{X}_{past}$ and $\widehat{X}_{fut}$ to determine the balanced state-space representation are computed by

$$\widehat{X}_{past} = L_{past}^{-1} \cdot X_{past}$$

$$\widehat{X}_{fut} = L_{fut}^{-1} \cdot X_{fut}.$$

A singular value decomposition is carried out (see e.g. de Cock 2002, and Section 4.1.3) to identify the orthonormal bases:

$$\left\{\frac{1}{N}\right\} \cdot \widehat{X}_{fut} \cdot \widehat{X}_{past}^{\mathrm{T}} = \Sigma_{yu} \cdot \widehat{V} \cdot \widehat{\Sigma} \cdot \widehat{U}.$$

Σ_{yu} denotes the cross-covariance between X_{fut} and X_{past}. We compute the state space by

$$\widehat{X}_t := \widehat{V}^{\mathrm{T}} \cdot \widehat{X}_{past} = \widehat{V}^{\mathrm{T}} \cdot L_{past}^{-1} \cdot X_{past}$$

and balance

$$\widehat{X}'_t = \widehat{\Sigma}^{\frac{1}{2}} \cdot \widehat{X}_t$$

such that for the covariance matrix it holds that

$$\left\{\frac{1}{N}\right\} \cdot \widehat{X}'_t \cdot \left(\widehat{X}'_t\right)^{\mathrm{T}} = \widehat{\Sigma}.$$

The requirement that the ρ_i's be nonnegative and ordered in decreasing magnitude makes the choice of bases unique if all canonical correlations are distinct. It is important to note that for a strict-sense stationary VAR(1) process $\{X_t\}$, only the p leading canonical correlations ρ_i of each pair $\left(X_{-\infty}^{-1}, X_0^{\infty}\right)$ of subprocesses are non-zero and therefore the cardinality of the base is equal to p (de Cock 2002; Boets et al. 2007). This is due to the simple fact that the process is Markovian and so the amount of information that the past provides about the future can always be encoded in the probability distribution over the p-dimensional present state (assuming an efficient coding mechanism is used). Furthermore, because of strict-sense stationarity, all ρ_i's are less than one.

3) EMC is a strictly increasing function of each of the canonical correlations. This property also follows directly from relation 265:

$$I\left[X_{-\infty}^{-1}; X_0^{\infty}\right] = -\frac{1}{2}\sum_{i=1}^{q} \log_2\left(1 - \rho_i^2\right). \tag{242}$$

4) The EMC is invariant under a transformation of the observations X_t by a nonsingular constant matrix $T \in \mathbb{R}^{p \times p}$. When we denote the transformed observations $Z_t = T \cdot X_t$, it holds that

$$I\left[Z_{-\infty}^{-1}; Z_0^{\infty}\right] = I\left[X_{-\infty}^{-1}; X_0^{\infty}\right]. \tag{243}$$

From the explicit result in Eq. 291 for the EMC of a process that is generated by a linear dynamical system with additive Gaussian noise, one can directly derive this invariance property. Similar to the notation in Section 4.2, $x_{t_1}^{t_2}$ denotes the vector obtained by stacking the observation sequence $X_{t_1}^{t_2}$ in a long vector of size $p(t_2 - t_1 + 1) \times 1$. We define the long vector $z_{t_1}^{t_2}$ in the same way. Then we can relate the transformed observations to the original ones via $Z_{t_1}^{t_2} = (I_{t_2-t_1+1} \otimes T) \cdot X_{t_1}^{t_2}$. The covariance of the history of transformed observations follows immediately and can be related to the covariance $(C_x)_{t_1}^{t_2} = E\left[X_{t_1}^{t_2}\left(X_{t_1}^{t_2}\right)^{\mathrm{T}}\right]$:

$$\begin{aligned}
(C_z)_{t_1}^{t_2} &= E\left[Z_{t_1}^{t_2}\left(Z_{t_1}^{t_2}\right)^{\mathrm{T}}\right] \\
&= E\left[(I_{t_2-t_1+1} \otimes T)X_{t_1}^{t_2}\left(X_{t_1}^{t_2}\right)^{\mathrm{T}}(I_{t_2-t_1+1} \otimes T)^{\mathrm{T}}\right] \\
&= (I_{t_2-t_1+1} \otimes T)(C_x)_{t_1}^{t_2}(I_{t_2-t_1+1} \otimes T)^{\mathrm{T}}.
\end{aligned}$$

It is straightforward to compute EMC by using the general expression for LDS (Eq. 291) with $H = I$ and $V = 0$ as

$$
\begin{aligned}
I\left[Z_{-\infty}^{-1};Z_0^{\infty}\right] &= \frac{1}{2}\log_2\frac{\mathrm{Det}(C_z)_{-\infty}^{-1}\mathrm{Det}(C_z)_0^{\infty}}{\mathrm{Det}(C_z)_{-\infty}^{\infty}} \\
&= \frac{1}{2}\lim_{\substack{t_p\to-\infty\\ t_f\to\infty}}\log_2\frac{\mathrm{Det}(C_z)_{t_p}^{-1}\mathrm{Det}(C_z)_0^{t_f}}{\mathrm{Det}(C_z)_{t_p}^{t_f}} \\
&= \frac{1}{2}\lim_{\substack{t_p\to-\infty\\ t_f\to\infty}}\left\{\log_2\mathrm{Det}\left(\left(I_{-t_p}\otimes T\right)(C_x)_{t_p}^{-1}\left(I_{-t_p}\otimes T\right)^{\mathrm{T}}\right)\right. \\
&\quad + \log_2\mathrm{Det}\left(\left(I_{t_f+1}\otimes T\right)(C_x)_0^{t_f}\left(I_{t_f+1}\otimes T\right)^{\mathrm{T}}\right) \\
&\quad \left. - \log_2\mathrm{Det}\left(\left(I_{t_f-t_p+1}\otimes T\right)(C_x)_{t_p}^{t_f}\left(I_{t_f-t_p+1}\otimes T\right)^{\mathrm{T}}\right)\right\} \\
&= \frac{1}{2}\lim_{\substack{t_p\to-\infty\\ t_f\to\infty}}\log_2\frac{\mathrm{Det}(C_x)_{t_p}^{-1}\mathrm{Det}(C_x)_0^{t_f}}{\mathrm{Det}(C_x)_{t_p}^{t_f}} \\
&= I\left[X_{-\infty}^{-1};X_0^{\infty}\right],
\end{aligned}
$$

where we have used the fact that $\mathrm{Det}(A\cdot B)=\mathrm{Det}(A)\cdot\mathrm{Det}(B)$ and that for matrices $A\in\mathbb{R}^{n\times n}, B\in\mathbb{R}^{m\times m}$ we have $\mathrm{Det}(A\otimes B)=(\mathrm{Det}(A))^n(\mathrm{Det}(B))^m$.

5) If the p-component vector of all observations X_t can be divided into two separate sets comprised of the vectors $X_t^{(1)}\in\mathbb{R}^{p_1}$ and $X_t^{(2)}\in\mathbb{R}^{p_2}$ with $p=p_1+p_2$, which are completely uncorrelated,

$$
C_{XX}(\tau)=E\left[X_tX_{t+\tau}^{\mathrm{T}}\right]=E\left[\begin{bmatrix}X_t^{(1)}\\ X_t^{(2)}\end{bmatrix}\begin{bmatrix}X_{t+\tau}^{(1)}\\ X_{t+\tau}^{(2)}\end{bmatrix}^{\mathrm{T}}\right]=\begin{bmatrix}C_{X^{(1)}X^{(1)}}(\tau) & 0\\ 0 & C_{X^{(2)}X^{(2)}}(\tau)\end{bmatrix},
$$

then the EMC of the whole sequence of observations equals the sum of the EMC of each set resulting from the partitioning:

$$
I\left[X_{-\infty}^{-1};X_0^{\infty}\right]=I\left[\left(X^{(1)}\right)_{-\infty}^{-1};\left(X^{(1)}\right)_0^{\infty}\right]+I\left[\left(X^{(2)}\right)_{-\infty}^{-1};\left(X^{(2)}\right)_0^{\infty}\right]. \tag{244}
$$

Since uncorrelated Gaussian random variables are independent, i.e. their joint *pdf* equals the product of the individual *pdfs*—in this case

$$
\begin{aligned}
f\left(x_{-\infty}^{-1},x_0^{\infty}\right) &= f\left[\left(x^{(1)}\right)_{-\infty}^{-1},\left(x^{(1)}\right)_0^{\infty},\left(x^{(2)}\right)_{-\infty}^{-1},\left(x^{(2)}\right)_0^{\infty}\right] \\
&= f\left[\left(x^{(1)}\right)_{-\infty}^{-1},\left(x^{(1)}\right)_0^{\infty}\right]\cdot f\left[\left(x^{(2)}\right)_{-\infty}^{-1},\left(x^{(2)}\right)_0^{\infty}\right],
\end{aligned}
$$

the above property of additivity for uncorrelated observations can be easily verified:

$$I\left[X_{-\infty}^{-1};X_0^{\infty}\right] = \int f\left[x_{-\infty}^{-1},x_0^{\infty}\right]\log_2\frac{f\left[x_{-\infty}^{-1},x_0^{\infty}\right]}{f\left[x_{-\infty}^{-1}\right]f\left[x_0^{\infty}\right]}dx_{-\infty}^{-1}dx_0^{\infty}$$

$$= \int f\left[x_{-\infty}^{-1},x_0^{\infty}\right]\log_2\frac{f\left[\left(x^{(1)}\right)_{-\infty}^{-1},\left(x^{(1)}\right)_0^{\infty}\right]f\left[\left(x^{(2)}\right)_{-\infty}^{-1},\left(x^{(2)}\right)_0^{\infty}\right]}{f\left[\left(x^{(1)}\right)_{-\infty}^{-1}\right]f\left[\left(x^{(1)}\right)_0^{\infty}\right]f\left[\left(x^{(2)}\right)_{-\infty}^{-1}\right]f\left[\left(x^{(2)}\right)_0^{\infty}\right]}dx_{-\infty}^{-1}dx_0^{\infty}$$

$$= \int f\left[\left(x^{(1)}\right)_{-\infty}^{-1},\left(x^{(1)}\right)_0^{\infty}\right]f\left[\left(x^{(2)}\right)_{-\infty}^{-1},\left(x^{(2)}\right)_0^{\infty}\right]$$

$$\log_2\frac{f\left[\left(x^{(1)}\right)_{-\infty}^{-1},\left(x^{(1)}\right)_0^{\infty}\right]}{f\left[\left(x^{(1)}\right)_{-\infty}^{-1}\right]f\left[\left(x^{(1)}\right)_0^{\infty}\right]}dx_{-\infty}^{-1}dx_0^{\infty}$$

$$+\int f\left[\left(x^{(1)}\right)_{-\infty}^{-1},\left(x^{(1)}\right)_0^{\infty}\right]f\left[\left(x^{(2)}\right)_{-\infty}^{-1},\left(x^{(2)}\right)_0^{\infty}\right]$$

$$\log_2\frac{f\left[\left(x^{(2)}\right)_{-\infty}^{-1},\left(x^{(2)}\right)_0^{\infty}\right]}{f\left[\left(x^{(2)}\right)_{-\infty}^{-1}\right]f\left[\left(x^{(2)}\right)_0^{\infty}\right]}dx_{-\infty}^{-1}dx_0^{\infty}.$$

In the first term, the integration with respect to $\left(x^{(2)}\right)_{-\infty}^{-1},\left(x^{(2)}\right)_0^{\infty}$ yields one, and analogously in the second term the integration with respect to the first variable set yields one. Ultimately, we obtain:

$$I\left[X_{-\infty}^{-1};X_0^{\infty}\right] = \int f\left[\left(x^{(1)}\right)_{-\infty}^{-1},\left(x^{(1)}\right)_0^{\infty}\right]\log_2\frac{f\left[\left(x^{(1)}\right)_{-\infty}^{-1},\left(x^{(1)}\right)_0^{\infty}\right]}{f\left[\left(x^{(1)}\right)_{-\infty}^{-1}\right]f\left[\left(x^{(1)}\right)_0^{\infty}\right]}d\left(x^{(1)}\right)_{-\infty}^{-1}d\left(x^{(1)}\right)_0^{\infty}$$

$$+\int f\left[\left(x^{(2)}\right)_{-\infty}^{-1},\left(x^{(2)}\right)_0^{\infty}\right]\log_2\frac{f\left[\left(x^{(2)}\right)_{-\infty}^{-1},\left(x^{(2)}\right)_0^{\infty}\right]}{f\left[\left(x^{(2)}\right)_{-\infty}^{-1}\right]f\left[\left(x^{(2)}\right)_0^{\infty}\right]}d\left(x^{(2)}\right)_{-\infty}^{-1}d\left(x^{(2)}\right)_0^{\infty}$$

$$= I\left[\left(X^{(1)}\right)_{-\infty}^{-1}\left(X^{(1)}\right)_0^{\infty}\right] + I\left[\left(X^{(2)}\right)_{-\infty}^{-1};\left(X^{(2)}\right)_0^{\infty}\right].$$

4.1.1 Closed-Form Solutions in Original State Space

To calculate the EMC on the basis of the generalized solution from Eq. 239 in the coordinates of the original state space $\mathbb{R}^p$, we must find the *pdf* of the generated stochastic process in the steady state. Let the p-dimensional random vector $X_{-\infty-\tau+1}$ be normally distributed with location $\mu_{-\infty-\tau+1} = A_0 \cdot x_{-\infty-\tau}$ and covariance

$\Sigma_{-\infty-\tau+1} = \Sigma_1$ (Eqs. 19 and 20), that is $X_{-\infty-\tau+1} \sim \mathcal{N}(x; A_0 \cdot x_{-\infty-\tau}, \Sigma_1)$. Starting with this random vector the project evolves according to state Eq. 8. As already shown in Section 2.2, the strictly stationary behavior for $t \to \infty$ means that a joint probability density is formed that is invariant under shifting the origin. Hence, for the locus we must have $\mu = A_0 \cdot \mu + E[\varepsilon_t] = A_0 \cdot \mu$, and for the covariance matrix the well-known Lyapunov criterion $\Sigma = A_0 \cdot \Sigma \cdot A_0^{\mathrm{T}} + \mathrm{Var}[\varepsilon_t] = A_0 \cdot \Sigma \cdot A_0^{\mathrm{T}} + C$ must be satisfied (Eqs. 4 and 27). It follows that μ must be an eigenvector corresponding to the eigenvalue 1 of the WTM A_0. Clearly, if the modeled project is asymptotically stable and the modulus of the largest eigenvalue of A_0 is less than 1, no such eigenvector can exist. Hence, the only vector that satisfies this equation is the zero vector 0_p, which indicates that there is no remaining work (Eq. 26).

Let $\lambda_1(A_0), \ldots, \lambda_p(A_0)$ be the eigenvalues of WTM A_0 ordered by magnitude. If $|\lambda_1(A_0)| < 1$, the solution of the Lyapunov Eq. 27 can be written as an infinite power series (Lancaster and Tismenetsky 1985):

$$\Sigma = \sum_{k=0}^{\infty} A_0^k \cdot C \cdot \left(A_0^{\mathrm{T}}\right)^k. \tag{245}$$

It can also be expressed using the Kronecker product $\otimes$:

$$vec[\Sigma] = \left[I_{p^2} - A_0 \otimes A_0\right]^{-1} vec[C] .$$

Σ is also positive-semidefinite. In the above equation it is assumed that $I_{p^2} - A_0 \otimes A_0$ is invertible, $vec[C\ ]$ is the vector function which was already used for the derivation of the least square estimators in Section 2.7, and I_{p^2} is the identity matrix of size $p^2 \times p^2$.

Under the assumption of Gaussian behavior, it is not difficult to find different closed-form solutions. Recalling that the random vector X_0 in steady state is normally distributed with location $\mu = 0_p$ and covariance Σ, it follows from textbooks (e.g. Cover and Thomas 1991) that the differential entropy as the second summand in Eq. 239 can be expressed as

$$\begin{aligned} -\int_{\mathbb{X}^p} f[x_0] \log_2 f[x_0] dx_0 &= -\int_{\mathbb{R}^p} \mathcal{N}(x_0; \mu, \Sigma) \log_2 \mathcal{N}(x_0; \mu, \Sigma) dx_0 \\ &= \frac{1}{2} \log_2 (2\pi e)^p \mathrm{Det}[\Sigma]. \end{aligned}$$

For the calculation of the conditional entropy (first summand in Eq. 239), the following insight is helpful. Given a value x_0, the distribution of X_1 is a normal distribution with location $A_0 \cdot x_0$ and covariance C. Hence, the conditional entropy is simply equal to minus the differential entropy of that distribution. For Gaussian distributions, the differential entropy is independent of the locus. Therefore, for the conditional entropy it holds that

$$\begin{aligned}
&\int_{\mathbb{X}^p}\int_{\mathbb{X}^p} f[x_1|x_0]f[x_0]\log_2 f[x_1|x_0]dx_0dx_1 \\
&= \int_{\mathbb{R}^p}\int_{\mathbb{R}^p} \mathcal{N}(x_1;A_0x_0,C)\mathcal{N}(x_0;\mu,\Sigma)\log_2\mathcal{N}(x_1;A_0x_0,C)dx_0dx_1 \\
&= \int_{\mathbb{R}^p} \mathcal{N}(x_1;A_0x_0,C)\ \log_2\mathcal{N}(x_1;A_0x_0,C)dx_1 \\
&= -\frac{1}{2}\log_2(2\pi e)^p\mathrm{Det}[C].
\end{aligned}$$

It follows for the VAR(1) model of cooperative work that

$$\begin{aligned}
\mathrm{EMC} &= \frac{1}{2}\log_2\left(\frac{\mathrm{Det}[\Sigma]}{\mathrm{Det}[C]}\right) = \frac{1}{2}\log_2\mathrm{Det}[\Sigma] - \frac{1}{2}\log_2\mathrm{Det}[C] \\
&= \frac{1}{2}\log_2\mathrm{Det}\left[\Sigma\cdot C^{-1}\right].
\end{aligned} \tag{246}$$

According to the above equation, the EMC can be decomposed additively into dynamic and pure-fluctuation parts. The dynamic part represents the variance of the process in steady state. If the fluctuations are isotropic, the dynamic part completely decouples from the fluctuations, as will be shown in Eqs. 250 and 251 (Ay et al. 2012). If the solution of the Lyapunov equation (Eq. 245) is substituted into the above equation, we can write the desired first closed-form solution as

$$\mathrm{EMC} = \frac{1}{2}\log_2\left(\frac{\mathrm{Det}\left[\sum_{k=0}^{\infty}A_0^k\cdot C\cdot\left(A_0^{\mathrm{T}}\right)^k\right]}{\mathrm{Det}[C]}\right). \tag{247}$$

The determinant Det[Σ] of the covariance matrix $\Sigma = \sum_{k=0}^{\infty}A_0^k\cdot C\cdot\left(A_0^{\mathrm{T}}\right)^k$ in the numerator of the solution above can be interpreted as a generalized variance of the stationary process. In the same manner Det[C] represents the generalized variance of the inherent fluctuations. The inverse C^{-1} is the so-called "concentration matrix" or "precision matrix" (Puri 2010). Det[C] can also be interpreted as the intrinsic (mean squared) one-step prediction error that cannot be underrun, even if we condition the observation on infinite past histories to build a maximally predictive model. An analogous interpretation of Det[Σ] is to consider it as the (mean squared) error for an infinite-step forecast of the VAR(1) model that is parameterized by the optimizing parameters x_0, A_0 and C (Lütkepohl 1985). In this sense, EMC is the logarithmic ratio related to the mean squared errors for infinite-step and one-step forecasts of the process state. Another interesting interpretation of Det[Σ] is produced if we do not refer to the predictions of a parameterized VAR(1) model over an infinite forecast horizon but instead to the one-step prediction error of a naïve

VAR(0) model whose predictions are based on the (zero) mean of the stationary process. It is evident that this kind of model completely lacks the ability to compress past project trajectories into a meaningful internal configuration to denote the state of the project and therefore has zero complexity. Hence, EMC can also be interpreted as the logarithmic ratio of the (mean squared) one-step prediction error of a naïve VAR(0) model with zero complexity and a standard VAR(1) model with non-negligible complexity due to procedural memory that incorporates an effective prediction mechanism. In this context "effective" means that the state should be formed in a way that the mean squared prediction error is minimized at fixed memory (sensu Still 2014). In terms of information theory, the generalized variance ratio can be interpreted as the entropy lost and information gained when the modeled project is in the steady state, and the state is observed by the project manager with predefined "error bars", which cannot be underrun because of the intrinsic prediction error (Bialek 2003).

The covariance matrices Σ and C are positive-semidefinite. Under the assumption that they are of full rank, the determinants are positive, and the range of the EMC is $\left[0,+\infty\right)$. This was already mentioned in the discussion of the essential properties of EMC (see Eq. 241).

Interestingly, we can reshape the above solution so that it can be interpreted in terms of Shannon's famous "Gaussian channel" (cf. Eq. 262 and the associated discussion) as

$$\mathrm{EMC}=\frac{1}{2}\log_2 \mathrm{Det}\left[I_p+\left(\sum_{k=1}^{\infty} A_0^k \cdot C \cdot \left(A_0^{\mathrm{T}}\right)^k\right)\cdot C^{-1}\right]. \tag{248}$$

If the covariance C is decomposed into an orthogonal forcing matrix K and a diagonal matrix Λ_K as shown in Eq. 22, the determinant in the denominator of Eq. 247 can be replaced by $\mathrm{Det}[C]=\mathrm{Det}[\Lambda_K]$.

We can also separate the noise component $K\cdot\Lambda_K\cdot K^{\mathrm{T}}$ in the sum and reshape the determinant in the numerator as follows:

$$\begin{aligned}
\mathrm{EMC}&=\frac{1}{2}\log_2\left(\frac{\mathrm{Det}\left[\sum_{k=0}^{\infty} A_0^k\cdot K\cdot\Lambda_K\cdot K^{\mathrm{T}}\cdot\left(A_0^{\mathrm{T}}\right)^k\right]}{\mathrm{Det}[\Lambda_K]}\right)\\
&=\frac{1}{2}\log_2\left(\frac{\mathrm{Det}\left[\sum_{k=1}^{\infty} A_0^k\cdot K\cdot\Lambda_K\cdot K^{\mathrm{T}}\cdot\left(A_0^{\mathrm{T}}\right)^k+K\cdot\Lambda_K\cdot K^{\mathrm{T}}\right]}{\mathrm{Det}[\Lambda_K]}\right)\\
&=\frac{1}{2}\log_2\left(\frac{\mathrm{Det}[K]\cdot\mathrm{Det}\left[K^{\mathrm{T}}\cdot\left(\sum_{k=1}^{\infty} A_0^k\cdot K\cdot\Lambda_K\cdot K^{\mathrm{T}}\cdot\left(A_0^{\mathrm{T}}\right)^k\right)\cdot K+\Lambda_K\right]\cdot\mathrm{Det}\left[K^{\mathrm{T}}\right]}{\mathrm{Det}[\Lambda_K]}\right)\\
&=\frac{1}{2}\log_2\left(\frac{\mathrm{Det}\left[K^{\mathrm{T}}\cdot\left(\sum_{k=1}^{\infty} A_0^k\cdot K\cdot\Lambda_K\cdot K^{\mathrm{T}}\cdot\left(A_0^{\mathrm{T}}\right)^k\right)\cdot K+\Lambda_K\right]}{\mathrm{Det}[\Lambda_K]}\right).
\end{aligned}$$

Moreover, because Λ_K is diagonal, taking $\mathrm{Tr}[\log_2(\Lambda_K)]$ is equivalent to $\log_2(\mathrm{Det}[\Lambda_K])$ and we have

$$\mathrm{EMC} = \frac{1}{2}\log_2 \mathrm{Det}\left[A_0^{\iota} + \Lambda_K\right] - \frac{1}{2}\sum_{i=1}^{p} \log_2 \lambda_i(C), \tag{249}$$

where $A_0^{\iota} = K^{\mathrm{T}} \cdot \left(\sum_{k=1}^{\infty} A_0^k \cdot K \cdot \Lambda_K \cdot K^{\mathrm{T}} \cdot \left(A_0^{\mathrm{T}}\right)^k\right) \cdot K$.

If the noise is isotropic, that is, the variances along the independent directions are equal $\left(C = \left\{\sigma^2\right\} \cdot I_p\right)$, and therefore correlations ρ_{ij} (Eq. 43) between performance fluctuations do not exist, we obtain a surprisingly simple solution:

$$\begin{aligned}
\mathrm{EMC} &= \frac{1}{2}\log_2 \mathrm{Det}\left[\sum_{k=0}^{\infty} A_0^k \cdot \left(A_0^{\mathrm{T}}\right)^k\right] \\
&= \frac{1}{2}\log_2 \mathrm{Det}\left[\left(I_p - A_0 \cdot A_0^{\mathrm{T}}\right)^{-1}\right] \\
&= -\frac{1}{2}\log_2 \mathrm{Det}\left[I_p - A_0 \cdot A_0^{\mathrm{T}}\right].
\end{aligned} \tag{250}$$

The above solution is based on the von Neumann series that generalizes the geometric series to matrices (cf. Section 2.2).

If the matrix A_0 is diagonalizable, it can be decomposed into eigenvectors $\vartheta_i(A_0)$ in the columns $S_{:i}$ of S (Eq. 35) and written as $A_0 = S \cdot \Lambda_S \cdot S^{-1}$. Λ_S is a diagonal matrix with eigenvalues $\lambda_i(A_0)$ along the principal diagonal. Hence, if $C = \left\{\sigma^2\right\} \cdot I_p$ and A_0 is diagonalizable, the EMC from Eq. 250 can be fully simplified:

$$\begin{aligned}
\mathrm{EMC} &= \frac{1}{2}\log_2 \prod_{i=1}^{p} \frac{1}{1 - \lambda_i(A_0)^2} \\
&= \frac{1}{2}\sum_{i=1}^{p} \log_2 \frac{1}{1 - \lambda_i(A_0)^2} \\
&= -\frac{1}{2}\sum_{i=1}^{p} \log_2\left(1 - \lambda_i(A_0)^2\right).
\end{aligned} \tag{251}$$

Both closed-form solutions that were obtained under the assumption of isotropic fluctuations only depend on the dynamical operator A_0, and therefore the dynamic part of the project can be seen to decouple completely from the unpredictable performance fluctuations. Under these circumstances the argument $\left(1 - \lambda_i(A_0)^2\right)$ of the binary logarithmic function can be interpreted as the damping coefficient of design mode $\phi_i = (\lambda_i(A_0), \vartheta_i(A_0))$ (see Section 2.1).

Similarly, for a project phase in which all p development tasks are processed independently at the same autonomous processing rate a, the dynamic part completely decouples from the performance fluctuations under arbitrary correlation coefficients. In this non-cooperative environment with minimum richness of temporal and structure-organizational dependencies, we simply have $A_0 = \text{Diag}[a, \ldots, a]$. For EMC, it therefore holds that

$$
\begin{aligned}
\text{EMC} &= \frac{1}{2}\log_2\left(\frac{\text{Det}\left[\sum_{k=0}^{\infty} (\text{Diag}[a, \ldots, a])^k \cdot C \cdot \left(\text{Diag}[a, \ldots, a]^{\text{T}}\right)^k\right]}{\text{Det}[C]}\right) \\
&= \frac{1}{2}\log_2\left(\frac{\text{Det}\left[C \cdot \sum_{k=0}^{\infty} (\text{Diag}[a, \ldots, a])^k \cdot (\text{Diag}[a, \ldots, a])^k\right]}{\text{Det}[C]}\right) \\
&= \frac{1}{2}\log_2 \text{Det}\left[\sum_{k=0}^{\infty} \left(\text{Diag}\left[a^2, \ldots, a^2\right]\right)^k\right] \\
&= \frac{1}{2}\log_2 \text{Det}\left[\text{Diag}\left[\frac{1}{1-a^2}, \ldots, \frac{1}{1-a^2}\right]\right] \\
&= -\frac{p}{2}\log_2\left(1-a^2\right).
\end{aligned} \tag{252}
$$

An additional closed-form solution in which the EMC can be expressed in terms of the dynamical operator A_0 and a so-called prewhitened operator W was formulated by DelSole and Tippett (2007) and Ay et al. (2012). Using $\text{Det}[A]/\text{Det}[B] = \text{Det}\left[A \cdot B^{-1}\right]$ and the Lyapunov Eq. 27 we can write

$$
\frac{\text{Det}[C]}{\text{Det}[\Sigma]} = \text{Det}\left[\left(\Sigma - A_0 \cdot \Sigma \cdot A_0^{\text{T}}\right) \cdot \Sigma^{-1}\right] = \text{Det}\left[I_p - A_0 \cdot \Sigma \cdot A_0^{\text{T}} \cdot \Sigma^{-1}\right].
$$

Defining

$$
W := \Sigma^{-\frac{1}{2}} \cdot A_0 \cdot \Sigma^{\frac{1}{2}}
$$

we obtain

$$
\frac{\text{Det}[C]}{\text{Det}[\Sigma]} = \text{Det}\left[I_p - W \cdot W^{\text{T}}\right],
$$

where $\text{Det}\left[I_p - A \cdot N \cdot A^{-1}\right] = \text{Det}\left[I_p - N\right]$ and $\Sigma = \Sigma^{\text{T}}$ were used. Hence, we obtain the EMC also as

$$\mathrm{EMC} = -\frac{1}{2}\log_2 \mathrm{Det}\left[I_p - W \cdot W^{\mathrm{T}}\right]. \tag{253}$$

According to DelSole and Tippett (2007) the application of the dynamical operator W can be regarded as a whitening transformation of the state-space coordinates of the dynamical operator A_0 by means of the covariance matrix Σ.

Concerning the evaluation of the persistent mutual information—represented by the variable EMC(τ)—of a vector autoregressive process, Section 3.2.4 showed that this can be expressed by the continuous-type mutual information $I[.;.]$ as

$$\begin{aligned}
\mathrm{EMC}(\tau) &= I\left[X_{-\infty}^{-1}; X_{\tau}^{\infty}\right] \\
&= \int_{\mathbb{X}^p} \cdots \int_{\mathbb{X}^p} f\left[x_{-\infty}^{-1}, x_{\tau}^{\infty}\right] \log_2 \frac{f\left[x_{-\infty}^{-1}, x_{\tau}^{\infty}\right]}{f\left[x_{-\infty}^{-1}\right] f\left[x_{\tau}^{\infty}\right]} \, dx_{-\infty}^{-1} \, dx_{\tau}^{\infty}.
\end{aligned}$$

The independent parameter $\tau \geq 0$ denotes the lead time. The term $f\left[x_{-\infty}^{-1}\right]$ designates the joint *pdf* of the infinite one-dimensional history of the stochastic process. Likewise, $f\left[x_{\tau}^{\infty}\right]$ designates the corresponding *pdf* of the infinite future from time τ onward. We used the shorthand notation $f\left[x_{-\infty}^{-1}, x_{\tau}^{\infty}\right] = f[x_{-\infty}, x_{-\infty+1}, \ldots, x_{-1}, x_{\tau}, x_{\tau+1}, , \ldots, x_{\infty-1}, x_{\infty}]$, $f\left[x_{-\infty}^{-1}\right] = f[x_{-\infty}, \ldots, x_{-1}]$, $f\left[x_{\tau}^{\infty}\right] = f[x_{\tau}, \ldots, x_{\infty}]$, $dx_{-\infty}^{-1} = dx_{-\infty} \ldots dx_{-1}$ and $dx_{\tau}^{\infty} = dx_{\tau} \ldots dx_{\infty}$. Informally, for positive lead times the term $I\left[X_{-\infty}^{-1}; X_{\tau}^{\infty}\right]$ can be interpreted as the information that is communicated from the past to the future ignoring the current length-τ sequence of observations $X_0^{\tau-1}$. Assuming strict stationarity, the joint *pdfs* are invariant under shifting the origin. Due to the Markov property of the VAR (1) model they can be factorized as follows:

$$\begin{aligned}
f\left[x_{-\infty}^{-1}, x_{\tau}^{\infty}\right] &= \int_{\mathbb{X}^p} \cdots \int_{\mathbb{X}^p} f\left[x_{-\infty}^{-1}, x_0^{\infty}\right] dx_0 \ldots dx_{\tau-1} \\
&= f[x_{-\infty}] f\left[x_{-\infty+1} \middle| x_{-\infty}\right] \ldots f\left[x_{-1} \middle| x_{-2}\right] f\left[x_{\tau+1} \middle| x_{\tau}\right] \ldots f\left[x_{\infty} \middle| x_{\infty-1}\right] \\
&\quad \times \int_{\mathbb{X}^p} \cdots \int_{\mathbb{X}^p} f\left[x_0 \middle| x_{-1}\right] \ldots f\left[x_{\tau} \middle| x_{\tau-1}\right] dx_0 \ldots dx_{\tau-1} \\
f\left[x_{-\infty}^{-1}\right] &= f[x_{-\infty}] f\left[x_{-\infty+1} \middle| x_{-\infty}\right] \ldots f\left[x_{-1} \middle| x_{-2}\right] \\
f\left[x_{\tau}^{\infty}\right] &= f[x_{\tau}] f\left[x_{\tau+1} \middle| x_{\tau}\right] \ldots f\left[x_{\infty} \middle| x_{\infty-1}\right]
\end{aligned}$$

Hence, we can simplify the mutual information as follows:

$$\begin{aligned}
I\left[X_{-\infty}^{-1}; X_{\tau}^{\infty}\right] &= \int_{\mathbb{X}^p} \cdots \int_{\mathbb{X}^p} f\left[x_{-\infty}^{-1}, x_{\tau}^{\infty}\right] \\
&\log_2 \frac{\int_{\mathbb{X}^p} \cdots \int_{\mathbb{X}^p} f\left[x_0 \middle| x_{-1}\right] \ldots f\left[x_{\tau} \middle| x_{\tau-1}\right] dx_0 \ldots dx_{\tau-1}}{f[x_{\tau}]} dx_{-\infty}^{-1} \, dx_{\tau}^{\infty}.
\end{aligned}$$

According to the famous Chapman-Kolmogorov equation (Papoulis and Pillai 2002) it holds that:

$$\int_{\mathbb{X}^p} \cdots \int_{\mathbb{X}^p} f\left[x_0 \middle| x_{-1}\right] \ldots f\left[x_\tau \middle| x_{\tau-1}\right] dx_0 \ldots dx_{\tau-1} = f\left[x_\tau \middle| x_{-1}\right]$$

Hence, we have

$$\begin{aligned}
I\left[X_{-\infty}^{-1}; X_\tau^\infty\right] &= \int_{\mathbb{X}^p}\int_{\mathbb{X}^p} \log_2 f\left[x_\tau \middle| x_{-1}\right] dx_{-1} dx_\tau \int_{\mathbb{X}^p} \cdots \int_{\mathbb{X}^p} f\left[x_{-\infty}^{-1}, x_\tau^\infty\right] dx_{-\infty}^{-2}\, dx_{\tau+1}^\infty \\
&\quad - \int_{\mathbb{X}^p} \log_2 f[x_\tau] dx_\tau \int_{\mathbb{X}^p} \cdots \int_{\mathbb{X}^p} f\left[x_{-\infty}^{-1}, x_\tau^\infty\right] dx_{-\infty}^{-1}\, dx_{\tau+1}^\infty \\
&= \int_{\mathbb{X}^p}\int_{\mathbb{X}^p} f[x_{-1}, x_\tau] \log_2 f\left[x_\tau \middle| x_{-1}\right] dx_{-1} dx_\tau - \int_{\mathbb{X}^p} f[x_\tau] \log_2 f[x_\tau]\, dx_\tau \\
&= \int_{\mathbb{X}^p} f[x_{-1}] dx_{-1} \int_{\mathbb{X}^p} f\left[x_\tau \middle| x_{-1}\right] \log_2 f\left[x_\tau \middle| x_{-1}\right] dx_\tau - \int_{\mathbb{X}^p} f[x_\tau] \log_2 f[x_\tau]\, dx_\tau.
\end{aligned}$$

For a VAR(1) process the transition function is defined as

$$f\left[x_\tau \middle| x_{-1}\right] = \mathcal{N}\left(x_\tau; A_0^\tau \cdot x_{-1}, C(\tau)\right),$$

with the lead-time dependent covariance

$$\begin{aligned}
C(\tau) &= A_0 \cdot C(\tau - 1) \cdot A_0^{\mathrm{T}} + C \\
&= \sum_{k=0}^{\tau} A_0^k \cdot C \cdot \left(A_0^{\mathrm{T}}\right)^k.
\end{aligned}$$

We find the solution

$$\begin{aligned}
\mathrm{EMC}(\tau) &= \frac{1}{2}\log_2 \mathrm{Det}[\Sigma] - \frac{1}{2}\log_2 \mathrm{Det}[C(\tau)] \\
&= \frac{1}{2}\log_2\left(\frac{\mathrm{Det}[\Sigma]}{\mathrm{Det}[C(\tau)]}\right) \\
&= \frac{1}{2}\log_2 \mathrm{Det}\left[\Sigma \cdot (C(\tau))^{-1}\right].
\end{aligned}$$

The solution can also be expressed as the logarithm of the variance ratio (Ay et al. 2012):

$$\mathrm{EMC}(\tau) = \frac{1}{2}\log_2\left(\frac{\mathrm{Det}[\Sigma]}{\mathrm{Det}\left[\Sigma - A_0^{\tau+1} \cdot \Sigma \cdot \left(A_0^{\mathrm{T}}\right)^{\tau+1}\right]}\right), \tag{254}$$

noting that $C = \Sigma - A_0 \cdot \Sigma \cdot A_0^{\mathrm{T}}$. As in Section 4.1 we can rewrite the above solution on the basis of the dynamical operator A_0 and lead-time dependent prewhitened operator $W(\tau)$ (Eq. 253; DelSole and Tippett 2007; Ay et al. 2012) as

$$\mathrm{EMC}(\tau) = -\frac{1}{2}\log_2 \mathrm{Det}\left[I_p - W(\tau)\cdot W(\tau)^{\mathrm{T}}\right],$$

with

$$W(\tau) = \Sigma^{-\frac{1}{2}} \cdot A_0^{\tau+1} \cdot \Sigma^{\frac{1}{2}}. \tag{255}$$

Following the same principles, a closed-form solution can be calculated for the elusive information $\sigma_\mu(\tau) = I\left[X_{-\infty}^{-1}; X_\tau^\infty \middle| X_o^{\tau-1}\right]$ from Eq. 231. As explained in Section 3.2.4, the elusive information is one of two essential pieces of the persistent mutual information and represents the Shannon information that is communicated from the past to the future by the stochastic process, but does not flow through the currently observed length-τ sequence $X_0^{\tau-1}$ (James et al. 2011). The key distinguishing feature of the persistent mutual information is that it is nonzero for $\tau \geq 1$ if a process necessarily has hidden states (Marzen and Crutchfield 2014). Conversely, due to the Markov property of the VAR(1) model, the elusive information completely vanishes for positive length τ.

This statement is easy to prove by using the definitions for the conditional mutual information from Eq. 214 and the conditional entropy from Eq. 213. Based on these definitions, the following relationship can be expressed:

$$\begin{aligned} I\left[X;Y\middle|Z\right] &= H\left[X\middle|Z\right] + H\left[Y\middle|Z\right] - H\left[X,Y\middle|Z\right] \\ &= H[X,Z] - H[Z] + H[Y,Z] - H[Z] - H[X,Y,Z] + H[Z] \\ &= H[X,Z] + H[Y,Z] - H[Z] - H[X,Y,Z]. \end{aligned}$$

As it holds

$$I[X;Z,Y] = H[X] + H[Z,Y] - H[X,Y,Z]$$

we find

$$\begin{aligned} I\left[X;Y\middle|Z\right] &= H[X,Z] - H[Z] - H[X] + I[X;Z,Y] \\ &= I[X;Z,Y] - I[X;Z]. \end{aligned}$$

In particular, we have

$$\begin{aligned} \sigma_\mu(\tau) &= I\left[X_{-\infty}^{-1}; X_\tau^\infty \middle| X_0^{\tau-1}\right] \\ &= I\left[X_{-\infty}^{-1}; X_0^{\tau-1}, X_\tau^\infty\right] - I\left[X_{-\infty}^{-1}; X_0^{\tau-1}\right] \\ &= I\left[X_{-\infty}^{-1}; X_0^\infty\right] - I\left[X_{-\infty}^{-1}; X_0^{\tau-1}\right]. \end{aligned}$$

Using the Markov property (Eq. 235), we see from the calculations Eq. 237–238 that the emergent complexity does not depend on the future of the autoregressive process beyond the lead time τ, i.e.

$$I\left[X_{-\infty}^{-1};X_0^{\infty}\right]=I\left[X_{-\infty}^{-1};X_0^{\tau-1}\right].$$

This proves that it holds for $\tau\geq 1$:

$$\sigma_\mu(\tau)=0.$$

This result is independent of the coordinate system in the vector autoregression model of cooperative work.

4.1.2 Closed-Form Solutions in the Spectral Basis

In this chapter, we calculate additional solutions in which the dependence of the EMC on the anisotropy of the performance fluctuations is made explicit. These solutions are much easier to interpret, and to derive them we work in the spectral basis (cf. Eq. 35). According to Neumaier and Schneider (2001), the steady-state covariance matrix Σ' in the spectral basis can be calculated on the basis of the transformed covariance matrix of the performance fluctuations $C'=S^{-1}\cdot C\cdot\left(\left[S^{\mathrm{T}}\right]^*\right)^{-1}$ (Eq. 41) as

$$\Sigma'=\begin{pmatrix}\frac{{c'_{11}}^2}{1-\lambda_1\overline{\lambda_1}} & \frac{\rho'_{12}c'_{11}c'_{22}}{1-\lambda_1\overline{\lambda_2}} & \cdots\\ \frac{\rho'_{12}c'_{11}c'_{22}}{1-\lambda_2\overline{\lambda_1}} & \frac{{c'_{22}}^2}{1-\lambda_2\overline{\lambda_2}} & \cdots\\ \vdots & \vdots & \ddots\end{pmatrix}. \tag{256}$$

In the above equation, the ρ'_{ij}'s are the transformed correlations, which were defined in Eq. 43 for a WTM A_0 with arbitrary structure and in Eq. 47 for A_0's that are symmetric. The ${c'_{ii}}^2$'s (cf. Eq. 10) and $\rho'_{ij}c'_{ii}c'_{jj}$'s (cf. Eq. 11) are the scalar-valued variance and covariance components of C' in the spectral basis:

$$C'=\begin{pmatrix}{c'_{11}}^2 & \rho'_{12}c'_{11}c'_{22} & \cdots\\ \rho'_{12}c'_{11}c'_{22} & {c'_{22}}^2 & \cdots\\ \vdots & \vdots & \ddots\end{pmatrix}. \tag{257}$$

The transformation into the spectral basis is a linear transformation of the state-space coordinates (see Eq. 41) and therefore does not change the mutual information being communicated from the infinite past into the infinite future by the stochastic process. Hence, the functional form of the closed-form solution from Eq. 246 holds, and the EMC can be calculated as the (logarithmic) variance ratio (Schneider and Griffies 1999; de Cock 2002):

$$\mathrm{EMC} = \frac{1}{2}\log_2\left(\frac{\mathrm{Det}[\Sigma']}{\mathrm{Det}[C']}\right) = \frac{1}{2}\log_2 \mathrm{Det}\left[\Sigma' \cdot C'^{-1}\right]. \tag{258}$$

The basis transformation does not change the positive-definiteness of the covariance matrices. Under the assumption that the matrices are of full rank, the determinants are positive. As already shown in Section 4.1.1., the determinant $\mathrm{Det}[\Sigma']$ of the covariance matrix Σ' can be interpreted as a generalized variance of the stationary process in the spectral basis, whereas $\mathrm{Det}[C']$ represents the generalized variance of the inherent performance fluctuations after the basis transformation. The variance ratio can also be interpreted in a geometrical framework (de Cock 2002). It is well known that the volume Vol[.] of the parallelepiped spanned by the rows or columns of a covariance matrix, e.g. Σ', is equal to the value of its determinant:

$$\mathrm{Vol}[\mathrm{parallelepiped}[\Sigma']] = \mathrm{Det}[\Sigma'].$$

In this sense the inverse variance ratio $\mathrm{Det}[C']/\mathrm{Det}[\Sigma']$ represents the factor by which the volume of the parallelepiped referring to the dynamical part of the process can be collapsed due to the state observation by the project manager leading to a certain information gain.

An important finding is that the scalar-valued variance and covariance components of the fluctuation part are not relevant for the calculation of the EMC. This follows from the definition of a determinant (see Eq. 267). The calculated determinants of Σ' and C' just give rise to the occurrence of the factor $\prod_{n=1}^{p} {c'_{nn}}^2$, which cancels out:

$$\mathrm{Det}\left[\Sigma' \cdot C'^{-1}\right] = \mathrm{Det}[\Sigma'] \cdot \mathrm{Det}\left[C'^{-1}\right] = \frac{\mathrm{Det}[\Sigma']}{\mathrm{Det}[C']} = \frac{\mathrm{Det}\left[\Sigma'_N\right]}{\mathrm{Det}\left[C'_N\right]}.$$

Hence, we can also calculate with the "normalized" covariance matrices Σ'_N and C'_N:

$$\Sigma'_N = \begin{pmatrix} \dfrac{1}{1-|\lambda_1|^2} & \dfrac{\rho'_{12}}{1-\lambda_1\overline{\lambda_2}} & \cdots \\ \dfrac{\rho'_{12}}{1-\lambda_2\overline{\lambda_1}} & \dfrac{1}{1-|\lambda_2|^2} & \cdots \\ \vdots & \vdots & \ddots \end{pmatrix} \tag{259}$$

$$C'_N = \begin{pmatrix} 1 & \rho'_{12} & \cdots \\ \rho'_{12} & 1 & \cdots \\ \vdots & \vdots & \ddots \end{pmatrix}. \tag{260}$$

It can be proved that the normalized covariance matrices are also positive-semidefinite. If they are furthermore not rank deficient, inconsistencies of the complexity measure do not occur. According to Shannon's classic information-theory findings about the capacity of a Gaussian channel (Cover and Thomas 1991), the normalized covariance matrix Σ'_N can be decomposed into summands as follows:

$$\Sigma'_N = C'_N + \begin{pmatrix} \frac{1}{1-|\lambda_1|^2} - 1 & \frac{\rho'_{12}}{1-\lambda_1\overline{\lambda_2}} - \rho'_{12} & \cdots \\ \frac{\rho'_{12}}{1-\lambda_2\overline{\lambda_1}} - \rho'_{12} & \frac{1}{1-|\lambda_2|^2} - 1 & \cdots \\ \vdots & \vdots & \ddots \end{pmatrix}.$$

The second summand in the above equation is defined as Σ''_N. This matrix can be simplified:

$$\Sigma''_N = \begin{pmatrix} \frac{|\lambda_1|^2}{1-|\lambda_1|^2} & \rho'_{12}\frac{\lambda_1\overline{\lambda_2}}{1-\lambda_1\overline{\lambda_2}} & \cdots \\ \rho'_{12}\frac{\lambda_2\overline{\lambda_1}}{1-\lambda_2\overline{\lambda_1}} & \frac{|\lambda_2|^2}{1-|\lambda_2|^2} & \cdots \\ \vdots & \vdots & \ddots \end{pmatrix}. \tag{261}$$

We obtain the most expressive closed-form solution based on the signal-to-noise ratio SNR $:= \Sigma''_N.{C'_N}^{-1}$:

$$\text{EMC} = \frac{1}{2}\log_2 \text{Det}\left[I_p + \Sigma''_N \cdot {C'_N}^{-1}\right]. \tag{262}$$

The SNR can be interpreted as the ratio of the variance Σ''_N of the signal in the spectral basis that is generated by cooperative task processing and the effective variance C'_N of the performance fluctuations. The variance of the signal drives the process to a certain extent and can be reinforced through the structural organization of the project. The effective fluctuations are in the same units as the input x_t. This is called "referring the noise to the input" and is a standard method in physics for characterizing detectors, amplifiers and other devices (Bialek 2003). Clearly, if one builds a photodetector it is not so useful to quote the noise level at the output in volts; one wants to know how this noise limits the ability to detect dim lights. Similarly, when we characterize a PD project that uses a stream of progress reports to document a quasicontinuous workflow, we don't want to know the variance in the absolute labor units; we want to know how variability in the performance of the developers limits precision in estimating the real work progress (signal), which amounts to defining an effective "noise level" in the units of the signal itself. In the present case, this is just a matter of "dividing" generalized variances, but in reality it is a fairly complex task. According to Sylvester's determinant theorem, we can swap the factors in the second summand:

$$\mathrm{Det}\left[I_p + \Sigma_N'' \cdot {C_N'}^{-1}\right] = \mathrm{Det}\left[I_p + {C_N'}^{-1} \cdot \Sigma_N''\right].$$

The obtained closed-form solution in the spectral basis has at most only $(p^2 - p)/2 + p = p(p+1)/2$ independent parameters, namely the eigenvalues $\lambda_i(A_0)$ of the WTM and the correlations ρ'_{ij} in the spectral basis, and not a maximum of the approximately $p^2 + (p^2 - p)/2 + p = p(3p+1)/2$ parameters encoded in both the WTM A_0 and the covariance matrix C (Eq. 248). In other words, through a transformation into the spectral basis we can identify the essential variables influencing emergent complexity in the sense of Grassberger's theory and reduce the dimensionality of the problem in many cases by the factor $(3p+1)/(p+1)$.

Furthermore, these independent parameters are easy to interpret, and at this point we can make a number of comments to stress the importance and usefulness of the analytical results. It is evident that the eigenvalues $\lambda_i(A_0)$ represent the essential temporal dependencies of the modeled project phase in terms of effective productivity rates on linearly independent scales determined by the eigenvectors $\vartheta_i(A_0)$ $(i = 1 \ldots p)$. The effective productivity rates depend only on the design modes ϕ_i of the WTM A_0 and therefore reflect the project's organizational design. The lower the effective productivity rates because of slow task processing or strong task couplings, the less the design modes are "damped," and hence the larger the project complexity. On the other hand, the correlations ρ'_{ij} model the essential dependencies between the unpredictable performance fluctuations in open organizational systems that can give rise to an excitation of the design modes and their interactions. This excitation can compensate for the damping factors of the design mode. The ρ'_{ij}'s scale linearly with the $\lambda_i(C)$ along each independent direction of the fluctuation variable ε'_t: the larger the $\lambda_i(C)$, the larger the correlations and the stronger the excitation (Eq. 43). However, the scale factors are determined not only by a linear interference between design modes ϕ_i and ϕ_j caused by cooperative task processing but also by the weighted interference with performance fluctuation modes Ψ_i and Ψ_j caused by correlations between performance variability (cf. Eqs. 43 and 47). In other words, the emergent complexity of the modeled project phase does not simply come from the least-damped design mode $\phi_1 = (\lambda_1(A_0), \vartheta_1(A_0))$ because this mode may not be sufficiently excited, but rather is caused (at least theoretically) by a complete interference between all design and performance fluctuation modes. Like the analytical considerations of Crutchfield et al. (2013) concerning stationary and ergodic stochastic processes whose measurement values cover a finite alphabet, the obtained closed-form solutions show that in a development process complexity is not just controlled by the "first spectral gap," i.e. the difference between the dominant eigenvalue and the eigenvalue with the second largest magnitude. Rather, the entire spectrum of eigenvalues is relevant and therefore all subspaces of the underlying causal-state process can contribute to emergent complexity (Crutchfield et al. 2013). In most practical case studies, only a few subspaces will dominate project dynamics. However, the closed-form solution from Eq. 262 in conjunction with Eqs. 260 and 261 shows that this is not generally the case. In Section 4.1.4, we will present fairly simple polynomial-based solutions for projects with only two or three tasks, and we will make the theoretical connections between the eigenvalues, the spectral gaps and the correlations very clear. The solution for two tasks will also allow us to identify simple scaling laws for real-valued eigenvalues. As a result, we see that emergent complexity in the sense of Grassberger's theory is a holistic property of the structure and process organization, and that, in most real cases, it cannot be reduced to singular properties of the project organizational design. This is a truly nonreductionist approach to complexity assessment insisting on the specific character of the organizational design as a whole.

Similarly to the previous chapter, we can obtain a closed-form solution for the persistent mutual information EMC(τ) in the spectral basis. The transformation into the spectral basis is a linear transformation of the state-space coordinates and therefore does not change the persistent mutual information communicated from the past into the future by the stochastic process. Hence, in analogy to Eq. 256 the variance ratio can also be calculated

$$\mathrm{EMC}(\tau) = \frac{1}{2}\log_2\left(\frac{\mathrm{Det}[\Sigma']}{\mathrm{Det}\left[\Sigma' - \Lambda_S^{\tau+1}\cdot\Sigma'\cdot\left(\left[\Lambda_S^{\mathrm{T}}\right]^*\right)^{\tau+1}\right]}\right)$$

in the spectral basis, where the diagonal matrix Λ_S is the dynamical operator (Eq. 39) as

$$\Lambda_S = \mathrm{Diag}[\lambda_i(A_0)] \quad 1 \le i \le p\,.$$

Because Λ_S is diagonal, the solution in the spectral basis can be simplified to

$$\begin{aligned}\mathrm{EMC}(\tau) &= -\frac{1}{2}\log_2\left(\frac{\mathrm{Det}\left[\Sigma' - \Lambda_S^{\tau+1}\cdot\Sigma'\cdot\left(\left[\Lambda_S^{\mathrm{T}}\right]^*\right)^{\tau+1}\right]}{\mathrm{Det}[\Sigma']}\right)\\ &= -\frac{1}{2}\log_2\left(\frac{\mathrm{Det}\left[\Sigma' - \Lambda_S^{\tau+1}\cdot\Sigma'\cdot\Lambda_S^{*\tau+1}\right]}{\mathrm{Det}[\Sigma']}\right)\\ &= -\frac{1}{2}\log_2\mathrm{Det}\left[I_p - \Lambda_S^{\tau+1}\cdot\Sigma'\cdot\Lambda_S^{*\tau+1}\cdot\Sigma'^{-1}\right]\\ &= -\frac{1}{2}\log_2\mathrm{Det}\left[I_p - \Sigma'(\tau)\cdot\Sigma'^{-1}\right],\end{aligned} \tag{263}$$

with $\Sigma'(\tau) = \Lambda_S^{\tau+1}\cdot\Sigma'\cdot\Lambda_S^{*\tau+1}$ $(\tau \geq 0)$.

As with the derivation of the expressive closed-form solution in Section 4.2, the generalized variance term $\Sigma' - \Lambda_S^{\tau+1}\cdot\Sigma'\cdot\left(\left[\Lambda_S^{\mathrm{T}}\right]^*\right)^{\tau+1} = \Sigma' - \Lambda_S^{\tau+1}\cdot\Sigma'\cdot\Lambda_S^{*\tau+1}$ in the denominator of the variance ratio can be written in an explicit matrix form:

$$\begin{aligned}&\Sigma' - \Lambda_S^{\tau+1}\cdot\Sigma'\cdot\Lambda_S^{*\tau+1}\\ &= \begin{pmatrix} \frac{c_{11}'^{\,2}}{1-\lambda_1\overline{\lambda_1}} - \frac{\lambda_1^{\tau+1}c_{11}'^{\,2}\left(\overline{\lambda_1}\right)^{\tau+1}}{1-\lambda_1\overline{\lambda_1}} & \frac{\rho_{12}'c_{11}'c_{22}'}{1-\lambda_1\overline{\lambda_2}} - \frac{\lambda_1^{\tau+1}\rho_{12}'c_{11}'c_{22}'\left(\overline{\lambda_2}\right)^{\tau+1}}{1-\lambda_1\overline{\lambda_2}} & \cdots \\ \frac{\rho_{12}'c_{11}'c_{22}'}{1-\lambda_2\overline{\lambda_1}} - \frac{\lambda_2^{\tau+1}\rho_{12}'c_{11}'c_{22}'\left(\overline{\lambda_1}\right)^{\tau+1}}{1-\lambda_2\overline{\lambda_1}} & \frac{c_{22}'^{\,2}}{1-\lambda_2\overline{\lambda_2}} - \frac{\lambda_2^{\tau+1}c_{22}'^{\,2}\left(\overline{\lambda_2}\right)^{\tau+1}}{1-\lambda_2\overline{\lambda_1}} & \cdots \\ \vdots & \vdots & \ddots \end{pmatrix}\\ &= \begin{pmatrix} \left(c_{11}'^{\,2}\right)\frac{1-|\lambda_1|^{2(\tau+1)}}{1-|\lambda_1|^2} & c_{11}'c_{22}'\frac{\rho_{12}'\left(1-\lambda_1^{\tau+1}\left(\overline{\lambda_2}\right)^{\tau+1}\right)}{1-\lambda_1\overline{\lambda_2}} & \cdots \\ c_{11}'c_{22}'\frac{\rho_{12}'\left(1-\lambda_2^{\tau+1}\left(\overline{\lambda_1}\right)^{\tau+1}\right)}{1-\lambda_2\overline{\lambda_1}} & \left(c_{22}'^{\,2}\right)\frac{1-|\lambda_1|^{2(\tau+1)}}{1-|\lambda_1|^2} & \cdots \\ \vdots & \vdots & \ddots \end{pmatrix}.\end{aligned}$$

It can be proved that the covariances c_{ij}' in the above matrix form are not relevant for the calculation of EMC(τ). This follows from the definition of a determinant (see Eq. 267). When calculating the determinants of Σ' and $\Sigma' - \Lambda_S^{\tau+1}\cdot\Sigma'\cdot\Lambda_S^{*\tau+1}$ they just give rise to the occurrence of a factor $\prod_{n=1}^{p} c_{nn}'^{\,2}$, which cancels out in the variance ratio. Therefore, the persistent mutual information can also be calculated using normalized covariance matrices. The normalized covariance matrix of Σ', termed Σ_N', was defined in Eq. 259. The normalized covariance matrix of $\Sigma' - \Lambda_S^{\tau+1}\cdot\Sigma'\cdot\Lambda_S^{*\tau+1}$ is simply

$$\Sigma'_N - \Lambda_S^{\tau+1} \cdot \Sigma'_N \cdot \Lambda_S^{*\tau+1}$$
$$= \begin{pmatrix} \frac{1-|\lambda_1|^{2(\tau+1)}}{1-|\lambda_1|^2} & \rho'_{12}\frac{\left(1-\lambda_1{}^{\tau+1}\left(\overline{\lambda_2}\right)^{\tau+1}\right)}{1-\lambda_1\overline{\lambda_2}} & \cdots \\ \rho'_{12}\frac{\left(1-\lambda_2{}^{\tau+1}\left(\overline{\lambda_1}\right)^{\tau+1}\right)}{1-\lambda_2\overline{\lambda_1}} & \frac{1-|\lambda_2|^{2(\tau+1)}}{1-|\lambda_2|^2} & \cdots \\ \vdots & \vdots & \ddots \end{pmatrix}.$$

Hence,

$$\begin{aligned} \mathrm{EMC}(\tau) &= -\frac{1}{2}\log_2\left(\frac{\mathrm{Det}\left[\Sigma'_N - \Lambda_S^{\tau+1} \cdot \Sigma'_N \cdot \Lambda_S^{*\tau+1}\right]}{\mathrm{Det}\left[\Sigma'_N\right]}\right) \\ &= -\frac{1}{2}\log_2\mathrm{Det}\left[I_p - \Lambda_S^{\tau+1} \cdot \Sigma'_N \cdot \Lambda_S^{*\tau+1} \cdot \Sigma'_N{}^{-1}\right] \\ &= -\frac{1}{2}\log_2\mathrm{Det}\left[I_p - \Sigma'_N(\tau) \cdot \Sigma'_N{}^{-1}\right], \end{aligned} \tag{264}$$

with $\Sigma'_N(\tau) = \Lambda_S^{\tau+1} \cdot \Sigma'_N \cdot \Lambda_S^{*\tau+1}\ (\tau \geq 0)$.

4.1.3 Closed-form Solution through Canonical Correlation Analysis

If the matrix C'_N representing the intrinsic prediction error in the spectral basis is diagonal in the same coordinate system as the normalized covariance matrix Σ'_N contributed by cooperative task processing, then the matrix product $\Sigma'_N \cdot C'_N{}^{-1} = \left(I_p + \Sigma''_N \cdot C'_N{}^{-1}\right)$ is diagonal, and simple reduction of emergent complexity to singular properties of the design modes $\phi_i = (\lambda_i(A_0), \vartheta_i(A_0))$ and performance fluctuation modes $\Psi_i = (\lambda_i(C), k_i(C))$ will work. In this case, the elements along the principal diagonal are the signal-to-noise ratios along each independent direction. Hence, the EMC is proportional to the sum of the log-transformed ratios, and these summands are the only independent parameters. However, in the general case we have to diagonalize the above matrix product in a first step to obtain an additional closed-form solution. This closed-form solution has the least number of independent parameters. In spite of its algebraic simplicity, the solution is not very expressive, because the spatiotemporal covariance structures of the open organizational system are not revealed. We will return to this point after presenting the solution.

Unfortunately, the diagonalization of the matrix product $\Sigma'_N \cdot C'_N{}^{-1}$ cannot be carried out through an eigendecomposition, because the product of two symmetric matrices is not necessarily symmetric itself. Therefore, the left and right eigenvectors can differ and do not form a set of mutually orthogonal vectors, as they would if the product was diagonal. Nevertheless, we can always rotate our coordinate system in the space of the output to make the matrix product diagonal (Schneider and Griffies 1999). To do this, we decompose $\Sigma'_N \cdot {C'_N}^{-1}$ into singular values (singular value decomposition, see e.g. de Cock 2002) as

$$\Sigma'_N \cdot {C'_N}^{-1} = U \cdot \Lambda_{UV} \cdot V^{\mathrm{T}},$$

where

$$U \cdot U^{\mathrm{T}} = I_p \quad \text{and} \quad V \cdot V^{\mathrm{T}} = I_p$$

and

$$\Lambda_{UV} = \mathrm{Diag}\left[\sigma'_i\right] \quad 1 \leq i \leq p\,.$$

The columns of U are the left singular vectors; those of V are the right singular vectors. The columns of V can be regarded as a set of orthonormal "input" basis vectors for $\Sigma'_N \cdot {C'_N}^{-1}$; the columns of U form a set of orthonormal "output" basis vectors. The diagonal values σ'_i in matrix Λ_{UV} are the singular values, which can be thought of as scalar "gain controls" by which each corresponding input is multiplied to give a corresponding output. The σ'_i's are the only independent parameters of the following closed-form solution (see). The relationship between the singular values σ'_i of $\Sigma'_N \cdot {C'_N}^{-1}$ and the canonical correlations ρ_i (see summary of properties of EMC at the end of Section 4.1) in our case is as follows (de Cock 2002):

$$\sigma'_i = \frac{1}{1-\rho_i^2} \quad 1 \leq i \leq p\,.$$

Under the assumption that $\mathrm{Det}\left[\Sigma'_N \cdot {C'_N}^{-1}\right] > 0$, it is possible to prove that $\mathrm{Det}[U] \cdot \mathrm{Det}[V] = 1$. We can obtain the desired closed-form solution as follows:

$$\begin{aligned}
\mathrm{EMC} &= \frac{1}{2}\log_2 \det\left[\Sigma'_N \cdot {C'_N}^{-1}\right] \\
&= \frac{1}{2}\log_2 \det\left[U \cdot \Lambda_{UV} \cdot V^{\mathrm{T}}\right] \\
&= \frac{1}{2}\log_2 (\mathrm{Det}[U] \cdot \mathrm{Det}[\Lambda_{UV}] \cdot \mathrm{Det}[V]) \\
&= \frac{1}{2}\log_2 \det[\Lambda_{UV}] \\
&= \frac{1}{2}\mathrm{Tr}[\log_2(\Lambda_{UV})] \\
&= \frac{1}{2}\sum_{i=1}^{p}\log_2 \sigma'_i \\
&= \frac{1}{2}\sum_{i=1}^{p}\log_2\left(\frac{1}{1-\rho_i^2}\right) \\
&= -\frac{1}{2}\sum_{i=1}^{p}\log_2\left(1-\rho_i^2\right).
\end{aligned} \tag{265}$$

In spite of its algebraic simplicity, a main disadvantage of this closed-form solution with only p parameters σ_i' or ρ_i^2 is that both the temporal dependencies of the modeled work process in terms of essential productivity rates (represented by the λ_i's), and the essential cooperative relationships exciting fluctuations (represented by the ρ_{ij}''s) are not explicit, but are compounded into correlation coefficients between the canonical variates. Therefore, it is impossible for the project manager to analyze and interpret the spatiotemporal covariance structures of the organizational system and to identify countermeasures for coping with emergent complexity.

A canonical correlation analysis over τ time steps leads to the following solution of the persistent mutual information:

$$\begin{aligned}
\mathrm{EMC}(\tau) &= \frac{1}{2}\log_2 \mathrm{Det}\left[\Sigma_N' \cdot \left(\Sigma_N' - \Lambda_S^{\tau+1} \cdot \Sigma_N' \cdot \Lambda_S^{*\tau+1}\right)^{-1}\right] \\
&= \frac{1}{2}\log_2 \mathrm{Det}\left[U(\tau) \cdot \Lambda_{UV}(\tau) \cdot V(\tau)^{\mathrm{T}}\right] \\
&= \frac{1}{2}\log_2 (\mathrm{Det}[U(\tau)] \cdot \mathrm{Det}[\Lambda_{UV}(\tau)] \cdot \mathrm{Det}[V(\tau)]) \\
&= \frac{1}{2}\log_2 \mathrm{Det}[\Lambda_{UV}(\tau)] \\
&= \frac{1}{2}\mathrm{Tr}[\log_2(\Lambda_{UV}(\tau))] \\
&= \frac{1}{2}\sum_{i=1}^{p}\log_2 \sigma_i'(\tau) \\
&= \frac{1}{2}\sum_{i=1}^{p}\log_2\left(\frac{1}{1-(\rho_i(\tau))^2}\right) \\
&= -\frac{1}{2}\sum_{i=1}^{p}\log_2\left(1-(\rho_i(\tau))^2\right).
\end{aligned} \tag{266}$$

The term $U(\tau) \cdot \Lambda_{UV}(\tau) \cdot V(\tau)^{\mathrm{T}}$ represents the product of the matrices resulting from a decomposition of $\Sigma_N' \cdot \left(\Sigma_N' - \Lambda_S^{\tau+1} \cdot \Sigma_N' \cdot (\Lambda_S{}^*)^{\tau+1}\right)^{-1}$ as a function of the lead time τ:

$$(U(\tau), \Lambda_{UV}(\tau), V(\tau)) = \mathrm{SVD}\left[\Sigma_N' \cdot \left(\Sigma_N' - \Lambda_S^{\tau+1} \cdot \Sigma_N' \cdot (\Lambda_S{}^*)^{\tau+1}\right)^{-1}\right],$$

where the matrix-valued function SVD[.] represents the singular value decomposition of the argument. The $\sigma_i'(\tau)$'s and $\rho_i(\tau)$'s represent, respectively, the singular values and canonical correlations given the lead time.

4.1.4 *Polynomial-Based Solutions for Processes with Two and Three Tasks*

We can also analyze the spatiotemporal covariance structure of Σ_N' (Eq. 259) in the spectral basis explicitly by recalling the definition of a determinant. If $B = (b_{ij})$ is a matrix of size p, then

$$\mathrm{Det}(B) = \sum_{\beta \in R_p} \mathrm{sgn}(\beta) \prod_{i=1}^{p} b_{i,\beta(i)} \tag{267}$$

holds. R_p is the set of all permutations of $\{1,\ldots,p\}$. Thus, because of the regular structure of the matrix Σ'_N, $\mathrm{Det}[\Sigma'_N]$ is a sum of $p!$ summands. Each of these summands is a fraction, because it is a product of elements from Σ'_N, where exactly one entry is chosen from each row and column. The denominator of those fractions is a product consisting of p factors of $1 - \lambda_i(\mathrm{A}_0)\lambda_i\overline{\lambda_j(\mathrm{A}_0)}$. The numerator is a product of $2, 3, \ldots, p$ factors ρ'_{ij}, or simply 1 if the permutation is the identity. (The case of one factor cannot occur, because the amount of factors equals the amount of numbers changed by the permutation β, and there is no permutation that changes just one number). The coefficients (i,j) of the factor $1 - \lambda_i(\mathrm{A}_0)\overline{\lambda_j(\mathrm{A}_0)}$ in the denominator correspond to the coefficients (k,l) of the factor ρ'_{kl} in the numerator, i.e. $i = l$ and $j = k$, if $i \neq k$ holds. Otherwise, in the case that $i = k$, no corresponding factor is multiplied in the numerator, because the appropriate entry of Σ'_N lies on the principal diagonal. Moreover, $1 - \lambda_i(A_0)\overline{\lambda_j(A_0)} = 1 - |\lambda_i(A_0)|^2$ holds in that case.

These circumstances are elucidated for project phases with only $p = 2$ and $p = 3$ fully interdependent tasks. For $p = 2$ we have

$$\Sigma'_N = \begin{pmatrix} \dfrac{1}{1-|\lambda_1|^2} & \dfrac{\rho'_{12}}{1-\lambda_1\overline{\lambda_2}} \\ \dfrac{\rho'_{12}}{1-\lambda_2\overline{\lambda_1}} & \dfrac{1}{1-|\lambda_2|^2} \end{pmatrix},$$

hence,

$$\mathrm{Det}\left[\Sigma'_N\right] = \frac{1}{\left(1-|\lambda_1|^2\right)\left(1-|\lambda_2|^2\right)} - \frac{{\rho'_{12}}^2}{\left(1-\lambda_2\overline{\lambda_1}\right)\left(1-\lambda_1\overline{\lambda_2}\right)}.$$

For $p = 3$ we have

$$\Sigma'_N = \begin{pmatrix} \dfrac{1}{1-|\lambda_1|^2} & \dfrac{\rho'_{12}}{1-\lambda_1\overline{\lambda_2}} & \dfrac{\rho'_{13}}{1-\lambda_1\overline{\lambda_3}} \\ \dfrac{\rho'_{12}}{1-\lambda_2\overline{\lambda_1}} & \dfrac{1}{1-|\lambda_2|^2} & \dfrac{\rho'_{23}}{1-\lambda_2\overline{\lambda_3}} \\ \dfrac{\rho'_{13}}{1-\lambda_3\overline{\lambda_1}} & \dfrac{\rho'_{23}}{1-\lambda_3\overline{\lambda_2}} & \dfrac{1}{1-|\lambda_3|^2} \end{pmatrix},$$

hence,

$$
\begin{aligned}
\mathrm{Det}\left[\Sigma'_N\right] = & \frac{1}{\left(1-|\lambda_1|^2\right)\left(1-|\lambda_2|^2\right)\left(1-|\lambda_3|^2\right)} \\
& - \frac{\rho'_{23}2}{\left(1-|\lambda_1|^2\right)\left(1-\lambda_3\overline{\lambda_2}\right)\left(1-\lambda_2\overline{\lambda_3}\right)} - \frac{\rho'_{13}2}{\left(1-|\lambda_2|^2\right)\left(1-\lambda_3\overline{\lambda_1}\right)\left(1-\lambda_1\overline{\lambda_3}\right)} \\
& - \frac{\rho'_{12}2}{\left(1-|\lambda_3|^2\right)\left(1-\lambda_1\overline{\lambda_2}\right)\left(1-\lambda_2\overline{\lambda_1}\right)} + \frac{\rho'_{12}\rho'_{13}\rho'_{23}}{\left(1-\lambda_1\overline{\lambda_2}\right)\left(1-\lambda_2\overline{\lambda_3}\right)\left(1-\lambda_3\overline{\lambda_1}\right)} \\
& + \frac{\rho'_{12}\rho'_{13}\rho'_{23}}{\left(1-\lambda_2\overline{\lambda_1}\right)\left(1-\lambda_3\overline{\lambda_2}\right)\left(1-\lambda_1\overline{\lambda_3}\right)}.
\end{aligned}
$$

The results for C'_N are much simpler. From Eqs. 259 and 260 it follows that the numerator is the same, whereas the denominator is simply 1.

For $p = 2$ we have

$$
C'_N = \begin{pmatrix} 1 & \rho'_{12} \\ \rho'_{12} & 1 \end{pmatrix},
$$

hence,

$$
\mathrm{Det}\left[C'_N\right] = 1 - {\rho'_{12}}^2.
$$

For $p = 3$ we have

$$
C'_N = \begin{pmatrix} 1 & \rho'_{12} & \rho'_{13} \\ \rho'_{12} & 1 & \rho'_{23} \\ \rho'_{13} & \rho'_{23} & 1 \end{pmatrix},
$$

hence,

$$
\mathrm{Det}\left[C'_N\right] = 1 + 2\rho'_{12}\rho'_{13}\rho'_{23} - {\rho'_{12}}^2 - {\rho'_{13}}^2 - {\rho'_{23}}^2.
$$

These results readily yield the closed-form expression

$$
\mathrm{EMC} = \frac{1}{2}\log_2\left[\frac{1}{1-{\rho'_{12}}^2}\left(\frac{1}{\left(1-|\lambda_1|^2\right)\left(1-|\lambda_2|^2\right)} - {\rho'_{12}}^2\frac{1}{\left(1-\lambda_2\overline{\lambda_1}\right)\left(1-\lambda_1\overline{\lambda_2}\right)}\right)\right] \tag{268}
$$

for $p = 2$ tasks and

$$\mathrm{EMC} = \frac{1}{2}\log_2\left[\frac{1}{1 + 2\rho'_{12}\rho'_{13}\rho'_{23} - {\rho'_{12}}^2 - {\rho'_{13}}^2 - {\rho'_{23}}^2}\mathrm{Det}\left[\Sigma'_N\right]\right] \tag{269}$$

for $p = 3$ tasks, where the simplified determinant $\mathrm{Det}[\Sigma'_N]$ of the normalized covariance matrix Σ'_N is given by

$$\begin{aligned}
\mathrm{Det}\left[\Sigma'_N\right] &= \frac{1}{\left(1 - |\lambda_1|^2\right)}\left(\frac{1}{\left(1 - |\lambda_2|^2\right)\left(1 - |\lambda_3|^2\right)} - \frac{\rho'_{23}}{\left(1 - \lambda_3\overline{\lambda_2}\right)\left(1 - \lambda_2\overline{\lambda_3}\right)}\right) \\
&+\rho'_{12}\rho'_{13}\rho'_{23}\left(\frac{1}{\left(1 - \lambda_1\overline{\lambda_2}\right)\left(1 - \lambda_2\overline{\lambda_3}\right)\left(1 - \lambda_3\overline{\lambda_1}\right)} + \frac{1}{\left(1 - \lambda_1\overline{\lambda_3}\right)\left(1 - \lambda_2\overline{\lambda_1}\right)\left(1 - \lambda_3\overline{\lambda_2}\right)}\right) \\
&- \frac{{\rho'_{12}}^2}{\left(1 - |\lambda_3|^2\right)\left(1 - \lambda_1\overline{\lambda_2}\right)\left(1 - \lambda_2\overline{\lambda_1}\right)} - \frac{{\rho'_{13}}^2}{\left(1 - |\lambda_2|^2\right)\left(1 - \lambda_3\overline{\lambda_1}\right)\left(1 - \lambda_1\overline{\lambda_3}\right)}.
\end{aligned} \tag{270}$$

Now, we suppose that all eigenvalues $\lambda_i(A_0)$ are real. Under this assumption EMC can be expressed by the spectral gaps $\left(\lambda_i - \lambda_j\right)_{i \neq j}$ between eigenvalues as

$$\begin{aligned}
\mathrm{EMC} &= \frac{1}{2}\log_2\left[\frac{1}{\left(1 - {\lambda_1}^2\right)\left(1 - {\lambda_2}^2\right)} + \frac{{\rho'_{12}}^2}{1 - {\rho'_{12}}^2}\frac{\left(\lambda_1 - \lambda_2\right)^2}{\left(1 - {\lambda_1}^2\right)\left(1 - {\lambda_2}^2\right)\left(1 - \lambda_1\lambda_2\right)^2}\right] \\
&= \frac{1}{2}\log_2\left[\frac{1}{\left(1 - {\lambda_1}^2\right)\left(1 - {\lambda_2}^2\right)}\left(1 + \frac{{\rho'_{12}}^2}{1 - {\rho'_{12}}^2}\frac{\left(\lambda_1 - \lambda_2\right)^2}{\left(1 - \lambda_1\lambda_2\right)^2}\right)\right] \\
&= -\frac{1}{2}\log_2\left[1 - {\lambda_1}^2\right] - \frac{1}{2}\log_2\left[1 - {\lambda_2}^2\right] + \log_2\left[1 + \frac{{\rho'_{12}}^2}{1 - {\rho'_{12}}^2}\frac{\left(\lambda_1 - \lambda_2\right)^2}{\left(1 - \lambda_1\lambda_2\right)^2}\right],
\end{aligned} \tag{271}$$

for $p = 2$, and as

$$\begin{aligned}
\mathrm{EMC} &= -\frac{1}{2}\log_2\left[1 - {\lambda_1}^2\right] - \frac{1}{2}\log_2\left[1 - {\lambda_2}^2\right] - \frac{1}{2}\log_2\left[1 - {\lambda_3}^2\right] \\
&+\frac{1}{2}\log_2\left[1 + \frac{{\rho'_{12}}^2}{\rho'}\frac{\left(\lambda_1 - \lambda_2\right)^2}{\left(1 - \lambda_1\lambda_2\right)^2} + \frac{{\rho'_{13}}^2}{\rho'}\frac{\left(\lambda_1 - \lambda_3\right)^2}{\left(1 - \lambda_1\lambda_3\right)^2} + \frac{{\rho'_{23}}^2}{\rho'}\frac{\left(\lambda_2 - \lambda_3\right)^2}{\left(1 - \lambda_2\lambda_3\right)^2}\right. \\
&\left. + \frac{2{\rho'_{12}}^2{\rho'_{13}}^2{\rho'_{23}}^2}{\rho'}\frac{\left(1 - {\lambda_1}^2\right)\left(1 - {\lambda_2}^2\right)\left(1 - {\lambda_3}^2\right) - (1 - \lambda_1\lambda_2)(1 - \lambda_1\lambda_3)(1 - \lambda_2\lambda_3)}{(1 - \lambda_1\lambda_2)(1 - \lambda_1\lambda_3)(1 - \lambda_2\lambda_3)}\right]
\end{aligned} \tag{272}$$

for $p = 3$ using analogous simplifications. The factor ρ' equals the determinant of the covariance matrix of a standard trivariate normal distribution taking variances $c_{11}^2 = c_{22}^2 = c_{33}^2 = 1$ and is given by

$$\rho' = 1 + 2\rho'_{12}\rho'_{13}\rho'_{23} - \rho'^{2}_{12} - \rho'^{2}_{13} - \rho'^{2}_{23}.$$

Hence, if the dynamical operator A_0 has only real eigenvalues $\lambda_i(A_0)$, EMC can be decomposed into simple additive complexity factors and the factor related to the correlations between the covariance components of C' in the spectral basis is a simple function of the spectral gap(s).

For a process with $p = 2$ tasks that is asymptotically stable in the sense of Lyapunov (Eq. 4), it is evident that the first, second and third summand in the last row of Eq. 271 can only take values in the range $[0, +\infty)$, and for different correlations $\rho'_{12} \in [-1; 1]$ the sum of the first and second summand $-1/2\log_2[1 - \lambda_1^2] - 1/2\log_2[1 - \lambda_2^2]$ is a lower bound. To gain additional insights into the scaling behavior of EMC in the spectral gap $\Delta\lambda = (\lambda_1 - \lambda_2)$ and the correlation coefficient ρ'_{12}, we define another variable $\varsigma = (\lambda_1 + \lambda_2)$ that is orthogonal to $\Delta\lambda$. The Taylor series expansion of EMC in the spectral gap $\Delta\lambda$ about the point $\Delta\lambda = 0$ to order $\Delta\lambda^2$ leads to:

$$\mathrm{EMC} = -\frac{1}{2}\log_2\left[\left(1 - \frac{\varsigma^2}{4}\right)^2\right] + \frac{4((1 + \rho'^{2}_{12}) + \varsigma^2 - 3\rho'^{2}_{12}\varsigma^2)}{(\rho'^{2}_{12} - 1)(\varsigma^2 - 4)\log_{10}(2)}\Delta\lambda^2 + o[\Delta\lambda]^3.$$

For the correlation coefficient ρ'_{12} we obtain the series expansion

$$\mathrm{EMC} = -\frac{1}{2}\log_2\left[\left(1 - \frac{(\varsigma - \Delta\lambda)^2}{4}\right)\left(1 - \frac{(\varsigma + \Delta\lambda)^2}{4}\right)\right] + \frac{2\Delta\lambda^2}{(4 + \Delta\lambda^2 - \varsigma^2)\log_{10}(2)}\rho'^{2}_{12} + o[\rho'_{12}]^3$$

about the point $\rho'_{12} = 0$ to order ρ'^{2}_{12}.

For $p = 3$ tasks it can also be proved that the fourth summand in Eq. 272 can only take values in the range $[0, +\infty)$ in view of the definition of the covariance matrix. The sum of the first, second and third summands is also a lower bound.

Interestingly, the coefficient $\rho'^{2}_{12}/(1 - \rho'^{2}_{12})$ in Eq. 271 is equivalent to Cohen's f^2, which is an effect size measure that is frequently used in the context of an F-test for ANOVA or multiple regression. By convention, in the behavioral sciences effect sizes of 0.02, 0.15, and 0.35 are termed small, medium, and large, respectively (Cohen 1988). The squared product-moment correlation ρ'^{2}_{12} can also be easily interpreted within the class of linear regression models. If an intercept is included in a linear regression model, then ρ'^{2}_{12} is equivalent to the well known coefficient of determination R^2. The coefficient of determination provides a measure of how well future outcomes are likely to be predicted by the statistical model.

Moreover, interesting questions arise from the identification of these lower bounds. The answers will improve the understanding of the unexpectedly rich dynamics that even small open organizational systems can generate. The identified lower bounds can be reached, if and only if either the performance fluctuations are isotropic, that is, for the corresponding covariance matrix in the original state-space coordinates the expression $C = \{\sigma^2\} \cdot I_p$ holds (see Eq. 250), or the dynamical operator A_0 is symmetric and the column vectors of the forcing matrix K are "aligned," in the sense that $A_0 = \{c\} \cdot K$ holds ($c \in \mathbb{R}$ or $c = \text{Diag}[c_i]$ in general). More details about the interrelationship between A_0 and K were presented earlier in Section 2.3. In the following, we focus on the question of how to identify the "optimal" spectrum of eigenvalues λ_i, in the sense that emergent complexity according to the metric $\text{EMC} = \frac{1}{2}\log_2 \text{Det}\left[\Sigma'_N \cdot C'^{-1}_N\right]$ is minimized subject to the constraint that the expected total amount of work $x_{tot} \in \mathbb{R}^+$ done over all tasks in the modeled project phase is constant. This constrained optimization problem will be solved under the assumptions that all eigenvalues $\lambda_i(A_0)$ are real, it holds that $\lambda_i(A_0) > 0$ and the performance fluctuations are isotropic, i.e. for a process consisting only of relaxators (see Fig. 2.6) and in which the design modes are excited as little as possible. We therefore need to find project organization designs that could, on average, process the same amount of work while leading to minimum emergent complexity. A closed-form solution of the mean vector $\bar{x}$ of the accumulated work for distinct tasks in an asymptotically stable process given the initial state x_0 can be calculated across an infinite time interval as $\bar{x} = \left(I_p - A_0\right)^{-1} \cdot x_0$ (see Section 2.2). The expected total amount of work $x_{tot} = \text{Total}[\bar{x}]$ is simply the sum of the vector components (Eq. 16). For two tasks, the above question can be formulated as the following constrained optimization problem:

$$\min_{(a_{11},a_{12},a_{21},a_{22})} \frac{1}{2}\log_2 \text{Det}\left[\left(\left(1-\lambda_1\left[\begin{pmatrix} a_{11} & a_{21} \\ a_{12} & a_{22} \end{pmatrix}\right]^2\right)\left(1-\lambda_2\left[\begin{pmatrix} a_{11} & a_{21} \\ a_{12} & a_{22} \end{pmatrix}\right]^2\right)\right)^{-1}\right]$$

$$\text{subject to Total}\left[\begin{pmatrix} 1-a_{11} & -a_{12} \\ -a_{21} & 1-a_{22} \end{pmatrix}^{-1} \cdot \begin{pmatrix} x_{01} \\ x_{02} \end{pmatrix}\right] = x_{tot}.$$

For three tasks, the corresponding formulation would be:

$$\min_{\left(\{a_{ij}\}_{(i,j)\epsilon\{1,2,3\}^2}\right)} \frac{1}{2}\log_2 \text{Det}\left[\prod_{i=1}^{3}\left(1-\left(\lambda_i\left[\begin{pmatrix} a_{11} & a_{12} & a_{13} \\ a_{21} & a_{22} & a_{23} \\ a_{31} & a_{32} & a_{33} \end{pmatrix}\right]^2\right)\right)^{-1}\right]$$

$$\text{subject to Total}\left[\begin{pmatrix} 1-a_{11} & -a_{12} & -a_{13} \\ -a_{21} & 1-a_{22} & -a_{23} \\ -a_{31} & -a_{32} & 1-a_{33} \end{pmatrix}^{-1} \cdot \begin{pmatrix} x_{01} \\ x_{02} \\ x_{03} \end{pmatrix}\right] = x_{tot}.$$

In these equations, $\lambda_i[.]$ represents the i-th eigenvalue of the argument matrix. To solve the constrained optimization problems, the method of Lagrange multipliers is used. Unfortunately, this method leads to simple and expressive closed-form solutions that this book can only present and discuss under additional constraints. The first additional constraint is that only two development tasks are processed. Furthermore, both tasks have to be "uncoupled" and the corresponding off-diagonal elements $a_{12} = 0$ and $a_{21} = 0$ indicate the absence of cooperative relationships. Finally, the initial state is constrained to a setting in which both tasks are 100% to be completed, that is $x_0 = [1\ 1]^{\mathrm{T}}$, and in this case the total amount of work must be larger than $2 (x_{tot} > 2)$. Under these constraints, it follows that the eigenvalues $\lambda_1(A_0)$ and $\lambda_2(A_0)$ are equal to the autonomous task processing rates:

$$\lambda_1\left[\begin{pmatrix} a_{11} & 0 \\ 0 & a_{22} \end{pmatrix}\right] = a_{11} \quad \text{and} \quad \lambda_2\left[\begin{pmatrix} a_{11} & 0 \\ 0 & a_{22} \end{pmatrix}\right] = a_{22}\,.$$

The closed-form solution of the constrained optimization problem is the piecewise-defined complexity function:

$$\mathrm{EMC}_{min} = \begin{cases} \log_2\left(\dfrac{{x_{tot}}^2}{(x_{tot}-1)}\right) - 2 & \text{if}\, 2 < x_{tot} \le 2+\sqrt{2} \\ \dfrac{1}{2}\log_2(2x_{tot}-1) - 1 & \text{if } 2+\sqrt{2} < x_{tot}. \end{cases}$$

The corresponding equations for the autonomous task processing rates (alias eigenvalues) are

$$a_{11}^{min} = \lambda_1^{min} = \begin{cases} \dfrac{x_{tot}-2}{x_{tot}} & \text{if } 2 < x_{tot} \le 2+\sqrt{2} \\ \dfrac{1}{x_{tot}-1-\sqrt{2+(x_{tot}-4)x_{tot}}} & \text{if } 2+\sqrt{2} < x_{tot} \end{cases}$$

$$a_{22}^{min} = \lambda_2^{min} = \begin{cases} \dfrac{x_{tot}-2}{x_{tot}} & \text{if } 2 < x_{tot} \le 2+\sqrt{2} \\ \dfrac{1}{x_{tot}-1+\sqrt{2+(x_{tot}-4)x_{tot}}} & \text{if } 2+\sqrt{2} < x_{tot}. \end{cases}$$

When we analyze the above solutions, an interesting finding is that the value $x_{tot}^1 = 2+\sqrt{2} \sim 3.414$ of the total amount of work indicates a kind of "bifurcation point" in the complexity landscape. Below that point, minimum complexity values are assigned for equal autonomous task processing rates (or eigenvalues); above it, minimum complexity values are attained, if and only if the difference between rates (the spectral gap $\Delta\lambda^{min} = \lambda_1^{min} - \lambda_2^{min}$) is

$$a_{11}^{min} - a_{22}^{min} = \lambda_1^{min} - \lambda_2^{min} = \frac{2\sqrt{2+(x_{tot}-4)x_{tot}}}{2x_{tot}-1}.$$

This bifurcation behavior of an open organizational system in which only two uncoupled tasks are concurrently processed was unexpected. Figure 4.1 shows the bifurcation point in detail.

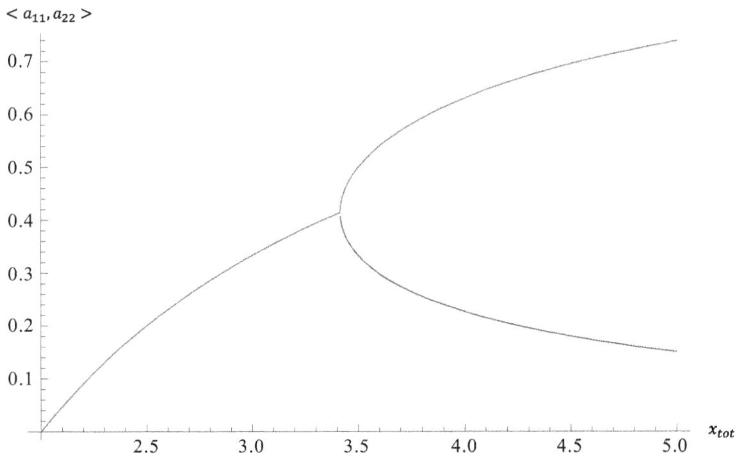

Fig. 4.1 Plot of autonomous task processing rates a_{11} and a_{22} leading to a minimum *EMC* subject to the constraint that the expected total amount of work x_{tot} is constant. The underlying closed-form solution was calculated based on Lagrange multipliers. Note that the solution only holds under the assumption that the tasks are uncoupled and the initial state is $x_0 = [1\ 1]^{\mathrm{T}}$, in which case x_{tot} must be larger than 2

We also found analytical results for the constrained optimization problem in the more general case of two uncoupled overlapping tasks, i.e. a bundle of independent tasks where, initially, only the second task has to be fully completed, while the first task is already completed to a level of $x\,\%$ and we therefore have an initial state $x_0 = [(1-x)/100\ 1]^{\mathrm{T}}$. However, the closed-form solutions are very complicated and, due to space limitations, cannot be presented here. It is important to note that the piecewise-defined complexity function and the corresponding bifurcation point are completely independent of the degree of task overlapping and only depend on the dynamics of task processing. This is a highly desirable property of the preferred complexity metric.

When we relax the constraint that both tasks have to be uncoupled, and consider all four matrix entries of the WTM A_0 as free parameters, we find another simple analytical solution to the constrained optimization problem. The initial state is constrained to be $x_0 = [1\ 1]^{\mathrm{T}}$ as before. However, the obtained solution is not very structurally informative, as all four elements of A_0 are supposed to be equal to $(x_{tot} - 2)/2x_{tot}$, and we have the symmetric matrix representation

$$A_0^{min} = \begin{pmatrix} \frac{1}{2} - \frac{1}{x_{tot}} & \frac{1}{2} - \frac{1}{x_{tot}} \\ \frac{1}{2} - \frac{1}{x_{tot}} & \frac{1}{2} - \frac{1}{x_{tot}} \end{pmatrix}.$$

From a practical point of view, this kind of project organizational design seems to be rather "pathological" because the relative couplings between tasks are extremely strong and one must expect a large amount of additional work in the iterations. The corresponding complexity solution is

$$\mathrm{EMC}_{min} = \frac{1}{2}\log_2\left(\frac{{x_{tot}}^2}{4(x_{tot}-1)}\right) \quad \text{if } 2 < x_{tot} \ .$$

It is evident that the above minimum of emergent complexity scales for $x_{tot} > 2.5$ almost linearly in the expected total amount of work.

4.1.5 Bounds on Effective Measure Complexity

To calculate the lower bounds on EMC for an arbitrary number of tasks we can make use of Oppenheim's inequality (see Horn and Johnson 1985). Let M and N be positive-semidefinite matrices and let $M \circ N$ be the entry-wise product of these matrices (so-called "Hadamard product"). The Hadamard product of two positive-semidefinite matrices is again positive-semidefinite. Furthermore, if M and N are positive-semidefinite, then the following equality based on Oppenheim holds:

$$\mathrm{Det}[M \circ N] \geq \left(\prod\nolimits_{i=1}^{p} M_{[[i,i]]}\right)\mathrm{Det}[N].$$

Let $M = \left(M_{[[i,j]]}\right) = \left(1/\left(1 - \lambda_i(A_0)\overline{\lambda_j(A_0)}\right)\right)$ be a Cauchy matrix $(1 \leq i, j \leq p)$. The elements along the principal diagonal of this matrix represent the "damping factor" $1 - |\lambda_i|^2$ of design mode ϕ_i, and the off-diagonal elements $1 - \lambda_i\overline{\lambda_i}$ are the damping factors between the interacting modes ϕ_i and ϕ_j. We follow the convention that the eigenvalues are ordered in decreasing magnitude in rows. Let $N = C'_N$ be the normalized covariance matrix of the noise, as defined in Eq. 260. Then the normalized covariance matrix of the signal Σ'_N from Eq. 259 can be written as the Hadamard product $\Sigma'_N = M \circ C'_N$. According to Oppenheim's inequality, the following inequality holds:

$$\begin{aligned}
\mathrm{EMC} = \frac{1}{2}\log_2\left(\frac{\mathrm{Det}\left[\Sigma'_N\right]}{\mathrm{Det}\left[C'_N\right]}\right) &= \frac{1}{2}\log_2\left(\frac{\mathrm{Det}\left[M \circ C'_N\right]}{\mathrm{Det}\left[C'_N\right]}\right) \geq \frac{1}{2}\log_2\left(\frac{\left(\prod_{i=1}^{p} M_{[[i,i]]}\right)\mathrm{Det}\left[C'_N\right]}{\mathrm{Det}\left[C'_N\right]}\right) \\
&= \frac{1}{2}\log_2\left(\prod\nolimits_{i=1}^{p}\frac{1}{1-|\lambda_i|^2}\right) \\
&= -\frac{1}{2}\sum_{i=1}^{p}\log_2\left(1-|\lambda_i|^2\right).
\end{aligned} \tag{273}$$

The lower bound according to the above equation is equal to the closed-form solution for EMC that was obtained under the assumptions of isotropic noise $\left(C = \{\sigma^2\} \cdot I_p\right)$ and A_0 being diagonalizable (see Eq. 251). In other words, emergent complexity in PD projects can be kept to a minimum, if the variances of the unpredictable performance fluctuations are equalized by purposeful interventions of the project manager and correlations between vector components are suppressed.

Next, because of the commutativity of the Hadamard product, it holds that

$$\mathrm{EMC} = \frac{1}{2}\log_2\left(\frac{\mathrm{Det}\left[\Sigma_N'\right]}{\mathrm{Det}\left[C_N'\right]}\right) = \frac{1}{2}\log_2\left(\frac{\mathrm{Det}\left[C_N' \circ M\right]}{\mathrm{Det}\left[C_N'\right]}\right) \geq \frac{1}{2}\log_2\left(\frac{\left(\prod_{i=1}^{p} C'_{N[[i,i]]}\right)\mathrm{Det}[M]}{\mathrm{Det}\left[C_N'\right]}\right)$$
$$= \frac{1}{2}\log_2\left(\frac{\mathrm{Det}[M]}{\mathrm{Det}\left[C_N'\right]}\right).$$

The determinant of the Cauchy matrix M in the numerator can be written as (Krattenthaler 2005)

$$\mathrm{Det}[M] = \mathrm{Det}\begin{bmatrix} \frac{1}{1-|\lambda_1|^2} & \frac{1}{1-\lambda_1\overline{\lambda_2}} & \cdots \\ \frac{1}{1-\lambda_2\overline{\lambda_1}} & \frac{1}{1-|\lambda_2|^2} & \cdots \\ \vdots & \vdots & \ddots \end{bmatrix} = \frac{\prod_{i<j}^{p}(\lambda_i-\lambda_j)\left(\overline{\lambda_i}-\overline{\lambda_j}\right)}{\prod_{i,j}^{p}\left(1-\lambda_i\overline{\lambda_j}\right)}.$$

Hence,

$$\begin{aligned}\mathrm{EMC} &= \frac{1}{2}\log_2\left(\frac{\mathrm{Det}\left[C_N' \circ M\right]}{\mathrm{Det}\left[C_N'\right]}\right)\\ &\geq \frac{1}{2}\log_2\left(\frac{\prod_{i<j}^{p}(\lambda_i-\lambda_j)\left(\overline{\lambda_i}-\overline{\lambda_j}\right)}{\prod_{i,j}^{p}\left(1-\lambda_i\overline{\lambda_j}\right)\mathrm{Det}\left[C_N'\right]}\right)\\ &= \frac{1}{2}\left(\sum_{i<j}^{p}\left(\log_2(\lambda_i-\lambda_j)+\log_2\left(\overline{\lambda_i}-\overline{\lambda_j}\right)\right)-\sum_{i,j}^{p}\log_2\left(1-\lambda_i\overline{\lambda_j}\right)-\log_2\mathrm{Det}\left[C_N'\right]\right).\end{aligned} \tag{274}$$

The lower bound on the EMC in the above equation is only defined for a dynamical operator A_0 with distinct eigenvalues. Under this assumption, a particularly interesting property of the bound is that it includes not only the damping factors $\left(1-\lambda_i\overline{\lambda_i}\right)$ inherent to the dynamical operator A_0 (as does the bound in Eq. 273) but also the spectral gap between eigenvalues $\left(\lambda_i-\lambda_j\right)$ and their complex conjugates $\left(\overline{\lambda_i}-\overline{\lambda_j}\right)$. We can draw the conclusion that under certain circumstances, differences among effective productivity rates (represented by the λ_i's) stimulate emergent complexity in PD (cf. Eqs. 271 and 272). Conversely, small complexity scores are assigned if the effective productivity rates are similar.

Additional analyses have shown that the lower bound defined in Eq. 273 is tighter when the eigenvalues of the dynamical operator A_0 are of similar magnitudes. Conversely, the lower bound defined in Eq. 274 comes closer to the true complexity values if the magnitudes of the eigenvalues are unevenly distributed.

Finally, it is also possible to put both upper and lower bounds on the EMC that are explicit functions of the dynamical operator A_0 and its dimension p. To find these bounds, we considered results for the determinant of the solution of the Lyapunov equation (Eq. 27, cf. Mori et al. 1982). Let Σ be the covariance matrix of the process in the steady state, and let the dominant eigenvalue $\rho(A_0) = \max|_i|$ of A_0 be less than 1 in magnitude (see Section 2.1). Then we have

$$\mathrm{Det}[\Sigma] \geq \frac{\mathrm{Det}[C]}{\left(1 - (\mathrm{Det}[A_0])^{\frac{2}{p}}\right)^p}.$$

Moreover, if A_0 is diagonalizable and $\rho\left(A_0^{\mathrm{T}} \cdot A_0\right) \cdot C - A_0 \cdot \Sigma \cdot A_0^{\mathrm{T}}$ is positive-semidefinite, then

$$\mathrm{Det}[\Sigma] \leq \frac{\mathrm{Det}[C]}{\left(1 - \rho\left(A_0^{\mathrm{T}} \cdot A_0\right)\right)^p},$$

where $\rho\left(A_0^{\mathrm{T}} \cdot A_0\right)$ denotes the dominant eigenvalue of $A_0^{\mathrm{T}} \cdot A_0$. Based on Eq. 246 we can calculate the following bounds:

$$-\frac{p}{2}\log_2\left(1 - (\mathrm{Det}[A_0])^{\frac{2}{p}}\right) \leq \mathrm{EMC} \leq -\frac{p}{2}\log_2\left(1 - \rho\left(A_0^{\mathrm{T}} \cdot A_0\right)\right). \tag{275}$$

The upper bound only holds if A_0 is diagonalizable and $\rho\left(A_0^{\mathrm{T}} \cdot A_0\right) \cdot C - A_0 \cdot \Sigma \cdot A_0^{\mathrm{T}}$ is positive-semidefinite. If C is diagonal, then $\rho\left(A_0^{\mathrm{T}} \cdot A_0\right) \cdot C - A_0 \cdot \Sigma \cdot A_0^{\mathrm{T}}$ is always positive-semidefinite. Both bounds grow strictly monotonically with the dimension of the dynamical operator A_0 and it is evident that the EMC assigns larger complexity values to projects with more tasks, if the task couplings are similar. One can also divide the measure by the dimension p of the state space and compare the complexity of project phases with different cardinalities.

4.1.6 *Closed-Form Solutions for Higher-Order Models*

It is also not difficult to calculate the EMC of stochastic processes in steady state that are generated by higher-order autoregressive models of cooperative work in PD projects. In Section 2.4, we said that a vector autoregression model of order n, abbreviated as VAR(n) model, without an intercept term is defined by the state equation (see Neumaier and Schneider 2001 or Lütkepohl 2005):

$$X_t = \sum_{i=0}^{n-1} A_i \cdot X_{t-i-1} + \varepsilon_t.$$

The probability density function of the vector ε_t of performance fluctuations is given in Eq. 13. It is evident that due to the autoregressive behavior involving n instances of the process in the past, the generated stochastic process $\{X_t\}$ does not possess the Markov property (cf. Eq. 18) and therefore neither the generalized complexity solution from Eq. 239 nor the closed-form solution for a VAR(1) process from Eq. 247 can be used to evaluate emergent complexity. However, as we showed in Section 2.5, we can make the stochastic process Markovian by "augmenting" the state vector and rewriting the state equation as a first-order recurrence relation (Eq. 59):

$$\tilde{X}_t = \tilde{A} \cdot \tilde{X}_{t-1} + \tilde{\varepsilon}_t \quad t = 1, \ldots, T, \tag{276}$$

where $\tilde{X}_t$ is the augmented state vector (Eq. 60)

$$\tilde{X}_t = \begin{pmatrix} X_t \\ X_{t-1} \\ \vdots \\ X_{t-n+1} \end{pmatrix},$$

$\tilde{\varepsilon}_t$ is the augmented noise vector (Eq. 61)

$$\tilde{\varepsilon}_t = \begin{pmatrix} \varepsilon_t \\ 0 \\ \vdots \\ 0 \end{pmatrix}$$

and $\tilde{A}$ is the extended dynamical operator (Eq. 62)

$$\tilde{A} = \begin{pmatrix} A_0 & A_1 & \cdots & A_{n-2} & A_{n-1} \\ I_p & 0 & \cdots & 0 & 0 \\ 0 & I_p & \cdots & 0 & 0 \\ & & & & \\ 0 & 0 & \ddots & 0 & 0 \\ 0 & 0 & \ldots & I_p & 0 \end{pmatrix}.$$

The covariance matrix $\tilde{C}$ can be written as

$$\tilde{C} = E\left[\tilde{\varepsilon}_t \tilde{\varepsilon}_t^{\mathrm{T}}\right] = \begin{pmatrix} C & 0 & \cdots & 0 \\ 0 & 0 & & 0 \\ \vdots & & \ddots & \vdots \\ 0 & 0 & \cdots & 0 \end{pmatrix}. \tag{277}$$

The partial covariance $C = E\left[\varepsilon_t \varepsilon_t^{\mathrm{T}}\right]$ represents the intrinsic one-step prediction error of the original autoregressive process.

In light of the mutual information that is communicated from the infinite past to the infinite future (by storing it in the present) the problem with this kind of order reduction by state-space augmentation is that the augmented state vector $\tilde{X}_t$ has vector components that are also included in the previous state vector $\tilde{X}_{t-1}$ and therefore the past and future are not completely shielded in information-theoretic terms, given the present state. To be able to apply the closed-form complexity solution from Eq. 247 directly to the higher-order model in the coordinates of the original state space $\mathbb{R}^p$, we have to find a state representation with disjoint vector components. This can be easily done by defining the combined future and present project state $\tilde{X}_{t+n-1}$ to be the block of random vectors

$$\tilde{X}_{t+n-1} = \begin{pmatrix} X_{t+n-1} \\ X_{t+n-2} \\ \vdots \\ X_t \end{pmatrix}$$

and the past project state $\tilde{X}_{t-1}$ to be the block of vectors

$$\tilde{X}_{t-1} = \begin{pmatrix} X_{t-1} \\ X_{t-2} \\ \vdots \\ X_{t-n} \end{pmatrix}.$$

The calculation of the n-th iterate of $\tilde{X}_{t+n-1}$ leads to the higher-order recurrence relation

$$\begin{aligned} \tilde{X}_{t+n-1} &= \tilde{A} \cdot \tilde{X}_{t+n-2} + \tilde{\varepsilon}_{t+n-1} \\ &= \tilde{A}\left(\tilde{A} \cdot \tilde{X}_{t+n-3} + \tilde{\varepsilon}_{t+n-2}\right) + \tilde{\varepsilon}_{t+n-1} \\ &= \tilde{A}^2 \cdot \tilde{X}_{t+n-3} + \tilde{A} \cdot \tilde{\varepsilon}_{t+n-2} + \tilde{\varepsilon}_{t+n-1} \\ &\vdots \\ &= \tilde{A}^n \cdot \tilde{X}_{t-1} + \sum_{i=1}^{n} \left(\tilde{A}\right)^{n-i} \cdot \tilde{\varepsilon}_{t+i-1} \quad t = 2-n, \ldots, T-n+1. \end{aligned} \tag{278}$$

Under the assumption of strictly stationary behavior of $\{\tilde{X}_t\}$ for $t \to \infty$, we can utilize the complexity solution from eq. 247 and express the mutual information that is communicated by the VAR(n) model from the infinite past to the infinite future through the present project state by the logarithmic generalized variance ratio as follows:

$$\mathrm{EMC} = \frac{1}{2}\log_2\left(\frac{\mathrm{Det}\left[\sum_{k=0}^{\infty}\left(\tilde{A}^n\right)^k \cdot \left(\sum_{i=1}^{n}\left(\tilde{A}\right)^{n-i} \cdot \tilde{C} \cdot \left(\left(\tilde{A}\right)^{n-i}\right)^{\mathrm{T}}\right) \cdot \left(\left(\tilde{A}^n\right)^{\mathrm{T}}\right)^k\right]}{\mathrm{Det}\left[\sum_{i=1}^{n}\left(\tilde{A}\right)^{n-i} \cdot \tilde{C} \cdot \left(\left(\tilde{A}\right)^{n-i}\right)^{\mathrm{T}}\right]}\right)$$

$$= \frac{1}{2}\log_2\left(\frac{\mathrm{Det}\left[\tilde{\Sigma}\right]}{\mathrm{Det}\left[\sum_{i=1}^{n}\left(\tilde{A}\right)^{n-i} \cdot \tilde{C} \cdot \left(\left(\tilde{A}\right)^{n-i}\right)^{\mathrm{T}}\right]}\right), \tag{279}$$

where the steady-state covariance $\tilde{\Sigma}$ in the denominator is given by the infinite sum

$$\tilde{\Sigma} = \sum_{k=0}^{\infty}\left(\tilde{A}^n\right)^k \cdot \left(\sum_{i=1}^{n}\left(\tilde{A}\right)^{n-i} \cdot \tilde{C} \cdot \left(\left(\tilde{A}\right)^{n-i}\right)^{\mathrm{T}}\right) \cdot \left(\left(\tilde{A}^n\right)^{\mathrm{T}}\right)^k.$$

As an alternative to this solution, we can calculate the mutual information between infinite past and future histories using the additive factors method of Li and Xie (1996). In this method, the total mutual information is decomposed into additive components which can be expressed as a ratio of conditional (auto)covariances of the steady-state process. This method is very appealing as it allows us to interpret the additive components in terms of the universal learning curve $\Lambda(m)$ that was formulated by Bialek et al. (2001, see Eq. 224) and is explained in detail in Section 3.2.4. EMC is simply the discrete integral of $\Lambda(m)$ with respect to the block length m, which controls the speed at which the mutual information converges to its limit (Crutchfield et al. 2010). When we use the block length as a natural order parameter of the additive components, we can also easily evaluate the speed of convergence. If convergence is slow, it is an indicator of emergent complexity (see discussion in Section 3.2.4).

Let

$$\boldsymbol{C}_m = \begin{pmatrix} C_{\tilde{X}\tilde{X}}(0) & C_{\tilde{X}\tilde{X}}(1) & \ldots & C_{\tilde{X}\tilde{X}}(m-1) \\ C_{\tilde{X}\tilde{X}}(1) & C_{\tilde{X}\tilde{X}}(0) & \ldots & C_{\tilde{X}\tilde{X}}(m-2) \\ \vdots & \vdots & \ddots & \vdots \\ C_{\tilde{X}\tilde{X}}(m-1) & C_{\tilde{X}\tilde{X}}(m-2) & \ldots & C_{\tilde{X}\tilde{X}}(0) \end{pmatrix} \tag{280}$$

be a $mp \times mp$ $(m \in \mathbb{N})$ Toeplitz matrix (Li and Xie 1996) storing the values of the autocovariance functions (Eq. 159)

$$\begin{aligned} C_{\tilde{X}\tilde{X}}(\tau) &= E\left[\left(\tilde{X}_t - \mu_{\tilde{X}}\right)\left(\tilde{X}_{t+\tau} - \mu_{\tilde{X}}\right)^{\mathrm{T}}\right] \\ &= E\left[\tilde{X}_t \tilde{X}_{t+\tau}^{\mathrm{T}}\right] - \mu_{\tilde{X}} \mu_{\tilde{X}}^{\mathrm{T}} \\ &= E\left[\tilde{X}_t \tilde{X}_{t+\tau}^{\mathrm{T}}\right] \end{aligned}$$

of the steady-state process generated by state Eq. 59 (and not Eq. 276) for lead times $\tau = 0, 1, \ldots, m-1$. We know from Section 2.9 that, in steady state, the autocovariance $C_{\tilde{X}\tilde{X}}(\tau)$ and the autocorrelation $R_{\tilde{X}\tilde{X}}(\tau)$ (Eq. 160) are equal and that we have $C_{\tilde{X}\tilde{X}}(\tau) = R_{\tilde{X}\tilde{X}}(\tau)$. Note that the matrix elements $C_{\tilde{X}\tilde{X}}(\tau)$ are defined to be $p \times p$ block autocovariance matrices of the corresponding subspaces. Furthermore, let

$$\tilde{\Sigma}_{\mu} = \mathrm{Det}\left[E\left[\left(\tilde{X}_t - E\left[\tilde{X}_t \middle| \tilde{X}_{t-1}, \ldots, \tilde{X}_{-\infty}\right]\right)\left(\tilde{X}_t - E\left[\tilde{X}_t \middle| \tilde{X}_{t-1}, \ldots, \tilde{X}_{-\infty}\right]\right)^{\mathrm{T}}\right]\right]$$

be the (mean squared) one-step prediction error with respect to the steady-state process and

$$\tilde{\Sigma}_{(m)} = \mathrm{Det}\left[E\left[\left(\tilde{X}_t - E\left[\tilde{X}_t \middle| \tilde{X}_{t-1}, \ldots, \tilde{X}_{t-m}\right]\right)\left(\tilde{X}_t - E\left[\tilde{X}_t \middle| \tilde{X}_{t-1}, \ldots, \tilde{X}_{t-m}\right]\right)^{\mathrm{T}}\right]\right] \tag{281}$$

be the one-step prediction error of order m (cf. Eq. 66 in Section 2.4). According to these definitions $\tilde{\Sigma}_{\mu}$ can be interpreted as the inherent prediction error of the process that cannot be underrun, even if we condition our observations on the infinite past to build a maximally predictive model. $\tilde{\Sigma}_{(m)}$ represents the prediction error resulting from conditioning the observations on only m past instances of the process to build a maximally predictive model, and not on all instances that were theoretically possible. In this sense a certain error component of $\tilde{\Sigma}_{(m)}$ does not result from the inherent unpredictability because of limited knowledge or chaotic behavior, but because of the unpredictability resulting from a limit of the length of the observation window on the state evolution. Under the assumption that $\boldsymbol{C}_m$ is invertible, the one-step prediction error $\tilde{\Sigma}_{(m)}$ of order m can be expressed as the generalized variance ratio (Li and Xie 1996):

$$\tilde{\Sigma}_{(m)} = \frac{\mathrm{Det}\left[\boldsymbol{C}_{m+1}\right]}{\mathrm{Det}\left[\boldsymbol{C}_m\right]}. \tag{282}$$

The zeroth-order prediction error can be derived from the autocovariance for zero lead time, and it holds that:

$$\tilde{\Sigma}_{(0)} = \mathrm{Det}\left[C_{\tilde{X}\tilde{X}}(0)\right]. \tag{283}$$

Following the information-theoretic considerations of a VAR(n) model that were carried out in Section 4.1 (cf. Eq. 238) it is not difficult to show that, for any autocovariance matrix representation $\boldsymbol{C}_m$ with $m \geq n$, it holds for the one-step prediction errors that

$$\tilde{\Sigma}_{(m)} = \tilde{\Sigma}_{(n)} = \tilde{\Sigma}_{\mu} \quad \forall m \geq n \, .$$

In other words, due to the limited "memory depth" of the generative VAR(n) model, conditioning the current observation on sequences larger than the regression order does not, on average, lead to further reductions of the one-step prediction error in steady state. Under these circumstances of severely limited procedural memory the prediction error of order n equals the intrinsic prediction error. As a consequence of this behavior, higher-dimensional matrix representations than $\boldsymbol{C}_n$ must not be considered when evaluating the past-future mutual information. An additional theoretical analysis of the vector autoregression model in the original state-space coordinates allows us to conclude that the inherent prediction error equals the determinant of the expectation $E[\varepsilon_t \varepsilon_t^{\mathrm{T}}]$ and that it can be simply expressed as

$$\tilde{\Sigma}_{\mu} = \mathrm{Det}\left[E\left[\varepsilon_t \varepsilon_t^{\mathrm{T}}\right]\right] = \mathrm{Det}[C].$$

Furthermore, in steady state the $np \times np$ matrix $\boldsymbol{C}_n$ storing all relevant autocovariances up to lead time $\tau = n - 1$ equals the steady-state covariance of the process generated by state Eq. 59, and we have (Eq. 245, Lancaster and Tismenetsky 1985):

$$\boldsymbol{C}_n = \sum_{k=0}^{\infty} \tilde{A}^{k} \cdot \tilde{C} \cdot \left(\tilde{A}^{\mathrm{T}}\right)^{k},$$

where $\tilde{A}$ is the extended dynamical operator from Eq. 62, and $\tilde{C}$ is the corresponding covariance matrix from Eq. 277. If needed, the autocovariances for smaller lead times can be easily extracted as block matrices from this large representation. Based on these theoretic considerations and the material of Li and Xie (1996), the mutual information between infinite past and future histories can be conveniently expressed by n additive components as

$$\begin{aligned} \mathrm{EMC} &= \frac{1}{2}\left(\sum_{i=0}^{n-1}\left(\log_2 \tilde{\Sigma}_{(i)} - \log_2 \tilde{\Sigma}_{\mu}\right)\right) \\ &= \frac{1}{2}\left(\sum_{i=0}^{n-1} \log_2 \tilde{\Sigma}_{(i)} - n \log_2 \tilde{\Sigma}_{\mu}\right) \\ &= \frac{1}{2}\sum_{i=0}^{n-1} \log_2 \tilde{\Sigma}_{(i)} - \frac{1}{2} n \log_2 \mathrm{Det}[C]. \end{aligned} \tag{284}$$

Each summand $1/2\left(\log_2 \tilde{\Sigma}_{(i)} - \log_2 \tilde{\Sigma}_{\mu}\right) = 1/2\left(\log_2 \tilde{\Sigma}_{(i)} - \log_2 \mathrm{Det}[C]\right)$ can be used to evaluate the local predictability of the process. The corresponding local "over-estimates" of the intrinsic prediction error allow us to define a universal learning curve $\Lambda(i)$ in the sense of Bialek et al. (2001) with respect to block length i as (cf. Eq. 224)

$$\Lambda(i) = \log_2 \tilde{\Sigma}_{(i-1)} - \log_2 \mathrm{Det}[C], \quad i = 1, 2, \ldots, n,$$

where the maximum block length is determined by the autoregression order of the generative model. As already explained in Section 3.1.4, in light of a learning curve, EMC measures the amount of apparent randomness at small order i, which can be "explained away" by considering correlations between sequences with increasing length $i+1,\ i+2, \ldots$.

Returning to state Eq. 276 for informationally separated instances of past and future histories, we can use the first-order recurrence relation to apply the solution principles that were introduced at the end of Section 4.1.1 and find a simple expression for the persistent mutual information EMC(τ) (Eq. 229) as a function of the lead time $\tau \geq 0$. Substituting the steady-state covariance and the dynamical operator in Eq. 254, we can express EMC(τ) as the logarithmic generalized variance ratio (Ay et al. 2012):

$$\mathrm{EMC}(\tau) = \frac{1}{2} \log_2 \left(\frac{\mathrm{Det}\left[\tilde{\Sigma}\right]}{\mathrm{Det}\left[\tilde{\Sigma} - \left(\tilde{A}^{n}\right)^{\tau+1} \cdot \tilde{\Sigma} \cdot \left(\left(\tilde{A}^{n}\right)^{\mathrm{T}}\right)^{\tau+1}\right]} \right), \tag{285}$$

As one would expect, the steady-state covariances in the numerators of the variance ratios related to both measures of emergent complexity are equal (Eqs. 279 and 285). We note that for the inherent one-step prediction error it holds that $E\left[\tilde{\varepsilon}_t \tilde{\varepsilon}_t^{\mathrm{T}}\right] = \tilde{C} = \tilde{\Sigma} - \tilde{A}^{n} \cdot \tilde{\Sigma} \cdot \left(\tilde{A}^{n}\right)^{\mathrm{T}}$.

Applying the principles and techniques introduced in Section 4.1.2 and 4.1.3, it is also not difficult to derive additional closed-form solutions in the spectral basis and other coordinate systems. We leave this as an exercise for the interested reader.

With the previous complexity considerations of higher-order autoregressive models of cooperative work in PD projects, it is possible to analyze in detail the differences between the EMC as originally developed by Grassberger (1986) and the persistent mutual information EMC(τ) according to Eq. 229, proposed recently by Ball et al. (2010) as a complexity measure. In order to clarify the differences between both measures we refer to the seminal work of Li (2006) and evaluate both the emergent complexity of a strict-sense stationary process $\{X_t\}$ generated by a VAR(n) model, and the emergent complexity related to the model in conjunction with a causal finite impulse response (FIR) filter (see e.g. Puri 2010) of order $m (m \geq 1)$. Each of the output sequences of such a filter is a weighted sum of the most recent m filter input values:

$$y_t = \sum_{i=0}^{m} b_i \cdot x_{t-i}.$$

The b_i's denotes the filter coefficients. The transfer function of the FIR filter is denoted by $H(z)$ (cf. Section 4.2.1). It is assumed that the filter has all its roots on the unit circle. We pass the VAR(n) model outputs x_t through the filter to obtain the output sequence y_t. If $m \geq 1$, according to Li (2006) it holds that the EMC^y related to the stationary filter output y_t is not finite:

$$\text{EMC}^y \rightarrow \infty.$$

However, the corresponding persistent mutual information $\text{EMC}^y(m)$ is finite and equal to the effective measure complexity EMC^x of the steady-state process that is filtered:

$$\text{EMC}^y(m) = \text{EMC}^x.$$

Li (2006) proved these properties for arbitrary stationary Gaussian processes. His theorems also show that zeros on the unit circle can easily cause EMC to be infinite. For instance, even for a simple first-order moving average process $\{X_t\}$ (a so-called MA(1) process, see Section 4.2.1) generated by state equation

$$X_t = \varepsilon_t - \varepsilon_{t-1}$$

the corresponding effective measure complexity

$$\text{EMC} \rightarrow \infty$$

grows over all given limits (Li 2006). Nevertheless, the persistent mutual information

$$\text{EMC}(1) < \infty$$

for lead time one is finite. Hence, in cases where we have a transfer function in the form of a polynomial of degree m that has all its roots on the unit circle, the persistent mutual information $\text{EMC}(\tau)$ according to Eq. 229 should be used instead of the original formulation of the EMC. However, these cases are extremely rare in project management.

4.2 Closed-Form Solutions of Effective Measure Complexity for Linear Dynamical System Models of Cooperative Work

4.2.1 Explicit Formulation

According to the analytical considerations set out at the beginning of Section 4.1, the EMC of a linear dynamical system (LDS, see Section 2.9.) as an advanced model of cooperative work in PD projects, which is defined by the system of equations

$$X_{t+1} = A_0 X_t + \varepsilon_t$$
$$Y_t = HX_t + \nu_t$$

with $\varepsilon_t = \mathcal{N}(\xi; 0_q, C)$ and $\nu_t = \mathcal{N}(\eta; 0_p, V)$, can be expressed by the continuous-type mutual information $I[.;.]$ as

$$\begin{aligned}\text{EMC} &= I\left[Y_{-\infty}^{-1}; Y_0^{\infty}\right] \\ &= \int f\left[y_{-\infty}^{-1}, y_0^{\infty}\right] \log_2 \frac{f\left[y_{-\infty}^{-1}, y_0^{\infty}\right]}{f\left[y_{-\infty}^{-1}\right] f\left[y_0^{\infty}\right]} dy_{-\infty}^{\infty} \ .\end{aligned} \tag{286}$$

In contrast to the previous chapters we have not written the multiplication symbol "." between a matrix and a vector explicitly in the above equations. We will use this more compact notation here and in the following chapter to save space and simplify the interpretation of longer terms. Their meaning should always be clear from the context.

The function $f\left[y_{-\infty}^{-1}\right]$ designates the joint *pdf* of the observable infinite one-dimensional history. Similarly, the function $f\left[y_0^{\infty}\right]$ represents the corresponding *pdf* of the observable infinite future.

It is important to point out that if, and only if, the joint *pdf* of the past $f\left[y_{-\infty}^{\infty}\right]$ and future $f\left[y_0^{\infty}\right]$ histories of observations reach the same steady state, the evaluation of the infinite-dimensional integral yields a finite value. Otherwise, the integral will diverge, as will become clear below. This is possible if the covariance for the initial state in the infinite past with *pdf* given by $f[x_{-\infty}] = (x_t; \mu, \Sigma_0)$ equals the one in the steady state Σ, i.e. the one that satisfies the Lyapunov criterion

$$\Sigma = A_0 \Sigma A_0^{\mathrm{T}} + C$$

from Eq. 27. If the initial state is in steady state, then its expected value is the zero vector $\mu = 0$.

In what follows we will therefore assume that the hidden Markov process $\{X_t\}$ is strict-sense stationary and that in steady state a stable distribution $f[x_\nu]$ is formed. From the state-space model, the following normal distributions can be deduced in steady state:

$$
\begin{aligned}
f[x_\nu] &= \mathcal{N}(x_\nu;\mu,\Sigma)\\
f[x_\nu|x_{\nu-1}] &= \mathcal{N}(x_\nu;A_0x_{\nu-1},C)\\
f[y_\nu|x_\nu] &= \mathcal{N}(y_\nu;Hx_\nu,V).
\end{aligned}
$$

Before we proceed, note of the following: if only the observations y_t are available, it is always possible to introduce an arbitrary invertible transform T so that the model for the observations $Y_t = HX_t + \nu_t$ remains unchanged if $H' = HT, X'_t = T^{-1}X_t$, as

$$
\begin{aligned}
Y_t &= H'X'_{\mathrm{t}} + \nu_t\\
&= HTT^{-1}X_t + \nu_t\\
&= HX_t + \nu_t.
\end{aligned}
\tag{287}
$$

For example, one could choose a whitening transform, cf. Eqs. 156, 157 and 158, $X'_t = \Lambda_u^{-1/2}U^{\mathrm{T}}X_t$ for the hidden-state process which leads to a covariance of the performance fluctuations equal to the identity matrix $C = I_q$. However, in the subsequent derivations, we will continue to use a general covariance C to clarify the interrelationships between the random performance fluctuations and emergent complexity. Following the notation introduced in Section 2.9, we will use the (long) vector $\boldsymbol{y}_{-\infty}^{\infty}$ of the stacked variables $y_{-\infty}^{\infty}$, i.e. $y_{-\infty}^{\infty} = \left(y_{-\infty}^T, \cdots, y_{\infty}^T\right)^T$ in what follows. The vectors $\boldsymbol{y}_{-\infty}^{-1}$ and $\boldsymbol{y}_0^{\infty}$ are defined accordingly. We also add subscripts and superscripts to the quantities $V, C, \Delta t, b$ to mark the corresponding time step. The three joint *pdf*'s in the general definition of the EMC are given for the Gaussian density model (see Eq. 134 in Section 2.9):

$$
f\left[\boldsymbol{y}_{-\infty}^{\infty}\right] = c_{y_{-\infty}^{\infty}}\,\mathrm{Exp}\left[-\frac{1}{2}\left(\boldsymbol{y}_{-\infty}^{\infty}\right)^{\mathrm{T}}\mathcal{V}_{-\infty}^{\infty}\boldsymbol{y}_{-\infty}^{\infty}\right]\frac{(2\pi)^{\Delta t_{-\infty}^{\infty}q/2}}{\sqrt{\mathrm{Det}C_{-\infty}^{\infty}}}\mathrm{Exp}\left[\frac{1}{2}\left(\boldsymbol{b}_{-\infty}^{\infty}\right)^{\mathrm{T}}\left(C_{-\infty}^{\infty}\right)^{-1}\boldsymbol{b}_{-\infty}^{\infty}\right]
\tag{288}
$$

$$
f\left[\boldsymbol{y}_{-\infty}^{-1}\right] = c_{y_{-\infty}^{-1}}\,\mathrm{Exp}\left[-\frac{1}{2}\left(\boldsymbol{y}_{-\infty}^{-1}\right)^{\mathrm{T}}\mathcal{V}_{-\infty}^{-1}\boldsymbol{y}_{-\infty}^{-1}\right]\frac{(2\pi)^{\Delta t_{-\infty}^{-1}q/2}}{\sqrt{\mathrm{Det}C_{-\infty}^{-1}}}\mathrm{Exp}\left[\frac{1}{2}\left(\boldsymbol{b}_{-\infty}^{-1}\right)^{\mathrm{T}}\left(C_{-\infty}^{-1}\right)^{-1}\boldsymbol{b}_{-\infty}^{-1}\right]
\tag{289}
$$

$$
f\left[\boldsymbol{y}_0^{\infty}\right] = c_{y_0^{\infty}}\,\mathrm{Exp}\left[-\frac{1}{2}\left(\boldsymbol{y}_0^{\infty}\right)^{\mathrm{T}}\mathcal{V}_0^{\infty}\boldsymbol{y}_0^{\infty}\right]\frac{(2\pi)^{\Delta t_0^{\infty}q/2}}{\sqrt{\mathrm{Det}C_0^{\infty}}}\mathrm{Exp}\left[\frac{1}{2}\left(\boldsymbol{b}_0^{\infty}\right)^{\mathrm{T}}\left(C_0^{\infty}\right)^{-1}\boldsymbol{b}_0^{\infty}\right].
\tag{290}
$$

Within a direct calculation of the EMC, given by the integral 286, here are two possible paths: One involves splitting the integral into two parts:

$$
\begin{aligned}
I\left[Y_{-\infty}^{-1};Y_0^{\infty}\right] &= \int f\left[y_{-\infty}^{-1},y_0^{\infty}\right]\log_2\frac{f\left[y_{-\infty}^{-1},y_0^{\infty}\right]}{f\left[y_{-\infty}^{-1}\right]f\left[y_0^{\infty}\right]}dy_{-\infty}^{\infty}\\
&= \int f\left[y_{-\infty}^{-1},y_0^{\infty}\right]\log_2 f\left[y_{-\infty}^{-1},y_0^{\infty}\right]dy_{-\infty}^{\infty}\\
&\quad -\int f\left[y_{-\infty}^{-1},y_0^{\infty}\right]\log_2 f\left[y_{-\infty}^{-1}\right]f\left[y_0^{\infty}\right]dy_{-\infty}^{\infty}.
\end{aligned}
$$

The other involves leaving the integral as a whole, computing first the ratio $f\left[y_{-\infty}^{\infty}\right]/\left(f\left[y_{-\infty}^{-1}\right]f\left[y_0^{\infty}\right]\right)$, and then carrying out the integration at the end. The latter approach will lead to an implicit formulation of the EMC. We will pursue this in Section 4.2.2.

For now, we will follow the first path, which will lead us to a result for the EMC in an expressive form given by the (logarithmic) ratio of the product of the determinants of the covariances of the joint *pdf*s of the past and future histories and the determinant of the covariance for the whole history. These covariances are infinite-dimensional in principle, but we will see, numerically, that low-dimensional approximations come very close to the asymptotic result. The smallest possible dimension, i.e. if only two time steps are involved, leads to a simple yet meaningful approximation, which will be discussed in more detail below.

For the first term we can use the result for the differential entropy of a Gaussian variable, see e.g. Cover and Thomas (1991),

$$
\int f\left[y_{-\infty}^{-1},y_0^{\infty}\right]\log_2 f\left[y_{-\infty}^{-1},y_0^{\infty}\right]dy_{-\infty}^{\infty} = -\frac{1}{2}\log_2\left((2\pi e)^{p\Delta t_{-\infty}^{\infty}}\right)-\frac{1}{2}\log_2\left(\mathrm{Det}\,(C_y)_{-\infty}^{\infty}\right).
$$

The second term can be computed as follows:

$$
\begin{aligned}
&\int f\left[y_{-\infty}^{-1},y_0^{\infty}\right]\log_2 f\left[y_{-\infty}^{-1}\right]f\left[y_0^{\infty}\right]dy_{-\infty}^{\infty}\\
&=\int f\left[y_{-\infty}^{-1},y_0^{\infty}\right]\log_2\frac{\mathrm{Exp}\left[-\frac{1}{2}(\mathbf{y}_{-\infty}^{-1})^{\mathrm{T}}\left((C_y)_{-\infty}^{-1}\right)^{-1}\mathbf{y}_{-\infty}^{-1}-\frac{1}{2}(\mathbf{y}_0^{\infty})^{\mathrm{T}}\left((C_y)_0^{\infty}\right)^{-1}\mathbf{y}_0^{\infty}\right]}{\sqrt{2\pi}^{p\Delta t_{-\infty}^{\infty}}\sqrt{\mathrm{Det}\,(C_y)_{-\infty}^{-1}\,\mathrm{Det}\,(C_y)_0^{\infty}}}dy_{-\infty}^{\infty}\\
&=-\frac{1}{2}\int f\left[y_{-\infty}^{-1},y_0^{\infty}\right]\frac{1}{\ln 2}(\mathbf{y}_{-\infty}^{\infty})^{\mathrm{T}}\widehat{C}\,\mathbf{y}_{-\infty}^{\infty}dy_{-\infty}^{\infty}-\frac{1}{2}\log_2\left((2\pi)^{p\Delta t_{-\infty}^{\infty}}\right)\\
&\quad-\frac{1}{2}\log_2\left(\mathrm{Det}\,(C_y)_{-\infty}^{-1}\,\mathrm{Det}\,(C_y)_0^{\infty}\right),
\end{aligned}
$$

with

$$
\widehat{C}=\begin{pmatrix}\left((C_y)_{-\infty}^{-1}\right)^{-1} & 0\\ 0 & \left((C_y)_0^{\infty}\right)^{-1}\end{pmatrix}.
$$

The integral in the first summand of the above equation yields

$$\begin{aligned}\frac{1}{2}\int f\left[y_{-\infty}^{-1}, y_0^{\infty}\right]\frac{1}{\ln 2}\left(\mathbf{y}_{-\infty}^{\infty}\right)^{\mathrm{T}}\widehat{\mathcal{C}}\,\mathbf{y}_{-\infty}^{\infty}dy_{-\infty}^{\infty} &= \frac{1}{2}\frac{1}{\ln 2}\mathrm{Tr}\left(\widehat{\mathcal{C}}\cdot\left(\mathcal{C}_y\right)_{-\infty}^{\infty}\right)\\ &= \frac{1}{2}\frac{1}{\ln 2}p\Delta t_{-\infty}^{\infty}\\ &= \frac{1}{2}\log_2 e^{p\Delta t_{-\infty}^{\infty}},\end{aligned}$$

where we used the fact that $\left(\mathcal{C}_y\right)_{-\infty}^{\infty}$ can be partitioned in a 2×2 block matrix, in which the upper left block equals $\left(\mathcal{C}_y\right)_{-\infty}^{-1}$ and the lower right block equals $\left(\mathcal{C}_y\right)_0^{\infty}$, as can be seen from the block Toeplitz structure of the covariance of the observations. The matrix product $\widehat{\mathcal{C}}\cdot\left(\mathcal{C}_y\right)_{-\infty}^{\infty}$ then has only ones on the diagonal and it is easy to evaluate the trace. Finally, by combining the individual results, we obtain

$$I\left[Y_{-\infty}^{-1}; Y_0^{\infty}\right] = \frac{1}{2}\log_2\frac{\mathrm{Det}\left(\mathcal{C}_y\right)_{-\infty}^{-1}\mathrm{Det}\left(\mathcal{C}_y\right)_0^{\infty}}{\mathrm{Det}\left(\mathcal{C}_y\right)_{-\infty}^{\infty}}. \tag{291}$$

Note that this result has been obtained in a more general context by de Cock (2002), see Eq. 295. The matrices are infinite dimensional, which makes this result impractical for direct use. However, we found in simulations that for a moderately small number of time steps $\Delta t = t_2 - t_1 + 1$ of either the past or the future (the total number of time steps involved is then $2\Delta t$), the value for the EMC tends to its asymptotic value (see Fig. 4.2).

As we have shown in Section 2.10, the likelihood of the observation sequence $\{y_t\}_{t_1}^{t_2}$ is invariant under an arbitrary invertible transform $\Psi\in\mathbb{R}^{q\times q}$ with $\mathrm{Det}(\Psi) = 1$ transforming the set of parameters as $x_t' = \Psi x_t$, $\pi_0' = \Psi\,\pi_0$, $A_0' = \Psi\,A_0\,\Psi^{-1}$, $C' = \Psi\,C\,\Psi^T$, $\Pi_0' = \Psi\,\Pi_0\,\Psi^T$, $H' = H\,\Psi^{-1}$ and $V' = V$. Therefore, the system matrices can not be identified uniquely.

However, the emergent complexity is invariant under this parameter transform, as easily proved by using the expression for the EMC from Eq. 291:

$$\begin{aligned}C_y' &= I_{\Delta t}\otimes V + (I_{\Delta t}\otimes H')C_x'(I_{\Delta t}\otimes H'^T)\\ &= I_{\Delta t}\otimes V + \left(I_{\Delta t}\otimes H\Psi^{-1}\right)(I_{\Delta t}\otimes\Psi)C_x\left(I_{\Delta t}\otimes\Psi^T\right)\left(I_{\Delta t}\otimes\Psi^{-T}H^T\right)\\ &= I_{\Delta t}\otimes V + (I_{\Delta t}\cdot I_{\Delta t})\otimes\left(H\Psi^{-1}\Psi\right)C_x(I_{\Delta t}\cdot I_{\Delta t})\otimes\left(\Psi^T\Psi^{-T}H^T\right)\\ &= I_{\Delta t}\otimes V + (I_{\Delta t}\otimes H)C_x\left(I_{\Delta t}\otimes H^T\right)\\ &= C_y.\end{aligned}$$

This general result holds for the covariance of any observation interval, in particular for $\left(C_y\right)_{-\infty}^{-1}$, $\left(C_y\right)_0^{\infty}$, $\left(C_y\right)_{-\infty}^{\infty}$, and, therefore, the EMC remains unchanged.

Surprisingly, the smallest possible value $\Delta t = 1$, i.e. if we consider that just

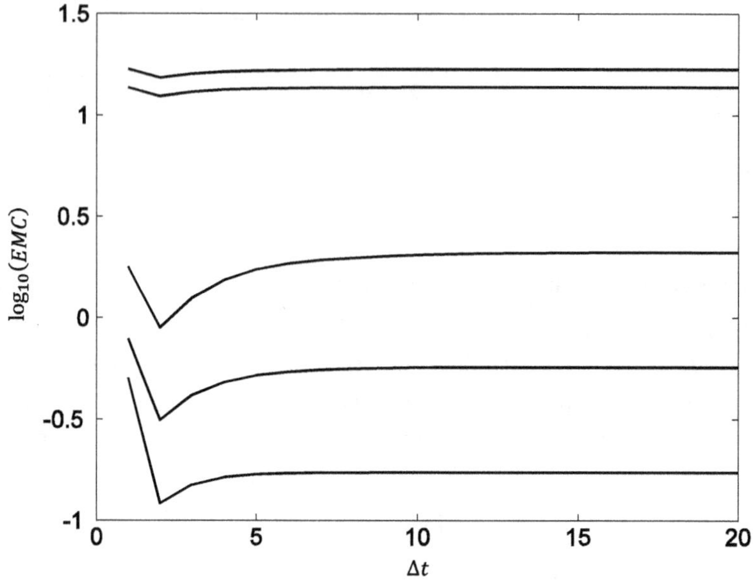

Fig. 4.2 Calculated values of $\log_{10}(EMC)$ for five different, randomly chosen system matrices A_0 and H and varying number of time steps Δt

$$\mathrm{EMC}^{(1)} = \frac{1}{2}\log_2 \frac{\mathrm{Det}\,(\mathcal{C}_y)_{-1}^{-1}\mathrm{Det}\,(\mathcal{C}_y)_0^0}{\mathrm{Det}\,(\mathcal{C}_y)_{-1}^0}$$

leads to a result that is very close to the asymptotic value (see Fig. 4.2). In this case, which we can call a first-order approximation, a very simple closed-form expression for the EMC can be derived. The covariances for a single time step are given by

$$\begin{aligned}(\mathcal{C}_y)_{-1}^{-1} = (\mathcal{C}_y)_0^0 &= I_1 \otimes V + (I_1 \otimes H)\mathcal{C}_x(I_1 \otimes H^{\mathrm{T}}) \\ &= V + H\mathcal{C}_x H^{\mathrm{T}} \\ &= V + H\Sigma H^{\mathrm{T}}.\end{aligned}$$

For two time steps we have

$$\begin{aligned}(\mathcal{C}_y)_{-1}^0 &= I_2 \otimes V + (I_2 \otimes H)\mathcal{C}_x(I_2 \otimes H^{\mathrm{T}}) \\ &= \begin{pmatrix} V & 0 \\ 0 & V \end{pmatrix} + \begin{pmatrix} H & 0 \\ 0 & H \end{pmatrix}\begin{pmatrix} \Sigma & \Sigma A_0^{\mathrm{T}} \\ A_0\Sigma & V \end{pmatrix}\begin{pmatrix} H^{\mathrm{T}} & 0 \\ 0 & H^{\mathrm{T}} \end{pmatrix} \\ &= \begin{pmatrix} V + H\Sigma H^{\mathrm{T}} & H\Sigma A_0^{\mathrm{T}} H^{\mathrm{T}} \\ HA_0\Sigma H^{\mathrm{T}} & V + H\Sigma H^{\mathrm{T}} \end{pmatrix}.\end{aligned}$$

The determinant of covariance of the two time steps can be simplified using a formula for the determinant of block-matrices,

$$\text{Det}\left(C_y\right)_{-1}^{0}=\text{Det}\left(V+H\Sigma H^{\text{T}}\right)\text{Det}\left(V+H\Sigma H^{\text{T}}-H\Sigma A_0^{\text{T}}H^{\text{T}}\left(V+H\Sigma H^{\text{T}}\right)^{-1}HA_0\Sigma H^{\text{T}}\right),$$

and we obtain the following first-order approximation for the EMC:

$$\begin{aligned}\text{EMC}^{(1)} &= \frac{1}{2}\log_2\frac{\text{Det}\left(V+H\Sigma H^{\text{T}}\right)}{\text{Det}\left(V+H\Sigma H^{\text{T}}-H\Sigma A_0^{\text{T}}H^{\text{T}}\left(V+H\Sigma H^{\text{T}}\right)^{-1}HA_0\Sigma H^{\text{T}}\right)}\\ &= -\frac{1}{2}\log_2\text{Det}\left(I_p-H\Sigma A_0^{\text{T}}H^{\text{T}}\left(V+H\Sigma H^{\text{T}}\right)^{-1}HA_0\Sigma H^{\text{T}}\left(V+H\Sigma H^{\text{T}}\right)^{-1}\right).\end{aligned} \tag{292}$$

Interestingly, this approximate result can be derived if one starts from the assumption that the Markov property holds in steady state for the observable process, and we have

$$\begin{aligned}f\left[y_{-\infty}^{-1},y_0^{\infty}\right]&=f[y_{-\infty}]f\left[y_{-\infty+1}\left|y_{-\infty}\right.\right]\cdots f[y_{-1}|y_{-2}]f[y_0|y_{-1}]f[y_1|y_0]\cdots f[y_\infty|y_{\infty-1}]\\ f\left[y_{-\infty}^{-1}\right]&=f[y_{-\infty}]f\left[y_{-\infty+1}\left|y_{-\infty}\right.\right]\cdots f[y_{-1}|y_{-2}]\\ f\left[y_0^{\infty}\right]&=f[y_0]f[y_1|y_0]\cdots f[y_\infty|y_{\infty-1}].\end{aligned}$$

Hence, the expression for the EMC reduces to

$$\begin{aligned}I\left[Y_{-\infty}^{-1};Y_0^{\infty}\right] &= \int_{\mathbb{Y}^p}\cdots\int_{\mathbb{Y}^p} f\left[y_{-\infty}^{-1},y_0^{\infty}\right]\log_2\frac{f[y_0|y_{-1}]}{f[y_0]}dy_{-\infty}^{-1}dy_0^{\infty}\\ &= \int_{\mathbb{Y}^p}\cdots\int_{\mathbb{Y}^p} f\left[y_{-\infty}^{-1},y_0^{\infty}\right]\log_2 f[y_0|y_{-1}]dy_{-\infty}^{-1}dy_0^{\infty}\\ &\quad -\int_{\mathbb{Y}^p}\cdots\int_{\mathbb{Y}^p} f\left[y_{-\infty}^{-1},y_0^{\infty}\right]\log_2 f[y_0]dy_{-\infty}^{-1}dy_0^{\infty}\\ &= \int_{\mathbb{Y}^p}\int_{\mathbb{Y}^p}\log_2 f[y_0|y_{-1}]dy_{-1}dy_0\left(\int_{\mathbb{Y}^p}\cdots\int_{\mathbb{Y}^p} f\left[y_{-\infty}^{-1},y_0^{\infty}\right]dy_{-\infty}\ldots dy_{-2}dy_1\ldots dy_\infty\right)\\ &\quad -\int_{\mathbb{Y}^p}\log_2 f[y_0]dy_0\left(\int_{\mathbb{Y}^p}\cdots\int_{\mathbb{Y}^p} f\left[y_{-\infty}^{-1},y_0^{\infty}\right]dy_{-\infty}\ldots dy_{-1}dy_1\ldots dy_\infty\right).\end{aligned}$$

Exploiting the relations for the marginal probability densities, we obtain:

$$\begin{aligned}I\left[Y_{-\infty}^{-1};Y_0^{\infty}\right] &= \int_{\mathbb{Y}^p} f[y_{-1},\, y_0]\log_2 f[y_0|y_{-1}]dy_{-1}dy_0-\int_{\mathbb{Y}^p} f[y_0]\log_2 f[y_0]dy_0\\ &= \int_{\mathbb{Y}^p} f[y_{-1}]dy_{-1}\int_{\mathbb{Y}^p} f[y_0|y_{-1}]\log_2 f[y_0|y_{-1}]dy_0-\int_{\mathbb{Y}^p} f[y_0]\log_2 f[y_0]dy_0.\end{aligned}$$

The probability density for the observable variable Y_t can then be expressed as:

$$\begin{aligned} f[y_t] &= \int_{\mathbb{X}^p} f[y_t, x_t] dx_t \\ &= \int_{\mathbb{X}^p} f[y_t|x_t] f[x_t] dx_t \\ &= \int_{\mathbb{X}^p} \mathcal{N}(y_t; Hx_t, V)\mathcal{N}(x_t; \mu, \Sigma) dx_t. \end{aligned}$$

In order to solve the above integral, it is useful to apply the following transformation formula for normal distributions:

$$\mathcal{N}(y; Hx, V)\mathcal{N}(x; \mu, \Sigma) = \mathcal{N}(y; H\mu, S)\mathcal{N}\left(x; \mu + W(y - H\mu), \Sigma - WSW^{\mathrm{T}}\right)$$

with

$$S = H\Sigma H^{\mathrm{T}} + V \text{ and } W = \Sigma H^{\mathrm{T}} S^{-1}.$$

Hence, we obtain:

$$f[y_t] = \mathcal{N}\left(y_t; H\mu, H\Sigma H^{\mathrm{T}} + V\right).$$

For the calculation of $f\left[y_0 \middle| y_{-1}\right] = f[y_{-1}, y_0]/f[y_{-1}]$ we insert the hidden states x_{-1} and x_0 and exploit the Markov property

$$\begin{aligned} f\left[y_0 \middle| y_{-1}\right] &= \int_{\mathbb{X}^q}\int_{\mathbb{X}^q} \frac{f[y_{-1}, y_0, x_{-1}, x_0]}{f[y_{-1}]} dx_{-1} dx_0 \\ &= \int_{\mathbb{X}^q}\int_{\mathbb{X}^q} f\left[x_{-1} \middle| y_{-1}\right] f[x_0|x_{-1}] f[y_0|x_0] dx_{-1} dx_0. \end{aligned}$$

Because of Bayes theorem

$$f\left[x_{-1} \middle| y_{-1}\right] = \frac{f[y_{-1}|x_{-1}] f[x_{-1}]}{f[y_{-1}]},$$

we find

$$\begin{aligned} & f\left[y_0 \middle| y_{-1}\right] \\ &= \frac{1}{f[y_{-1}]} \int_{\mathbb{X}^q}\int_{\mathbb{X}^q} f[x_{-1}] f[y_{-1}|x_{-1}] f[x_0|x_{-1}] f[y_0|x_0] dx_{-1} dx_0 \\ &= \frac{1}{f[y_{-1}]} \int_{\mathbb{R}^q}\int_{\mathbb{R}^q} \mathcal{N}(x_{-1};\mu,\Sigma)\mathcal{N}(y_{-1};Hx_{-1},V)\mathcal{N}(x_0;A_0x_{-1},C)\mathcal{N}(y_0;Hx_0,V) dx_{-1} dx_0. \end{aligned}$$

First, we transform the first two Gaussians as:

$$\mathcal{N}(x_{-1};\mu,\Sigma)\mathcal{N}(y_{-1};Hx_{-1},V) = \mathcal{N}\left(y_{-1};H\mu,H\Sigma H^{\mathrm{T}}+V\right)$$
$$\times\mathcal{N}\left(x_{-1};\mu+W(y_{-1}-H\mu),\Sigma-WSW^{\mathrm{T}}\right),$$

with

$$S = H\Sigma H^{\mathrm{T}}+V \text{ and } W = \Sigma H^{\mathrm{T}}S^{-1}.$$

The first Gaussian on the right hand side cancels $f[y_{-1}]$. The second Gaussian on the right hand side together with the third Gaussian $\mathcal{N}(x_0;A_0x_{-1},C)$ from the previous expression for $f\left[y_0|y_{-1}\right]$ yields:

$$\begin{aligned}&\mathcal{N}(x_0;A_0x_{-1},C)\mathcal{N}\left(x_{-1};\mu+W(y_{-1}-H\mu),\Sigma-WSW^{\mathrm{T}}\right)\\&=\mathcal{N}\left(x_0;A_0(\mu+W(y_{-1}-H\mu)),\ A_0(\Sigma-WSW^{\mathrm{T}})A_0^{\mathrm{T}}+C\right)\mathcal{N}(x_{-1};\overline{x}_{-1},C')\end{aligned}$$

with some inconsequential mean $\overline{x}_{-1}$ and covariance C'. After integration with respect to x_{-1} we have:

$$f\left[y_0|y_{-1}\right]=\int_{\mathbb{R}^q}\mathcal{N}\left(x_0;A_0(\mu+W(y_{-1}-H\mu)),A_0(\Sigma-WSW^{\mathrm{T}})A_0^{\mathrm{T}}+C\right)\mathcal{N}(y_0;Hx_0,V)dx_0.$$

Again, by transforming the two Gaussians we can carry out easily the integration with respect to x_0 and obtain:

$$f\left[y_0|y_{-1}\right]=\mathcal{N}\left(y_0;HA_0(\mu+W(y_{-1}-H\mu)),H\left(A_0\left(\Sigma-WSW^{T}\right)A_0^{\mathrm{T}}+C\right)H^{\mathrm{T}}+V\right).$$

Using the fact that the differential entropy of a multivariate Gaussian distribution $\mathcal{N}(x;\mu,C)$ is given by

$$-\int_{\mathbb{R}^q}\mathcal{N}(x;\mu,C)\log_2\mathcal{N}(x;\mu,C)dx=\frac{1}{2}\log_2(2\pi e)^p\mathrm{Det}[C],$$

we arrive at the known first-order approximation from Eq. 292 for the EMC:

$$\mathrm{EMC}^{(1)}=\frac{1}{2}\log_2\mathrm{Det}\left[H\Sigma H^{\mathrm{T}}+V\right]-\frac{1}{2}\log_2\mathrm{Det}[D],$$

with

$$D=V+H\Sigma H^{\mathrm{T}}-H\Sigma A_0^{\mathrm{T}}H^{\mathrm{T}}\left(V+H\Sigma H^{\mathrm{T}}\right)^{-1}HA_0\Sigma H^{\mathrm{T}}.$$

It is evident that in the case of $H=I$ and $V=\{0\}I_q$, we obtain the same result as we did for the VAR(1) model (see Eq. 246).

For small covariance V, i.e. if the eigenvalues of $\left(H\Sigma H^{\mathrm{T}}\right)^{-1}V$ lie inside the unit circle, we can expand

$$\begin{aligned}\left(H\Sigma H^{\mathrm{T}}+V\right)^{-\mathrm{T}} &= \left(H\Sigma H^{\mathrm{T}}\right)^{-1}\left(I_p+\left(H\Sigma H^{\mathrm{T}}\right)^{-1}V\right)^{-1}\\ &\approx \left(H\Sigma H^{\mathrm{T}}\right)^{-1}\left(I_p-\left(H\Sigma H^{\mathrm{T}}\right)^{-1}V\right)\end{aligned}$$

and arrive at an approximate expression for D:

$$D=H\left(A_0H^{-1}H^{-\mathrm{T}}\Sigma^{-1}H^{-1}VH\Sigma A_0+C\right)H^{\mathrm{T}}+V.$$

Assuming furthermore $V=\{\sigma_v^2\}I_p$, we obtain:

$$D=H\left(\{\sigma_v^2\}A_0H^{-1}H^{-\mathrm{T}}A_0+C\right)H^{\mathrm{T}}+\{\sigma_v^2\}I_p.$$

Following the procedure from Section 4.1.2, we can also express the first-order approximation for the EMC as the signal-to-noise ratio:

$$\begin{aligned}\mathrm{EMC}^{(1)} &= -\frac{1}{2}\log_2\mathrm{Det}\left[I_p-HA_0\Sigma H^{\mathrm{T}}\left(H\Sigma H^{\mathrm{T}}+V\right)^{-\mathrm{T}}H\Sigma^{\mathrm{T}}A_0^{\mathrm{T}}H^{\mathrm{T}}\left(H\Sigma H^{\mathrm{T}}+V\right)^{-1}\right]\\ &= \frac{1}{2}\log_2\left(\mathrm{Det}\left[I_p-HA_0\Sigma H^{\mathrm{T}}\left(H\Sigma H^{\mathrm{T}}+V\right)^{-\mathrm{T}}H\Sigma^{\mathrm{T}}A_0^{\mathrm{T}}H^{\mathrm{T}}\left(H\Sigma H^{\mathrm{T}}+V\right)^{-1}\right]\right)^{-1}\\ &= \frac{1}{2}\log_2\mathrm{Det}\left[\left(I_p-HA_0\Sigma H^{\mathrm{T}}\left(H\Sigma H^{\mathrm{T}}+V\right)^{-\mathrm{T}}H\Sigma^{\mathrm{T}}A_0^{\mathrm{T}}H^{\mathrm{T}}\left(H\Sigma H^{\mathrm{T}}+V\right)^{-1}\right)^{-1}\right]\\ &= \frac{1}{2}\log_2\mathrm{Det}\left[\sum_{k=0}^{\infty}\left(HA_0\Sigma H^{\mathrm{T}}\left(H\Sigma H^{\mathrm{T}}+V\right)^{-\mathrm{T}}H\Sigma^{\mathrm{T}}A_0^{\mathrm{T}}H^{\mathrm{T}}\left(H\Sigma H^{\mathrm{T}}+V\right)^{-1}\right)^k\right]\\ &= \frac{1}{2}\log_2\mathrm{Det}\left[I_p+\sum_{k=1}^{\infty}\left(HA_0\Sigma H^{\mathrm{T}}\left(H\Sigma H^{\mathrm{T}}+V\right)^{-\mathrm{T}}H\Sigma^{\mathrm{T}}A_0^{\mathrm{T}}H^{\mathrm{T}}\left(H\Sigma H^{\mathrm{T}}+V\right)^{-1}\right)^k\right].\end{aligned} \tag{293}$$

The above derivation is based on the von Neumann series generated by the operator $HA_0\Sigma H^{\mathrm{T}}\left(H\Sigma H^{\mathrm{T}}+V\right)^{-\mathrm{T}}H\Sigma^{\mathrm{T}}A_0^{\mathrm{T}}H^{\mathrm{T}}\left(H\Sigma H^{\mathrm{T}}+V\right)^{-1}$. The von Neumann series generalizes the geometric series (cf. Section 2.1). The infinite sum represents the signal-to-noise ratio.

The closed-form solution from Eq. 286 also allows us to develop homologous vector autoregression models for linear dynamical systems. For $t\to\infty$ these models generate stochastic processes with equivalent effective measure complexity, but the state variables are completely observable. In this sense, the homologous models reveal all correlations and dynamical dependency structures during the observation time and do not possess any kind of crypticity (Ellison et al. 2009). To make this possible we usually have to use a higher dimensionality $p>q$. We start by focusing on homologous VAR(1) models with dynamical operator A_0^h and covariance matrix C^h that are defined over a p-dimensional space $\mathbb{R}^p$ of observable states X_t^h:

$$X_t^h = A_0^h X_{t-1}^h + \varepsilon_t^h \quad t = 1, \ldots, T,$$

with

$$\varepsilon_t^h \sim \mathcal{N}(0_p, C^h).$$

Assuming that the performance fluctuations represented by the homologous model are isotropic and temporally uncorrelated, i.e. $\varepsilon_t^h \sim \mathcal{N}(0_p, \{\sigma_v^2\} I_p)$ and $E\left[\varepsilon_t^h (\varepsilon_s^h)^{\mathrm{T}}\right] = C^h \delta_{ts}$, we can construct a dynamical operator A_0^h representing a large variety of cooperative relationships. The preferred structure of relationships must be determined in the specific application context of complexity evaluation. According to the analysis in Section 4.1.1 only two constraints must be satisfied: (1) A_0^h must be diagonalizable and (2) for the weighted sum of eigenvalues $\lambda_i(A_0^h)$, it must hold that (cf. Eq. 251):

$$\begin{aligned} -\frac{1}{2}\sum_{i=1}^{p} \log_2\left(1 - \lambda_i\left(A_0^h\right)^2\right) &= \frac{1}{2}\log_2 \frac{\mathrm{Det}\ (\mathcal{C}_y)_{-\infty}^{-1} \mathrm{Det}\ (\mathcal{C}_y)_0^{\infty}}{\mathrm{Det}\ (\mathcal{C}_y)_{-\infty}^{\infty}} \\ &= \frac{1}{2}\left(\log_2 \mathrm{Det}\ (\mathcal{C}_y)_{-\infty}^{-1} \mathrm{Det}\ (\mathcal{C}_y)_0^{\infty} - \log_2 \mathrm{Det}\ (\mathcal{C}_y)_{-\infty}^{\infty}\right). \end{aligned} \tag{294}$$

It is evident that the most simple homologous model can be constructed by setting the autonomous task processing rates as diagonal elements of A_0^h to the same rate a, i.e. $A_0^h = \mathrm{Diag}[a, \ldots, a]$. For this structurally non-informative model, the corresponding stationary stochastic process communicates the same amount of information from the infinite past to the infinite future, if

$$a = \sqrt{1 - 2^{-\frac{1}{p}\left(\log_2 \mathrm{Det}\ (\mathcal{C}_y)_{-\infty}^{-1} \mathrm{Det}\ (\mathcal{C}_y)_0^{\infty} - \log_2 \mathrm{Det}\ (\mathcal{C}_y)_{-\infty}^{\infty}\right)}}.$$

The above equation also holds for homologous models with non-isotropic fluctuations, because all tasks are processed at the same time scale.

Finally, we can develop a homologous model that is defined over a one-dimensional state space. This model is termed an auto-regressive moving average (ARMA) model and is characterized by the following linear difference equation (see e.g. Puri 2010):

$$Y_t = \sum_{i=1}^{p} a_i Y_{t-i} + \sum_{j=1}^{q} b_j U_{t-j}.$$

The input of the model is Gaussian white noise with variance $\sigma^2 = 1$, i.e. $U \sim \mathcal{N}(0, 1)$. This model is notated ARMA(p, q) in the literature (note that in this notation the variable q does not denote the dimensionality of the observation vectors Y_t; it denotes the number of inputs U_{t-j} driving the process). It is evident that an ARMA(p,q) model can be rewritten as either a VAR(p) model of order p (Section 2.4) or an LDS$(p,1)$ model (Section 2.9) (see e.g. de Cock 2002). It is not difficult to show that for a stable and strictly minimum phase ARMA(p,q) model the effective measure complexity is given by

$$\mathrm{EMC} = \frac{1}{2}\log_2 \frac{\prod_{i,\,j=1}^{p,q} \left|1-\alpha_i\overline{\beta}_j\right|}{\prod_{i,\,j=1}^{p} \left|1-\alpha_i\overline{\alpha}_j\right| \prod_{i,\,j=1}^{q} \left|1-\beta_i\overline{\beta}_j\right|}$$
$$= \frac{1}{2}\left(\sum_{i,\,j}^{p,q} \log_2\left|1-\alpha_i\overline{\beta}_j\right| - \sum_{i,\,j}^{p} \log_2\left|1-\alpha_i\overline{\alpha}_j\right| + \sum_{i,\,j}^{q} \log_2\left|1-\beta_i\overline{\beta}_j\right|\right),$$

where the variables $\alpha_1,\ldots,\alpha_p$ denote the roots of the polynomial $a(z) = z^p + a_1 z^{p-1} + \ldots + a_p$ and $\beta_1,\ldots,\beta_q$ the roots of the polynomial $b(z) = z^q + b_1 z^{q-1} + \ldots + b_q$ (see e.g. de Cock 2002). These polynomials are the results of the z-transform of the difference equation of the ARMA(p,q) model. The well-known transfer function $H(z)$ from control theory is the quotient of these polynomials. Since the polynomials are real, the roots are all real or come in conjugate pairs. Hence, for the poles $\alpha_1,\ldots,\alpha_p$ and the zeros $\beta_1,\ldots,\beta_q$ of the transfer function $H(z)$ of the homologous ARMA(p,q) model, it must hold that

$$\frac{1}{2}\left(\sum_{i,\,j}^{p,q} \log_2\left|1-\alpha_i\overline{\beta}_j\right| - \sum_{i,\,j}^{p} \log_2\left|1-\alpha_i\overline{\alpha}_j\right| + \sum_{i,\,j}^{q} \log_2\left|1-\beta_i\overline{\beta}_j\right|\right)$$
$$= \frac{1}{2}\left(\log_2 \mathrm{Det}\,(\mathcal{C}_y)_{-\infty}^{-1}\mathrm{Det}\,(\mathcal{C}_y)_0^{\infty} - \log_2 \mathrm{Det}\,(\mathcal{C}_y)_{-\infty}^{\infty}\right).$$

4.2.2 Implicit Formulation

Interestingly, the sophisticated closed-form solution of EMC from Eq. 286 that was obtained through the evaluation of the infinite-dimensional integral of the continuous-type mutual information (Eq. 286) can also be written in a structurally rich implicit form. This form is based on the seminal work of de Cock (2002). The implicit form is especially easy to interpret because its independent parameters can be derived from solutions of fundamental equations. In order to derive the implicit form of de Cock (2002) we work with the "forward innovation model" from Section 2.9 (Eqs. 164 and 165):

$$X_{t+1}^f = A_0 X_t^f + K\eta_t$$
$$Y_t = H X_t^f + \eta_t.$$

According to de Cock (2002), the effective measure complexity can be expressed as

$$\begin{aligned}\mathrm{EMC} &= I\left[Y_{-\infty}^{-1}; Y_0^{\infty}\right] \\ &= -\frac{1}{2}\log_2 \mathrm{Det}\left[I_q - \Sigma_f\left(G_z^{-1} + \Sigma^f\right)^{-1}\right].\end{aligned} \tag{295}$$

The covariance matrix Σ^f is the solution of the Lyapunov equation (cf. Eq. 167)

$$\Sigma^f = A_0 \Sigma^f A_0^{\mathrm{T}} + KSK^{\mathrm{T}}.$$

In the above Lyapunov equation

$$K = \left(G^f - A_0 \Sigma^f H^{\mathrm{T}}\right)\left(C_{YY}(0) - H\Sigma^f H^{\mathrm{T}}\right)^{-1}$$

is the Kalman gain (Eq. 169) and

$$S = C_{YY}(0) - H\Sigma^f H^{\mathrm{T}}.$$

is the covariance $S_{t+1|t}$ (Eq. 168) of the single-source performance fluctuations η_t for $t \to \infty$. Hence, we have the following algebraic Riccati equation for Σ^f (van Overschee and de Moor 1996):

$$\Sigma^f = A_0 \Sigma^f A_0^{\mathrm{T}} + \left(G^f - A_0 \Sigma^f H^{\mathrm{T}}\right)\left(C_{YY}(0) - H\Sigma^f H^{\mathrm{T}}\right)^{-1}\left(\left(G^f\right)^{\mathrm{T}} - H\Sigma^f H^{\mathrm{T}}\right). \tag{296}$$

The additional covariance matrix G_z from Eq. 295 satisfies the Lyapunov equation

$$G_z = (A_0 - KH)^{\mathrm{T}} G_z (A_0 - KH) + H^{\mathrm{T}} S^{-1} H. \tag{297}$$

An important finding of de Cock (2002) is that the inverse aggregated covariance matrix $\left(G_z^{-1} + \Sigma^f\right)^{-1}$ is the solution of another Lyapunov equation

$$\begin{aligned}\overline{\Sigma}^b &= \overline{A}_0 \overline{\Sigma}^b \overline{A}_0^{\mathrm{T}} + \overline{K}\,\overline{S}\,\overline{K}^{\mathrm{T}} \\ &= A_0^{\mathrm{T}} \overline{\Sigma}^b A_0 + \overline{K}\,\overline{S}\,\overline{K}^{\mathrm{T}},\end{aligned}$$

which is related to the backward innovation representation of the corresponding backward model (Eqs. 178 and 179):

$$\begin{aligned}\overline{X}_{t-1}^b &= \overline{A}_0 \overline{X}_t^b + \overline{K}\overline{\eta}_t \\ Y_t &= \overline{H}\,\overline{X}_t^b + \overline{\eta}_t.\end{aligned}$$

Substituting the Kalman gain $\overline{K}$ (Eq. 181) and the fluctuations covariance $\overline{S} = E\left[\overline{\eta}_t \overline{\eta}_t^{\mathrm{T}}\right]$ (Eq. 182) in the Lyapunov equation for the backward innovation representation leads to the following algebraic Riccati equation for the backward state covariance matrix:

$$\begin{aligned}\overline{\Sigma}^b &= A_0^{\mathrm{T}}\overline{\Sigma}^b A_0 + \left(H^{\mathrm{T}} - A_0^{\mathrm{T}}\overline{\Sigma}^b G\right)\left(\left(H^{\mathrm{T}} - A_0^{\mathrm{T}}\overline{\Sigma}^b G\right)\left(C_{YY}(0) - G^{\mathrm{T}}\overline{\Sigma}^b G\right)^{-1}\right)^{\mathrm{T}} \\ &= A_0^{\mathrm{T}}\overline{\Sigma}^b A_0 + \left(H^{\mathrm{T}} - A_0^{\mathrm{T}}\overline{\Sigma}^b G\right)\left(C_{YY}(0) - G^{\mathrm{T}}\overline{\Sigma}^b G\right)^{-\mathrm{T}}\left(H^{\mathrm{T}} - A_0^{\mathrm{T}}\overline{\Sigma}^b G\right)^{\mathrm{T}} \\ &= A_0^{\mathrm{T}}\overline{\Sigma}^b A_0 + \left(H^{\mathrm{T}} - A_0^{\mathrm{T}}\overline{\Sigma}^b G\right)\left(C_{YY}(0) - G^{\mathrm{T}}\overline{\Sigma}^b G\right)^{-1}\left(H - G^{\mathrm{T}}\overline{\Sigma}^b A_0\right).\end{aligned} \tag{298}$$

Hence, the most intuitive solution is obtained (de Cock 2002):

$$\begin{aligned}I\left[Y_{-\infty}^{-1}; Y_0^{\infty}\right] &= -\frac{1}{2}\log_2 \mathrm{Det}\left[I_q - \Sigma^f\left(G_z^{-1} + \Sigma^f\right)^{-1}\right] \\ &= -\frac{1}{2}\log_2 \mathrm{Det}\left[I_q - \Sigma^f\overline{\Sigma}^b\right].\end{aligned} \tag{299}$$

According to Sylvester's determinant theorem, this solution can equivalently be expressed based on the signal-to-noise ratio $\mathrm{SNR} = G_z\left(\left(\Sigma^f\right)^{-1}\right)^{-1}$, and we have:

$$\begin{aligned}I\left[Y_{-\infty}^{-1}; Y_0^{\infty}\right] &= -\frac{1}{2}\log_2 \mathrm{Det}\left[I_q - \overline{\Sigma}^b\Sigma^f\right] \\ &= \frac{1}{2}\log_2 \mathrm{Det}\left[I_q + G_z\Sigma^f\right] \\ &= \frac{1}{2}\log_2 \mathrm{Det}\left[I_q + \Sigma^f G_z\right].\end{aligned}$$

The standard numerical approach to solve the forward Riccati Eq. 296 is to solve the generalized eigenvalue problem

$$\begin{pmatrix} A_0^{\mathrm{T}} - H^{\mathrm{T}}(C_{YY}(0))^{-1}G^f & 0 \\ -G^f(C_{YY}(0))^{-1}(G^f)^{\mathrm{T}} & I_q \end{pmatrix}\begin{pmatrix} W_1 \\ W_2 \end{pmatrix} = \begin{pmatrix} I_q & -H^{\mathrm{T}}(C_{YY}(0))^{-1}H \\ 0 & A_0 - G^f(C_{YY}(0))^{-1}H \end{pmatrix}\begin{pmatrix} W_1 \\ W_2 \end{pmatrix}\Lambda$$

and compute the covariance matrix Σ^f as

$$\Sigma^f = W_2 W_1^{-1},$$

see, e.g. van Overschee and de Moor (1996). The complementary backward Riccati Eq. 298 can be tackled by solving

$$\begin{pmatrix} A_0 - G(C_{YY}(0))^{-1}H & 0 \\ -H^{\mathrm{T}}(C_{YY}(0))^{-1}H & I_q \end{pmatrix}\begin{pmatrix} W_1 \\ W_2 \end{pmatrix} = \begin{pmatrix} I_q & -G(C_{YY}(0))^{-1}H^{\mathrm{T}} \\ 0 & A_0^{\mathrm{T}} - H^{\mathrm{T}}(C_{YY}(0))^{-1}(G^f)^{\mathrm{T}} \end{pmatrix}\begin{pmatrix} W_1 \\ W_2 \end{pmatrix}\Lambda$$

and computing the covariance matrix $\overline{\Sigma}_b$ as

$$\overline{\Sigma}^b = W_2 W_1^{-1}.$$

It is evident that the same numerical function can be used in the preferred programming language to solve the above generalized eigenvalue problems. This function must be called for the forward Riccati equation with the argument $(A_0, H, G^f, C_{YY}(0))$, whilst for the backward Riccati equation the argument must be $(A_0^{\mathrm{T}}, (G^f)^{\mathrm{T}}, H^{\mathrm{T}}, C_{YY}(0))$.

Similar to the canonical correlation analysis of the basic VAR(1) process in Section 4.1.3, we can diagonalize the forward and backward state covariance matrices obtained by solving the algebraic Riccati Eqs. 296 and 298 simultaneously and bring them in a form called "stochastic balanced realization" (Desai and Pal 1984). A stochastic balanced representation is an innovations representation with state covariance matrix equal to the canonical correlation coefficient matrix for the sequence of observations. Let the eigendecomposition (cf. Eq. 22) of the product of the state covariance matrices $\Sigma^f \overline{\Sigma}^b$ be given by the representation

$$\Sigma^f \overline{\Sigma}^b = M \Lambda_M^2 M^{-1}$$
$$\Lambda_M^2 = \mathrm{Diag}\left[\lambda_i\left(\Sigma^f \overline{\Sigma}^b\right)\right] \quad 1 \le i \le q,$$

where the eigenvector matrix M is picked as

$$M = U_{\overline{\Sigma}^b} \Lambda_{\overline{\Sigma}^b}^{-1/2} U_{\Sigma^f} \Lambda_{\mathrm{M}}^{1/2}.$$

The matrices $U_{\overline{\Sigma}^b}$ and $\Lambda_{\overline{\Sigma}^b}$ can be specified by the eigendecomposition of $\overline{\Sigma}_b$ as

$$U_{\overline{\Sigma}^b} \Lambda_{\overline{\Sigma}^b} U_{\overline{\Sigma}^b}^{-1} = \overline{\Sigma}^b,$$

and for U_{Σ^f} it holds that

$$U_{\Sigma^f} \Lambda_M^2 U_{\Sigma^f}^{-1} = \Lambda_{\overline{\Sigma}^b}^{1/2} U_{\overline{\Sigma}^b}^{-1} \Sigma^f \Lambda_{\overline{\Sigma}^b}^{1/2}.$$

Furthermore, let the forward state that is subject to the simultaneous diagonalization of the state covariance matrices be

$$X_t^d = T X_t^f$$

and the corresponding backward state be

$$\overline{X}_t^d = T^{-1} \overline{X}_t^b$$

with the coefficient of the similarity transformation

$$T = M^{\mathrm{T}},$$

then in steady state it holds for the expectations (Desai and Pal 1984) that:

$$E\left[X_t^d \left(X_t^d\right)^{\mathrm{T}}\right] = \Lambda_M = E\left[\overline{X}_t^d \left(\overline{X}_t^d\right)^{\mathrm{T}}\right].$$

Hence, the stochastic balanced representation allows us to make the dependency of the effective measure complexity on the eigenvalues of the product of the state covariance matrices $\Sigma^f \overline{\Sigma}^b$ explicit:

$$\begin{aligned} I\left[Y_{-\infty}^{-1}; Y_0^{\infty}\right] &= -\frac{1}{2}\log_2 \mathrm{Det}\left[I_p - \Sigma^f \overline{\Sigma}^b\right] \\ &= -\frac{1}{2}\log_2 \mathrm{Det}\left[I_q - \Lambda_M^2\right] \\ &= -\frac{1}{2}\log_2 \prod_{i=1}^{q}\left(1 - \lambda_i\left(\Sigma^f \overline{\Sigma}^b\right)\right) \\ &= -\frac{1}{2}\log_2 \prod_{i=1}^{q}\left(1 - \rho_i^2\right) \\ &= -\frac{1}{2}\sum_{i=1}^{q}\log_2\left(1 - \rho_i^2\right). \end{aligned} \tag{300}$$

In the last line of the above equation the ρ_i's represent the canonical correlations, which were already introduced in Section 4.1.3 (cf. Eq. 265) to analyze emergent complexity based on a reduced number of independent parameters. In other words, the eigenvalues of $\Sigma^f \overline{\Sigma}^b$ are simply the squares of the canonical correlation coefficients between the canonical variates. However, it is important to note that in contrast to Section 4.1.3 the infinite random sequences representing the past $X_{past} = \left(X_{-\infty}^{\mathrm{T}} \ \cdots \ X_{-2}^{\mathrm{T}} \ X_{-1}^{\mathrm{T}}\right)^{\mathrm{T}}$ and future $X_{fut} = \left(X_0^{\mathrm{T}} \ X_1^{\mathrm{T}} \ \cdots \ X_{\infty}^{\mathrm{T}}\right)^{\mathrm{T}}$ histories of the hidden state process are not the subject of the canonical correlation analysis, but rather the canonical correlations between the pair $\left(\left(Y_{-\infty}^{\mathrm{T}} \ \cdots \ Y_{-2}^{\mathrm{T}} \ Y_{-1}^{\mathrm{T}}\right)^{\mathrm{T}}, \left(Y_0^{\mathrm{T}} \ Y_1^{\mathrm{T}} \ \cdots \ Y_{\infty}^{\mathrm{T}}\right)^{\mathrm{T}}\right)$ of past and future histories of the observation process $\{Y_t\}$ are considered to evaluate complexity explicitly. Due to the potentially higher dimensionality of the state space of the hidden state process $(q > p)$, all q complexity-shaping summands $\log_2(1 - \rho_i^2)$ that can give rise to correlations between observations of the project state must therefore be considered. The reduced dimension of the observation process is usually not sufficient, because apart from organizationally retarded cases not only the p but also the q leading canonical correlations are non-zero. The observation process is not necessarily Markovian and therefore the amount of information that the past provides about the future usually cannot be "stored" in the p-dimensional present state. However, because of strict-sense stationarity of the state process, all ρ_i's are less than one. The canonical correlations ρ_i's should not be confused with the ordinary correlations ρ_{ij} and ρ'_{ij}, which were introduced in Chapter 2.

As an alternative to the use of the stochastic balanced representation of Desai and Pal (1984), a minimum phase balancing based on the scheme of McGinnie (1994) could be carried out. The minimum phase balancing scheme allows us to find a forward innovation form of the LDS model in which the state covariance matrix Σ^f (Eq. 296) and the covariance matrix G_z (Eq. 297) are equal and diagonal. Let

$$\Lambda_P = \mathrm{Diag}[\sigma_i] \quad 1 \leq i \leq q$$

be this diagonal matrix and σ_i the minimum phase singular values of the dynamical system. Under these circumstances, we simply have

$$I\left[Y_{-\infty}^{-1}; Y_0^{\infty}\right] = -\frac{1}{2}\sum_{i=1}^{q} \log_2\left(1 - {\sigma_i}^2\right). \tag{301}$$

A structurally different implicit formulation for the EMC can be obained when we compute the integral in formula 286 directly. Plugging the results for the joint *pdf*s of the past, the future and the whole observation sequence into the general expression for the EMC from Eq. 286, the ratio of the whole *pdf* to the ones of the past and the future histories is given by:

$$\begin{aligned}\frac{f\left[y_{-\infty}^{\infty}\right]}{f\left[y_{-\infty}^{-1}\right]f\left[y_0^{\infty}\right]} &= \frac{c_{y_{-\infty}^{\infty}}}{c_{y_{-\infty}^{-1}}c_{y_0^{\infty}}} \cdot \frac{\sqrt{\mathrm{Det}\, C_{-\infty}^{-1}}\sqrt{\mathrm{Det}\, C_0^{\infty}}}{\sqrt{\mathrm{Det}\, C_{-\infty}^{\infty}}}\\ &\cdot \mathrm{Exp}\left[\frac{1}{2}(\mathbf{b}_{-\infty}^{\infty})^{\mathrm{T}}(C_{-\infty}^{\infty})^{-1}\mathbf{b}_{-\infty}^{\infty} - \frac{1}{2}(\mathbf{b}_{-\infty}^{-1})^{\mathrm{T}}(C_{-\infty}^{-1})^{-1}\mathbf{b}_{-\infty}^{-1} - \frac{1}{2}(\mathbf{b}_0^{\infty})^{\mathrm{T}}(C_0^{\infty})^{-1}\mathbf{b}_0^{\infty}\right]\\ &= c_1 \cdot \mathrm{Exp}\left[\frac{1}{2}(\mathbf{y}_{-\infty}^{\infty})^{\mathrm{T}}\mathcal{B}\mathbf{y}_{-\infty}^{\infty}\right].\end{aligned} \tag{302}$$

The constant c_1 is defined accordingly. As we can write

$$\mathbf{b}_{t_1}^{t_2} = \left(I \otimes H^{\mathrm{T}}V^{-1}\right)\mathbf{y}_{t_1}^{t_2},$$

we defined the covariance matrix

$$\mathcal{B} = \left(I \otimes V^{-1}H\right)\left(\left(C_{-\infty}^{\infty}\right)^{-1} - \begin{bmatrix}\left(C_{-\infty}^{-1}\right)^{-1} & 0\\ 0 & \left(C_0^{\infty}\right)^{-1}\end{bmatrix}\right)\left(I \otimes H^{\mathrm{T}}V^{-1}\right).$$

Inserting Eq. 302 into the general Eq. 286 leads to

$$\begin{aligned} I\left[Y_{-\infty}^{-1};Y_0^{\infty}\right] &= \int f\left[y_{-\infty}^{-1},y_0^{\infty}\right]\log_2 c_1 \mathrm{Exp}\left[\frac{1}{2}\left(\mathbf{y}_{-\infty}^{\infty}\right)^{\mathrm{T}}\mathcal{B}\mathbf{y}_{-\infty}^{\infty}\right]dy_{-\infty}^{\infty} \\ &= \int f\left[y_{-\infty}^{-1},y_0^{\infty}\right]\log_2 c_1 dy_{-\infty}^{\infty} + \frac{1}{\ln 2}\int f\left[y_{-\infty}^{-1},y_0^{\infty}\right]\left(\mathbf{y}_{-\infty}^{\infty}\right)^{\mathrm{T}}\mathcal{B}\mathbf{y}_{-\infty}^{\infty}dy_{-\infty}^{\infty}. \end{aligned}$$

Using the fact that the joint *pdf* is normalized to one and some well-known results for Gaussian integrals, we obtain

$$I\left[Y_{-\infty}^{-1};Y_0^{\infty}\right] = \log_2 c_1 + \frac{1}{\ln 2}\mathrm{Tr}\left[\mathcal{B}\left(\mathcal{V}_{-\infty}^{\infty} - \mathcal{B}_{-\infty}^{\infty}\right)^{-1}\right], \tag{303}$$

where

$$\mathcal{B}_{-\infty}^{\infty} = \left(I \otimes V^{-1}H\right)\left(\mathcal{C}_{-\infty}^{\infty}\right)^{-1}\left(I \otimes H^{\mathrm{T}}V^{-1}\right).$$

Using the Woodbury matrix identity (Higham 2002, Eq. 148)

$$(A + UCV)^{-1} = A^{-1} - A^{-1}U\left(C^{-1} + VA^{-1}U\right)^{-1}VA^{-1},$$

we can calculate

$$\begin{aligned} \left(\mathcal{V}_{-\infty}^{\infty} - \mathcal{B}_{-\infty}^{\infty}\right)^{-1} &= \left(\left(I \otimes V^{-1}\right) - \left(I \otimes V^{-1}H\right)\left(\mathcal{C}_{-\infty}^{\infty}\right)^{-1}\left(I \otimes H^{\mathrm{T}}V^{-1}\right)\right)^{-1} \\ &= \left(I \otimes V^{-1}\right)^{-1} - \left(I \otimes V^{-1}\right)^{-1}\left(I \otimes V^{-1}H\right) \\ &\cdot\left(-\mathcal{C}_{-\infty}^{\infty} + \left(I \otimes H^{\mathrm{T}}V^{-1}\right)\left(I \otimes V^{-1}\right)^{-1}\left(I \otimes V^{-1}H\right)\right)^{-1} \\ &\cdot\left(I \otimes H^{\mathrm{T}}V^{-1}\right)\left(I \otimes V^{-1}\right)^{-1}. \end{aligned}$$

Using the identities $(A \otimes B)^{-1} = A^{-1} \otimes B^{-1}$ and $(A \otimes B)(C \otimes D) = AC \otimes BD$, we find

$$\begin{aligned} \left(\mathcal{V}_{-\infty}^{\infty} - \mathcal{B}_{-\infty}^{\infty}\right)^{-1} &= I \otimes V + (I \otimes H)\left(\mathcal{C}_{-\infty}^{\infty} - I \otimes H^{T}V^{-1}H\right)^{-1}\left(I \otimes H^{\mathrm{T}}\right) \\ &= I \otimes V + (I \otimes H)\left((\mathcal{C}_2)_{-\infty}^{\infty}\right)^{-1}\left(I \otimes H^{\mathrm{T}}\right). \end{aligned}$$

We note that the above expression equals the inverse of the covariance of the observed states $y_{-\infty}^{\infty}$.

The first part of the constant c_1 can be evaluated directly:

$$c_1 = \sqrt{\frac{\mathrm{Det}\ \Sigma}{\mathrm{Det}\ C}} \cdot \frac{\sqrt{\mathrm{Det}\ \mathcal{C}_{-\infty}^{-1}}\sqrt{\mathrm{Det}\ \mathcal{C}_0^{\infty}}}{\sqrt{\mathrm{Det}\ \mathcal{C}_{-\infty}^{\infty}}}, \tag{304}$$

whereas the second part containing the determinants can be solved as follows: first, we observe that we have

$$\sqrt{\text{Det } C_{-\infty}^{-1}}\,\sqrt{\text{Det } C_0^{\infty}} = \sqrt{\text{Det } \tilde{C}_{-\infty}^{\infty}},$$

where the covariance $\tilde{C}_{-\infty}^{\infty}$ is given by

$$\tilde{C}_{-\infty}^{\infty} = \begin{pmatrix} C_{-\infty}^{-1} & \begin{matrix} 0 & \cdots & 0 \\ 0 & & \vdots \\ A & 0 & 0 \end{matrix} \\ 0 & C_0^{\infty} \end{pmatrix}$$

$$= \begin{pmatrix} B_1 & A & & & & & & & & \\ A^{\mathrm{T}} & B & \ddots & & & & & & & \\ & \ddots & \ddots & \ddots & & & & & & \\ & & \ddots & B & A & & & & & \\ & & & A^{\mathrm{T}} & B_{\Delta t} & A & & & & \\ & & & & & B_1 & A & & & \\ & & & & & A^{\mathrm{T}} & B & \ddots & & \\ & & & & & & \ddots & \ddots & \ddots & \\ & & & & & & & \ddots & B & A \\ & & & & & & & & A^{\mathrm{T}} & B_{\Delta t} \end{pmatrix}.$$

$\tilde{C}_{-\infty}^{\infty}$ differs from $C_{-\infty}^{\infty}$ given by

$$C_{-\infty}^{\infty} = \begin{pmatrix} B_1 & A & & & & & & & \\ A^{\mathrm{T}} & B & \ddots & & & & & & \\ & \ddots & \ddots & \ddots & & & & & \\ & & \ddots & B & A & & & & \\ & & & A^{\mathrm{T}} & B & A & & & \\ & & & & A^{\mathrm{T}} & B & A & & \\ & & & & & A^{\mathrm{T}} & B & \ddots & \\ & & & & & & \ddots & \ddots & \ddots \\ & & & & & & & \ddots & B & A \\ & & & & & & & & A^{\mathrm{T}} & B_{\Delta t} \end{pmatrix}$$

only in three blocks in the center. If we introduce a vector $e_t = (0\cdots 010\cdots 0)^{\mathrm{T}}$ which has only a one at the position corresponding to time step t, then $e_i e_j^{\mathrm{T}}$ is a matrix that has zeros everywhere except at postion i,j. Accordingly, $\left(e_i e_j^{\mathrm{T}}\right) \otimes A$ is a block-matrix where only the block i,j contains the matrix A. With these prerequisites we can write

$$\tilde{C}^{\infty}_{-\infty} = C^{\infty}_{-\infty} + \left(e_0 e_0^{\mathrm{T}}\right) \otimes \begin{pmatrix} (B_{\Delta t} - B) & 0 \\ -A^{\mathrm{T}} & B_1 - B \end{pmatrix},$$

where e_0 is chosen to select the central four blocks of $\tilde{C}^{\infty}_{-\infty}$. Using the identity for Kronecker products

$$AC \otimes BD = (A \otimes B)(\mathrm{C} \otimes \mathrm{D})$$

we get

$$\tilde{C}^{\infty}_{-\infty} = C^{\infty}_{-\infty} + \left(e_0 \otimes \begin{pmatrix} (B_{\Delta t} - B) & 0 \\ -A^{\mathrm{T}} & B_1 - B \end{pmatrix}\right)\left(e_0^{\mathrm{T}} \otimes I_{2q}\right).$$

Using Sylvester's determinant theorem, which states that for matrices $A \in \mathbb{R}^{m\times n}$, $B \in \mathbb{R}^{n\times m}, X \in \mathbb{R}^{m\times m}$ it holds that

$$\mathrm{Det}\ (X + AB) = \mathrm{Det}\ (X)\mathrm{Det}\ \left(I_n + BX^{-1}A\right),$$

we obtain

$$\mathrm{Det}\left(\tilde{C}^{\infty}_{-\infty}\right) = \mathrm{Det}\left(C^{\infty}_{-\infty}\right)\mathrm{Det}\left(I_{2q} + \left(e_0^{\mathrm{T}} \otimes I_{2q}\right)\left(C^{\infty}_{-\infty}\right)^{-1}\left(e_0 \otimes \begin{pmatrix} (B_{\Delta t} - B) & 0 \\ -A^{\mathrm{T}} & B_1 - B \end{pmatrix}\right)\right).$$

In the second term of the right determinant, only the central blocks of $C^{\infty}_{-\infty}$ contribute. They are denoted as

$$\left(\left(C^{\infty}_{-\infty}\right)^{-1}\right)_{i=\{-1,0\},\, j=\{-1,0\}} := \begin{pmatrix} X_0 & X_1 \\ X_1^{\mathrm{T}} & X_0 \end{pmatrix}.$$

Note that $C^{\infty}_{-\infty}$ is symmetric and that the blocks along the diagonal are contant in the asymptotic regime. Finally, we get

$$\frac{\sqrt{\mathrm{Det}\ C^{-1}_{-\infty}}\sqrt{\mathrm{Det}\ C^{\infty}_{0}}}{\sqrt{\mathrm{Det}\ C^{\infty}_{-\infty}}} = \sqrt{\mathrm{Det}\left(I_{2q} + \begin{pmatrix} X_0 & X_1 \\ X_1^{\mathrm{T}} & X_0 \end{pmatrix}\begin{pmatrix} B_{\Delta t} - B & 0 \\ -A^{\mathrm{T}} & B_1 - B \end{pmatrix}\right)}. \tag{305}$$

Using the fact that $B_1 - B = \Sigma^{-1} - C^{-1}, B_{\Delta t} - B = -A_0^{\mathrm{T}} C^{-1} A_0$, and $A^{\mathrm{T}} = -C^{-1} A_0$ an expression for the constant c_1 in terms of the system matrices then reads:

$$c_1 = \sqrt{\frac{\mathrm{Det}\,\Sigma}{\mathrm{Det}\,C}} \sqrt{\mathrm{Det}\left(I_{2q} + \begin{pmatrix} (X_1 - X_0 A_0^{\mathrm{T}})C^{-1}A_0 & X_1(\Sigma^{-1} - C^{-1}) \\ (X_0 - X_1^{\mathrm{T}} A_0^{\mathrm{T}})C^{-1}A_0 & X_0(\Sigma^{-1} - C^{-1}) \end{pmatrix}\right)}. \quad (306)$$

Next, we turn to the second term in the expression for the EMC, Eq. 303. The matrix $\mathcal{B}$ consists of the difference of $(\mathcal{C}_{-\infty}^{\infty})^{-1}$ for the whole time axis and $(\mathcal{C}_{-\infty}^{-1})^{-1}$ $+(\mathcal{C}_0^{\infty})^{-1}$ for the past and the future of the observed process. Evidently, the matrix for the whole time history $(\mathcal{C}_{-\infty}^{\infty})^{-1}$ coincides with the one for the past $(\mathcal{C}_{-\infty}^{\infty})^{-1}$ at least from the infinite past until a certain point of time in the past, where the later matrix elements are still in the asymptotic regime. For later times until $t = -1$, the matrix $(\mathcal{C}_{-\infty}^{-1})^{-1}$ is characterized by the transition regime due to the transient phase in the recursions for $\mathbf{N}_j$. This is illustrated in Fig. 4.3 where the transition regimes are shown in yellow. Similarly, the matrix corresponding to the future observations $(\mathcal{C}_0^{\infty})^{-1}$ deviates significantly from the one for the whole time history only in the beginning and up to some finite point in time in the future (we assumed that the whole process is in steady state and that the covariances of the initial states are equal). Furthermore, the matrix elements decay exponentially in the direction perpendicular to the diagonal. Therefore, the contributions to the sum in Eq. 303 only come from the finite area enclosed by the red lines in Fig. 4.3.

Altogether, in order to numerically compute the EMC given the system matrices, one has to calculate the invers matrices of $\mathcal{C}$ and $\mathcal{C}_2$ for some sufficiently large order Δt, where the asymptotic regime has been reached in the center. The corresponding matrix elements can then be plugged directly into the final result.

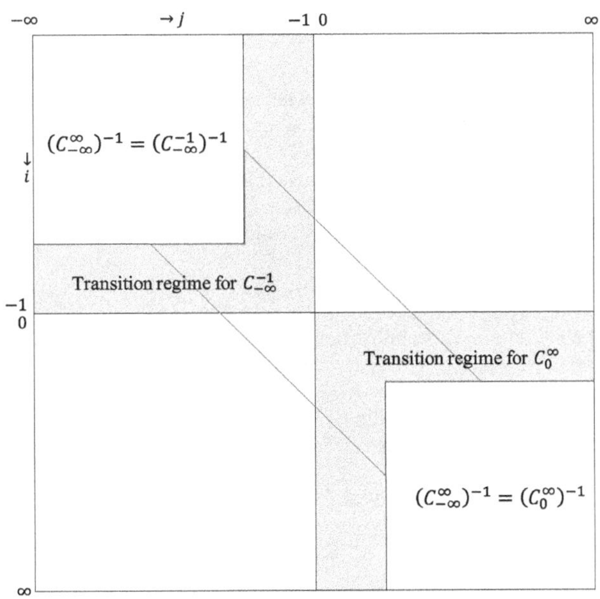

Fig. 4.3 Structure of inverses of $\mathcal{C}_{-\infty}^{\infty}$, $\mathcal{C}_{-\infty}^{-1}$, and $\mathcal{C}_0^{\infty}$: only in the area enclosed by the red lines do the inverses differ and contribute to the *EMC*

Using some of the results obtained so far, we can now simplify the general result Eq. 286: We use the results for the normalization of the joint *pdf*, which we obtained by integrating over the hidden states. The ratio of the normalization constants of the joint *pdf*'s was denoted as

$$c_1 = \frac{c_{y_{-\infty}^{\infty}}}{c_{y_{-\infty}^{-1}} c_{y_0^{\infty}}} \cdot \frac{\sqrt{\text{Det } C_{-\infty}^{-1}} \sqrt{\text{Det } C_0^{\infty}}}{\sqrt{\text{Det } C_{-\infty}^{\infty}}}.$$

Alternatively, the ratio can be expressed directly in terms of the determinants of the corresponding covariances:

$$c_1 = \frac{\sqrt{\text{Det } (C_y)_{-\infty}^{-1}} \sqrt{\text{Det } (C_y)_0^{\infty}}}{\sqrt{\text{Det } (C_y)_{-\infty}^{\infty}}}. \tag{307}$$

Therefore, we can finally write the closed-form solution as

$$\begin{aligned} I\left[Y_{-\infty}^{-1}; Y_0^{\infty}\right] &= \log_2 c_1 \\ &= \log_2 \sqrt{\frac{\text{Det } \Sigma}{\text{Det } C}} \sqrt{\text{Det } \left(I_{2q} + \begin{bmatrix} (X_1 - X_0 A_0^{\mathrm{T}}) C^{-1} A_0 & X_1 (\Sigma^{-1} - C^{-1}) \\ (X_0 - X_1^{\mathrm{T}} A_0^{\mathrm{T}}) C^{-1} A_0 & X_0 (\Sigma^{-1} - C^{-1}) \end{bmatrix}\right)} \\ &= \frac{1}{2} \log_2 \frac{\text{Det } \Sigma}{\text{Det } C} \text{Det } \left(I_q + X_0 (\Sigma^{-1} - C^{-1})\right) \\ &\cdot \text{Det } \left(I_q + (X_1 - X_0 A_0^{\mathrm{T}}) C^{-1} A_0 - X_1 \left(X_0 + (\Sigma^{-1} - C^{-1})\right)^{-1} (X_0 - X_1^{\mathrm{T}} A_0^{\mathrm{T}}) C^{-1} A_0\right). \end{aligned} \tag{308}$$

References

Ay, N., Bernigau, H., Der, R., Prokopenko, M.: Information driven self-organization: the dynamic system approach to autonomous robot behavior. Theory Biosci. Special issue on Guided Self-Organisation (GSO-2010), **131**, 161–179 (2012)

Ball, R.C., Diakonova, M., MacKay, R.S.: Quantifying emergence in terms of persistent mutual information. (2010). Advances in Complex Systems **13**(1), 327–338 (2010)

Bialek, W.: Some background on information theory. Unpublished Working Paper. Princeton University, (2003)

Bialek, W., Nemenman, I., Tishby, N.: Predictability, complexity and learning. Neural Comput. **13** (1), 2409–2463 (2001)

Billingsley, P.: Probability and Measure, 3rd edn. Wiley, New York, NY (1995)

Boets, J., de Cock, K., de Moor, B.: A mutual information based distance for multivariate Gaussian processes. In: Chiuo, A., et al. (eds.) Modeling, Estimation and Control, LNCIS 364, pp. 15–33. Springer, Berlin (2007)

Brockwell, P.J., Davis, R.A.: Time Series: Theory and Methods, 2nd edn. Springer, New York, NY (1991)

Cohen, J.: Statistical Power Analysis for the Behavioral Sciences, 2nd edn. Lawrence Erlbaum Associates, Mahwah, NJ (1988)

Cover, T.M., Thomas, J.A.: Elements of Information Theory. John Wiley and Sons, New York, NY (1991)

Creutzig, F.: Sufficient encoding of dynamical systems—from the grasshopper auditory system to general principles. Ph.D. thesis, Humboldt-Universität zu Berlin (2008)

Crutchfield, J.P., Ellison, C.J., James, R.G., Mahoney, J.R.: Synchronization and control in intrinsic and designed computation: an information-theoretic analysis of competing models of stochastic computation. Santa Fe Institute Working Paper 2010-08-015, (2010)

Crutchfield, J.P., Ellison, C.J., Riechers, P.M.: Exact complexity: the spectral decomposition of intrinsic computation. Santa Fe Institute Working Paper 2013-09-028, (2013)

de Cock, K.D.: Principal angles in system theory, information theory and signal processing. Ph.D. thesis, Katholieke Universiteit Leuven (2002)

DelSole, T., Tippett, M.K.: Predictability: recent insights from information theory. Rev. Geophys. **45**, RG4002 (2007)

Desai, U.N., Pal, D.: A transformation approach to stochastic model reduction. IEEE Trans. Autom. Control **29**(12), 1097–1100 (1984)

Ellison, C.J., Mahoney, J.R., Crutchfield, J.P.: Prediction, retrodiction, and the amount of information stored in the present. Santa Fe Institute Working Paper 2009-05-017, (2009)

Grassberger, P.: Toward a quantitative theory of self-generated complexity. Int. J. Theor. Phys. **25** (9), 907–938 (1986)

Higham, N.J.: Accuracy and Stability of Numerical Algorithms, 2nd edn. Society for Industrial and Applied Mathematics (SIAM), Philadelphia, PA (2002)

Horn, R.A., Johnson, C.R.: Matrix Analysis. Cambridge University Press, Cambridge U.K (1985)

Hotelling, H.: The most predictable criterion. J. Educ. Psychol. **26**, 139–142 (1935)

James, R., Ellison, C.J., Crutchfield, J.P.: Anatomy of a bit: information in a time series observation. Santa Fe Institute Working Paper 2011-05-019, (2011)

Kraskov, A., Stögbauer, H., Grassberger, P.: Estimating Mutual Information. (2004). arXiv:cond-mat/0305641v1 [cond-mat.stat-mech]

Krattenthaler, C.: Advanced determinant calculus: A complement. Linear Algebra Appl. **411**(2), 68–166 (2005)

Lancaster, P., Tismenetsky, M.: The Theory of Matrices, 2nd edn. Academic, Orlando, FL (1985)

Li, L.: Some notes on mutual information between past and future. J. Time Ser. Anal. **27**(2), 309–322 (2006)

Li, L., Xie, Z.: Model selection and order determination for time series by information between the past and the future. J. Time Ser. Anal. **17**(1), 65–84 (1996)

Lütkepohl, H.: Comparison of criteria for estimating the order of a vector autoregressive process. J. Time Ser. Anal. **6**, 35–52 (1985)

Lütkepohl, H.: New Introduction to Multiple Time Series Analysis. Springer, Berlin (2005)

Marzen, S., Crutchfield, J.P.: Circumventing the curse of dimensionality in prediction: causal rate-distortion for infinite-order Markov processes. Santa Fe Institute Working Paper 2014-12-047, (2014)

McGinnie, B.P.: A balanced view of system identification. PhD thesis, Cambridge University, Cambridge, UK (1994)

Mori, T., Fukuma, N., Kuwahara, M.: On the discrete Lyapunov matrix equation. IEEE Trans. Autom. Control **27**(2), 463–464 (1982)

Neumaier, A., Schneider, T.: Estimation of parameters and Eigenmodes of multivariate autoregressive models. ACM Trans. Math. Softw. **27**, 27–57 (2001)

Papoulis, A., Pillai, S.U.: Probability, Random Variables and Stochastic Processes. McGraw-Hill, Boston, MA (2002)

Puri, N.N.: Fundamentals of Linear Systems for Physical Scientists and Engineers. CRC Press, Boca Raton, FL (2010)

Schneider, T., Griffies, S.M.: A conceptual framework for predictability studies. J. Clim. **12**, 3133–3155 (1999)

Still, S.: Information bottleneck approach to predictive inference. Entropy **16**(1), 968–989 (2014)

Van Overschee, P., de Moor, B.: Subspace Identification for Linear Systems: Theory, Implementations, Applications. Kluwer Academic Publishers, Boston, MA (1996)

Chapter 5
Validity Analysis of Selected Closed-Form Solutions for Effective Measure Complexity

Following our comprehensive and unified treatment of emergent complexity based on information theory and the application of information-theoretic methods associated with complexity measures, we will now analyze the validity of closed-form solutions for the effective measure complexity (EMC) that were obtained for vector autoregression models as the basic mathematical representation of cooperative work in PD projects (see Sections 2.2, 2.3, 2.4 and 2.6 in conjunction with Section 4.1). In the validation studies we not only investigated "flat" project organization forms but also analyzed work processes with periodically correlated components due to a multi-level hierarchical coordination structure. It is well established that validity is one of the most influential concepts in industrial engineering and engineering management because questions concerning its nature and scope influence everything from the design of project organization for a PD project to the application and evaluation of specific design criteria. In this context we follow classic validity theory and distinguish between criterion-related, content-related, and construct-related validity (Salkind and Rasmussen 2007). Criterion-related validity refers to the extent to which a measure—EMC in our case—predicts the values of another measure, for instance the total time taken to complete particular development activities in a PD project (Eaves and Woods-Grooves 2007, cf. Section 5.2). The first measure is usually called the predictor variable. We have dubbed the second measure the criterion variable because our extensive analysis in Section 3.2.4 has already shown that EMC is theoretically valid. The literature distinguishes between two types of criterion-related validity (see e.g. Eaves and Woods-Grooves 2007): (1) predictive validity and (2) concurrent validity. The distinctive factor here is the time interval between obtaining the first and the second set of measurements. For predictive validity, the data related to the criterion variable is collected some time after the data for the predictor variable. For concurrent validity, the data from both variables is collected at about the same time in the same experiment.

© Springer International Publishing Switzerland 2016

C. Schlick, B. Demissie, *Product Development Projects*, Understanding Complex Systems, DOI 10.1007/978-3-319-21717-8_5

5.1 Selection of Predictive Models of Cooperative Work

First, we focus on concurrent validity. Before presenting the methods and results of the studies of concurrent validity in Section 5.1.2, we must first lay additional theoretical foundations on model selection. These foundations build primarily on the work of Li and Xie (1996) on the principle of "minimal mutual information." Although this principle was developed independently from the complexity-theoretic measures of Grassberger (1986) and other researchers in this area, it is very closely related to EMC as the mutual information communicated from the infinite past to the infinite future by the stochastic process (see Section 3.2.4) is evaluated to select the class of parametric models under certain constraints. Furthermore, the same principle can be used to formulate a model selection criterion that is based on EMC estimates from data and can be directly applied to identify the optimal model order within a preferred class (Section 5.1.1). The principle and its instantiations have universal reach and its scope of application extends beyond selecting an optimal model for project management activities. The validation studies themselves are much narrower in their focus and based on Monte Carlo experiments. As already demonstrated in the simulation study from Section 2.8, Monte Carlo experiments are a special class of algorithms that are based on repeated random sampling of the work processes to compute their outcomes. The repeated random sampling was carried out in a Mathematica® modeling and simulation environment. After presenting the methods used in the validation studies (Section 5.1.2), we will present and discuss the results of two Monte Carlo studies (Section 5.1.3). The studies had the same overall objective but used different parametric model forms. The overall objective was to compare the accuracy of the EMC-based model selection criterion with standard criteria like the (original and bias corrected) Akaike information criterion (see Sections 2.4 and 3.2.2) and the Schwarz-Bayes criterion (see Sections 2.3, 2.10 and 3.2.2). We hypothesized that model selection based on EMC allows us to select the true model order with high accuracy and that the histogram distributions of the selected model orders are not significantly different from the distributions obtained under the alternative criteria. To evaluate this hypothesis we focused on the class of VAR models, which were analyzed and discussed in detail in Sections 2.2, 2.4 and 4.1. The parametric model forms were not only derived from field data of a PD project in a small industrial company, but were also synthetically generated so that we could systematically compare the concurrent validity of the different model selection criteria. In the first study, we used the specific VAR(1) and VAR(2) model representations from Section 2.5 to simulate the cooperative task processing of the developers in a sensor development project (Schlick et al. 2013). In the second study, we partially replicated Lütkepohl's experimental setup (Lütkepohl 1985) and analyzed 1000 bivariate VAR(1) and 1000 bivariate VAR(2) models with complex conjugate and distinct real characteristic roots. We also investigated 400 trivariate (i.e. three-dimensional) VAR(1) models. By systematically evaluating a total of 2400 models we were able to consider a large variety of parameter sets corresponding to a wide range of characteristic roots in different regions of the unit circle and thus generalize the model selection results (Schlick et al. 2013).

5.1.1 Principles for Model Selection

As already discussed in Sections 2.4 and 3.2.2 in relation to VAR models and Markov chains, a central aspect in model selection tasks within a predefined class of statistical models is to find a good trade-off between the predictive accuracy gained by increasing the number of freely estimated parameters, and the danger of overfitting the model to non-predictable fluctuations and not regularities in the observed processes that generalize to other datasets. As mentioned earlier in the literature review in Section 3.2.2, Akaike (1973, 1974) developed the first rigorous approach along this line of thought based on the expected Kullback-Leibler information. His famous AIC criterion (see Sections 2.4 and 3.2.2) represents the asymptotic bias correction term of the maximized log-likelihood from each approximating model to full reality. Akaike's fundamental ideas were further developed by Rissanen in a series of papers and books starting from 1978 (see Section 3.2.2). He emphasizes that fitting a statistical model to data is equivalent to finding an efficient encoding. We therefore need to measure both the code length required to describe the deviations in the data from the model's predictions, and the code length required to specify the model's structure and independent parameters (Bialek et al. 2001). Without explicit links to complexity theory, Schwarz developed his now famous criterion (Schwarz-Bayes criterion, or Bayesian information criterion, BIC, see Sections 2.4, 2.10 and 3.2.2) as an asymptotic approximation to a transformation of the Bayesian posterior probability of an approximating model. When the sample size is large, this criterion favors a model that ideally corresponds to the approximating model, which is most probable a posteriori.

To underline the specific properties of EMC, we have to leave the pragmatic considerations from Sections 2.4, 2.7 and 2.10 behind and interpret model selection from a theoretical perspective developed in the classic works of Jaynes (1957, comprehensively described in his posthumous book edited by Bretthorst and published in 2003), Li and Xie (1996) and others on universal information-theoretic principles. Their works show that model selection has not one but two central aspects: (1) selection of the model class based on the preferred principle, (2) selection of the optimal model order within that class using the same or at least similar principles. Our previous considerations focused exclusively on the second aspect and did not address model class selection as this is, in a sense, a more abstract decision problem for finding a theoretically optimal representation of the regularities of an entity that can be generalized without posing unjustified constraints on the internal configuration. Jaynes' famous principle of maximum entropy can be applied in the first step to select the model class that maximizes the entropy subject to given constraints on certain statistics on the observations from the process such as the expectation values, autocovariances (see e.g. 309), or any other function for which the samples provide a reliable estimate. The principle assumes that we base our decision on exactly stated prior information and that the set of all trial probability distributions is considered that would adequately encode the prior information. From all continuous prior distributions we seek a model probability density which is normalized and will exactly reproduce the measured expectation values (or autocovariances etc.), but will otherwise be as unstructured, or random, as possible—hence maximum entropy (Tkačik and Bialek 2014). According to the principle, of the densities considered, the one with the maximal entropy is the proper density and therefore the differential entropy according to Eq. 232 is to be maximized. As such, the principle selects a probability density function that makes the "least claim" to being informative beyond the stated prior and therefore admits the most ignorance about the dynamical dependency structure of the process beyond the stated prior (sensu Jaynes 2003). In practice, this amounts to solving a variational problem for the model probability density. The resulting maximum entropy distribution has an exponential form (see e.g. Tkačik and Bialek, 2014). The principle of maximum entropy does not correspond to a single model for the process, but rather provides a systematic method for building a hierarchy of models that provide increasingly good approximations to the true distribution. At each step in the hierarchy, the models are parsimonious, having the minimal dynamical dependency structure required to reproduce certain statistics on the data that one is trying to match (Tkačik and Bialek, 2014).

An analogous principle is the principle of "minimum mutual information" (MMI) which is based not on the differential entropy but on the mutual information (see definition in Eq. 234) and was formulated later by Li and Xie (1996). For time series of state vectors, the principle of MMI selects the model that minimizes EMC as an indicator of mutual information between the infinite past-future histories of the process subject to given constraints on autocovariances or other important parameters. If, for example, the values $C_{XX}(0),\ldots,C_{XX}(n)$ of the first $n+1$ autocovariance functions

$$\boldsymbol{C}_{n+1} = \begin{pmatrix} C_{XX}(0) & C_{XX}(1) & \cdots & C_{XX}(n) \\ C_{XX}(1) & C_{XX}(0) & \cdots & C_{XX}(n-1) \\ \vdots & \vdots & \ddots & \vdots \\ C_{XX}(n) & C_{XX}(n-1) & \cdots & C_{XX}(0) \end{pmatrix} \tag{309}$$

of a p-dimensional stationary Gaussian process $\{X_t\}$ are given in the form of a $(n+1)p \times (n+1)p$ $(n \in \mathbb{N})$ Toeplitz matrix $\boldsymbol{C}_{n+1}$ (Eq. 280), then under the principle of MMI the state vector X_t has to satisfy a VAR(n) model (Li and Xie 1996) such as

$$X_t = \sum_{i=0}^{n-1} A_i \cdot X_{t-i-1} + \varepsilon_t.$$

(cf. state equation 58). In other words, the VAR(n) model minimizes EMC with respect to given autocovariances. Note that the matrix elements $C_{XX}(\tau)$ in the Toeplitz matrix are defined as $p \times p$ block autocovariance matrices of the corresponding subspaces for lead times $\tau = 0, 1, \ldots, n$. Using Jaynes' principle of maximum entropy for model class selection under the same constraints on autocovariances will also result in a VAR(n) model. The conceptual difference between both principles is that the principle of MMI favors model classes that do not impose more constraints on the canonical correlations between the infinite past-future histories than are justified by the informational structure of the process independently of the preferred coordinate system (see, e.g., the beginning of Chapter 4 and Section 4.1.3), whilst the principle of maximum entropy lacks invariance under change of variables $x_t \rightarrow y(x_t)$. To do something similar to this principle, a definite group of transformation must be specified, which requires additional knowledge of the human model builder (Jaynes 2003). If, in another case, the l cepstrum coefficients (see de Cock 2002) are given instead of the autocovariances, the MMI principle selects the Bloomfield model BL(l) (see Li and Xie 1996).

In Section 3.2.4 we introduced two key invariants of stochastic processes—Grassberger's EMC (Eq. 216) and Shannon's source entropy rate h_μ (Eq. 218). We also saw that Bialek et al.'s (2001) analysis of the predictive information $I_{pred}(n)$ (Eq. 226) uncovered interesting interrelationships between these two key invariants. Let us recall that the predictive information $I_{pred}(n)$ (Eq. 226) indicates for a strict-sense stationary process the distribution of EMC over the block length n and can be interpreted as the subextensive component of the block entropy $H(n)$ (Eq. 227). The extensive component of the block entropy is driven by the source entropy rate h_μ and it holds that

$$H(n) = nh_\mu + I_{pred}(n).$$

To analyze this relationship from a different perspective, we refer to a theorem of Li (2006) and express the log-likelihood $\ln f_\theta[x_1, \ldots, x_T]$ of T observations from an arbitrary Gaussian stationary process $\{X_t\}$ with density function f_θ with the following sum:

$$\ln f_\theta[x_1, \ldots, x_T] = -\frac{1}{2}\Big(Tp(\ln 2\pi) + \ln\mathrm{Det}[\boldsymbol{C}_T] + (x_1, \ldots, x_T)^{\mathrm{T}} \boldsymbol{C}_T^{-1} (x_1, \ldots, x_T)\Big).$$

$\boldsymbol{C}_T$ is a $Tp \times Tp$ Toeplitz matrix (cf. Eqs. 280 and 309) encoding the autocovariances of the stationary process over T time steps. Li (2006) shows that the entropy rate and the effective measure complexity are—apart from a constant—simply the first and second term in the expansion of $\ln \mathrm{Det}[\boldsymbol{C}_T]$:

$$\ln\mathrm{Det}[\boldsymbol{C}_T] + Tp(\ln 2\pi) = 2Th_\mu + 2EMC + o(1). \tag{310}$$

The term $o(1)$ denotes an arbitrary sequence that remains bounded. It is not difficult to see that the entropy rate and the effective measure complexity denote the extensive and subextensive components of the log-likelihood of the observation sequence in the same manner as the entropy rate and the predictive information represent the extensive and subextensive parts of the block entropy. Hence, EMC and h_μ can be regarded as complementary key invariants not only with respect to the block length but also in the time domain. We will now illustrate their complementarity by applying them to model order selection.

In our information-theoretic considerations on stochastic complexity in Section 3.2.2 we pointed out that Akaike's AIC was originally developed to identify the candidate model that minimizes the expected Kullback-Leibler divergence between the candidate model and the true model. From an engineering perspective, we can interpret the AIC statistic as a penalty term of the entropy rate

$$h_\mu = \lim_{\eta \to \infty} \frac{H[X^{\eta=\eta}]}{\eta}$$

of the stationary process (Eq. 218) and rewrite the criterion informally as

$$AIC(n) = 2\hat{h}_\mu(n) + \text{penalty}, \tag{311}$$

where $\hat{h}_\mu(n)$ denotes the estimate of the entropy rate under the fitted model of order n (Li 2006). Using the closed-form solutions developed in Section 4.1.6, we can now express the estimate of the entropy rate in the class of VAR models explicitly as

$$\hat{h}_\mu(n) = \frac{1}{2}\log_2 \mathrm{Det}\left[\hat{\Sigma}_{(n)}\right] + \frac{p}{2}(\log_2 2\pi e),$$

where

$$\begin{aligned}\hat{\Sigma}_{(n)} &= \left\{\frac{(T-n_p)}{T}\right\} \cdot \hat{C}\\ &= \left\{\frac{1}{T}\right\} \cdot R_{22}^{\mathrm{T}} \cdot R_{22}\end{aligned}$$

denotes, according to Eq. 66, the one-step prediction error matrix of a VAR(n) model that was fitted to data based on least squares estimation for a time series of state vectors X_t indexed by $t = 1 - n, \ldots T$. Alternatively, the maximum likelihood parameters could be used. The index n denotes the model order and p the dimensionality of the state vectors. It holds that $n_p = np$.

In the same manner, Li and Xie (1996) constructed penalized estimates of the EMC. In their work they refer to the mutual information between the infinite past-future histories of the stationary process as simply "past-future information" and denote it with the symbol I_{p-f} (cf. Eq. 216). For finite complexity values we have the following identities (Eqs. 216 and 226)

$$\text{EMC} = I_{p-f} = I\left[X_{-\infty}^{-1}; X_0^{\infty}\right] = \lim_{\eta \to \infty} I_{pred}(n = \eta).$$

A new complexity-based criterion for order selection can be derived by replacing the entropy rate $\hat{h}_\mu(n)$ estimated under an n-th order VAR model by the estimated effective measure complexity $\widehat{EMC}(n)$ and properly adjusting the penalty term. Li and Xie's (1996) development of the criterion leads to an alternative statistic, termed the mutual information criterion, which can be informally written as:

$$MIC(n) = \widehat{EMC}(n) + \text{penalty}. \tag{312}$$

Analogously to $AIC(n)$ and $BIC(n)$ (see Eqs. 67 and 71 for VAR models), the penalty term should take the sample size and the effective number of parameters into account.

Li and Xie (1996) developed the mutual information criterion for one-dimensional autoregressive processes. Their formulation can be easily generalized to p-dimensional VAR(n) models. We arrive at

$$MIC(n) = \begin{cases} 0 & \text{for } n = 0 \\ -2\ \widehat{EMC}(n) + \dfrac{\log_2 T}{T} \dfrac{np(np+1)}{2} & \text{for } n \geq 1. \end{cases} \tag{313}$$

In information-theoretic terms, the order n is considered to be the optimal one if it satisfies:

$$n_{opt} = \text{argmin}_n MIC(n). \tag{314}$$

In Eq. 313 $\widehat{EMC}(n)$ denotes the complexity value that is assigned to the fitted nth-order VAR model. T denotes the last time step of the state vectors indexed by $t = 1 - n, \ldots T$ which are used to estimate the parameters. The second summand in the definition of $MIC(n)$ for $n \geq 1$ is the penalty for models that are unnecessarily complex with respect to the finite sample size. This term serves the same purpose as the penalties formulated for $AIC(n)$ in Eq. 67 and for $BIC(n)$ in Eq. 71. Similarly to the alternative formulations of model selection criteria, the penalty only depends on the sample size, the model order n and the effective number $np(np+1)/2$ of parameters. The effective number of parameters corresponds to the number of freely estimated parameters in the Toeplitz matrix $\boldsymbol{C}_n$ (Eq. 309).

To make the model selection mechanism fully operational, we can apply the closed-form solutions from Section 4.1.6 and express the complexity estimate in matrix form (Eq. 279 in conjunction with Eqs. 62 and 277) as

$$\widehat{EMC}(n) = \frac{1}{2}\log_2\left(\frac{\mathrm{Det}\left[\hat{\Sigma}\right]}{\mathrm{Det}\left[\sum_{i=1}^{n}\left(\hat{A}\right)^{n-i}\cdot\widetilde{C}\cdot\left(\left(\hat{A}\right)^{n-i}\right)^{\mathrm{T}}\right]}\right),$$

where the estimated steady-state covariance matrix $\hat{\Sigma}$ in the numerator is given by the infinite sum

$$\hat{\Sigma} = \sum_{k=0}^{\infty}\left(\hat{A}^{n}\right)^{k}\cdot\left(\sum_{i=1}^{n}\left(\hat{A}\right)^{n-i}\cdot\hat{C}\cdot\left(\left(\hat{A}\right)^{n-i}\right)^{\mathrm{T}}\right)\cdot\left(\left(\hat{A}^{n}\right)^{\mathrm{T}}\right)^{k}.$$

An alternative, more intuitive expression can be obtained on the basis of the (mean squared) one-step prediction errors $\widetilde{\Sigma}_{(i)}$ (Eq. 284):

$$\widehat{EMC}(n) = \frac{1}{2}\sum_{i=0}^{n-1}\log_2\widehat{\widetilde{\Sigma}}_{(i)} - \frac{1}{2}n\log_2\widehat{\widetilde{\Sigma}}_{(n)},$$

where the estimated one-step prediction errors for $i > 0$ are given by (Eq. 282)

$$\widehat{\widetilde{\Sigma}}_{(i)} = \frac{\mathrm{Det}\left[\widehat{\boldsymbol{C}}_{i+1}\right]}{\mathrm{Det}\left[\widehat{\boldsymbol{C}}_{i}\right]}.$$

The zeroth-order prediction error equals the determinant of the estimated autocovariance for zero lead time (Eq. 283):

$$\widehat{\widetilde{\Sigma}}_{(0)} = \mathrm{Det}\left[\widehat{C}_{\widetilde{X}\widetilde{X}}(0)\right].$$

As shown in Section 4.1.6, the alternative expression can also be interpreted using an information-theoretic learning curve $\widehat{\Lambda}(i)$ $(i = 1, 2, \ldots, n)$ in which each summand $1/2\left(\log_2\widehat{\widetilde{\Sigma}}_{(i-1)} - \log_2\widehat{\widetilde{\Sigma}}_{(n)}\right)$ can be used to evaluate the amount of apparent randomness at small model order i, which can be "explained away" by considering parameterized models with increasing order $i+1,\ i+2, \ldots$.

At first glance, the objective function formulated in Eq. 314 in conjunction with Eq. 313 to select the optimal order within the class of VAR models seems to contradict the MMI principle, because it is designed to select the model that maximizes emergent complexity in the meaning of EMC and not a representation that corresponds to minimum mutual information. However, this is only seemingly in contradiction, because each nth-order model defines a map from the process past history $(x_0, x_1, \ldots, x_t)$ to a Gaussian state variable S_t summarizing the properties. We are interested in a state representation that summarizes the properties in such a way that it enables us to make good predictions (sensu Still 2014). As the VAR model's memory only reaches back n time steps into the past because the state is the weighted value of the last n observations, we need efficient mapping to capture as much information about the past as possible to improve predictions of possible future histories $(X_{t+1}, X_{t+2}, \ldots, X_{\infty})$. This mapping can be carried out by either maximizing the predictive information $I_{pred}(n \to \infty)$ (Eq. 226), or equivalently EMC. However, our information-theoretic considerations in Section 4.1.6 have shown that the captured predictive information for VAR(n) models cannot on average be larger than the amount of information corresponding to the model order and it holds that $I_{pred}(n) = I_{pred}(n+1) = \ldots$. In other words, a VAR($n$) model has a limited capacity to store the predictive information communicated over an infinite time horizon. Maximizing the predictive information over all fitted models up to order n thus corresponds to maximizing the predictive accuracy of possible future histories within the range of model capacities which, in turn, requires maximum effective measure complexity values as stated in the objective function. With a finite sample size, the complexity values must be adjusted in the same way as the penalty term.

Li and Xie (1996) also formulated an alternative criterion, called the "least-information criterion," abbreviated as $LIC(n)$. In our own analysis we found that this criterion is not only less rigorous in its derivation but also often leads to lower accuracy in model selection tasks. It is therefore not considered below. For more information, the corresponding proofs and additional background information on both criteria are presented in Li and Xie (1996).

5.1.2 Methods

Two Monte Carlo studies were carried out to investigate the concurrent validity of the mutual information criterion MIC according to Eq. 314 in conjunction with Eq. 313 within the class of VAR models in different model selection tasks. We speak of concurrent validity, because the data for the calculation of the criterion and the estimates of the underlying effective measure complexity are collected simultaneously in the same Monte Carlo experiment. In the first validation study, we used the field data that were acquired in the industrial case study as input for the experiments and followed the procedure that was introduced in Section 2.5 to model and simulate the task processing. Within the industrial multiproject environment, we focused on Project A, the largest project, with ten partially overlapping development tasks spanning the project phases from conceptual design to product documentation. Here, we concentrate on the first two overlapping tasks of this project, and only modeled their overlapping range (see Section 2.5). We used the least square method developed by Neumaier and Schneider (2001) to estimate the parameters of the corresponding bivariate VAR(n) models of different order n. The model selection procedure based on the classic criteria showed that the Bayesian information criterion BIC according to Eq. 71 is minimal for a VAR(1) model (see

Section 2.5). For this model, we obtained the following optimizing parameters (Eqs. 73, 74, 75):

$$\hat{x}_0 = \begin{pmatrix} 0.6016 \\ 1.0000 \end{pmatrix}$$
$$\hat{A}_0 = \begin{pmatrix} 0.9406 & -0.0017 \\ 0.0085 & 0.8720 \end{pmatrix}$$
$$\hat{C} = \begin{pmatrix} (0.0135)^2 & -0.38 \cdot 0.0135 \cdot 0.0416 \\ -0.38 \cdot 0.0135 \cdot 0.0416 & (0.0416)^2 \end{pmatrix}.$$

When using the corrected Akaike information criterion AIC_c according to Eq. 69 instead of the Bayesian information criterion, a VAR(2) model is assigned minimum scores. We focused on AIC_c in the first Monte Carlo study, because it is known to have superior bias properties for small samples. Burnham and Anderson (2002) recommend applying the corrected criterion instead of the classic AIC when it holds that $T/k < 40$ (Eq. 70), as in our case. The impact of the sample size on the bias properties and on selection accuracy was investigated in the second Monte Carlo study. The optimizing parameters for the VAR(2) model are given by the representation (Eqs. 73, 76, 77, 78 and 79):

$$\hat{x}_0 = \begin{pmatrix} 0.6016 \\ 1.0000 \end{pmatrix}$$
$$\hat{x}_1 = \begin{pmatrix} 0.6016 \\ 0.7154 \end{pmatrix}$$
$$\hat{A}_0 = \begin{pmatrix} 1.1884 & -0.1476 \\ 0.0470 & 1.1496 \end{pmatrix}$$
$$\hat{A}_1 = \begin{pmatrix} -0.2418 & 0.1344 \\ -0.0554 & -0.2622 \end{pmatrix}$$
$$\hat{C} = \begin{pmatrix} (0.0116)^2 & -0.013 \cdot 0.0116 \cdot 0.0257 \\ -0.013 \cdot 0.0116 \cdot 0.0257 & (0.0257)^2 \end{pmatrix}.$$

The VAR(1) and VAR(2) models were then used to simulate the processing of the first two development tasks in repeated trials and to generate the data sets required to investigate concurrent validity. To ensure robust parameter estimation, two independent time series of 100 time steps were generated in each trial. These time series were used to estimate the parameters of candidate VAR(n) models of different order based on the cited least square method. The model orders considered were in the range $n_{min} = 1$ and $n_{max} = 6$. The candidate VAR(n) models were then evaluated based on the mutual information criterion $MIC(n)$ from Eq. 313 and the alternative $BIC(n)$ and $AIC_c(n)$ criteria. We hypothesized that the complexity-based $MIC(n)$ is highly accurate and can identify the true model order (either $n = 1$ for the VAR(1) model with parameters according to Eqs. 73, 74, 75, or $n = 2$ for the VAR (2) model represented by Eqs. 73, 76, 77, 78 and 79) from the generated data sets with a level of accuracy that is not significantly different from the alternative criteria in statistical terms. 1000 independent trials were computed for each model to obtain a good statistic for the pairwise comparison of the histogram distributions of the model orders which were selected by the three different criteria. Pearson's chi squared test (see e.g. Field 2009) was used to investigate whether there were significant differences in the histogram distribution. The level of significance in all tests was set to $\alpha = 0.05$.

In the second study of concurrent validity, we partially replicated Lütkepohl's (1985) experimental setup and analyzed a total of 1000 bivariate VAR(1) and 1000 bivariate VAR(2) models with both complex conjugate and distinct real characteristic roots as well as 400 trivariate (i.e. three-dimensional) VAR(1) models with distinct real characteristic roots. In contrast to Lütkepohl's (1985) study, we only allowed for zero means and therefore did not include intercept terms in the models. Furthermore, with respect to the theoretical analysis of cooperative work in PD projects in Chapter 2, we excluded moving average models and focused on purely autoregressive recurrence relations. In general, bivariate VAR models can be used to simulate the processing of two development tasks, whilst three-dimensional models can simulate the processing of three tasks in a compact statistical representation form. The basic idea underlying Lütkepohl's (1985) design of the Monte Carlo experiments is to consider a large variety of parameter sets corresponding to a wide range of characteristic roots in different regions of the unit circle in the data generation. To simplify the interpretation of the results, the data can be regarded as instantiations of different cooperative relationships between developers processing either two or three tasks under varying regimes of cooperative problem solving processes. The aggregated evaluation of the task processing indicates which criterion has the highest accuracy of model selection where no prior information about the location and covariance of the true parameters is available. Note that even if only stable bivariate and trivariate autoregressive processes are considered, the parameter space is theoretically unbounded and therefore only a very small area can be investigated.

The investigated bivariate VAR(1) models were defined by the well-known state equation (cf. Eq. 2):

$$X_t = A_0 \cdot X_{t-1} + \varepsilon_t.$$

Ignoring the error term ε_t and using a parametric representation of the dynamical operator

$$A_0 = \begin{pmatrix} a_{11} & a_{12} \\ a_{21} & a_{11} + \Delta a \end{pmatrix}$$

according to Eq. 54, where $\{a_{11}, \Delta a, a_{12}, a_{21}\} \in \mathbb{R}^+$, we can express the eigenvalues (Eqs. 56 and 57) as roots of the characteristic polynomial by

Table 5.1 Parameters and parameter values of the bivariate $VAR(1)$ and $VAR(2)$ processes investigated in the Monte Carlo experiments (after Lütkepohl 1985)

Bivariate processes					
(a) Processes with complex conjugate characteristic roots					
Parameter	Parameter values				
a_{11}	-1	-0.5	0	0.5	1
a_{21}	-1.5	-0.5	0.5	1.5	
θ		$2\pi k/10$	$k = 0, 1, \ldots, 9$		
r	0.8	0.6	0.4	0.2	
(b) Processes with distinct real characteristic roots					
Parameter	Parameter values				
a_{11}	-1	-0.5	0	0.5	1
a_{21}	-0.5	0.5			
λ_1	-0.8	-0.4	0	0.4	0.8
λ_2	-0.6	-0.2	0.2	0.6	
Covariance matrices $C = I_2$					

$$\begin{aligned}\lambda_1 &= \frac{1}{2}\left(2a_{11} + \Delta a - \sqrt{\Delta a^2 + 4a_{12}a_{21}}\right)\\ \lambda_2 &= \frac{1}{2}\left(2a_{11} + \Delta a + \sqrt{\Delta a^2 + 4a_{12}a_{21}}\right).\end{aligned}$$

Theoretically, all parameter values $\{a_{11}, \Delta a, a_{12}, a_{21}\} \in \mathbb{R}$ for which it holds that $|\lambda_1| < 1$ and $|\lambda_2| < 1$ lead to asymptotically stable processes (see Section 2.1). In the Monte Carlo study, we concentrated on Lütkepohl's (1985) parameter values corresponding to a wide range of characteristic roots in different regions of the unit circle. These values are summarized in Table 5.1.

If the eigenvalues are complex, they can be written as complex conjugate pairs (cf. Eq. 53) and interpreted as a stochastically driven oscillator with non-minimum period (see Section 2.3). Based on the typical notation

$$\begin{aligned}\lambda_1 &= r(\cos(\theta) + i\ \sin(\theta))\\ \lambda_2 &= r(\cos(\theta) - i\ \sin(\theta)),\end{aligned}$$

it is easy to see that the matrix entries Δa and a_{12} can be expressed as

$$\begin{aligned}\Delta a &= 2r\cos(\theta)\\ a_{12} &= \frac{2a_{12}r\cos(\theta) - a_{11}^2 - r^2}{a_{21}}\end{aligned}$$

provided $a_{21} \neq 0$ (Lütkepohl 1985). If, on the other hand, both eigenvalues are real, the matrix entries Δa and a_{12} are given by

$$\begin{aligned}\Delta a &= \lambda_1 + \lambda_2 - 2a_{11}\\ a_{12} &= \frac{\lambda_1^2 - \lambda_1(2a_{11} + \Delta a) + a_{11}(a_{11} + \Delta a)}{a_{21}}\end{aligned}$$

provided $a_{21} \neq 0$ (Lütkepohl 1985). In this case the eigenvalues represent either relaxators or stochastically driven damped oscillator with minimum period (see Section 2.3). Once the values of λ_1, λ_2, a_{11} and a_{21} are given, the matrix entries Δa and a_{12} can be calculated using the above equations.

The complete set of parameters and parameter values for all 1000 bivariate VAR (1) processes is shown in Table 5.1. As shown in Table 5.1, 800 VAR(1) models with complex conjugate characteristic roots and 200 VAR(1) models with real distinct characteristic roots were investigated in the second Monte Carlo study.

The Monte Carlo experiments for the bivariate VAR(2) models were carried out in a similar way. To ensure comparability, the bivariate VAR(2) models were defined by a recurrence relation with the same dynamical operator A_0 and a state variable X_t that is determined by the prior state of the process reaching back two time steps into the past and ignoring the previous time step:

$$X_t = A_0 \cdot X_{t-2} + \varepsilon_t.$$

The same parameter values from Table 5.1 were applied to the VAR(2) model representation, meaning that 800 VAR(2) models with complex conjugate characteristic roots and 200 VAR(2) models with real distinct characteristic roots were investigated.

In all Monte Carlo experiments the error process ε_t corresponded to a white noise process. The covariance matrix of ε_t was equal to the identity matrix, so that the variances and standard deviations equal 1 and the covariance equals zero:

$$C = \begin{pmatrix} 1 & 0 \\ 0 & 1 \end{pmatrix}.$$

The investigated three-dimensional VAR(1) models with distinct real characteristic roots are based on the parametric representation

$$A_0 = \begin{pmatrix} a_{11} & a_{12} & a_{13} \\ a_{21} & a_{22} & a_{23} \\ a_{31} & a_{32} & a_{33} \end{pmatrix}$$

of the dynamical operator, where $\{a_{ij}\} \in \mathbb{R}$ $(1 \leq i, j \leq 3)$. The eigenvalues of A_0 are the roots of the characteristic polynomial

$$\lambda^3 + \lambda^2(-a_{11} - a_{22} - a_{33}) + \lambda(-a_{12}a_{21} + a_{11}a_{22} - a_{13}a_{31} - a_{23}a_{32} + a_{11}a_{33} + a_{22}a_{33}) \\ + a_{13}a_{22}a_{31} - a_{12}a_{23}a_{31} - a_{13}a_{21}a_{32} + a_{11}a_{23}a_{32} + a_{12}a_{21}a_{33} - a_{11}a_{22}a_{33}.$$

All parameter values $\{a_{ij}\} \in \mathbb{R}$ for which it holds that $|\lambda_1| < 1$, $|\lambda_2| < 1$ and $|\lambda_3| < 1$ correspond to asymptotically stable processes (see Section 2.1). As before, we used Lütkepohl's (1985) parametric representation and parameter values in the Monte Carlo study to cover a wide range of characteristic roots in different regions of the unit circle. The parameters and parameter values are summarized in Table 5.2.

Table 5.2 Parameters and parameter values of the trivariate *VAR*(1) processes investigated in the Monte Carlo experiments (after Lütkepohl 1985)

Three-dimensional processes (only real characteristic roots $\lambda_1, \lambda_2, \lambda_3$)					
Parameter	Parameter values				
a_{11}	−0.5				
a_{12}	0.5				
a_{21}	0.5				
a_{22}	0.5				
a_{31}	−0.5	0.5			
a_{32}	−1	−0.5	0	0.5	1
λ_1	−0.6	−0.2	0.2	0.6	
λ_2	−0.8	−0.4	0	0.4	0.8
λ_3	−0.5	0.5			
Covariance matrix $C = I_3$					

In the parametric representation of Table 5.2, the matrix entries a_{13}, a_{23} and a_{33} can be determined from given eigenvalues λ_1, λ_2 and λ_3 on the basis of the following three equations:

$$\begin{aligned}
a_{13} = 1/\big(-a_{12}\,a_{31}^2 + a_{32}((a_{11}-a_{22})a_{31} + a_{21}\,a_{32})\big)\big[&a_{12}^2\,a_{21}\,a_{31} - a_{11}^3\,a_{32} + a_{32}\,\lambda_1\,\lambda_2\,\lambda_3 \\
&+a_{11}^2(a_{12}\,a_{31} + a_{32}(\lambda_1+\lambda_2+\lambda_3)) \\
&+a_{11}\,(-a_{32}(\lambda_2\,\lambda_3 + \lambda_1\,(\lambda_2+\lambda_3)) \\
&+a_{12}\big((a_{22}\,a_{31} - 2\,a_{21}\,a_{32} - a_{31}\,(\lambda_1+\lambda_2+\lambda_3)\big)\big) \\
&+a_{12}\,\big(a_{22}^2\,a_{31} + a_{21}\,a_{32}\,(\lambda_1+\lambda_2+\lambda_3) - a_{31}\,(\lambda_1+\lambda_2+(\lambda_1+\lambda_2)\,\lambda_3) \\
&-a_{22}\,(a_{21}\,a_{32} + a_{31}\,(\lambda_1+\lambda_2+\lambda_3)))\big]
\end{aligned}$$

$$\begin{aligned}
a_{23} = 1/\big(a_{12}\ a_{31}^2 - a_{32}\ ((a_{11}-a_{22})\ a_{31} + a_{21}\ a_{32})\big)\big[&a_{11}^2\ a_{21}\ a_{32} \\
&-(a_{22}\ a_{31} - a_{21}\ a_{32})(a_{22}-\lambda_1)(a_{22}-\lambda_2) \\
&+((a_{22}\ a_{31} - a_{21}\ a_{32})(a_{22}-\lambda_1) + (a_{21}\ a_{32} + a_{31}(-a_{22}+\lambda_1))\lambda_2)\lambda_3 \\
&+a_{12}\ a_{21}\ (a_{21}\ a_{32} + a_{31}\ (-2\ a_{22}+\lambda_1+\lambda_2+\lambda_3)) \\
&-a_{11}\ a_{21}\ (a_{12}\ a_{31} + a_{32}(-a_{22}+\lambda_1+\lambda_2+\lambda_3))\big]
\end{aligned}$$

$$a_{33} = \lambda_1 + \lambda_2 + \lambda_3 - a_{11} - a_{22}.$$

Similarly to the investigation of the bivariate models, the Monte Carlo experiments were carried out under the condition that the error process ε_t corresponded to a white noise process, such that the variances and standard deviations equal 1 and the covariance equals zero:

$$C = \begin{pmatrix} 1 & 0 & 0 \\ 0 & 1 & 0 \\ 0 & 0 & 1 \end{pmatrix}.$$

A total of 400 three-dimensional VAR(1) models were investigated.

A single time series was generated for each model. To investigate a broad range of execution conditions, the length was varied systematically between $T_1 = 40$ and $T_{17} = 200$ time steps. Starting with $T_1 = 40$ time steps, the number of time steps was increased by 10 after each simulation run, so that in total 19 different lengths were analyzed in independent runs. A pre-sample warm-up interval with 100 time steps was calculated to reduce the impact of starting-up values. From the time series generated, the independent parameters were estimated based on the cited least square method. As before, the order was varied systematically between first- and sixth-order models, the corresponding likelihood and penalty terms were calculated for each parameterized model, and the order was estimated based on the minimization principle introduced above. We carried out ten separate repeated and independent trials so that we could make an inferential statistical comparison of model order selection accuracy for all five investigated types of models from Tables 5.1 and 5.2. Regarding the penalty terms in the model selection procedure, we did not only use the corrected Akaike information criterion $AIC_c(n)$ from Eq. 69, but also included Akaike's classic $AIC(n)$ criterion according to Eq. 67 to investigate whether the heuristic decision rule advocated by Burnham and Anderson (2002) leads to good selection decisions. This decision rule recommends using AIC_c instead of AIC when it holds that $T/k < 40$ (Eq. 70). T denotes the sample size and k is the number of freely estimated parameters in the dynamical operator. For two-dimensional VAR(1) models there are $k = 4$ freely estimated parameters so for a sample size of $T_{13} = 160$ time steps or larger switching from the corrected criterion to the original formulation should be considered to obtain an optimal selection accuracy. For two-dimensional VAR(2) and three-dimensional VAR (1) models the number of freely estimated parameters is $k = 8$ and $k = 9$, respectively. With respect to the decision rule, in both cases even the maximum sample size of $T_{17} = 200$ time steps can regarded as small and hence the superior bias properties of AIC$_c$ should produce a higher accuracy level in model selection. If the sample size exceeds 500 time steps under the given experimental conditions, AIC and AIC$_c$ will strongly tend to consistently select the same model order.

In a similar way to the first Monte Carlo study, Pearson's chi squared test was used for each time series of given length to investigate whether there are significant differences in the histogram distribution of the empirically selected model orders between the $AIC(n)$, $AIC_c(n)$, $BIC(n)$ and $MIC(n)$ criteria for all types of models listed in Tables 5.1 and 5.2 over all ten trials. Where significant differences were found, the values in the histogram distribution corresponding to the true model order were compared against each other to decide which criterion leads to highest accuracy in the model selection task. The level of significance was set to $\alpha = 0.05$ for all tests. To focus our presentation and discussion of the results of the second Monte Carlo study on the essential effects, we have not included a report on the individual chi squared tests in the following chapters. Instead, we will present the relative frequencies f_{cs} of correct model order selection for the five types of models investigated in the form of error list plots and only report characteristic parameter settings in which significant differences in the histogram distribution of the empirically selected model orders were found.

5.1.3 Results and Discussion

The results of the first Monte Carlo study show that the complexity-based mutual information criterion $MIC(n)$ identifies the correct order of the VAR(1) model of the first two overlapping tasks with optimizing parameters according to Eqs. 73, 74, 75 in 933 of 1000 trials. In 64 trials a wrong estimate of $n = 2$ was obtained, and in only 4 trials an estimate of $n = 3$. Higher model orders were never assigned minimum scores and were therefore never selected. Comparing this histogram distribution with the corresponding distributions of the alternative criteria, we can see that the Bayesian information criterion BIC is the only one that leads to a higher level of accuracy. For the VAR(1) model, BIC assigned the correct model order in 999 cases. Based on AIC_c, the first model order was selected correctly in 928 trials. For the VAR(2) model represented by Eqs. 77, 78 and 79, the accuracy of selecting the true model order based on $MIC(n)$ led to similar results, with a correct selection of the model order in 930 of 1000 trials. A third-order model was considered to be the true model in only 15 trials, whilst the first order was wrongly selected in 55 trials. Other model orders never led to minimum scores and were therefore never selected. The number of correct identifications of the VAR(2) model for BIC and AIC_C was much lower, at 813 and 892, respectively. Table 5.3 shows the relative frequency of the selected model order for the VAR(1) model for all three criteria. The corresponding frequency distribution for the VAR(2) model is shown in Table 5.4.

The datasets in Tables 5.3 and 5.4 show that the complexity-based mutual information criterion MIC enables the selection of the true model orders in approximately 93% of all trials and therefore leads to the highest overall accuracy level. However, we have to differentiate between first- and second-order models. In the case of a first-order model, the null hypothesis that the datasets related $MIC(n)$ and $BIC(n)$ have the same distribution is rejected at the $\alpha = 0.05$ level (test statistic $x_0 = 11.61$, $p < 0.0404$). This test result indicates that, in terms of the histogram distribution, the novel criterion is inferior to the classic Bayesian approach. Compared to AIC_c, we did not find any significant differences in accuracy ($x_0 = 0.84$, $p < 0.9738$). Regarding the second-order model, the chi squared tests show that the corresponding null hypotheses related to the distributions of $MIC(n)$ and $BIC(n)$ as well as $MIC(n)$ and $AIC_C(n)$ are both rejected at the $\alpha = 0.05$ level (MIC - BIC: $x_0 = 55.54$, $p < 0.0001$; MIC $-$ AIC_c: $x_0 = 19.16$, $p < 0.0018$). Due to a higher frequency of correctly identified model orders (Table 5.4), we can conclude that

Table 5.3 Relative frequency of identified model order for VAR(1) model based on the corrected Akaike information criterion (AIC_c, Eq. 69), the Bayesian information criterion (*BIC*, Eq. 71), and the mutual information criterion (*MIC*, Eq. 313)

Model order	1	2	3	4	5	6
AIC_c	0.928	0.062	0.009	0.001	0	0
BIC	0.999	0.001	0	0	0	0
MIC	0.933	0.640	0.003	0	0	0

Table 5.4 Relative frequency of identified model order for VAR(2) model based on the corrected Akaike information criterion (AIC_c, Eq. 69), the Bayesian information criterion (*BIC*, Eq. 71), and the mutual information criterion (*MIC*, Eq. 313)

Model order	1	2	3	4	5	6
AIC_c	0.015	0.892	0.070	0.015	0.008	0
BIC	0.185	0.813	0.002	0	0	0
MIC	0.055	0.930	0.015	0	0	0

under the given experimental conditions the novel criterion leads to superior accuracy for processes with larger memory depth. The results of the second Monte Carlo study will give us additional evidence to decide whether this conclusion can be generalized to model selection tasks in much larger areas of the parameters space, or whether it only applies to the two specific two-dimensional model representations of PD environments investigated.

The relative frequencies f_{cs} of correct model order selection in the second Monte Carlo study for bivariate VAR(1) processes based on AIC, AIC_C, BIC and MIC are shown in the error list plots in Fig. 5.1. Each plot shows the means as note points, and the 95% confidence intervals as error bars. The note points have been slightly offset to make it is easier to see the error bars. The 95% confidence intervals were calculated under the assumption of a normal distribution and therefore correspond to ± 1.96 standard deviations. Part (a) of the figure shows the relative frequencies for first-order processes with complex conjugate characteristic roots (see parameter and parameter values in part (a) of Table 5.1), whilst part (b) presents the frequencies for the case of distinct real characteristic roots (see parameters and parameter values in part (b) of Table 5.1). By comparing both parts of the figure, we can see that the means and 95% confidence intervals of the relative frequencies are very similar. A detailed analysis of the data sets also revealed great similarity in the results of the statistical tests. We will therefore interpret them in a single discussion. According to the error list plots in both parts of Fig. 5.1, the Bayesian information criterion BIC leads to the highest accuracy on average and the smallest 95 % confidence intervals and is therefore the empirically optimal criterion for the selection of bivariate first-order models under all investigated parameters and parameter values. Both parts of Fig. 5.1 also show that for time series of length equal or larger than $T_2 = 50$ time steps the novel mutual information criterion MIC leads to second highest means and second smallest 95 % confidence intervals. The corrected Akaike information criterion AIC_C only revealed higher average accuracies than MIC and similarly large 95 % confidence intervals in the minimum sample size of $T_1 = 40$ time steps. The differences in accuracy here are also significant as shown by the chi squared tests of the histogram distributions ($x_0 = 51.28$, $p < 0.0342$ for first-order processes with complex conjugate characteristic roots; $x_0 = 12.04$, $p < 0.0342$ in the case of distinct real characteristic roots). For time series of greater length, MIC produces significantly higher accuracies than AIC_C. This is due to the fact that AIC and AIC_c strongly tend to consistently select the same model order as the sample size increases, in this case on a consistently lower

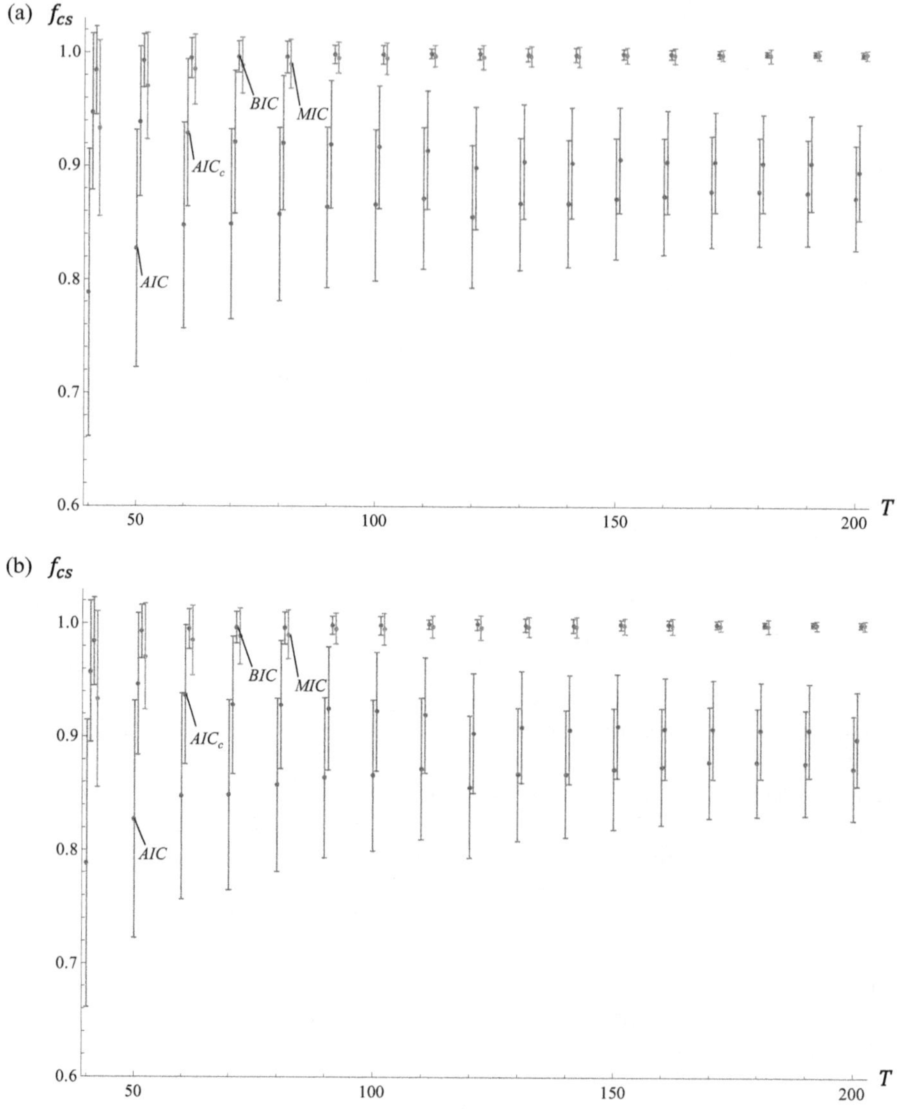

Fig. 5.1 Error list plots of relative frequencies f_{cs} of correct model order selection for bivariate VAR(1) processes based on Akaike's classic information criterion (*AIC*, Eq. 67), the corrected Akaike information criterion (*AIC*$_c$, Eq. 69), the Bayesian information criterion (*BIC*, Eq. 71), and the mutual information criterion (*MIC*, Eq. 313). *T* denotes the length of the time series that was used to estimate the model order and parameters. A pre-sample warm-up interval with 100 time steps was calculated. A total of 10 separate and independent trials were computed to obtain the statistic. The plots show mean frequencies as note points and 95 % confidence intervals as error bars. Note points have been offset to distinguish the error bars. Part (**a**) of the figure (*top*) shows the relative frequencies for VAR(1) processes with complex conjugate characteristic roots. The parameters and parameter values of these 800 processes are given in Table 5.1, part (**a**). Part (**b**) of the figure (*bottom*) shows the corresponding frequencies for VAR(1) processes with distinct real characteristic roots. The parameters and parameter values of these 200 processes are given in Table 5.1, part (**b**)

accuracy level (Fig. 5.1). An interesting finding when comparing BIC and MIC is that there were no significant differences in the histogram distributions for time series of length equal or larger than $T_3 = 60$ time steps—neither for models with complex conjugate characteristic roots nor for representations with distinct real characteristic roots. In other words, the empirical optimality of BIC can only be proved statistically for the smallest and second smallest sample size. For larger sample sizes BIC and MIC must be regarded as equally accurate in the class of bivariate first-order models under the predefined parameters and parameter values. Both parts of Fig. 5.1 also show that Akaike's classic criterion does not lead to an accuracy level that is comparable to the other criteria. For all investigated lengths of the time series the means were much lower. The chi squared tests demonstrate that even for the largest sample size of $T_{17} = 200$ time steps the differences in the histogram distributions are significant for the 800 models with complex conjugate characteristic roots ($x_0 = 14.45, p < 0.0115$). In contrast, the 200 models with distinct real characteristic roots do not lead to significant differences for the largest sample size considered ($x_0 = 5.14, p < 0.3991$). We can conclude that the heuristic decision rule introduced by Burnham and Anderson (2002) to use AIC_c instead of AIC when it holds that $T/k < 40$ (Eq. 70) is too conservative for the investigated first-order models and leads to suboptimal accuracy. We can still find superior bias properties of AIC_c for $T/k < 50$.

In an analogous manner to Fig. 5.1, Fig. 5.2 shows the means and 95 % confidence intervals of relative frequencies f_{cs} of correct model order selection for the investigated bivariate VAR(2) processes. Part (a) of the figure shows the note points and error bars of the frequencies corresponding to the 800 models with complex conjugate characteristic roots. Part (b) presents the results for the 200 generated processes with distinct real characteristic roots. As to be expected after the investigation of the first-order models, the means and 95 % confidence intervals of relative frequencies evolve very similarly over the time series length T under both conditions and we can discuss the data sets independently from of whether they have complex conjugate or distinct real characteristic roots. As can be seen in the error list plots in both parts of Fig. 5.2, for time series of length equal to or larger than $T_2 = 50$ time steps the Bayesian information criterion BIC leads to the highest mean accuracy and the smallest 95 % confidence intervals. This makes it the optimal criterion for the selection of bivariate second-order model under the given experimental conditions. Parts (a) and (b) of Fig. 5.2 also show that for sample sizes equal to or larger than $T_3 = 60$ time steps the mutual information criterion MIC results in the second highest means of selection accuracy and second smallest 95 % confidence intervals. The corrected Akaike information criterion AIC_c only leads to highest average accuracy and the smallest 95 % confidence interval in the smallest sample sizes of $T_1 = 40$ time steps. In this case there were significant differences in the histogram distributions for AIC_c and MIC and for AIC_c and BIC ($\mathrm{AIC}_c - \mathrm{MIC}$: $x_0 = 88.16$, $p < 0.0001$ for complex conjugate characteristic roots and $x_0 = 25.82$, $p < 0.0001$ for distinct real characteristic roots; BIC − MIC: $x_0 = 5483.16$, $p < 0.0001$ for complex conjugate characteristic roots and $x_0 = 60.68$, $p < 0.0001$ for distinct real characteristic roots). For a sample

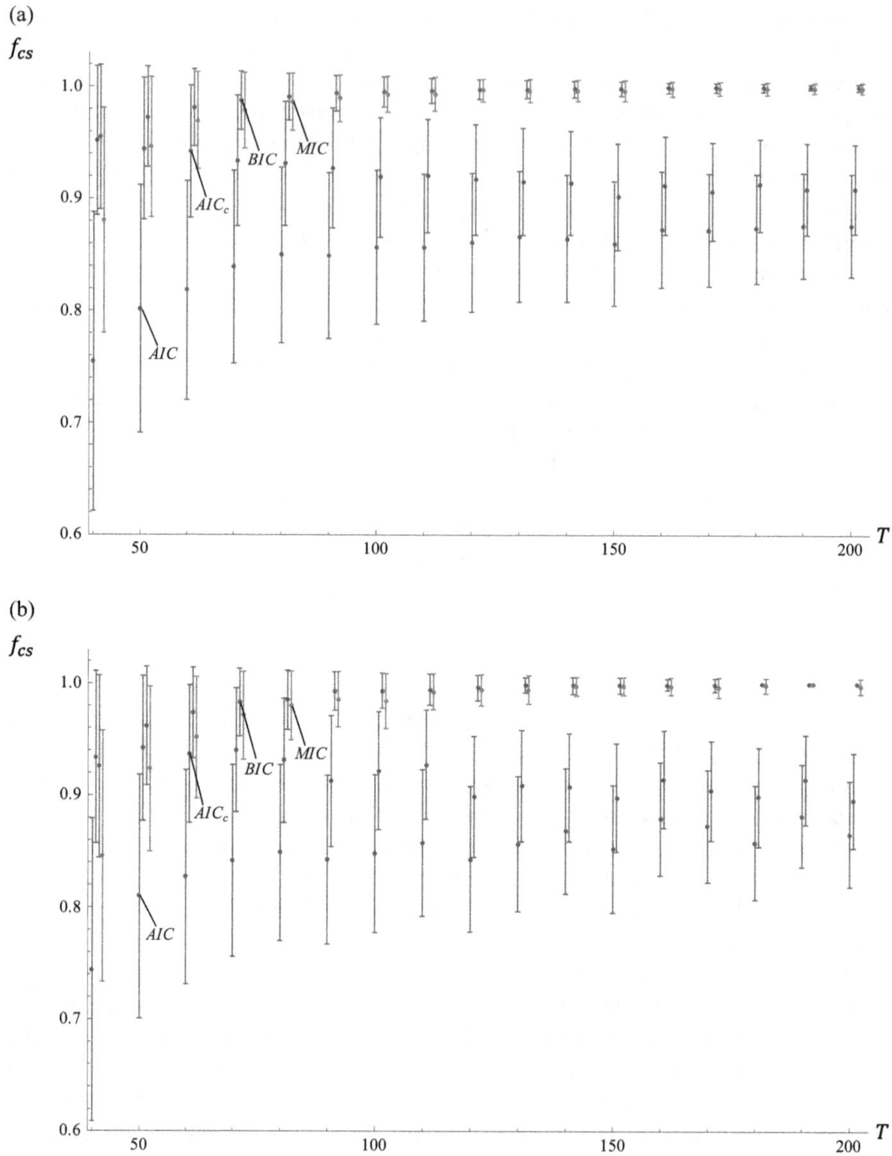

Fig. 5.2 Error list plots of relative frequencies f_{cs} of correct model order selection for bivariate VAR(2) processes based on Akaike's classic information criterion (*AIC*, Eq. 67), the corrected Akaike information criterion (AIC_c, Eq. 69), the Bayesian information criterion (*BIC*, Eq. 71), and the mutual information criterion (*MIC*, Eq. 313). *T* denotes the length of the time series that was used to estimate the model order and parameters. The experimental conditions were the same as in Fig. 5.1. Part (**a**) of the figure (*top*) shows the relative frequencies for VAR(2) processes with complex conjugate characteristic roots. The parameters and parameter values of these 800 processes are given in Table 5.1, part (**a**). Part (**b**) of the figure (*bottom*) shows the corresponding frequencies for VAR(1) processes with distinct real characteristic roots. The parameters and parameter values of these 200 processes are given in Table 5.1, part (**b**)

size of $T_2 = 50$ time steps the corrected Akaike information criterion AIC_C also leads to higher average accuracies than MIC and to slightly smaller 95 % confidence intervals. This difference in accuracy is again significant ($x_0 = 38.00$, $p < 0.0001$ for complex conjugate characteristic roots and $x_0 = 12.48$, $p < 0.0287$ for distinct real characteristic roots). Similar to the investigation of the first-order models, the more data points we have at hand to estimate the parameters, the higher the accuracy of MIC becomes in comparison to AIC_c (and the lower the corresponding p-values in the chi squared tests) as AIC and AIC_c strongly tend to consistently select the same model order (Fig. 5.2). When we compare BIC and MIC, the statistical tests show that for the 800 second-order models with complex conjugate characteristic roots a sample size of equal to or larger than $T_5 = 80$ time steps is needed to render the differences in histogram distributions insignificant. In contrast, for the 200 model representations with distinct real characteristic roots a sample size equal to or larger than $T_3 = 60$ time steps is already sufficient to lead to insignificant differences in accuracy. Weighting this evidence conservatively, we can demonstrate empirical optimality of BIC only under the constraint that we have a sample size of equal to or smaller than $T_4 = 70$ time steps. For larger sample sizes BIC and MIC have proven to be equally accurate under the parametric representation that were defined in parts (a) and (b) of Table 5.1. Similar to the first-order processes, parts (a) and (b) of Fig. 5.2 show that when AIC is used for model selection, it is impossible to reach an accuracy level that is comparable to that of the other criteria. For all investigated lengths of the time series the means are much lower. The chi squared tests show that even for the largest sample size of $T_{17} = 200$ time steps there are significant differences in the histogram distributions for models with complex conjugate characteristic roots ($x_0 = 28.11$, $p < 0.0001$). For models with distinct real characteristic roots, however, there were no significant differences ($x_0 = 4.93$, $p < 0.4241$). Due to the fact that we have eight and not only four freely estimated parameters in the bivariate VAR(2) model representation, the cited decision rule of Burnham and Anderson (2002) leads to the highest possible accuracy level over all VAR(2) models under evaluation and therefore seems to be adequate. Comparing Figs. 5.1 and 5.2, we find that the selection accuracy of the four criteria is quite similar for bivariate VAR(1) and for VAR(2) processes. However, the larger the model order and hence the memory depth of the process, the more data points are needed for the estimation of parameters to reach comparable accuracy levels. From a more qualitative perspective, we consistently find that the larger the sample size, the larger the differences in accuracy between BIC and MIC on the one hand, and AIC_c and AIC on the other. Another consistent finding is that the classic BIC criterion outperforms the other criteria on average accuracy. In spite of this fact, even for small sample sizes, the novel MIC criterion leads to histogram distributions of selected model orders that are not significantly different from the distributions produced by BIC on the $\alpha = 0.05$ level.

Finally, Fig. 5.3 shows the means and 95 % confidence intervals of relative frequencies f_{cs} of correct model order selection for the evaluated three-dimensional VAR(1) processes. Compared to the two bivariate model representations the results are unexpected. With the smallest sample size of $T_1 = 40$ time steps, the corrected

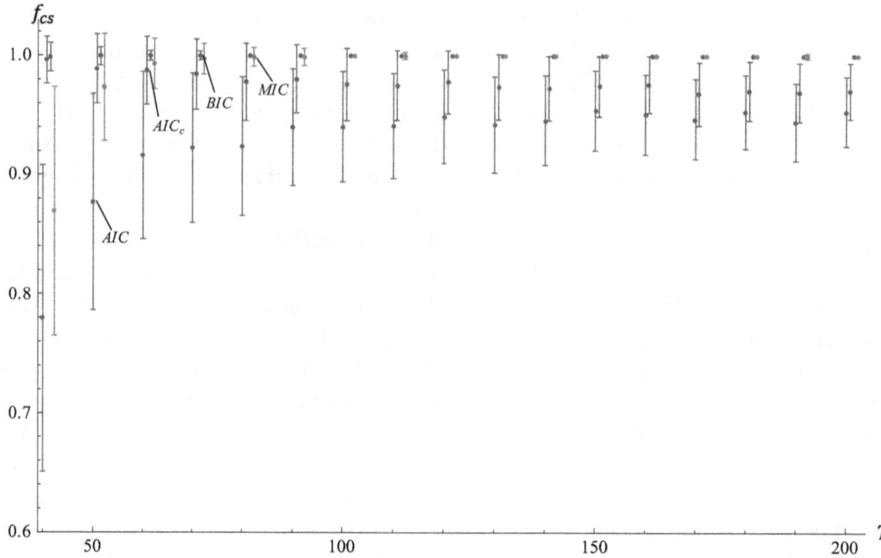

Fig. 5.3 Error list plots of relative frequencies f_{cs} of correct model order selection for three-dimensional VAR(1) processes based on Akaike's classic information criterion (*AIC*, Eq. 67), the corrected Akaike information criterion (AIC_c, Eq. 69), the Bayesian information criterion (*BIC*, Eq. 71), and the mutual information criterion (*MIC*, Eq. 313). The experimental conditions were the same as in Fig. 5.1. Only processes with real characteristic roots were investigated. The parameters and parameter values of these 400 processes are given in Table 5.2

Akaike information criterion AIC_c already leads to a relative frequency of more than 0.99 and is therefore extremely accurate. This level can only be matched by using the Bayesian information criterion BIC in the same manner (Fig. 5.3). Differences in the corresponding histogram distributions were not found $(x_0 = 1.05, p < 0.9588)$. The accuracy of the novel mutual information criterion MIC is significantly lower $(\text{AIC}_c - \text{MIC}:\ x_0 = 95.15,\ p < 0.0001;\ \text{MIC} - \text{BIC}:\ x_0 = 98.12,\ p < 0.0001)$. However, when the time series data is extended to $T_3 = 60$ or more time steps, the accuracy of MIC becomes significantly higher than AIC_c as we would expect for consistency reasons (Fig. 5.3). Similar to the bivariate first-order processes, the chi squared tests of the histogram distributions related to BIC and MIC show that a sample size of equal or larger than $T_3 = 60$ time steps is sufficient to render the differences in distributions insignificant. For equal or more $T_3 = 60$ data points MIC also leads to significantly higher selection accuracies than AIC_C. Finally, the chi squared tests show that for the largest sample size of $T_{17} = 200$ time steps there are no significant differences in the histogram distributions of AIC_C and AIC $(x_0 = 4.89, p < 0.43)$. Hence, the heuristic decision rule of Burnham and Anderson (2002) can be safely applied.

The results of both Monte Carlo studies show that the novel complexity-based mutual information criterion MIC is not only very effective for making model selection decisions in specific PD environments but that it appears to be a highly accurate universal quantity for model selection in the class of vector autoregression models. Although the results of the more comprehensive second Monte Carlo study do not confirm the preliminary conclusion from the first study that MIC has a significantly higher accuracy for second-order processes than AIC_c and BIC, we can conclude that MIC is, on average, only outperformed by the Bayesian information criterion BIC and leads to the second highest overall accuracy. The 95 % confidence intervals are also second smallest. Furthermore, the statistical tests of the histogram distributions show that a sample size of only $T_3 = 60$ time steps is sufficient in four out of five types of models investigated (see Tables 5.1 and 5.2) to render the differences between MIC and BIC in distributions insignificant. The only exception are second-order processes with complex conjugate characteristic roots. In this case a sample size of $T_5 = 80$ time steps is sufficiently large. Additional Monte Carlo studies replicating the setups of Mantalos et al. (2010) and Pereira et al. (2012) have shown that the mutual information criterion MIC is also highly accurate in selecting the true model order for larger memory depths. Furthermore, it is important to mention that the formulation of the criterion according to Eq. 314 in conjunction with Eq. 313 can be generalized to the broader class of linear dynamical systems (see Section 2.9). To do so, we only have to count the free parameters in the equations generating the state and the observation processes (Eqs. 136 and 137) and (re-)estimate the effective number of parameters by using the introduced MMI principle.

5.2 Optimization of Project Organization

Following this analysis of concurrent validity of the preferred operationalization of emergent complexity in model selection tasks based on Monte Carlo experiments, we will now move our focus on to predictive validity and focus on more application-oriented problems in project management. In this context, predictive validity means the degree to which the information-theoretic metric EMC can predict (or correlate with) standard key performance indicators of PD projects, such as the total time to complete the process (duration) and total amount of work done in the process (effort), which are measured at some point in the future during the (simulated) execution of a project. In order to analyze the predictive validity, we systematically manipulate different independent variables (such as differences in productivity between developers and development teams, and the release period of information about geometrical/topological entities between teams on different levels of the hierarchy) to see what effect it has on emergent complexity and on the other cited indicators of the same construct. The analysis is based on simple parameter studies and additional Monte Carlo experiments. The repeated random sampling was carried out within a self-developed simulation environment. We will present and discuss the results of two validation studies with different objectives and different mathematical models of cooperative work. In the first study we used the least complex first-order vector autoregression model (VAR(1) model, see Sections 2.2 and 2.3) to represent the direct cooperative relationships between the developers and the corresponding development teams (Schlick et al. 2009). In the second study we also allowed the occurrence of periodically correlated work processes (PVAR model, see Section 2.6) that are caused by a two-level hierarchical coordination structure in which subsystem-level and component-level design teams cooperate directly and indirectly (Schlick et al. 2011).

The objective of the first study is to design the project organization of an PD project subjected to concurrent engineering for minimal emergent complexity. To evaluate emergent complexity the information-theoretic metric EMC is used in the spectral basis (see Section 2.3) and different settings of cooperative task processing that can be represented by the basic VAR(1) model are considered. To simplify the calculations we developed efficient numerical functions based on the most expressive closed-form solution from Eq. 262. Organizational optimization based on a formal complexity metric in conjunction with vector autoregression models of cooperative work is an application area that is particularly interesting, because complex sociotechnical systems can be purposefully designed, and established management principles and heuristics can be objectively evaluated. Especially in PD projects requiring intensive cooperation, the classical principles and heuristics (e.g. constructing self-contained systems, Peters 1991; striving for decoupled design with minimum information content, Suh 2005; etc.) can fall short because they focus on the formalized design problem and product and tend to underestimate the effects of the cooperative problem solving process. As shown in the previous chapters, the iterative and closely interacting work processes can induce unexpected variability and generate effects that cannot be trivially reduced to singular properties of the constituent tasks. These effects emerge as a result of higher-order interactions and can lead to critical phenomena of emergent complexity such as the cited "design churns" (Yassine et al. 2003) or "problem-solving oscillations" (Mihm et al. 2003; Mihm and Loch 2006). Moreover, from a theoretical point of view, it is interesting to analyze whether EMC is not only valid for stochastic processes in the steady state but can also assess the "preasymptotic" behavior of project dynamics. It is evident that different projects can have different preasymptotics, according to the speed and kind of convergence to that asymptote. Some properties that hold in the preasymptote of a complex project can be significantly different from those that take place in the long run, and we want to investigate in the following chapters whether the relevant features, in terms of duration and effort, are captured by the complexity metric. Moreover, a nonnegligible percentage of PD projects in industry show divergent work remaining that does not have any asymptote at all. Maintaining the terminology of Section 2.1 we refer to these projects as unstable, because the work remaining then exceeds all given limits. If a project is unstable, a complete redesign of tasks and their interactions is necessary. Although divergent behavior of projects is critical from a practical point of view, it can be predicted easily within the framework of the developed theory of cooperative work and need not be analyzed further. This is because EMC simply assigns infinite complexity values to a divergent process as one would expect. For instance, for first-order models defined in Eq. 8 it is not difficult to see that the infinite sum in Eq. 247 diverges if the dominant eigenvalue $\lambda_1(A_0)$ of WTM A_0 has a magnitude larger than 1. The equation $|\lambda_1(A_0)| = 1$ determines the boundary between stable and asymptotically stable regimes. Recall from Section 2.2 that the first-order model is asymptotically stable in the sense of Lyapunov (Eq. 4) if and only if the spectral radius of WTM A_0 is strictly less than one and the matrix in Eq. 4 is positive definite. A first-order model without diffusion and unit spectral radius $\rho(A_0) = 1$ would steadily move away from the equilibrium state x_e and therefore only be marginally stable (Halanay and Rasvan 2000).

5.2.1 *Unconstrained Optimization*

We start the studies on optimizing project organization by formulating an unconstrained optimization problem and solving it through a complete enumeration of organization designs satisfying certain boundary conditions. In a second step, a constrained optimization problem is formulated and solved by applying the same principle. The constraint is that the expected total amount of work x_{tot} according to Eq. 17 is constant among the experimental conditions.

5.2.1.1 Methods

The developed objective function in our first study quantifies the complexity of a given organization design under the dynamic regime of the introduced state equations (Eq. 8 for original state space coordinates and Eq. 39 for spectral basis). We seek to minimize emergent complexity by systematically choosing the optimal project organization from within an allowed set. The elements of the set are distinct project organization designs that satisfy boundary conditions on productivity, cooperative relationships and performance variability. The set is complete in the sense that valid alternative organization designs with different asymptotic behavior do not exist. To simplify the problem formulation, the elements of the set are represented by WTMs and the corresponding covariance matrices. The covariance matrices are linear functions of the WTMs.

In the first optimization study, we consider a small but complete PD project that is organized according to the management concept of concurrent engineering (CE, see definition of Winner et al., 1988, in Chapter 1) and involves different teams. We decided to model and simulate a complete project and not only a distinct phase in order to demonstrate the introduced modeling concepts in a holistic manner. This also simplifies the interpretation of the results, because, in this case, the time to complete the process corresponds to the known project duration and the amount of work done in the process corresponds to the project effort involved in completing all tasks. The project duration is also termed the finishing time in the following. We focus on three CE teams in the project whose work is coordinated by a system-integration engineer. Each CE team has three members, with each team member i processing one development task i with an autonomous task processing rate a_{ii}. Tasks 1, 2 and 3 are processed by the members in team 1, tasks 4, 5 and 6 are processed by team 2 and tasks 7, 8 and 9 by team 3. The teams work on a component design level (see the V-model of the systems engineering process from Fig. 2.11). A work transformation from one time step to the next represents 1 week of development. Due to the complexity of the system to be developed, the finishing time of the complete project is on average more than 4 years of continuous development even under the most favorable organizational conditions (cf. Section 5.2.1.2). Similar to the modeling example in Section 2.8, the vector components of the state variable X_t represent the relative number of labor units required to complete the tasks. The tasks in each team are "fully coupled" with respect to the components to be designed (fully interdependent tasks, see Section 5.2), and the corresponding off-diagonal elements $a_{ij}(i \neq j)$ of the WTM indicate a symmetric intensity of cooperative relationships that is encoded by the independent parameter $f_1 > 0$. To avoid additional reinforcement loops, it is assumed that the three CE teams are not directly cooperating. The average task processing rate of the developers is represented by the independent parameter $a \in (0; 1]$. The individual task processing rates must not be equal but can vary around the mean by an offset $\Delta a > 0$. There are three distinct productivity levels: (1) the most productive developers were able to process their tasks at rate $a_{ii} = a - \Delta a$; (2) the least productive developers processed their tasks at rate $a_{ii} = a + \Delta a$; and (3) averagely productive developers processed their tasks at rate $a_{ii} = a$. Because the three CE teams are not directly cooperating, boundary-spanning activities have to be coordinated by a 10th system-integration engineer ($i = 10$) who exchanged information directly with all nine developers. The productivity of this engineer is average, and it holds that $a_{10,10} = a$ for task 10. The additional independent parameters $0 < f_2 \ll a$ and $0 < f_3 \ll a$ represent the strength of the forward and backward informational couplings between the nine developers and the system-integration engineer. In real projects, for instance in the German automotive industry, the system-integration engineer is usually a member of a superordinate subsystem-level or module-level team coordinating the development work on large scale (e.g. powertrain, door module). However, for the sake of simplicity, in what follows we ignore this additional hierarchical coordination structure (see systems engineering considerations in Section 2.6). We also do not consider other technical or organizational interfaces between teams. In addition to the cited boundary conditions holding on individual and team levels, we assumed that the mean task processing rate $\bar{a}$ of all individuals in the entire project is a. We also assume that the project is asymptotically stable and that the means converge to the fix point of no remaining work for all tasks. In order to guarantee asymptotic stability, the values of the independent parameters a, $\Delta a, f_1, f_2$ and f_3 must be carefully chosen, so that for all feasible project organization designs the dominant eigenvalue $\lambda_1[.]$ of the corresponding WTM has a magnitude smaller than 1. By doing so, only finite complexity values are assigned.

The optimization of project organization aims to assign team members with different productivity levels ($a - \Delta a$, $a + \Delta a$ or a) to the three CE teams such that the emergent complexity in the sense of the EMC metric can be kept to a minimum. Under the given boundary conditions a total of 40,320 assignments of team members can be distinguished. However, due the symmetry of cooperative relationships within teams, these assignments can be reduced to eight essential assignments and therefore the allowed set consists of only eight distinct work transformation matrices (WTMs). The additional assignments are simply permutations of the eight basic WTMs and therefore lead to identical complexity values. The eight distinct WTMs can be ordered according to the total variance of the productivity rates over all three teams. In the following we also term the total variance the "diversity" of the organization design, because it represents the accumulated deviation of the individual productivity rates from the mean rate a. Equation 315 shows the first WTM A_{01} from the allowed set, where the total variance of productivity rates is maximal and therefore represents a organization design with maximum diversity. This is due to the fact that the mean task-processing rate a holds not only for the entire project but also on the level of the three CE teams. For each team the variance of productivity rates is Δa^2. Hence, the total variance is $3\Delta a^2$.

$$A_{01} = \begin{pmatrix} a-\Delta a & f_1 & f_1 & 0 & 0 & 0 & 0 & 0 & 0 & f_3 \\ f_1 & a & f_1 & 0 & 0 & 0 & 0 & 0 & 0 & f_3 \\ f_1 & f_1 & a+\Delta a & 0 & 0 & 0 & 0 & 0 & 0 & f_3 \\ 0 & 0 & 0 & a-\Delta a & f_1 & f_1 & 0 & 0 & 0 & f_3 \\ 0 & 0 & 0 & f_1 & a & f_1 & 0 & 0 & 0 & f_3 \\ 0 & 0 & 0 & f_1 & f_1 & a+\Delta a & 0 & 0 & 0 & f_3 \\ 0 & 0 & 0 & 0 & 0 & 0 & a-\Delta a & f_1 & f_1 & f_3 \\ 0 & 0 & 0 & 0 & 0 & 0 & f_1 & a & f_1 & f_3 \\ 0 & 0 & 0 & 0 & 0 & 0 & f_1 & f_1 & a+\Delta a & f_3 \\ f_2 & f_2 & f_2 & f_2 & f_2 & f_2 & f_2 & f_2 & f_2 & a \end{pmatrix} \quad (315)$$

Equation 316 shows the WTM A_{08} as the last element of the allowed set.

$$A_{08} = \begin{pmatrix} a-\Delta a & f_1 & f_1 & 0 & 0 & 0 & 0 & 0 & 0 & f_3 \\ f_1 & a-\Delta a & f_1 & 0 & 0 & 0 & 0 & 0 & 0 & f_3 \\ f_1 & f_1 & a-\Delta a & 0 & 0 & 0 & 0 & 0 & 0 & f_3 \\ 0 & 0 & 0 & a & f_1 & f_1 & 0 & 0 & 0 & f_3 \\ 0 & 0 & 0 & f_1 & a & f_1 & 0 & 0 & 0 & f_3 \\ 0 & 0 & 0 & f_1 & f_1 & a & 0 & 0 & 0 & f_3 \\ 0 & 0 & 0 & 0 & 0 & 0 & a+\Delta a & f_1 & f_1 & f_3 \\ 0 & 0 & 0 & 0 & 0 & 0 & f_1 & a+\Delta a & f_1 & f_3 \\ 0 & 0 & 0 & 0 & 0 & 0 & f_1 & f_1 & a+\Delta a & f_3 \\ f_2 & f_2 & f_2 & f_2 & f_2 & f_2 & f_2 & f_2 & f_2 & a \end{pmatrix} \quad (316)$$

WTM A_{08} represents an organization design with zero productivity variance on team level and therefore also zero total variance. It is evident that this design has minimum diversity. In terms of human-centered organization design and management we have an extreme kind of "selective" team organization because CE team 1 only includes team members with maximum productivity, whilst team 3 consist only of persons with low productivity. To conserve space we do not show all eight

Table 5.5 Overview of the eight distinct assignments of team members with different productivity levels $(a - \Delta a,\ a + \Delta a\ or\ a)$ to the three CE teams

	mean productivity			variance productivity		
WTM	team 1	team 2	team 3	team 1	team 2	team 3
A_{01}	a	a	a	Δa^2	Δa^2	Δa^2
A_{02}	a	$a - 1/3\Delta a$	$a + 1/3\Delta a$	Δa^2	$4/3\Delta a^2$	$1/3\Delta a^2$
A_{03}	$a - 1/3\Delta a$	a	$a + 1/3\Delta a$	$4/3\Delta a^2$	0	$4/3\Delta a^2$
A_{04}	a	$a - 1/3\Delta a$	$a + 1/3\Delta a$	Δa^2	$1/3\Delta a^2$	$4/3\Delta a^2$
A_{05}	$a + 1/3\Delta a$	$a + 1/3\Delta a$	$a - 2/3\Delta a$	$4/3\Delta a^2$	$1/3\Delta a^2$	$1/3\Delta a^2$
A_{06}	$a - 2/3\Delta a$	a	$a + 2/3\Delta a$	$1/3\Delta a^2$	Δa^2	$1/3\Delta a^2$
A_{07}	$a - 2/3\Delta a$	$a - 1/3\Delta a$	$a + \Delta a$	$1/3\Delta a^2$	$1/3\Delta a^2$	0
A_{08}	$a - \Delta a$	a	$a + \Delta a$	0	0	0

The WTMs A_{01} and A_{08} representing project organization designs with maximum total variance $3\Delta a^2$ and zero total variance of autonomous productivity rates are shown explicitly in Eqs. 315 and 316. The mean productivity rate over all three CE teams is always a

distinct WTMs explicitly but present the essential properties of the team organization in parametric form in Table 5.5.

For all eight organization designs in Table 5.5, we assumed that the standard deviation c_{ii} of performance fluctuations (Eq. 10) influencing task i in the project is proportional to the task processing rate $a_{ii}(\mathrm{i} = 1, \ldots, 10)$: the faster a task is processed, the less it is perturbed, and the smaller the standard deviation $(c_{ii} \sim a_{ii})$. The c_{ii}^2's are the elements along the principal diagonal of covariance matrix $C_{j \in \{1,\ldots,8\}}$ (see Eq. 10). The proportionality constant is $r = 0.02$. We also assumed that correlations between the performance variability of tasks do not exist and that the covariance matrices are diagonal. We therefore have the following covariance matrices related to the WTMs A_{01} and A_{08} that represent organization designs with maximum and minimum diversity:

$$C_1 = \{r^2\}.\begin{pmatrix} (a-\Delta a)^2 & 0 & 0 & 0 & 0 & 0 & 0 & 0 & 0 & 0 \\ 0 & a^2 & 0 & 0 & 0 & 0 & 0 & 0 & 0 & 0 \\ 0 & 0 & (a+\Delta a)^2 & 0 & 0 & 0 & 0 & 0 & 0 & 0 \\ 0 & 0 & 0 & (a-\Delta a)^2 & 0 & 0 & 0 & 0 & 0 & 0 \\ 0 & 0 & 0 & 0 & a^2 & 0 & 0 & 0 & 0 & 0 \\ 0 & 0 & 0 & 0 & 0 & (a+\Delta a)^2 & 0 & 0 & 0 & 0 \\ 0 & 0 & 0 & 0 & 0 & 0 & (a-\Delta a)^2 & 0 & 0 & 0 \\ 0 & 0 & 0 & 0 & 0 & 0 & 0 & a^2 & 0 & 0 \\ 0 & 0 & 0 & 0 & 0 & 0 & 0 & 0 & (a+\Delta a)^2 & 0 \\ 0 & 0 & 0 & 0 & 0 & 0 & 0 & 0 & 0 & a^2 \end{pmatrix} \tag{317}$$

C_8

$$
=\{r^2\}\cdot\begin{pmatrix}
(a-\Delta a)^2 & 0 & 0 & 0 & 0 & 0 & 0 & 0 & 0 & 0\\
0 & (a-\Delta a)^2 & 0 & 0 & 0 & 0 & 0 & 0 & 0 & 0\\
0 & 0 & (a-\Delta a)^2 & 0 & 0 & 0 & 0 & 0 & 0 & 0\\
0 & 0 & 0 & a^2 & 0 & 0 & 0 & 0 & 0 & 0\\
0 & 0 & 0 & 0 & a^2 & 0 & 0 & 0 & 0 & 0\\
0 & 0 & 0 & 0 & 0 & a^2 & 0 & 0 & 0 & 0\\
0 & 0 & 0 & 0 & 0 & 0 & (a+\Delta a)^2 & 0 & 0 & 0\\
0 & 0 & 0 & 0 & 0 & 0 & 0 & (a+\Delta a)^2 & 0 & 0\\
0 & 0 & 0 & 0 & 0 & 0 & 0 & 0 & (a+\Delta a)^2 & 0\\
0 & 0 & 0 & 0 & 0 & 0 & 0 & 0 & 0 & a^2
\end{pmatrix}. \tag{318}
$$

We assumed that all ten parallel tasks were initially 100 % incomplete and the initial state is

$$
x_0=\begin{pmatrix}1\\1\\1\\1\\1\\1\\1\\1\\1\\1\end{pmatrix}. \tag{319}
$$

In conformity with the basic principle of diversity management, namely to manage teams in organizations so that the potential advantages of diversity are maximized while its potential disadvantages are minimized (Cox 1994), it is hypothesized that for significant individual differences between developers attributable to their skills, abilities or access to information (Lazear 1998, 1999), "productivity balancing" at the team level minimizes emergent complexity. Productivity balancing at the team level means that in each of the three CE teams, members with high productivity $(a - \Delta a)$, low productivity $(a + \Delta a)$ and average productivity (a) directly cooperate and that the average task processing rate a does not only hold for the whole project but also on the team level. Such an assignment was shown in WTM A_{01} (Eq. 315). Productivity balancing is a self-developed concept that borrows some highly effective elements of the popular concept of production leveling (see e.g. Liker 2004) and transfers them from the domain of manufacturing systems to PD projects and knowledge-intensive service processes in such a way that cooperative work is also designed on human terms. Production leveling, also known as production smoothing, is a rather pragmatic concept for improving efficiency and not the working conditions. It was vital to developing production efficiency in the Toyota Production System and lean production (Liker 2004). The goal is to produce parts, components and modules at a constant rate so that further processing and assembly can also be carried out at a constant rate with small variance. If a later process step varies its output in terms of timing and quality, the variability of these variables increases in a demand-driven system as we move up the line towards the earlier processes and therefore tends to excite demand fluctuations. In textbooks, this phenomenon is termed demand amplification (see e.g. Liker 2004). It can also spill over into the complete supply chain, leading to the well-known bullwhip effect (see e.g. Sterman 2000). For this reason, demand amplification induced variability of internal or external ordering patterns must be reduced as far as possible to improve overall productivity in manufacturing systems. The concept of productivity balancing also aims to increase the productivity of an organization and the performance of work teams and individuals. The general idea, however, is not to standardize the work and process the work tasks at a constant rate, which would severely limit the possible scope of action and stifle creative expression in product development, but rather to make the cited basic principle of diversity management (Cox 1994) operational and find optimal (or near optimal) assignments of team members with different productivity levels due to individual differences. This allows the project work to be carried out effectively and efficiently without the cited "design churns" (Yassine et al. 2003) or similar critical emergent phenomena of complex sociotechnical systems that lead to unacceptably high levels of stress and create an unbearable workload. Productivity balancing also makes it possible to keep developing human knowledge, skills and abilities through cooperation and communication. Productivity balancing is not a demand-driven concept in the way that production leveling is. Rather, it is a holistic approach to systematically designing interactions between humans, tasks and products/services that considers performance fluctuations as an opportunity to innovate and learn. The objective is to increase awareness of emergent phenomena that are characteristic to open organizational systems and to leverage from them the greatest advantage for the individuals and the work teams. In large-scale development organizations, productivity balancing is most effective if the incentive systems support the coherence of the work teams and if the individual differences in skills and abilities are not too large. Furthermore, the productivity goals that are set by the management must be attainable and realistic and the corresponding action plans must be compatible. In the light of the concept of productivity balancing, we can reinterpret the above "productivity balancing" hypothesis and formulate a complementary "team diversity" hypothesis positing that maximum productivity diversity in teams leads to working conditions with minimum self-generated complexity and therefore reduces the potential risks of stress caused by an excessive workload and a narrow scope of action. In the framework of the developed theory and models of cooperative work, both aspects are just two sides of the same coin.

To verify these complementary hypotheses, all distinct eight assignments of team members to the three CE teams were analyzed. For each assignment the value of the complexity measure EMC was calculated on the basis of Eq. 262. The base set of independent parameters was $\theta_1 = [a = 0.9\ f_1 = 0.01\ f_2 = 0.01\ f_3 = 0.005]$. The productivity offset Δa was varied systematically on levels $\Delta a_1 = 0.001$ (small difference) and $\Delta a_2 = 0.01$ (large difference). The small productivity difference was primarily of theoretical interest and served as the baseline condition. As mentioned earlier in Section 5.2, in addition to EMC as an innovative information-theory key performance indicator (KPI) the project duration T_δ and the total amount of work x_{tot} involved in completing all tasks were considered to be conventional KPIs based on a sample of 10,000 independent Monte Carlo runs for all valid assignments. The means and standard deviations were calculated and evaluated for both KPIs. According to the model formulation from Chapter 2, the project duration $T_{\delta,l}$ in each run l (also termed replication l in what follows) is determined on the basis of the decision rule that the work remaining is at most 100δ percent for all p tasks and therefore that the one-dimensional stopping criterion δ is met (see Section 2.1). In the Monte Carlo experiments, we worked with a stopping criterion of $\delta = 0.05$. Similarly, the total effort $x_{tot,l}$ involved in completing the tasks in each run l is determined by accumulating the work remaining over all time steps and all ten tasks until the cited stopping criterion is satisfied. The time units are [weeks], the effort units are [work measurement units], abbreviated as [wmu]. The [wmu] refer to the units of the vector components of the state variable X_t and therefore represent the relative number of labor units required to complete the tasks. To simplify the discussion in the following chapter, we refer to the mean total amount of work $\overline{x}_{tot}$ done over all tasks as the mean total amount of work. Clearly, the larger the mean project duration $\overline{T}$ or mean total amount of work $\overline{x}_{tot}$ and the corresponding standard deviations, the lower the performance under the given organizational boundary conditions. We also calculated the expected total amount of work x_{tot} done over all tasks analytically according to Eq. 17. We accumulated the work over an infinite past history and therefore did not take the stopping criterion of the Monte Carlo experiments into account. Furthermore, the expected duration $\overline{T}_{\sigma=0}$ of the project was considered under the assumption that the processing of the development tasks is unperturbed and therefore the variances and covariances incorporated in the covariance matrix of unpredictable performance fluctuations are all zero. The expected duration was determined by inspecting the unperturbed state vectors for increasing time intervals T (Eq. 15), until all vector components are smaller than the stopping criterion of $\delta = 0.05$.

In addition, we analyzed the data sets with the conventional KPIs which were calculated in the Monte Carlo experiments with respect to the assumption that they come from a log-normal distribution function. The log-normal distribution is a reasonable tool for modeling stochastic execution times of work processes because it possesses the following properties (Baker and Trietsch 2009; Trietsch et al., 2012): (1) it is strictly positive, (2) its coefficient of variation is not restricted, (3) it can approximate sums of positive random variables, (4) it can represent the relationship between (limited) work capacity and execution time and (5) it can also represent the ratio between actual and estimated execution time. The last property greatly facilitates parameter estimation by regression in applied studies. Similar arguments hold true when we want to model the cumulative effort expended. The stochastic model developed by Huberman and Wilkinson (2005) leads to log-normally distributed finishing times for the modeled design project. The most important effect represented by a log-normal distribution function, regardless of whether we are modeling execution times or cumulative effort, is to shift the average dynamics away from the unperturbed course leading to skewed data sets. This effect was not only observed by Huberman and Wilkinson (2005) in Monte Carlo experiments but was also investigated in detail through analytical considerations. However, the Huberman-Wilkinson model incorporates multiplicative "noise" instead of non-predictable performance fluctuations having an additive effect (cf. Eq. 8). The main question in our context is therefore whether the developed basic model of cooperative work in conjunction with the predefined stopping criterion can simulate the special kind of dynamics that leads to log-normally distributed data sets. To evaluate whether the data sets on time and effort were coming from the log-normal distribution, quantile-quantile (Q-Q) plots (see e.g. Field 2009) were computed for all experimental conditions. In order to simplify interpretation of the scales, the natural logarithm of the data points was calculated and plotted against the quantiles of a normal distribution. For each condition, we estimated the parameters using a maximum likelihood estimator (see e.g. Kundu and Raqab 2007). If the Q-Q plots showed that the quantiles of the theoretical and data distributions agree and therefore the plotted points fall on or are near the line $y = x$, the log-likelihood functions of the simulated time and effort values were calculated on the basis of the formula given in Kundu and Raqab (2007). Note that in this formula the natural logarithm is used as in the definition of the log-likelihood function for a LDS from Eq. 183 and not the binary logarithm. The computed log-likelihood values are denoted by LL in the following. Furthermore, as the log-normal distribution shares many properties with the generalized Rayleigh distribution and both can be used effectively to analyze skewed data sets, we used the likelihood ratio test developed by Kundu and Raqab (2007) to discriminate between the two distribution functions. The test statistic of this test is based on the difference between the log-likelihood values for the maximizing parameters. The generalized Rayleigh distribution was introduced recently by Surles and Padgett (2001). It does not only have a scale parameter λ, like the conventional Rayleigh distribution, but also a shape parameter α. The shape of the density function only depends on α. It is known that for $\alpha \leq 1/2$, the density function is strictly decreasing and for $\alpha > 1/2$, it is unimodal (Kundu and Raqab 2007). The density functions of a generalized Rayleigh distribution are always right skewed. For $\alpha = 1$ the generalized Rayleigh distribution corresponds to the conventional Rayleigh distribution which is quite popular in project management to estimate cost and effort. The conventional Rayleigh distribution was initially proposed by Norden (1970) as a statistical model of the manpower utilizations during a project. Putnam (1978) was the first to apply this model to software development projects. Lee, Hogue and Hoffmann (1993) evaluated the agreement of the model with observed outlays in a wide variety of defense acquisition programs which were in the development phase. Additional studies by Lee and others (see e.g. Lee et al. 2002) using data from Earned Value Management systems on research and development projects have shown that the cost can be accurately estimated using the Rayleigh distribution. Because of similar teamwork mechanisms in engineering projects, Bennett and Ho (2014) and others advocate modeling the cumulative effort to complete a phase of work by a Rayleigh distribution. Following the procedure of Kundu and Raqab (2007), we calculated the probability of correct selection of the log-normal and generalized Rayleigh distribution functions in each case through additional Monte Carlo simulations. We replicated the selection process 100 times to obtain an estimate of the probability of correct selection.

The Mathematica software package from Wolfram Research was used to carry out the Monte Carlo experiments and to compute the dependent variables.

5.2.1.2 Results and Discussion

To get an impression of the dynamics of concurrent task processing and performance variability, Fig. 5.4 shows the results of a typical run of the Monte Carlo simulation for WTM A_{01} (Eq. 315) representing the project organization design with maximum productivity diversity in teams. The initial state x_0 is given in Eq. 319. A large productivity offset was simulated and therefore the complete parameter vector is

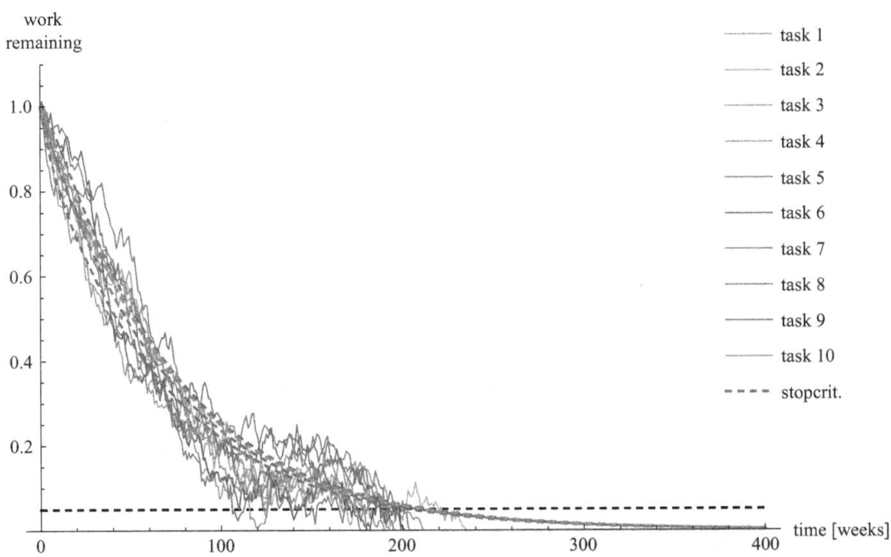

Fig. 5.4 List plot of work remaining in a simulated product development project, in which the mean task-processing rate a holds not only for the entire project but also on the level of the three CE teams. This project organizational design leads to maximum diversity of autonomous task processing rates within teams (see WTM A_{01} in Eq. 315). The concurrent processing of all ten development tasks is shown. The data are based on a single run of the Monte Carlo experiment with initial state x_0 (Eq. 319). The plot also shows the expected work remaining as dashed curves. The Monte Carlo experiment was based on state equation 8. A large productivity offset was simulated and therefore the parameters were $a = 0.9$, $f_1 = 0.04$, $f_2 = 0.01$, $f_3 = 0.005$ and $\Delta a = 0.01$. The stopping criterion of 5 % is marked by a *dashed line* at the bottom of the plot

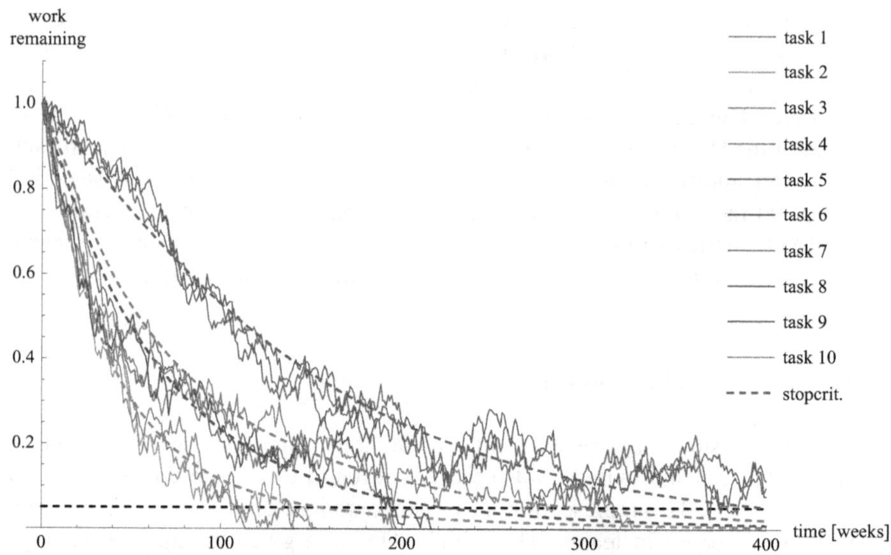

Fig. 5.5 List plot of work remaining in a simulated product development project with minimum diversity of autonomous task processing rates within teams (see WTM A_{08} in Eq. 316). The concurrent processing of all ten development tasks is shown. The data are based on a single run of the Monte Carlo experiment. The plot also shows the expected work remaining as *dashed curves*. The other simulation conditions and parameters were the same as in Fig. 5.4. The stopping criterion of 5 % is marked by a *dashed line* at the bottom of the plot

$\theta_1 = [a=0.9\ f_1=0.04\ f_2=0.01\ f_3=0.005\ \Delta a = \Delta a_2 = 0.01]$. Note that in the run shown, the task processing was simulated in a way that meant negative values of work remaining could not occur. If the unpredictable performance fluctuations for a certain task had led to a negative value of work remaining in the next time step, the corresponding vector component was set to zero for the next and all following time steps. In this sense, the zero state was "absorbing" for all state vector components. The same procedure was also used to calculate the other typical runs given in Figs. 5.5, 5.17, 5.18 and 5.19.

Even though the chosen productivity offset $\Delta a_2 = 0.01$ is large in the run shown in Fig. 5.4 and the intensity of cooperative relationships $(f_1 = 0.04)$ is high, the time series of work remaining decay quite smoothly to the fixpoint of zero work remaining and do not show heavy performance fluctuations around mean values. In contrast to this, Fig. 5.5 shows a typical simulation run for WTM A_{08} according to Eq. 316 representing the organization design with minimum diversity. The same initial state and parameter vector were used. As can be seen in Fig. 5.5, this kind of "unbalanced" organization design with minimum diversity within development teams leads to larger performance fluctuations, especially in tasks 7, 8 and 9, which were processed slowly.

Knowing that it is not very meaningful to discuss individual traces of work remaining and their deviation from the means for different organization designs

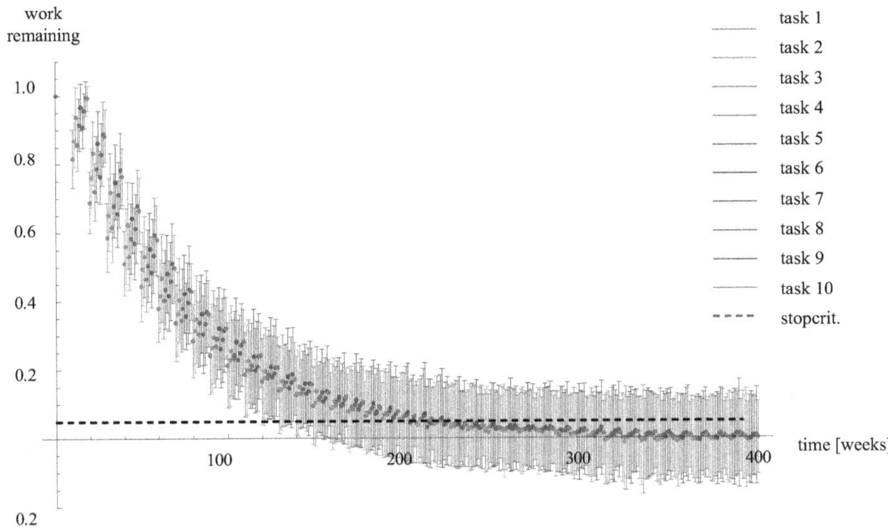

Fig. 5.6 Error list plot of work remaining in simulated PD projects with maximum diversity of autonomous task processing rates within teams (see WTM A_{01} in Eq. 315). The other simulation conditions and parameters are the same as in Fig. 5.4. A total of 100 separate and independent runs were calculated. The plot shows means of work remaining as note points and 95 % confidence intervals as error bars. Note points have been offset to distinguish the error bars

under the given initial and boundary conditions, we also calculated a statistic for both WTMs A_{01} and A_{08} based on 100 simulated projects and visualized them in the form of error list plots. These plots are given in Figs. 5.6 and 5.7. They show the mean values as note points for each time step, and the 95% confidence intervals as error bars. To simplify the visual analysis, the 95% confidence intervals were calculated under the assumption of a normal distribution and therefore correspond to ± 1.96 standard deviations as in the previous figures.

A comparison of the error list plots from Figs. 5.6 and 5.7 shows that the organization design with minimum diversity (encoded by WTM A_{08}) leads to significantly slower processing of development tasks 7, 8 and 9, which is to be expected as the least productive developers are all in team 3. It also leads to much stronger growth in the performance variability of that team over the first 300 time steps. Furthermore, this increased performance variability spills over into the other design teams and leads to slightly enhanced correlations between all work processes.

The analytical analyses show that for a small productivity offset ($\Delta a_1 = 0.001$) but high intensity of cooperative relationships ($f_1 = 0.04$), the organization design has little influence on complexity. The lowest complexity value under these conditions is $\text{EMC}(A_{01}) = 14.205$ and the largest value is $\text{EMC}(A_{01}) = 14.2068$. The corresponding expected total amount of work is $x_{tot}(A_{01}) = 680.851$ and $x_{tot}(A_{08}) = 682.092$. The maximum difference in the complexity variable EMC among the valid assignments is only 0.00163. Surprisingly, this holds, although the

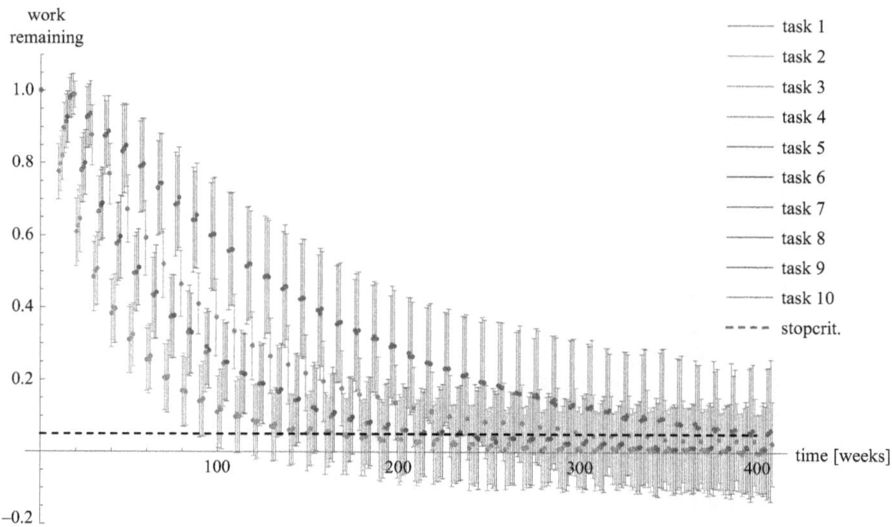

Fig. 5.7 Error list plot of work remaining in simulated PD projects with minimum diversity of autonomous task processing rates within teams (see WTM A_{08} in Eq. 316). The other simulation conditions and parameters are the same as in Fig. 5.4. A total of 100 separate and independent runs were calculated. The plot shows means of work remaining as note points and 95 % confidence intervals as error bars. Note points have been offset to distinguish the error bars

intensity of cooperative relationships is close to the bound of project divergence. Interestingly, in case of such a small productivity offset the Monte Carlo experiments show that the mean project duration differs by only 0.65 % among the eight valid assignments. The shortest mean project duration $\overline{T}(\mathrm{A}_{01})$ is 213.019 [weeks]. It is obtained for WTM A_{01}, which was assigned the minimum complexity value. The standard deviation is 36.50 [weeks]. The largest mean project duration is $\overline{T}(\mathrm{A}_{08})$ = 213.73 [weeks]. As expected, it is obtained for WTM A_{08} representing a process with maximum complexity. The standard deviation is 37.05 [weeks]. In the case of unperturbed task processing with zero performance fluctuations the differences in the project duration are a little larger and differ by 0.98% among the eight valid assignments. As expected, the shortest project duration of $\overline{T}_{\sigma=0}(\mathrm{A}_{01}) = 206$ [weeks] is obtained for WTM A_{01}, while the longest finishing time of $\overline{T}_{\sigma=0}(\mathrm{A}_{08}) = 211$ [weeks] occurs in the organization design represented by WTM A_{08}. We can conclude that if the project manager is able to "balance" the productivity of all team members through a good design of the work processes and the fair sharing of resources with such a small offset, the organization design has little effect. This finding holds for all considered KPIs and all mean task processing rates a that guarantee the asymptotic stability of the process (see Section 2.1), because all tasks are processed on very similar time scales.

However, when the productivity offset is increased to $\Delta a_2 = 0.01$—ceteris paribus—the complexity differences among the eight distinct assignments of team members grow significantly. The corresponding EMC values are shown in

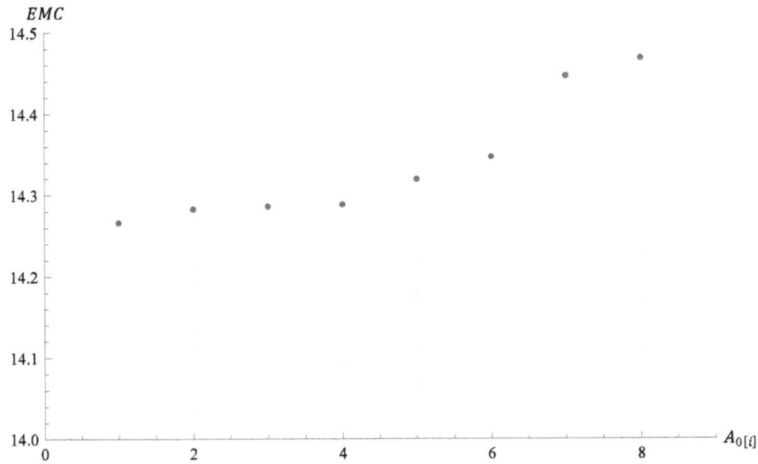

Fig. 5.8 Effective measure complexity for the eight organization designs that are encoded by the WTMs A_{01} to A_{08} from Table 5.5. The complexity values were calculated under the assumption that the productivity offset between team members is large and it holds that $\Delta a = \Delta a_2 = 0.01$. The additional parameters were $a = 0.9$, $f_1 = 0.04$, $f_2 = 0.01$ and $f_3 = 0.005$

Table 5.6 Means and standard deviations (SD) of the project duration and total amount of work that were calculated in the Monte Carlo experiments for the eight distinct assignments of team members with different productivity levels to the three CE teams

		Project duration		Total amount of work	
WTM	EMC	Mean	SD	Mean	SD
A_{01}	14.266	220.435	38.45	675.999	55.99
A_{02}	14.283	230.344	43.80	689.649	59.33
A_{03}	14.286	232.692	45.00	692.033	60.30
A_{04}	14.288	234.233	45.18	694.933	60.71
A_{05}	14.319	246.673	47.97	716.940	64.41
A_{06}	14.347	268.479	59.11	742.150	71.87
A_{07}	14.446	341.481	87.21	840.662	99.10
A_{08}	14.468	344.091	86.34	855.171	99.93

These assignments are encoded by the WTMs A_{01} to A_{08} according to Table 5.5. The experiments are based on state equation 8. The sample consisted of 10,000 independent runs. In these runs all tasks were initially 100 % incomplete. Simulation conditions and parameters are the same as in Fig. 5.9. In addition, the *EMC* values that were obtained on the basis of the closed form solution from Eq. 262 are shown in the second column

Fig. 5.8 and are given in numeric form in the second column of Table 5.6. Interestingly, the EMC values increase monotonically with the total variance of autonomous productivity rates over all three teams (see Table 5.5). Hence, the complexity metric partially acts as a scale parameter of the autonomous task processing rates.

The most important finding from the analysis of the eight organization designs that are encoded by the WTMs A_{01} to A_{08} from Table 5.5 is that an assignment with

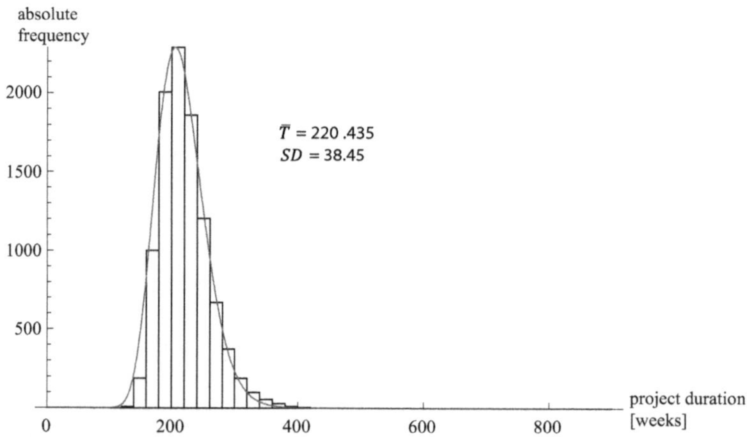

Fig. 5.9 Histogram of the project duration calculated for the organization design with maximum diversity of autonomous task processing rates within teams. This design is encoded by WTM A_{01} in Eq. 315. The total variance of autonomous productivity rates is $3\Delta a^2$ (Table 5.5). The sample consisted of 10,000 independent runs. In these runs all tasks were initially 100 % incomplete. The Monte Carlo experiment was based on state equation 8. The parameters were $a = 0.9, f_1 = 0.04, f_2 = 0.01, f_3 = 0.005$ and $\Delta a = 0.01$. The effective measure complexity is $EMC(A_{01}) = 14.266$. We overlaid the probability density function of a log-normal distribution for comparison

balanced productivity at the team level and therefore maximum diversity of productivity within teams leads to minimal complexity and it holds that $\text{EMC}(\text{A}_{01}) = 14.266$. This supports the cited "productivity balancing" hypothesis and the complementary "team diversity" hypothesis. The expected total amount of work is $x_{tot}(\text{A}_{01}) = 701.939$. In the case of unperturbed task processing with zero performance fluctuations, a finishing time of $\overline{T}_{\sigma=0}(\text{A}_{01}) = 215$ [weeks] was calculated. The corresponding histogram of the durations that were computed on the basis of a sample of 10,000 simulated projects is shown in Fig. 5.9. In this and the following histograms we overlay the probability density function of a log-normal distribution for comparison. The sample mean is $\overline{T}(A_{01}) = 220.435$ [weeks]. The standard deviation is 38.45 [weeks]. The corresponding histogram of the total amount of work x_{tot} in the simulated projects is shown in Fig. 5.10. The sample mean is $\overline{x}_{tot}(A_{01}) = 675.888$ [weeks] and the standard deviation is 55.988 [weeks].

Conversely, CE team building toward low diversity (above- or below-average productivity) at the team level significantly increases emergent complexity. An extreme example of organization design with zero total variance of autonomous task processing rates and therefore minimum diversity within teams was shown in WTM A_{08} (Eq. 316 and Table 5.5). In this case, the calculated complexity value is at a maximum, with $\text{EMC}(\text{A}_{08}) = 14.468$. The corresponding expected total amount of work is $x_{tot}(\text{A}_{08}) = 886.207$. Under the assumption of unperturbed task processing, a project duration of $\overline{T}_{\sigma=0}(\text{A}_{08}) = 404$ [weeks] is obtained. The histogram of the project duration that was calculated in the Monte Carlo

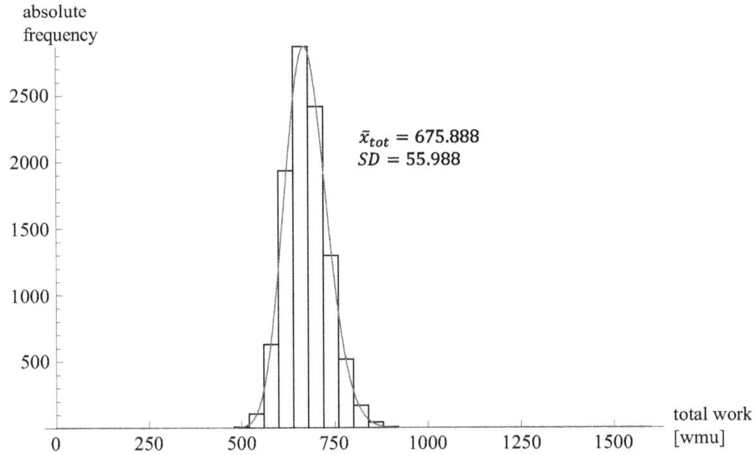

Fig. 5.10 Histogram of the total amount of work calculated for the organization design with maximum diversity of autonomous task processing rates within teams. This design is encoded by WTM A_{01} in Eq. 315. The sample consisted of 10,000 independent runs. In these runs all tasks were initially 100 % incomplete. The Monte Carlo experiment was based on state equation 8. Simulation conditions and parameters are the same as in Fig. 5.9. The effective measure complexity is $EMC(A_{01}) = 14.266$. The expected total amount of work is $x_{tot}(A_{01}) = 680.851$. We overlaid the probability density function of a log-normal distribution for comparison

experiments is shown in Fig. 5.11. The mean grows from $\overline{T}(A_{01}) = 220.467$ [weeks] in the case of maximum diversity (Fig. 5.9) to $\overline{T}(A_{08}) = 344.091$ [weeks] in the extremely nondiverse case shown. Furthermore, the standard deviation increases from 38.449 to 86.34 [weeks], and therefore the risk of schedule overruns grows significantly. The growth of means and standard deviation of the project duration in the Monte Carlo experiments is not unexpected because teams 1 and 2 and the system-integration engineer have to wait for the members of team 3 to finish their work. Therefore, in spite of performance fluctuations the project duration is largely determined by the least productive team. The corresponding histogram of the total amount of work x_{tot} is shown in Fig. 5.11. The sample mean is $\overline{x}_{tot}(A_{08}) = 855.171$ [weeks] and the standard deviation is 99.93 [weeks]. For the other six WTMs representing cases of team diversity in between the extremes, the means and standard deviations of the project duration, as well as the total amount of work grow monotonically with EMC. The detailed values are shown in Table 5.6. Hence, the complexity metric is a good predictor for both KPIs.

Detailed analyses have shown that the larger the emergent complexity, the more the evolution toward an equilibrium state of work remaining can differ from the unperturbed process. As a result, projects that in the absence of unpredictable performance fluctuations would converge smoothly to the desired goal state of zero remaining work can deviate significantly from this path. When emergent complexity is low, convergence to zero remaining work is smooth, and the project duration as well as the total amount of work can be statistically accurately modeled

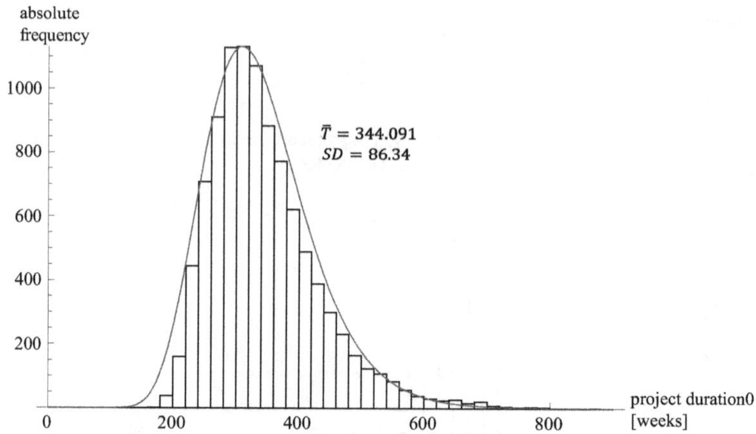

Fig. 5.11 Histogram of the project duration calculated for the organization design with minimum diversity of autonomous task processing rates within teams. This design is encoded by WTM A_{08} in Eq. 316. The total variance of autonomous task processing rates within teams is zero (Table 5.5). Simulation conditions and parameters were the same as in Fig. 5.9. The effective measure complexity is $EMC(A_{08}) = 14.468$. We overlaid the probability density function of a log-normal distribution for comparison

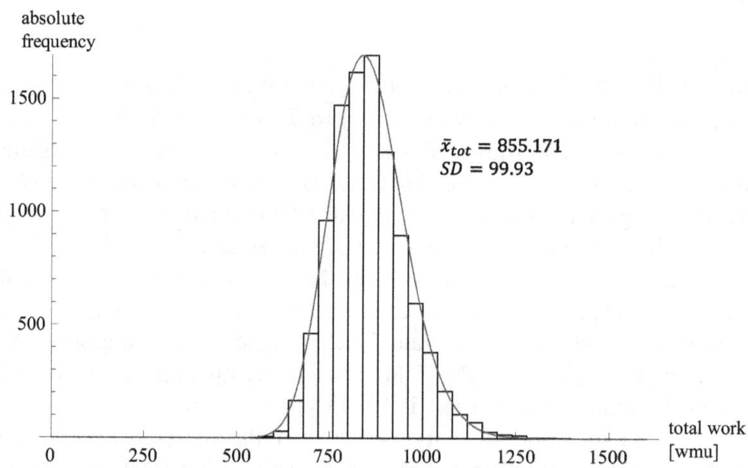

Fig. 5.12 Histogram of the total amount of work calculated for the organization design with minimum diversity of autonomous task processing rates within teams. This design is encoded by WTM A_{08} in Eq. 316. Simulation conditions and parameters were the same as in Fig. 5.9. The effective measure complexity is $EMC(A_{08}) = 14.468$. We overlaid the probability density function of a log-normal distribution for comparison

by log-normal distributions with small variance and therefore "short tails" (see Figs. 5.9 and 5.10). Above certain complexity thresholds, however, the distributions undergo a visible transition to a long-tailed log-normal form (see Figs. 5.11 and 5.12) and the correlations between tasks processed imply a possible execution time and cumulative effort that can significantly deviate from the average unperturbed course. Concerning the statistical data analysis, in all cases the Q-Q plots of the data sets showed the fit of the log-normal to be very good. Only for the tails on the far left and the far right sides of the theoretical distributions did the plotted quantile points deviate a little bit from the corresponding straight line. However, the maximum deviation from the straight line is in all cases smaller than 0.2. This corresponds to an estimated p-value for a two-tailed test of less than 0.05. The log-likelihood functions were thus calculated to allow for a comparison of the different organizational conditions. The evaluation of the log-likelihood functions of the simulated time and effort data under the assumption that they come from a log-normal distribution consistently produced high values. For a small productivity offset ($\Delta a_1 = 0.001$), the log-likelihood of the simulated project durations is between $LL_{min} = -49.716$ and $LL_{max} = -49.520$. For the total amount of work, log-likelihood values between $LL_{min} = -53.969$ and $LL_{max} = -53.814$ were obtained. If the productivity offset is large ($\Delta a_2 = 0.01$), the log-likelihood of the simulated project durations decreases to values between $L_{min} = -57.819$ and $LL_{max} = -50.149$. Similarly, the log-likelihood values of the generated total amount of work decrease to the range between $LL_{min} = -59.983$ and $LL_{max} = -54.340$. For a large productivity offset, the log-likelihood values decrease monotonically with the total variance of autonomous productivity rates of all three teams (see variances in Table 5.5). The high goodness of fit of the log-normal distribution as represented by high log-likelihood values is somehow counterintuitive as the model of the work processes is based on a linear stochastic difference equation and does not incorporate multiplicative noise as the one developed by Huberman and Wilkinson (2005), which by design leads to log-normally distributed time behavior. However, because of the necessary stopping criterion that must be assigned by the project manager, significant deviations from normality can occur, and lead times as well as amounts of work far from the average unperturbed process are quite likely. The careful reader may have noticed that under certain circumstances large deviations from normality can also contribute to an "accelerated" processing of the tasks. Accelerated processing means that the mean project duration $\overline{T}$ in the Monte Carlo experiments is shorter than the expected duration $\overline{T}_{\sigma=0}$. Recall that the expected duration was determined under the assumption that the processing of the development tasks is unperturbed. This "acceleration effect" can be found for WTM A_{08}: for the whole process an expected duration of $\overline{T}_{\sigma=0}(\mathrm{A}_{08}) = 404$ [weeks] is calculated, whilst the sample mean is only $\overline{T}(A_{08}) = 344.091$ [weeks] (see Fig. 5.11 and Table 5.6). However, it is very difficult to make use of this effect in applied cases, because the 95 % confidence intervals also grow monotonically with the expected duration, which significantly increases the risk of not meeting the schedule. Due to the limited space in this book, we can only report this interesting tradeoff and cannot carry out additional computational analyses. This has to be subject of future work. Finally, the results of the likelihood ratio tests according to Kundu and Raqab (2007) indicate that, independently of the productivity offset, for the majority of the investigated project organization designs the project duration can be modeled more accurately by generalized Rayleigh distribution functions. The estimated probability of correct selection is in all cases larger than 0.95. In contrast, the test results show that the log-normal distribution should be preferred in the majority of organizational settings to model the total amount of work under uncertainty. For this KPI, the probability of correct selection in all Monte Carlo simulations is also larger than 0.95. However, the values of the test statistic are only in an interval of [−171;69] and are therefore very small.

5.2.2 *Constrained Optimization*

After the presentation and discussion of the results of the basic unconstrained optimization problem in project organization, we move on to formulating and solving an associated constrained optimization problem. The constraint is that the expected total amount of work x_{tot} remains on a constant level among the different assignments of individuals to the three CE teams. The constraint is satisfied by systematic intervention in the strength f_3 of the backward informational couplings between the nine developers and the system-integration engineer. We only considered a setting in which the productivity offset Δa was large $(\Delta a = \Delta a_2 = 0.01)$. The base set of independent parameters therefore was $\theta_2 = [a = 0.9 \quad f_1 = 0.04 \quad f_2 = 0.01 \quad \Delta a = 0.01]$. The WTMs A_{01} to A_{08} were arranged in order of emergent complexity as before (see Table 5.6). Hence, WTM A_{01} represents the organization design that leads to minimum emergent complexity in the sense of the EMC metric and WTM A_{08} to maximum complexity. This order corresponds to an ordering by the total variance of autonomous task processing rates over all three design teams (see Table 5.5). We start by presenting analytical complexity results and go on to present the results of the Monte Carlo experiments.

5.2.2.1 Methods

As in the previous study, the developed objective function in the constraint optimization represents the emergent complexity of a given project organization under the dynamic regime of the state equations 8 and 39. We seek to minimize complexity by systematically choosing the organization design from the introduced eight assignments under the constraint that the expected total amount of work x_{tot} according Eq. 17 equals 701.939 [wmu]. This expected total amount of work corresponds to the minimum value that was identified in the previous study for the organization design with minimum emergent complexity. This design is encoded by WTM A_{01} (Eq. 315) and is characterized by a maximum diversity of autonomous productivity rates in the three teams. Starting with the base level $f_3 = 0.005$ of the strength of the backward informational couplings between the nine developers and the system-integration engineer, the feedback strength was reduced incrementally for organization designs with less diversity of autonomous productivity rates until the required total amount of work $x_{tot} = 701.939$ was reached. In other words, the independent parameter was adjusted by systematic algorithmic intervention of the experimenter so that the total expected effort in the project did not change under the eight distinct organization designs. To keep the total amount of work constant the independent parameter f_3 was adjusted by a self-developed iterative method so that it did not deviate more than 10^{-6}[wmu] from the correct value $x_{tot} = x_{tot}(A_{01}) = 701.939$. The time scale was not modified.

Following the procedure in Section 5.2.1.1, we assumed that the standard deviation c_{ii} of performance fluctuations (Eq. 10), which influence task i in the project is proportional to the task processing rate a_{ii} with proportionality constant $r = 0.02$. Hence, the covariance matrices must not be modified (see Eq. 317 for organization design encoded by WTM A_{01} and Eq. 318 for organization design encoded by WTM A_{08}). Other correlations between vector components were not considered. The initial state was not changed and is given by Eq. 319.

The Mathematica software package from Wolfram Research was used to carry out the analytical calculations and the Monte Carlo experiments. The stopping criterion was that if a maximum of 5 % of work remained for all tasks the simulated project was terminated. In addition to EMC as an innovative information-theory KPI, the project duration and total amount of work were used to evaluate performance in the same way as in the unconstrained optimization problem from the previous chapter. To calculate these KPIs, 10,000 independent runs were considered for each organization design. Furthermore, the expected duration $\overline{T}_{\sigma=0}$ was calculated under the assumption that the processing of the development tasks is unperturbed.

The results of the Monte Carlo experiments were analyzed in detail by descriptive and inferential statistical methods. To evaluate whether the data sets of both KPIs conform to the log-normal distribution, we followed the same procedure as in Section 5.2.1.1 and computed Q-Q plots for all cases using a maximum likelihood estimator. If the Q-Q plots showed that the quantiles of the theoretical and data distributions agree, the log-likelihood functions of the simulated time and effort values were calculated. Furthermore, likelihood ratio tests according to Kundu and Raqab (2007) were carried out to discriminate between log-normal and generalized Rayleigh distribution functions. The probability of correct selection was determined by the same procedure as in Section 5.2.1.1. To carry out an additional inferential statistical analysis of the organization designs, additional samples based on 100 independent Monte Carlo runs were drawn and the corresponding test statistics for the project duration and total amount of work were calculated. To simplify the analysis, only the organization designs with maximum diversity of autonomous task processing rates within teams (see WTM A_{01} in Eq. 299) and zero diversity (see WTM A_{08} in Eq. 316) were considered as before. We hypothesized that lower values of the complexity metric EMC lead to a significantly lower project duration. To evaluate this hypothesis the Kruskal-Wallis (see e.g. Field 2009) location equivalence test was used. The level of significance in the test was set to $\alpha = 0.05$. The Kruskal-Wallis test performs a hypothesis test on the project duration data with null hypothesis $H_{0,T}$ that the true location parameters of the samples are equal, i.e. $\mu_T(A_{01}) = \mu_T(A_{08})$, and alternative hypothesis $H_{a,T}$ that at least one is different. The test is a non-parametric method and is based on ranks. We decided to use a non-parametric test as the previous Monte Carlo study has shown that depending on the specific project organization design, either a log-normal or a generalized Rayleigh distribution function is more accurate to obtain statistical models for both KPIs. Under the specified execution conditions it is therefore not possible to specify a consistent parametric null distribution. In addition to the cited hypothesis on the project duration, we hypothesized that different values of the complexity metric do not lead to significantly different means of the total amount of work in the Monte Carlo experiments. The rationale behind this (possibly slightly counterintuitive) hypothesis is that we have formulated a constrained optimization problem, in which the analytically obtained expected total amount of work x_{tot} is deliberately kept constant under the different organizational conditions and this systematic intervention should not lead to significant differences of the total effort in the simulated projects. Hence, we formulate the null hypothesis $H_{0,x_{tot}}$ that the true location parameters of the samples are equal, i.e. $\mu_{x_{tot}}(A_{01}) = \mu_{x_{tot}}(A_{08})$.

We also carried out goodness-of-fit hypothesis tests to evaluate the differences between the distributions of performance data for both organization designs. The null hypothesis $H_{0,gof}$ was that performance data drawn from a sample with maximum diversity in autonomous task processing rates do not come from a different distribution than the data that was obtained for zero diversity. The alternative hypothesis $H_{a,gof}$ is that the data comes from a different distribution. The well-known Kolmogorov-Smirnov test was used to evaluate the hypothesis (see e.g. Field 2009). The level of significance was also set to $\alpha = 0.05$.

5.2.2.2 Results and Discussion

In order to satisfy the constraint $x_{tot} = 701.939$ that was imposed on the total effort involved in completing the deliverables, the strength f_3 of the backward informational couplings between the nine developers and the system-integration engineer had to be reduced by a minimum value of 0.00027 for an organization design in which the mean productivity of team 1 is average, the mean autonomous task processing rates of team 2 are $1/3\Delta a$ below average and the mean autonomous task processing rates of team 3 are $1/3\Delta a$ above average. The means and variance in productivity of the associated WTM A_{02} are shown in Table 5.5. Due to limited space, the complete matrix representation is not given but readers can easily construct it themselves. Interestingly, the maximum reduction of the backward coupling strength was necessary for the organization design with zero diversity (see WTM A_{08} in Eq. 316). In this case the reduction was 0.00303. A list plot of the reductions of the backward coupling strength is shown in Fig. 5.13. The results

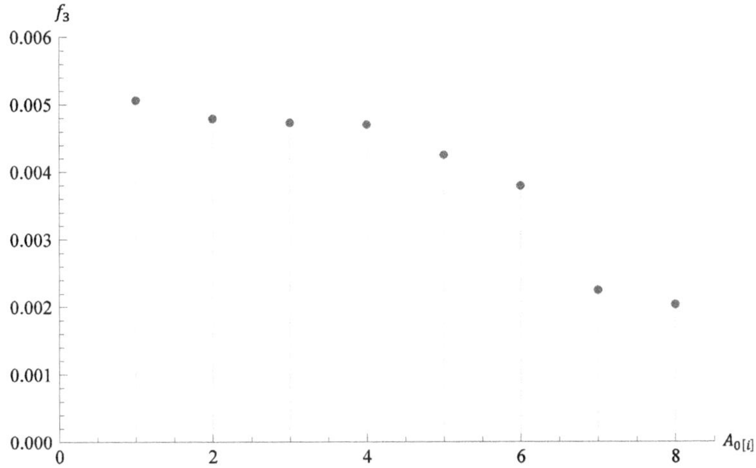

Fig. 5.13 Adjustments of original coupling parameter $f_3 = 0.005$ (see for instance the WTMs in Eqs. 315 and 316) that were necessary to satisfy the constraint on the expected total amount of work $x_{tot}(A_{01}) = \ldots = x_{tot}(A_{08}) = 701.939$ for the eight distinct project organization designs represented by the WTMs $A_{0[i]}(i = 1, \ldots, 8)$. An organizational setting was considered in which the productivity offset was large and it holds that $\Delta a = 0.01$. The additional parameters were $a = 0.9$, $f_1 = 0.04$ and $f_2 = 0.01$

show that the lower the productivity diversity within teams (following the order of the WTMs A_{01} to A_{08} from left to right in the figure), the more the backward coupling strength must be reduced to satisfy the constraint.

The values of the complexity metric EMC that correspond to the reduction of the backward coupling strength f_3 are visualized in Fig. 5.14. The numerical values can be found in the second column of Table 5.7. Interestingly, a comparison of Figs. 5.8 and 5.14 shows that although the backward informational couplings between the nine developers and the system-integration engineer are reduced in strength step-by-step (Fig. 5.13) and therefore tend to decrease the emergent complexity of the process, the increase in total variance of autonomous task processing rates of the ordered WTMs (Table 5.5) overcompensates this effect. The net effect is that the constrained optimization still leads to complexity values that grow monotonically with the total variance of autonomous task processing rates and do not shrink with decreasing backward coupling strength f_3. Hence, the consistent ordering of the organization designs by total variance of autonomous task processing rates as well as emergent complexity that was found in the previous chapter does not change after imposing the constraint on the total effort (cf. Fig. 5.8). However, the adjustments of the original coupling parameter $f_3 = 0.005$ that were necessary to satisfy the constraint lead to considerably lower complexity values for WTMs A_{02} to A_{08} and therefore the complexity metric acts less intensively as a scale parameter of the autonomous task processing rates (see Figs. 5.8 and 5.14).

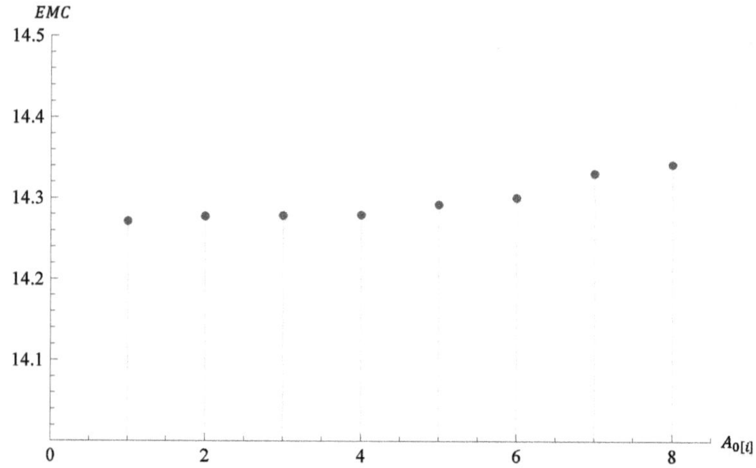

Fig. 5.14 Effective measure complexity for the investigated eight project organization designs under the constraint that the expected total amount of work is kept on constant level, i.e. $x_{tot}(A_{01}) = \ldots = x_{tot}(A_{08}) = 701.939$. To keep the expected total amount of work constant, the original coupling parameter $f_3 = 0.005$ was adjusted according to Fig. 5.13. The additional independent parameters are the same as in Fig. 5.13

Table 5.7 Means and standard deviations (SD) of the project duration and total amount of work obtained in the Monte Carlo experiments for the eight distinct assignments of team members with different productivity levels to the three CE teams under the constraint that the expected total amount of work remains on the level $x_{tot} = 701.939$

		Project duration		Total amount of work	
WTM	EMC	Mean	SD	Mean	SD
A_{01}	14.266	220.170	34.45	676.506	56.42
A_{02}	14.272	226.444	42.93	676.272	58.12
A_{03}	14.274	228.767	43.63	677.338	58.22
A_{04}	14.274	229.371	44.31	677.707	58.83
A_{05}	14.287	235.878	46.59	678.210	60.56
A_{06}	14.296	249.771	54.75	679.208	62.99
A_{07}	14.326	289.127	75.05	680.362	74.90
A_{08}	14.337	286.438	72.37	679.318	73.48

These assignments are encoded by the WTMs A_{01} to A_{08} according to Table 5.5. The experiments are based on state equation 8. The sample consisted of 10,000 independent runs. In these runs all tasks were initially 100 % incomplete. The coupling strength $f_3 = 0.005$ was adjusted according to Fig. 5.13 to satisfy the constraint on the total amount of work. The additional independent parameters are the same as in Fig. 5.13. In addition, the *EMC* values that were obtained on the basis of the closed form solution from Eq. 262 are shown in the second column

For unperturbed task processing with zero performance fluctuations the shortest finishing time of $\overline{T}_{\sigma=0}(A_{01}) = 215$ [weeks] is obtained, as expected, for WTM A_{01}. Interestingly, under these conditions the longest project duration of $\overline{T}_{\sigma=0}(A_{07}) =$ 336 [weeks] occurs in the case of the organization design represented by WTM A_{07}

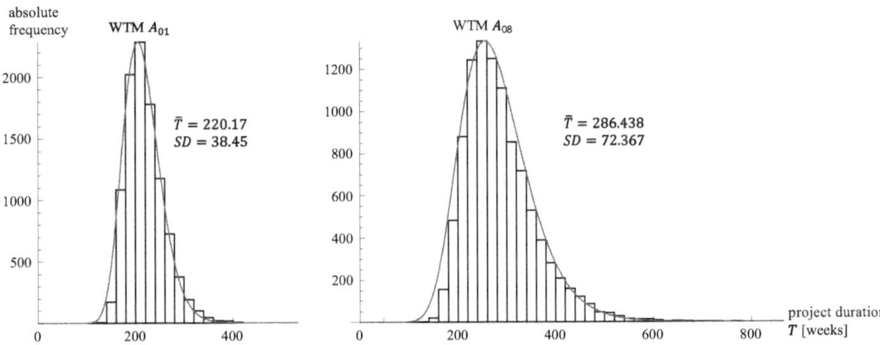

Fig. 5.15 Histograms of the project duration obtained for organization designs with maximum diversity (encoded by WTM A_{01}) and minimum diversity (encoded by WTM A_{08}) of autonomous task processing rates within teams. We computed 10,000 independent runs. In these runs all tasks were initially 100 % incomplete. The stopping criterion for the simulated projects was that a maximum of 5 % of work remained for all tasks. The Monte Carlo experiments were based on state equation 8. The coupling strength $f_3 = 0.005$ was adjusted according to Fig. 5.13 to satisfy the constraint on the total amount of work. The additional parameters are the same as in Fig. 5.13. We overlaid the probability density function of a log-normal distribution for comparison

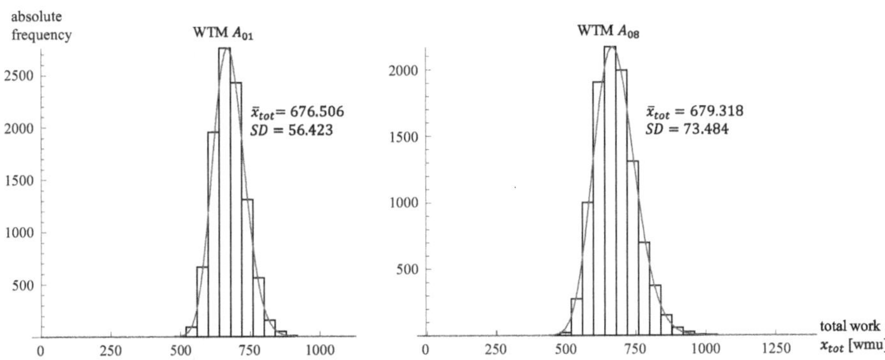

Fig. 5.16 Histograms of the total amount of work obtained for organization designs with maximum diversity (encoded by WTM A_{01}) and minimum diversity (encoded by WTM A_{08}) of autonomous task processing rates within teams. We computed 10,000 independent runs. In these runs all tasks were initially 100 % incomplete. The stopping criterion for the simulated projects was that a maximum of 5 % of work remained for all tasks. The Monte Carlo experiments were based on state equation 8. The coupling strength $f_3 = 0.005$ was adjusted according to Fig. 5.13 to satisfy the constraint on the total amount of work. The additional parameters are the same as in Fig. 5.13. We overlaid the probability density function of a log-normal distribution for comparison

and not WTM A_{08} as before. However, for WTM A_{08} the project duration is $\overline{T}_{\sigma=0}(A_{08}) = 333$ [weeks] and thus only slightly shorter.

Selected results of the Monte Carlo experiments are shown in Figs. 5.15 and 5.16. Figure 5.15 shows the histogram of the simulated project duration for organization designs with maximum and minimum diversity of autonomous

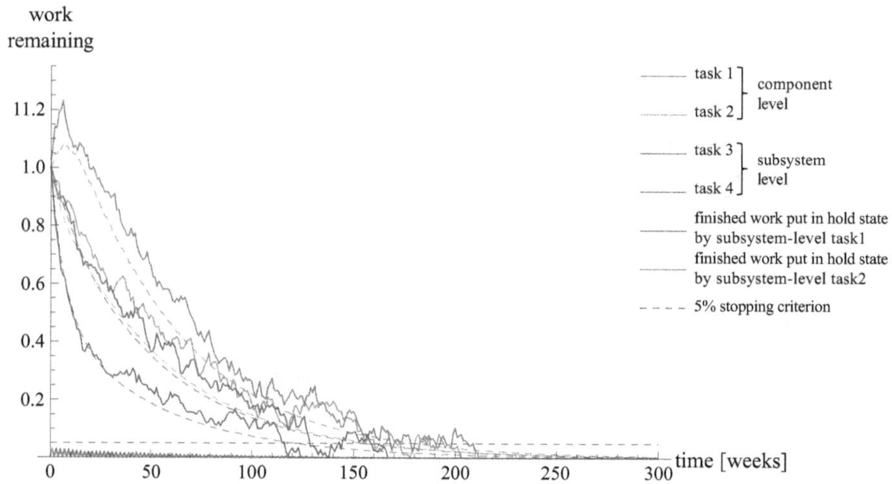

Fig. 5.17 List plot of work remaining in a simulated product development project with correlated work processes. It shows the simultaneous processing of all four development tasks. The release period is $s = 2$ [weeks]. The data is based on a single run of the Monte Carlo experiment with initial state x_0^* (Eq. 327). The plot also shows the means as dashed curves. The Monte Carlo experiment was based on state equation 89. The parameters are given by Eqs. 321–326. The stopping criterion of 5 % is marked by a *dashed line* at the bottom of the plot

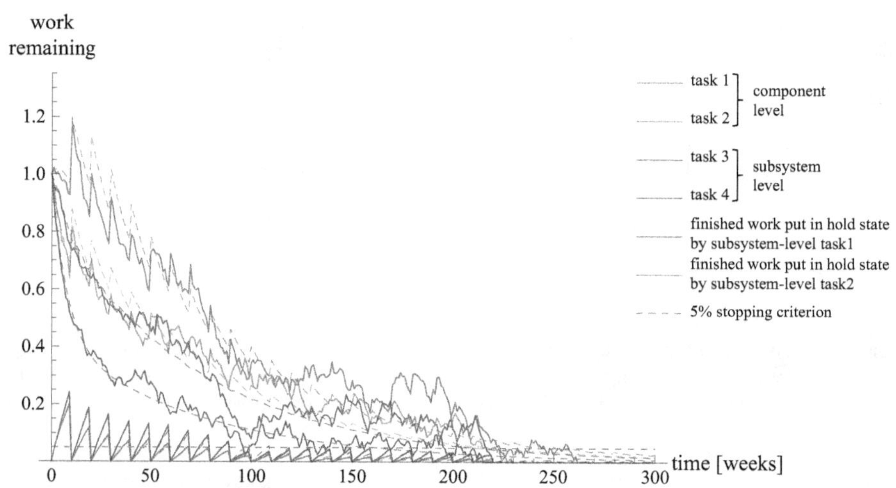

Fig. 5.18 List plot of work remaining in a simulated product development project with correlated work processes. The release period is $s = 10$ [weeks]. The other simulation conditions and parameters are the same as in Fig. 5.17

productivity rates within teams. In Fig. 5.16, the histograms for the total amount of work are given. To simplify the interpretation of the results, the figures also show the means and standard deviations.

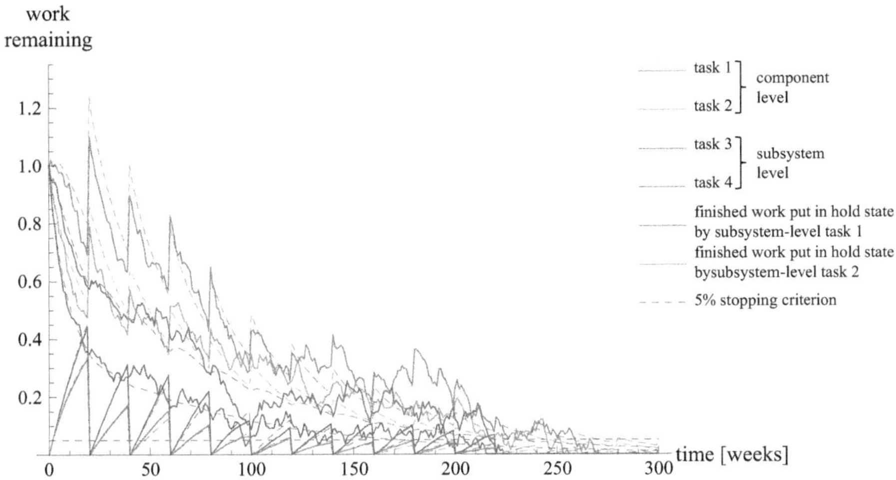

Fig. 5.19 List plot of work remaining in a simulated product development project with correlated work processes. The release period is $s = 20$ [weeks]. The data are based on a single run of the Monte Carlo experiment. The other simulation conditions and parameters are the same as in Fig. 5.17

According to Fig. 5.15, the organization design with maximum diversity of autonomous task processing rates within teams that is encoded by WTM A_{01} leads to an average project duration of $\overline{T}(A_{01}) = 220.17$ [weeks]. If the team members are assigned so that diversity is minimum within teams, the average project duration is extended to $\overline{T}(\mathrm{A}_{01}) = 286.439$ [weeks]. Furthermore, the standard deviation is more than twice as large. In contrast to these findings, the difference in the mean total amount of work for both organizational conditions is less than 1 % (Fig. 5.16). The constraint optimization based on the analytically obtained expected total amount of work therefore leads to very similar mean efforts in the Monte Carlo experiments.

Similarly to the unconstrained optimization of project organization, the means and standard deviations of the project duration monotonically grow with EMC. The values are summarized in Table 5.7. Hence, the complexity metric is also a good predictor for the project duration under the constraint that the expected total amount of work x_{tot} remains on an (almost) constant level. To verify the adjustments of the original coupling parameter $f_3 = 0.005$ that were necessary to satisfy this constraint, the means and standard deviations of the total amount of work are also given in Table 5.7, which shows that the adjustments are effective and lead to mean values with an average deviation of less than 0.5%. Note that the amounts of work shown in Table 5.7 are approximately 4% lower than the value of the constraint $x_{tot} = 701.939$, because we worked with a stopping criterion of $\delta = 0.05$ on the project duration.

Regarding the comparison of the quantiles of the theoretical distribution with the data distributions of both KPIs, in all cases the Q-Q plots showed the fit of the log-normal to be very good. Similar to the unconstrained optimization, the plotted quantile points only deviated a little from the corresponding straight line for the parts of the theoretical distributions that are on the far left and right sides of the mean. The maximum deviation from the straight line is in all cases smaller than 0.2. The evaluation of the log-likelihood functions of the simulated time and effort data consistently led to high values. The log-likelihood of the simulated project durations is between $LL_{min} = -56.075$ and $LL_{max} = -50.139$. For the total amount of work, values between $LL_{min} = -56.937$ and $LL_{max} = -54.408$ were obtained. The likelihood ratio tests according to Kundu and Raqab (2007) lead to very similar results as in the unconstrained optimization and show that for the majority of the investigated organization designs the project duration can be modeled more accurately by a generalized Raleigh distribution function. In contrast, the tests indicated that the log-normal distribution should be preferred by the modeler in the majority of organizational settings to represent the total amount of work. For both KPIs the probability of correct selection is also larger than 0.95. However, similar to the unconstrained optimization the values of the test statistic are very small (in the range of [−185;55]).

For 100 additional independent runs, the Kruskal-Wallis test on the project duration data shows that the location differences between both organization designs are significant $\left(K_T = 68.25, p = 8.35 \cdot 10^{-20}\right)$. Hence, the null hypothesis $H_{0,T}$ that the true location parameters of the samples are equal can be rejected on the significance level of $\alpha = 0.05$. The Kruskal-Wallis test on the total amount of work data comes to a different result. It shows that the locations between both organization designs are not significantly different $\left(K_{x_{tot}} = 1.96, p = 0.16\right)$. The null hypothesis $H_{0,x_{tot}}$ that the true location parameters of the samples are equal cannot be rejected on a significance level of $\alpha = 0.05$. Hence, the constraint imposed on the objective function is effective in the Monte Carlo experiments and leads to very small and insignificant differences in total effort. The goodness-of-fit hypothesis test between both organization designs also indicates that the differences in the distributions from which the total amount of work data were drawn are not significant. The Kolmogorov-Smirnov test statistic is $D_T = 0.0688$. The associated p-value is $p = 0.705$. The slight differences in the complexity metric according to Fig. 5.14 therefore do not lead to significant differences in probability distributions of the total amount of work if the work processes are systematically reorganized by the project manager to satisfy the constraint. As expected, the additional distribution fit test of the project duration data shows significant differences among the organizational conditions. The corresponding test statistic is $D_{x_{tot}} = 0.374$ $\left(p = 7.25 \cdot 10^{-13}\right)$.

The combined theoretical and computational analyses provide some evidence that the information-theory complexity metric EMC is not only a theoretically highly satisfactory quantity for the evaluation of emergent complexity of vector autoregressive processes as statistical models of cooperative work in PD, but under the investigated initial and boundary conditions it is also a good predictor of the mean and standard deviation of classic KPIs such as the project duration and total effort involved in completing the deliverables. Moreover, the results show that the self-developed concept of productivity balancing and designing organizations on the team level for diversity are promising for optimizing cooperative work in PD projects subjected to concurrent engineering.

5.3 Optimization of Release Period of Finished Work Between Design Teams at the Subsystem- and Component-Levels

The second study aims to optimize the release period using EMC as optimization objective function in projects where information about design, integration and tests of geometric/topological entities is deliberately withheld by systems engineering teams and not released to design teams working at the component level (Schlick et al. 2011; cf. Yassine et al. 2003). Based on the systems engineering considerations from Section 2.6, we focus on the subsystem and component levels of product design (levels 3 and 4 in V-model of the systems engineering process, see Fig. 2.11). According to Section 2.6, the outcome of finished work on subsystem level is "hidden" between the releases, and work in the subordinate component-level teams is based on product and process knowledge from the previous release period. This kind of noncooperative behavior is justified by the aim to improve the implementation of the product architecture through better subsystem-level design and validation and thus release only those designs that have a sufficient level of maturity. This can significantly reduce the overall amount of coordination. Optimizing the release period by using a formal complexity metric in conjunction with a mathematical model of periodically correlated work processes is an especially interesting application area because cross-hierarchical teamwork in large-scale PD projects can be designed systematically and unnecessary coordination efforts can be avoided. As in the previous chapter, it is also theoretically interesting to analyze whether, in addition to being valid for steady-state processes, EMC can be used to evaluate the preasymptotic range of the modeled project (phase). We start by formulating the unconstrained optimization problem and the presentation of its solution based on a complete enumeration of the release period. We then formulate a constrained optimization problem and solve it by applying the same principle. The constraint is that the expected total amount of work x_{tot} done over all tasks in the limit $n \rightarrow \infty$ (Eq. 91) is constant across the different release periods.

5.3.1 Unconstrained Optimization

5.3.1.1 Methods

The objective function developed in the second study quantifies the complexity of periodically correlated work processes in PD as a function of the release period under the dynamic regime of the state equation (Eq. 89). We seek to minimize emergent complexity. The release period was varied systematically in the range

[2;20] by increments of 1 week. The analytical calculations and Monte Carlo experiments consider different correlation lengths and simulate the work processes accordingly. The time scale is [weeks]. Using state equation 89 in conjunction with the closed-form solution from Eq. 247, we can express the EMC of the generated process in the original state space coordinates as

$$\mathrm{EMC_{PVAR}} = \frac{1}{2}\log_2\left(\frac{\mathrm{Det}\left[\sum_{k=0}^{\infty}\left(\left(\Phi_0^*\right)^{-1}\cdot\Phi_1^*\right)^k\cdot\left(\left(\Phi_0^*\right)^{-1}\cdot C^*\cdot\left(\Phi_0^*\right)^{-\mathrm{T}}\right)\cdot\left(\left(\Phi_0^*\right)^{-\mathrm{T}}\cdot\left(\Phi_1^*\right)^{\mathrm{T}}\right)^k\right]}{\mathrm{Det}\left[\left(\Phi_0^*\right)^{-1}\cdot C^*\cdot\left(\Phi_0^*\right)^{-\mathrm{T}}\right]}\right). \tag{320}$$

The transformed matrix Φ_0^* is defined in Eq. 85. Its inverse is given by the representation in Eq. 90. The autoregressive coefficient matrix Φ_1^* was defined in Eq. 86. The above definition of the complexity metric is an implicit function of the release period s as the dimension of both matrices Φ_0^* and Φ_1^* scales linearly with the period (cf. Eq. 86).

To ensure comparability between the first and second validation study, we also modeled a complete PD project that involves different teams. As before, the total time to complete the process corresponds to the project duration and the total amount of work done in the process to the project effort. For the Monte Carlo experiments, we developed an example project that includes two component-level and two subsystem-level tasks. The example project is based on the work of McDaniel et al. (1996). This work was developed into a complete PD project with periodically correlated work processes (Schlick et al. 2011). In the project, different teams process the tasks simultaneously. Every team is assigned a specific complex design task. Individual task processing is not considered. The component-level tasks aim to design and develop components of a high-end instrument panel for a completely new vehicle, including a drive-by-wire steering wheel, completely digital instrument panel cluster, navigation infotainment systems, gesture control for intuitive interaction etc. The subsystem-level tasks deal with panel design as a whole and integration testing of components. Due to the many components of the panel that must be newly developed, their multiple interfaces and the complex software functions, the finishing time of the complete project, even under the most favorable organizational conditions, is on average more than three and a half years of continuous development (cf. Sections 5.2.1.2 and 5.2.2.2). As in the previous validation study, the vector components of the state variable X_{ns+v} that are related to processing the subsystem-level and component-level tasks represent the relative number of labor units required to complete the tasks. We assume that both component-level design teams work at the same autonomous task processing rate $a_{11}^{\mathrm{C}} = a_{22}^{\mathrm{C}} = 0.90$. The tasks are coupled with symmetric strength and we have $a_{12}^{\mathrm{C}} = a_{21}^{\mathrm{C}} = 0.05$. Similarly, the teams responsible for subsystems design, validation and integration testing both work at the same (but slightly lower) autonomous task processing rate $a_{11}^{\mathrm{SS}} = a_{22}^{\mathrm{SS}} = 0.85$. The tasks are coupled at the same (but slightly higher) strength $a_{12}^{\mathrm{SS}} = a_{21}^{\mathrm{SS}} = 0.07$. Both subsystem level tasks generate 3% of finished work at each short iteration that is put in hold state until it is released at time step

ns $(n \in \mathbb{N})$. Hence, $a_{11}^{\mathrm{SSH}} = a_{22}^{\mathrm{SSH}} = 0.03$. Furthermore, the first component-level task generates 6% of finished work at each iteration for the first system-level task and vice versa. Hence, we have $a_{11}^{\mathrm{CSS}} = a_{11}^{\mathrm{SSC}} = 0.06$. The accumulated development issues of the subsystem-level teams are released to component-level teams in the form of a reworked subsystem design at the end of the period $\left(a_{11}^{\mathrm{HC}} = a_{22}^{\mathrm{HC}} = 1\right)$. Additional dynamical dependencies were not considered and therefore all other matrix entries were defined to be zero.

The complete representation for state equation 89 is as follows:

Combined dynamical operator $A_0^* = \left(\Phi_0^*\right)^{-1} \cdot \Phi_1^*$:

$$\left(\Phi_0^*\right)^{-1} \cdot \Phi_1^* = \begin{pmatrix} \Phi_1(s)\left(\Phi_1(1)\right)^{s-1} & 0 & 0 & \cdots & 0 \\ \left(\Phi_1(1)\right)^{s-1} & 0 & 0 & \cdots & 0 \\ \left(\Phi_1(1)\right)^{s-2} & 0 & 0 & \cdots & 0 \\ \vdots & \vdots & 0 & \ddots & \vdots \\ \Phi_1(1) & 0 & 0 & \cdots & 0 \end{pmatrix}$$

Work transformation sub-matrices:

$$A_0^{\mathrm{C}} = \begin{pmatrix} 0.90 & 0.05 \\ 0.05 & 0.90 \end{pmatrix} \tag{321}$$

$$A_0^{\mathrm{SS}} = \begin{pmatrix} 0.85 & 0.07 \\ 0.07 & 0.85 \end{pmatrix} \tag{322}$$

$$A_0^{\mathrm{CSS}} = \begin{pmatrix} 0.06 & 0 \\ 0 & 0 \end{pmatrix} \tag{323}$$

$$A_0^{\mathrm{SSC}} = \begin{pmatrix} 0.06 & 0 \\ 0 & 0 \end{pmatrix} \tag{324}$$

$$A_0^{\mathrm{SSH}} = \begin{pmatrix} 0.03 & 0 \\ 0 & 0.03 \end{pmatrix} \tag{325}$$

$$A_0^{\mathrm{HC}} = \begin{pmatrix} 1 & 0 \\ 0 & 1 \end{pmatrix} \tag{326}$$

Transformation matrices:

$$\Phi_1(1) = \begin{pmatrix} A_0^{\mathrm{C}} & A_0^{\mathrm{SSC}} & 0 \\ A_0^{\mathrm{CSS}} & A_0^{\mathrm{SS}} & 0 \\ 0 & A_0^{\mathrm{SSH}} & \{1-\varepsilon\} \cdot I_2 \end{pmatrix}$$

$$\Phi_1(s) = \begin{pmatrix} A_0^{\mathrm{C}} & A_0^{\mathrm{SSC}} & A_0^{\mathrm{HC}} \\ A_0^{\mathrm{CSS}} & A_0^{\mathrm{SS}} & 0 \\ 0 & 0 & \{\varepsilon\} \cdot I_2 \end{pmatrix}.$$

As explained in Section 2.7, the variable ε is necessary for an explicit complexity evaluation. $\mathrm{EMC_{PVAR}}$ then contains an correction term that scales linearly with ε, i.e. it holds that $\text{EMC}_{\text{PVAR}} = \text{EMC}_{\text{PVAR}}(\varepsilon = 0) + \varepsilon \cdot h\left(\Phi_0^*, \Phi_1^*, C^*\right)$ (cf. Eq. 320). We calculated with $\varepsilon = 10^{-4}$. By doing so, the finished work after release is set back to a nonzero but negligible amount in terms of productivity.

The initial state x_0^* was defined based on the assumption that all parallel tasks are initially to be fully completed and that no work is in hold state. Hence, for the minimum release period $s_{min} = 2$, we have:

$$x_0^* = \begin{pmatrix} 1 \\ 1 \\ 1 \\ 1 \\ 0 \\ 0 \\ 0 \\ 0 \\ 0 \\ 0 \\ 0 \\ 0 \end{pmatrix}, \tag{327}$$

For larger release periods, additional zeros were appended to the initial state.

Following the procedure of the first study (Section 5.1.1), we assumed that the standard deviation c_{ii} of performance fluctuations (Eq. 10) influencing task i in the project is proportional to the task processing rate. The proportionality constant is $r = 0.02$. Other correlations among vector components were not considered. Furthermore, we assumed that the variance of the fluctuations related to the finished work put in a hold state by system-level design teams is reduced by the factor $\varepsilon' = 10^{-4}$ and is therefore numerically negligible in the Monte Carlo experiments. Hence, we have the covariance matrix $C^* = E\left[\varepsilon_n^* \varepsilon_n^{*\text{T}}\right]$

$$C^* = \begin{pmatrix} C_s & 0 & 0 & 0 \\ 0 & C_1 & 0 & 0 \\ 0 & 0 & \ddots & 0 \\ 0 & 0 & 0 & C_1 \end{pmatrix},$$

where the submatrices are given by

$$C_1 = \{r^2\} \cdot \begin{pmatrix} \left(a_{11}^{\mathrm{C}}\right)^2 & 0 & 0 & 0 & 0 & 0 \\ 0 & \left(a_{22}^{\mathrm{C}}\right)^2 & 0 & 0 & 0 & 0 \\ 0 & 0 & \left(a_{11}^{\mathrm{SS}}\right)^2 & 0 & 0 & 0 \\ 0 & 0 & 0 & \left(a_{22}^{\mathrm{SS}}\right)^2 & 0 & 0 \\ 0 & 0 & 0 & 0 & \varepsilon'(1-\varepsilon)^2 & 0 \\ 0 & 0 & 0 & 0 & 0 & \varepsilon'(1-\varepsilon)^2 \end{pmatrix}$$

$$= \{0.02^2\} \cdot \begin{pmatrix} 0.9^2 & 0 & 0 & 0 & 0 & 0 \\ 0 & 0.9^2 & 0 & 0 & 0 & 0 \\ 0 & 0 & 0.85^2 & 0 & 0 & 0 \\ 0 & 0 & 0 & 0.85^2 & 0 & 0 \\ 0 & 0 & 0 & 0 & 10^{-4}\left(1-10^{-4}\right)^2 & 0 \\ 0 & 0 & 0 & 0 & 0 & 10^{-4}\left(1-10^{-4}\right)^2 \end{pmatrix} \tag{328}$$

and

$$C_s = \{r^2\} \cdot \begin{pmatrix} \left(a_{11}^{\mathrm{C}}\right)^2 & 0 & 0 & 0 & 0 & 0 \\ 0 & \left(a_{22}^{\mathrm{C}}\right)^2 & 0 & 0 & 0 & 0 \\ 0 & 0 & \left(a_{11}^{\mathrm{SS}}\right)^2 & 0 & 0 & 0 \\ 0 & 0 & 0 & \left(a_{22}^{\mathrm{SS}}\right)^2 & 0 & 0 \\ 0 & 0 & 0 & 0 & \varepsilon'\varepsilon^2 & 0 \\ 0 & 0 & 0 & 0 & 0 & \varepsilon'\varepsilon^2 \end{pmatrix}$$

$$= \{0.02^2\} \cdot \begin{pmatrix} 0.9^2 & 0 & 0 & 0 & 0 & 0 \\ 0 & 0.9^2 & 0 & 0 & 0 & 0 \\ 0 & 0 & 0.85^2 & 0 & 0 & 0 \\ 0 & 0 & 0 & 0.85^2 & 0 & 0 \\ 0 & 0 & 0 & 0 & 10^{-12} & 0 \\ 0 & 0 & 0 & 0 & 0 & 10^{-12} \end{pmatrix}. \tag{329}$$

The covariance matrix of the transformed error vector $\varepsilon_n^\star$ is given by $C^\star = \left(\Phi_0^*\right)^{-1} \cdot C^* \cdot \left(\Phi_0^*\right)^{-\mathrm{T}}$. Due to space limitations, we do not show this matrix.

The Mathematica software package from Wolfram Research was used to carry out the analytical calculations and the Monte Carlo experiments. The stopping criterion for the simulated projects was that a maximum of 5% of work remained for all tasks. The classic KPIs "project duration" and "total effort" were used in addition to EMC. To calculate these KPIs, we generated samples of 10,000 independent runs for each release period. We calculated the expected total amount of work x_{tot} analytically according to Eq. 91. Furthermore, the expected duration $\overline{T}_{\sigma=0}$ was calculated under the assumption that the processing of the development tasks is unperturbed.

We analyzed the data sets generated in the Monte Carlo experiments in the same manner as in the previous studies and used Q-Q plots to evaluate whether they come from a log-normal distribution (see Section 5.2.1.1). If the Q-Q plots showed that the quantiles of the theoretical and data distributions agree, the log-likelihood functions of the simulated time and effort values were calculated. As before, the log-likelihood values are denoted by *LL*. Following the procedure from Sections 5.2.1.1 and 5.2.2.1, we used the likelihood ratio test developed by Kundu and Raqab (2007) to discriminate between log-normal and generalized Rayleigh distribution functions.

5.3.1.2 Results and Discussion

Figure 5.17 shows the results of a typical run of the Monte Carlo simulation for the parameterized project model with initial state x_0^* assuming a minimal release period of $s = s_{min} = 2$ [weeks]. The finished work that was put in hold state when processing both subsystem-level tasks at each short iteration is also shown in the list plot around the abscissa. Additional typical time series for extended release periods with $s = 10$ and $s = s_{max} = 20$ [weeks] are shown in Figs. 5.18 and 5.19, respectively.

As in Section 5.1, we calculated a separate statistic for the three release periods $s \in \{2, 10, 20\}$ [weeks] based on 100 simulated projects and visualized them in the form of error list plots. These plots are shown in Figs. 5.20, 5.21 and 5.22 to give a comprehensive overview of how the release period length affects the means and 95% confidence intervals of work remaining. The comparison of Figs. 5.20, 5.21 and 5.22 shows that an extension of the release period from 2 weeks to 10 or 20 weeks increases the average work remaining before the stopping criterion of 5% is met. Furthermore, it is not difficult to see that, for the development task on the component level, the magnitude of the performance fluctuations is inversely proportional to the period length: the longer the release period, the more single instances of task processing deviate from the mean (unperturbed) work remaining and the larger the average performance variability. For all three release periods significant deviations from the means occur as early as in week 20 and proceed until the project is finished. Figures 5.21 and 5.22 also clearly show that the "sawtooth" behavior of finished work that is put in hold state by the teams processing the subsystem-level tasks spills over to the component-level tasks and is exacerbated by the unpredictable performance fluctuations. Conversely, under the given boundary conditions, the processing of the subsystem-level tasks is relatively fast and smooth.

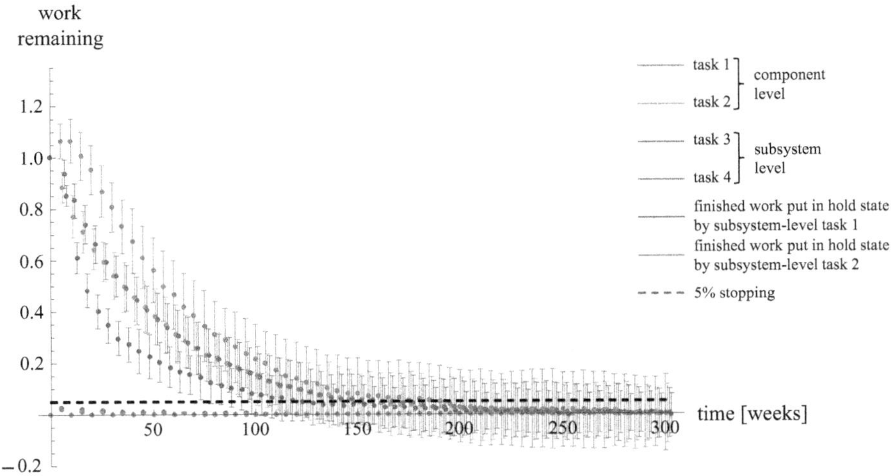

Fig. 5.20 Error list plot of work remaining in simulated product development projects with correlated work processes. The release period is $s = 2$ [weeks]. A total of 100 separate and independent runs were calculated. The plot shows means of work remaining as note points and 95 % confidence intervals as error bars. Note points have been offset to distinguish the error bars. The other simulation conditions and parameters are the same as in Fig. 5.17

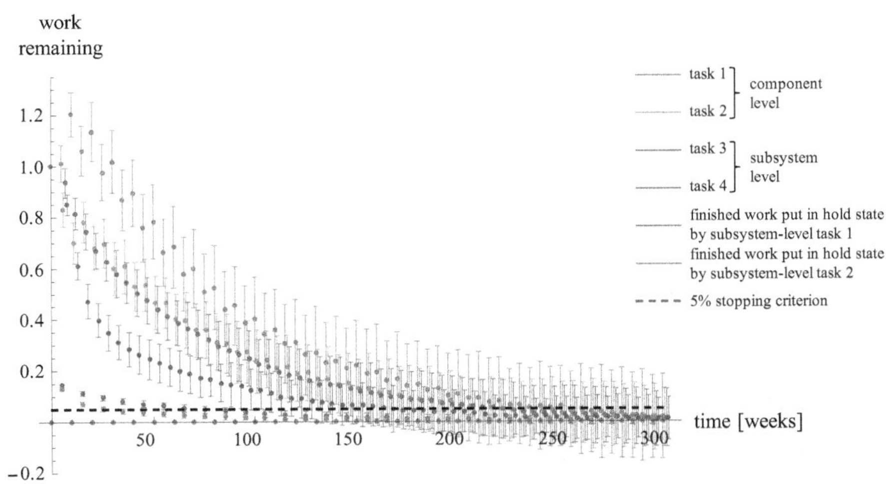

Fig. 5.21 Error list plot of work remaining in simulated product development projects with correlated work processes. The release period is $s = 10$ [weeks]. The other simulation conditions and parameters are the same as in Fig. 5.17

When the development tasks are processed deterministically without performance fluctuations, the shortest project duration of $\overline{T}_{\sigma=0}(s = 2) = 182$ [weeks] is obtained, as one must expect, for the minimum release period of 2 weeks. The

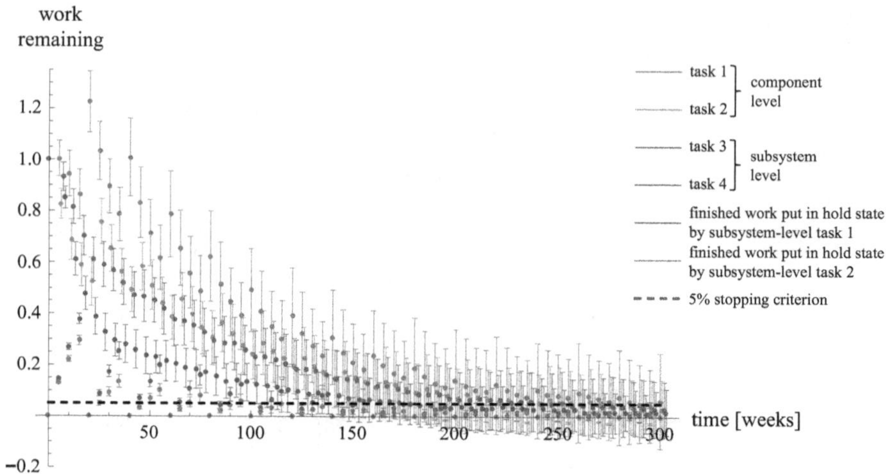

Fig. 5.22 Error list plot of work remaining in simulated product development projects with correlated work processes. The release period is $s = 20$ [weeks]. The other simulation conditions and parameters are the same as in Fig. 5.17

longest finishing time of $\overline{T}_{\sigma=0}(s = 2) = 259$ [weeks] occurs in the case of the maximum release period of 20 weeks.

The 10,000 runs that were computed for each release period show that the minimum release period of 2 weeks leads to a mean project duration of $\overline{T}(s = 2) = 158.553$ [weeks]. The standard deviation is $SD(s = 2) = 34.234$ [weeks]. Both the mean project duration and the standard deviation are minimal within the sample. If the release period of finished work is extended to 10 weeks, the mean project duration increases to $\overline{T}(s = 10) = 212.828$ [weeks] and the standard deviation to $SD(s = 10) = 51.181$ [weeks]. An additional extension of the release period to the maximum of 20 weeks further increases the mean project duration and standard deviation, and we have $\overline{T}(s = 20) = 226.333$ [weeks] and $SD(s = 20) = 54.000$ [weeks]. Figure 5.23 shows the histograms of the calculated project duration for the three considered release periods. For the longest release period of 20 weeks the histogram shows quite heavy oscillations of the distribution of the probability mass for project durations of longer than 175 weeks. This effect is due to the long time span between average release points. The oscillation period follows the release period.

The means and standard deviations of the total amount of work x_{tot} (see Section 2.7) in the simulated projects follows a similar pattern of growth (Fig. 5.24). However, the accumulation of work reduces the intrinsic periodic correlations, and an oscillation of the distribution of the probability mass does not occur for the release period of 20 weeks.

We also calculated analytically the expected total amount of work x_{tot} in the modeled project for different release periods. The expected total amount of work is by definition accumulated over an infinite past history and therefore does not take

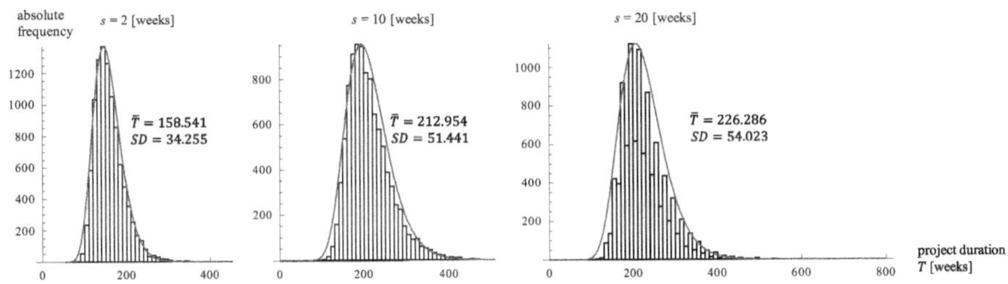

Fig. 5.23 Histograms of the project duration obtained for three different release periods $s = 2$, 10 and 20 [weeks]. We computed 10,000 independent runs. In these runs all tasks were initially 100 % incomplete. The stopping criterion for the simulated projects was that a maximum of 5 % of work remained for all tasks. The Monte Carlo experiments were based on state equation 89. The parameters are given by Eqs. 321–326. We overlaid the probability density function of a log-normal distribution for comparison

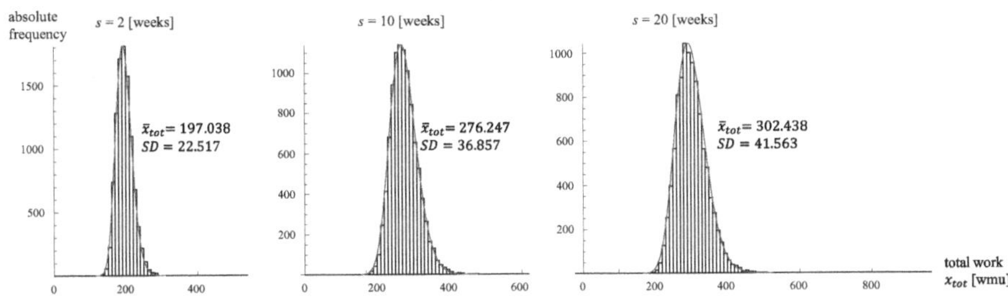

Fig. 5.24 Histograms of the total amount of work obtained for three different release periods $s = 2$, 10 and 20 [weeks]. We computed 10,000 independent runs. In these runs all tasks were initially 100 % incomplete. The stopping criterion for the simulated projects was that a maximum of 5 % of work remained for all tasks. The Monte Carlo experiments were based on state equation 89. The parameters are given by Eqs. 321–326. We overlaid the probability density function of a log-normal distribution for comparison

the stopping criterion of the Monte Carlo experiments into account. The results are presented in Fig. 5.25, which shows that the total amount of work is smallest for the shortest period length $s_{min} = 2$ and grows sublinearly with the release period.

The complexity values EMC_{PVAR} (Eq. 320) that were obtained for different period lengths are shown in Fig. 5.26.

The comparison of Figs. 5.25 and 5.26 shows that the complexity metric EMC_{PVAR} closely resembles the functional behavior of the expected total amount of work x_{tot} in the modeled project over different periods. Moreover—and most importantly in view of the objective of the study—the smallest complexity values are assigned to periodically correlated work processes with minimum period $s_{opt} = 2$ [weeks]. In other words, for the given initial and boundary conditions it makes sense to minimize the period in which information about system design and integration testing of geometric/topological entities is deliberately withheld by subsystem-level teams and not released to component-level teams. By minimizing

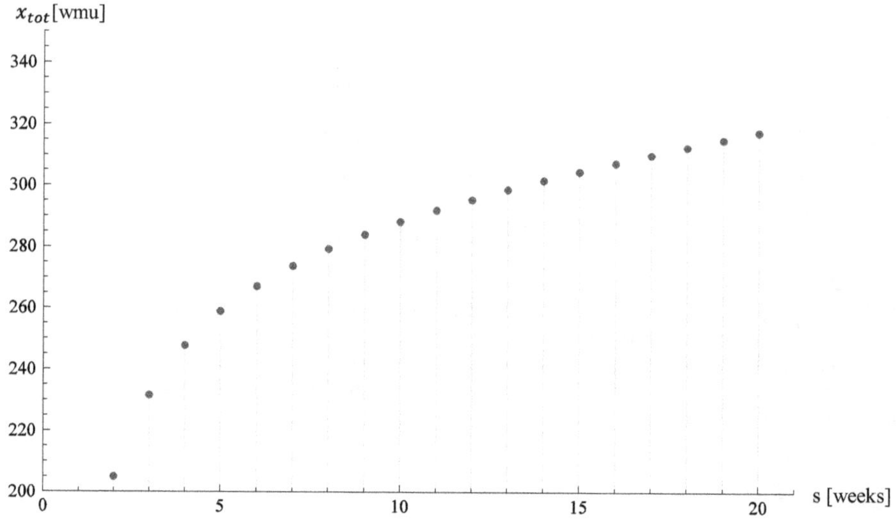

Fig. 5.25 Expected total amount of work x_{tot} in the modeled project according to Eq. 91. The units are work measurement units [wmu] that refer to the definition of the state of work remaining in the project. The parameters are given by Eqs. 321–326

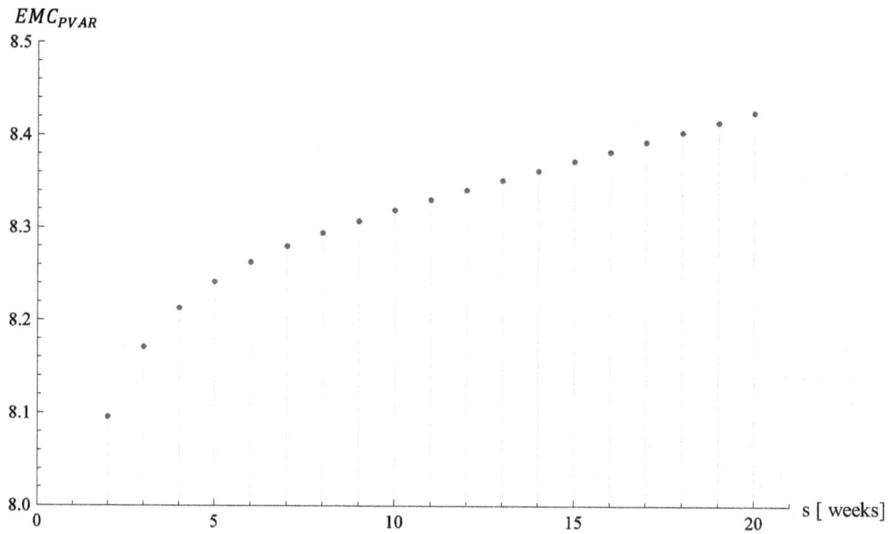

Fig. 5.26 Effective measure complexity EMC_{PVAR} in the modeled project according to Eq. 320. The parameters are given by Eqs. 321–326

the period length, the emergent complexity can be kept to a minimum and the total effort involved in the project can be reduced as far as possible. This recommendation is also fully supported by the results of the Monte Carlo experiments because they show that, as the period length increases, the means and standard deviation of the project duration and the total amount of work also increase (see Figs. 5.23 and 5.24).

Similar to the optimization of project organization in Sections 5.2.1.1 and 5.2.2.1, the descriptive statistical analysis based on the Q-Q plots showed for all periods that the fit of the log-normal distribution is very good and noticeable deviations from the corresponding straight line occurred only for the tails on the far left and the far right sides. The maximum deviation from the straight line is in all cases smaller than 0.2. This corresponds to an estimated p-value for a two-tailed test of less than 0.05. Interestingly, although for the longest release period of 20 weeks the histogram in Fig. 5.23 shows significant oscillations of the distribution of the probability mass for project durations of longer than 175 weeks, these oscillations only have a very small effect on the displacement of data points in the Q-Q plots. Hence, even for periodically correlated work processes with large periods, the choice of this type of distribution seems to be appropriate. The evaluation of the log-likelihood functions of the time and effort variables also consistently led to high values. The log-likelihood of the simulated project duration decreases monotonically with the period and is between $LL_{max} = -48.767$ (s = 2) and $LL_{min} = -53.201$ (s = 20). For the total amount of work the values are much larger, but follow the same pattern. We have $LL_{max} = -45.138$ for $s = 2$ and $LL_{min} = -51.177$ for $s = 20$. Therefore, as one must expect from the results of the previous Monte Carlo studies, significant deviations from normality can also occur for periodically correlated work processes, and lead times as well as amounts of work far from the average unperturbed process are quite likely. In contrast to the optimization of project organization, the results of the likelihood ratio tests according to Kundu and Raqab (2007) show that for all periods the project duration can be modeled more accurately by a log-normal than by a generalized Raleigh distribution function. On the other hand, the tests indicate that the generalized Rayleigh distribution should be preferred to represent the total amount of work: for both KPIs the probability of correct selection is larger than 0.95. However, compared with the previous Monte Carlo studies the values of the test statistic are in an even narrower range of [−137;33] and therefore indicate a low discriminative power.

5.3.2 *Constrained Optimization*

After presenting and discussing the results of the basic unconstrained optimization problem, we move on to formulating and solving a corresponding constrained optimization problem. This is done in a similar manner as in Section 5.1.2. The constraint is that the expected total amount of work x_{tot} remains on a constant level among the different release periods and does not vary with the period as in the previous chapter. To satisfy this constraint, systematic interventions were carried out on the subsystem-level of cooperative development. We start by presenting analytical complexity results and go on to present results of the Monte Carlo experiments.

5.3.2.1 Methods

The objective function for the constrained optimization quantifies the emergent complexity of periodically correlated work processes in PD in the same manner as in the previous chapter, and is given by Eq. 320. We seek to minimize complexity under the constraint that the expected total amount of work x_{tot} according to Eq. 91 is equal to 204.897 [wmu] for different release periods. This expected total effort corresponds to the minimum value that was identified in the previous study. Recall that this minimum expected total amount of work is obtained for the minimum release period $s_{min} = 2$ [weeks] and for a parameter vector according to Eqs. 321–329. Starting with the minimum period, the release period was extended by increments of 1 week until the maximum release period $s_{max} = 20$ [weeks] was reached. The analytical considerations and Monte Carlo experiments must therefore not only consider the different correlation lengths, they must also adjust the independent parameters by systematic algorithmic intervention of the experimenter so that the total expected effort in the project does not change under the different release conditions. To keep x_{tot} constant, the independent parameters a_{11}^{SSH} and a_{22}^{SSH} were adjusted (see Eq. 325). These parameters represent the fraction of work that is put in hold state by the subsystem design teams at each short iteration before it is released at the end of period s. The parameter adjustment was done by a self-developed iterative method in which the reference values $a_{11}^{\mathrm{SSH}} = 0.03$ and $a_{22}^{\mathrm{SSH}} = 0.03$ were reduced incrementally until the expected value $x_{tot}(3 \leq s < 20)$ did not deviate more than 10^{-6}[wmu] from the correct value $x_{tot}(s = 2) = 204.897$. Both reference values were reduced by the same amount and it always held $a_{11}^{\mathrm{SSH}} = a_{22}^{\mathrm{SSH}}$. The time scale was not modified.

Following the previous procedures, we assumed that the standard deviation c_{ii} of performance fluctuations (Eq. 10) influencing task i in the project is proportional to the task processing rate with proportionality constant $r = 0.02$. Other correlations between vector components were not considered. The variance of the fluctuations related to the finished work that is put in hold state is again reduced by the factor 10^{-3}.

The Mathematica software package from Wolfram Research was used to carry out the analytical calculations and the Monte Carlo experiments. The stopping criterion for the Monte Carlo experiments was that a maximum of 5 % of work remained for all tasks in the simulated projects. It was assumed that all development tasks on component- and subsystem-levels were initially fully incomplete. In addition to EMC, we used the KPIs "project duration" and "total effort" to evaluate performance as before. To calculate these KPIs, 10,000 independent runs were considered for each release period. To allow a systematic comparison of the data, the expected duration $\overline{T}_{\sigma=0}$ was calculated under the assumption that the processing of the development tasks is unperturbed.

The data sets were analyzed by the same descriptive and inferential statistical methods that were described in Sections 5.2.2.1 and 5.3.1.1, respectively. If the Q-Q plots in the descriptive analysis showed that the quantiles of the theoretical and data distributions agreed, the log-likelihood functions of the simulated time and effort values were calculated. As before, the log-likelihood values are denoted by *LL*. The likelihood ratio test developed by Kundu and Raqab (2007) and the corresponding simulation techniques were used to discriminate between log-normal and generalized Rayleigh distribution functions. For a better comparability of results, we applied the same non-parametric location equivalence test as in Section 5.2.2.1 in the inferential statistical analysis. To carry out this test, we drew additional samples based on 100 independent runs. To simplify the interpretation and discussion of the data, we only considered three release periods, namely $s_{min} = 2$, $s_{med} = 10$ and $s_{max} = 20$, in order to guarantee a sufficient coverage of the complete interval. We hypothesized that lower values of the complexity metric EMC_{PVAR} lead to a significantly lower expected project duration. We also hypothesized that different levels of the complexity metric EMC_{PVAR} do not correspond to significantly different means of the total amount of work in the Monte Carlo experiments. The null hypotheses $H_{0,T}$ and $H_{0,x_{tot}}$ were formulated accordingly. To evaluate these hypotheses the Kruskal-Wallis location equivalence test was used. The level of significance was set to $\alpha = 0.05$.

Following the procedure from Section 5.1.2, additional goodness-of-fit hypothesis tests were carried out to evaluate the differences between the distributions of performance data for the three release periods. The focus was on paired comparisons between the minimum release period $s_{min} = 2$ and the other periods. The null hypothesis $H_{0,gof}$ was always that performance data drawn from a sample with release period $s_{med} = 10$ or $s_{max} = 20$ do not come from a different distribution than the data obtained for the minimum release period $s_{min} = 2$. The alternative hypothesis $H_{a,gof}$ is that the data come from a different distribution. The Kolmogorov-Smirnov test was used to evaluate the hypotheses. The level of significance was also set to $\alpha = 0.05$.

5.3.2.2 Results and Discussion

In order to satisfy the constraint imposed for the total amount of work, the fractions $a_{11}^{\text{SSH}} = 0.03$ and $a_{22}^{\text{SSH}} = 0.03$ of work that are put in hold state by the teams working on subsystems-level at each short iteration had to be reduced by a minimum value of 0.00769 for release period $s = 3$ and a maximum value of 0.01646 for release period $s_{max} = 20$. A list plot of the necessary reductions of a_{11}^{SSH} and a_{22}^{SSH} is shown in Fig. 5.27. It shows that the longer the release period, the more the fraction of work put in hold state at each short iteration must be reduced in order to satisfy the constraint. The reduction scales sublinearly with the length of the release period.

The corresponding values of the complexity metric EMC_{PVAR} are shown in Fig. 5.28. Interestingly, the constrained optimization of the release period leads to complexity values that strictly decrease as the period increases. Hence, the release period minimizing emergent complexity in the sense of the complexity metric under the constraint $x_{tot}(2) = \ldots = x_{tot}(20) = 204.897$ is the maximum

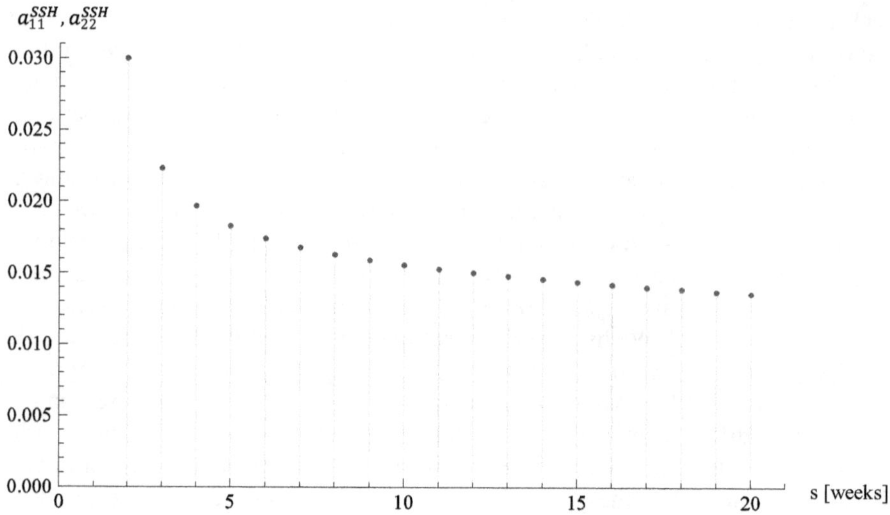

Fig. 5.27 Adjustments of original parameters $a_{11}^{SSH} = 0.03$ and $a_{22}^{SSH} = 0.03$ (Eq. 325) that were made to satisfy the constraint on the expected total amount of work $x_{tot}(2) = \ldots = x_{tot}(20) = 204.897$ in the modeled projects with periodically correlated work processes. The additional parameters were not changed and are given by Eqs. 321–326

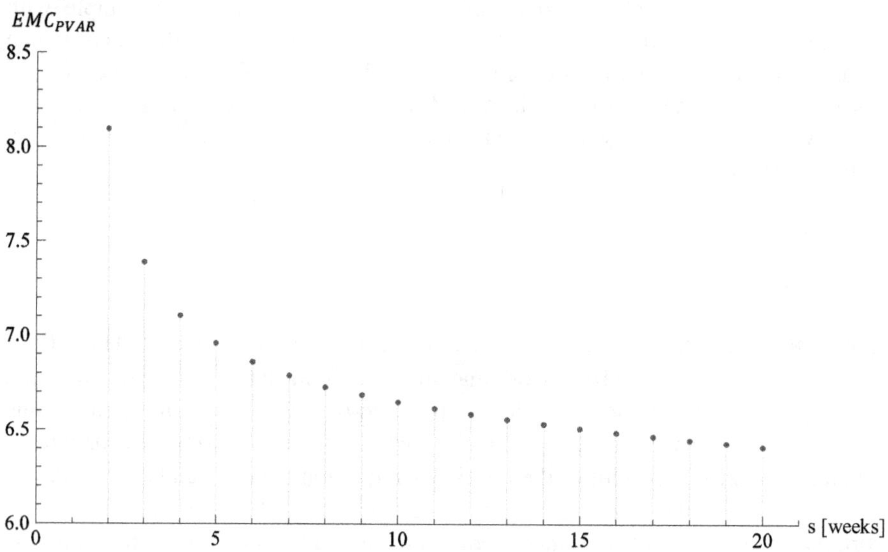

Fig. 5.28 Effective measure complexity EMC_{PVAR} in the modeled project according to Eq. 320 under the constraint that the expected total amount of work is kept on a constant level, i.e. $x_{tot}(2) = \ldots = x_{tot}(20) = 204.897$. To keep the expected total amount of work constant, the original parameters $a_{11}^{SSH} = 0.03$ and $a_{22}^{SSH} = 0.03$ (Eq. 325) were adjusted by the same value (see Fig. 5.27). The additional parameters were not changed and are given by Eqs. 321–326

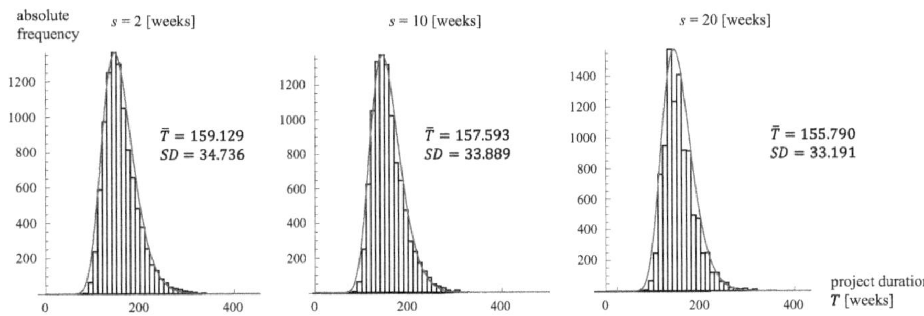

Fig. 5.29 Histograms of the project duration obtained for the three release periods $s = 2,\ 10$ and 20 [weeks] under the constraint $x_{tot}(2) = x_{tot}(10) = x_{tot}(20) = 204.897$. We computed 10,000 independent runs. In these runs all tasks were initially 100 % incomplete. The stopping criterion for the simulated projects was that a maximum of 5 % of work remained for all tasks. The Monte Carlo experiments are based on state equation 89. The parameters are given by Eqs. 321–326. The adjustments of parameters a_{11}^{SSH} and a_{22}^{SSH} follows the list plot from Fig. 5.27. We overlaid the probability density function of a log-normal distribution for comparison

period $s_{opt}^{c} = 20$ [weeks]. This result is in stark contrast to the solution of the unconstrained optimization problem, in which minimum complexity values were assigned to periodically correlated work processes with minimum period and we had $s_{opt} = 2$ [weeks]. Thus, for the given constraint on the total effort, the theory recommends to maximizing the period in which information about system design and integration testing of geometric/topological entities is deliberately withheld by subsystem-level teams and not released to component-level teams. When extending the period length to the largest possible value, the emergent complexity in the sense of Grassberger's measure can be kept to a minimum. Note that this recommendation, in principle, is only valid if the work processes can be organized in a way that for large release periods it is possible to significantly reduce the fractions of work that are put in hold state by the teams working on subsystems-level at each short iteration. Significant means in this context reducing the fractions of work by at least 25 %. In applied project management this is extremely difficult to achieve.

Interestingly, through the simulated reduction of the fractions of work that are put in hold state, the deterministic processing of the development tasks leads to finishing times that only slightly fluctuate between $\overline{T}_{\sigma=0}(s = 16) = 175$ and $\overline{T}_{\sigma=0}(s = 3) = 181$ [weeks].

The corresponding results of the Monte Carlo experiments for minimum and maximum release periods are shown in Fig. 5.29. The results for a release period of 10 weeks are also included (cf. Fig. 5.23). The associated histograms of the total amount of work are shown in Fig. 5.30.

The analysis of project duration and total amount of work from Figs. 5.29 and 5.30 shows that, once again, the values of the complexity metric EMC_{PVAR} that were given in Fig. 5.28 are predictive for the means and standard deviations of these classic KPIs. The smaller the complexity values assigned to a given release period, the shorter the resulting mean project duration and mean total amount of work, and

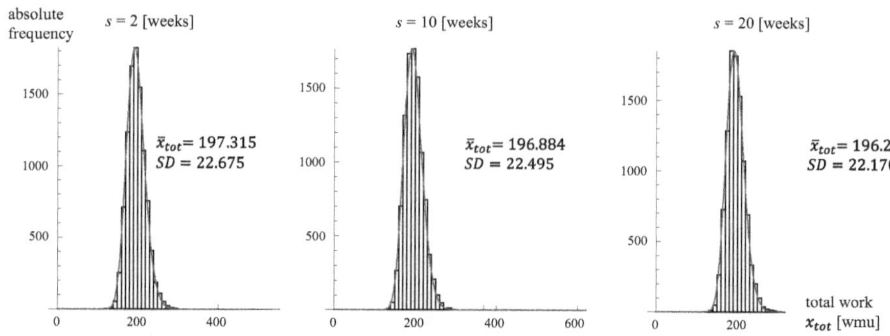

Fig. 5.30 Histograms of the total amount of work obtained for the three release periods $s = 2$, 10 and 20 [weeks] under the constraint $x_{tot}(2) = x_{tot}(10) = x_{tot}(20) = 204.897$. We computed 10,000 independent runs. In these runs all tasks were initially 100 % incomplete. The stopping criterion for the simulated projects was that a maximum of 5 % of work remained for all tasks. The Monte Carlo experiments are based on state equation 89. The parameters are given by Eqs. 321–326. The adjustments of parameters a_{11}^{SSH} and a_{22}^{SSH} follows the list plot from Fig. 5.27. We overlaid the probability density function of a log-normal distribution for comparison

the smaller the corresponding standard deviations. However, the performance differences are rather small compared to the previous analysis. The question is therefore whether these small differences are significant or not.

As in Sections 5.2.1.2, 5.2.2.2 and 5.3.1.2, the descriptive statistical analysis revealed a very good fit of the log-normal distribution for all release periods. Noticeable deviations from the corresponding straight line occurred in the Q-Q plots only for the tails on the far left and the far right sides of the theoretical distributions. However, the maximum deviation from the straight line is in all cases smaller than 0.2. Similar to the unconstrained optimization, although for the longest period of 20 weeks the histogram in Fig. 5.29 shows significant oscillations of the distribution of the probability mass for project durations of longer than 175 weeks, the deviation of the data points in the Q-Q plots from the straight line is very small and therefore the choice of a log-normal distribution seems to be appropriate. The evaluation of the log-likelihood functions consistently leads to satisfactorily high values. However, the log-likelihood of the simulated project durations grows monotonically with the period and is between $LL_{min} = -48.924 \quad (s = 2)$ and $LL_{max} = -44.416 \quad (s = 20)$. This result is in contrast to the solution of the unconstrained optimization problem, in which minimum likelihood values were assigned to work processes with maximum period. The evaluation of the total amount of work leads to the same monotonic growth of the log-likelihood with the release period and values between $LL_{min} = -45.226 \ (s = 2)$ and $LL_{max} = -44.985 \ (s = 20)$ are obtained. Due to the constraint on the total amount of work the range of the log-likelihood values is much smaller. We can conclude that under this execution condition, too, significant deviations from normality can occur, and lead times as well as amounts of work far from the average unperturbed process are quite likely. The likelihood ratio tests according to Kundu and Raqab (2007) lead to very similar results as in the unconstrained optimization problem: for all periods the project duration can be modeled more accurately by a log-normal than by a generalized Raleigh distribution function, while the generalized Rayleigh distribution should be preferred to model the total amount of work under uncertainty. For both KPIs the Monte Carlo simulations reveal estimated probabilities of correct selection that are larger than 0.95. The values of the test statistic are in the range of [−75;4] and therefore, as before, indicate only a low discriminative power.

The final question in this chapter is whether the comparably small differences in performance as shown in the histograms of Figs. 5.29 and 5.30 are significant or not. For 100 additional independent runs, the Kruskal-Wallis test on the project duration data shows that the location differences between the three release periods are not significant ($K_T = 2.5157, p = 0.2852$). Hence, the null hypothesis $H_{0,T}$ that the true location parameters of the samples are equal cannot be rejected on the significance level of $\alpha = 0.05$. The Kruskal-Wallis test on the total amount of work data comes to a similar conclusion and shows that the location differences between the three release periods are not significant ($K_{x_{tot}} = 1.2870, p = 0.5270$). The null hypothesis $H_{0,x_{tot}}$ that the true location parameters of the samples are equal also cannot be rejected on the significance level of $\alpha = 0.05$.

In spite of these insignificant differences in mean performance, the goodness-of-fit hypothesis test between minimum release period $s_{min} = 2$ and maximum period $s_{max} = 20$ indicates significant differences in the distributions of the project duration from which data were drawn. In this case, the Kolmogorov-Smirnov test statistic is $D_{T,2-20} = 0.138$. The associated p-value is $p = 0.0401$. Hence, the differences in the complexity metric EMC_{PVAR} according to the interval bounds shown in Fig. 5.28 also lead to significant differences in probability distributions. The additional distribution fit tests of the project duration and total effort data do not show significant differences among the experimental conditions. The test statistic related to the project duration data for release periods $s_{min} = 2$ and $s_{med} = 10$ is $D_{T,2-10} = 0.096$ ($p = 0.2961$). For the simulated total amount of work the test statistic for release periods $s_{min} = 2$ and $s_{med} = 10$ is $D_{x_{tot},2-10} = 0.0945$ ($p = 0.3135$) and for periods $s_{min} = 2$ and $s_{max} = 20$ it is $D_{x_{tot},2-20} = 0.1038$ ($p = 0.2159$).

The combined theoretical and computational analyses also provide, for the periodically correlated work processes, some evidence that the information-theory complexity metric EMC is a theoretically very satisfactory quantity and that, under the investigated organizational conditions, it is also a good predictor of the mean and standard deviation of the project duration for different release periods and different formulations of the optimization problem—either unconstrained in the total amount of work (see Fig. 5.26 in conjunction with Figs. 5.23 and 5.24) or constrained (see Fig. 5.28 in conjunction with Figs. 5.29 and 5.30). The predictive power is much larger for the unconstrained optimization of the release period. In summary, inconsistencies in performance predictions or disordinal interactions between the independent parameters did not occur and prove the predictive validity of the closed-form solutions for the investigated class of models of cooperative work.

References

Akaike, H.: Information theory and an extension of the maximum likelihood principle. In: Petrov, B.N., Csaki, F. (eds.) Second International Symposium of Information Theory, pp. 267–281. Akademia Kiado, Budapest (1973)

Akaike, H.: A new look at the statistical model identification. IEEE Trans. Automat. Contr. **19**, 716–723 (1974)

Baker, K.R., Trietsch, D.: Principles of Sequencing and Scheduling. Wiley, New York, NY (2009)

Bennett, J.M., Ho, D.S.K.: Project Management for Engineers. World Scientific, Singapore (2014)

Bialek, W., Nemenman, I., Tishby, N.: Predictability, complexity and learning. Neural Computation **13**(1), 2409–2463 (2001)

Burnham, K.P., Anderson, D.R.: Model Selection and Multimodel Inference: A Practical Information-Theoretic Approach. Springer, New York, NY (2002)

Cox, T.: Cultural Diversity in Organizations: Theory, Research and Practice. Berrett-Koehler, San Francisco, CA (1994)

de Cock, K.D.: Principal angles in system theory, information theory and signal processing. Ph.D. thesis, Katholieke Universiteit Leuven (2002)

Eaves, R., Woods-Grooves, S.: Criterion validity. In: Salkind, N. (ed.) Encyclopedia of Measurement and Statistics, pp. 201–203. Sage, Thousand Oaks, CA (2007)

Field, A.: Discovering Statistics Using SPSS, 3rd edn. Sage, London (2009)

Grassberger, P.: Toward a quantitative theory of self-generated complexity. Int. J. Theor. Phys. **25** (9), 907–938 (1986)

Halanay, A., Rasvan, V.: Stability and Stable Oscillations in Discrete Time Systems. CRC, Boca Raton (2000)

Huberman, B.A., Wilkinson, D.M.: Performance variability and project dynamics. Comput. Math. Org. Theory **11**(4), 307–332 (2005)

Jaynes, E.T.: Information theory and statistical mechanics. Phys. Rev. **106**, 620–630; **108**, 171–190 (1957)

Jaynes, E.T.: Probability theory. In: Larry Bretthorst, G. (ed.) The Logic of Science. Cambridge University Press, Cambridge, UK (2003)

Kundu, D., Raqab, M.: Discriminating between the generalized rayleigh and log-normal distribution. Statistics **41**(6), 505–515 (2007)

Lazear, E.P.: Personnel Economics for Managers. Wiley, New York, NY (1998)

Lazear, E.P.: Globalization and the market for team-mates. Econ. J. **104**(454), 15–40 (1999)

Lee, D.A., Hogue, M., Hoffman, D.: Time histories of expenditures for defense acquisition programs in the development phase. In: Proceedings of the 1993 Annual Meeting of the International Society for Parametric Analysis, 1993

Lee, D.A.: Norden-Rayleigh Analysis: A Useful Tool for EVM in Development Projects. Logistics Management Institute, The Measurable News, March 2002

Li, L.: Some notes on mutual information between past and future. J. Time Ser. Anal. **27**(2), 309–322 (2006)

Li, L., Xie, Z.: Model selection and order determination for time series by information between the past and the future. J. Time Ser. Anal. **17**(1), 65–84 (1996)

Liker, J.K.: The Toyota Way: 14 Management Principles from the World's Greatest Manufacturer. McGraw-Hill, Maidenhead, Berkshire (2004)

Lütkepohl, H.: Comparison of criteria for estimating the order of a vector autoregressive process. J. Time Ser. Anal. **6**, 35–52 (1985)

Mantalos, P., Mattheou, K., Karagrigoriou, A.: Vector autoregressive order selection and forecasting via the modified divergence information criterion. Int. J. Comput. Econ. Econom. **1** (3/4), 254–377 (2010)

McDaniel, C.D.: A linear systems framework for analyzing the automotive appearance design process. Master's Thesis (Management/Electrical Engineering), MIT, Cambridge, MA (1996)

Mihm, J., Loch, C.: Spiraling out of control: problem-solving dynamics in complex distributed engineering projects. In: Braha, D., Minai, A.A., Bar-Yam, Y. (eds.) Complex Engineered Systems: Science Meets Technology, pp. 141–158. Springer, Berlin (2006)

Mihm, J., Loch, C., Huchzermeier, A.: Problem-solving oscillations in complex engineering. Manag. Sci. **46**(6), 733–750 (2003)

Neumaier, A., Schneider, T.: Estimation of parameters and eigenmodes of multivariate autoregressive models. ACM Trans. Math. Softw. **27**, 27–57 (2001)

Norden, P.V.: Useful tools for project management. In: Starr, M.K. (ed.) Management of Production. Penguin Books, Baltimore, MA (1970)

Pereira, D.C., Vilas, D., Monteil, N.R., Prado, R.R., del Valle, A.G.: Autocorrelation effects in manufacturing systems performance: a simulation analysis. In: Proceedings of the Winter Simulation Conference (WSC 2012), 2012

Peters, T.: Get innovative or get dead. California Management Review **33**(2), 9–23 (1991)

Putnam, L.H.: A general empirical solution to the macro software sizing and estimating problem. IEEE Trans. Softw. Eng. **4**(4), 345–361 (1978)

Salkind, N., Rasmussen, K.: Validity theory. In: Salkind, N. (ed.) Encyclopedia of Measurement and Statistics. Sage, Thousand Oaks, CA (2007)

Schlick, C.M., Duckwitz, S., Gärtner, T., Tackenberg, S.: Optimization of concurrent engineering projects using an information-theoretic complexity metric. In: Proceedings of the 11th International DSM Conference, 2009, pp. 53–64

Schlick, C.M., Schneider, S., Duckwitz, S.: Modeling of periodically correlated work processes in large-scale concurrent engineering projects based on the DSM. In: Proceedings of the 13th International Dependency and Structure Modeling Conference, DSM 2011, pp. 273–290

Schlick, C.M., Schneider, S., Duckwitz, S.: A universal complexity criterion for model selection in dynamic models of cooperative work based on the DSM. In: Proceedings of the 15th International Dependency and Structure Modeling Conference, DSM 2013, pp. 99–105

Sterman, J.D.: Business Dynamics: Systems Thinking and Modeling for a Complex World. McGraw-Hill Higher Education, Burr Ridge, IL (2000)

Still, S.: Information bottleneck approach to predictive inference. Entropy **16**(1), 968–989 (2014)

Suh, N.P.: Complexity—Theory and Applications. Oxford University Press, Oxford (2005)

Surles, J.G., Padgett, W.J.: Inference for reliability and stress-strength for a scaled Burr Type X distribution. Lifetime Data Anal. **7**(1), 187–200 (2001)

Tkačik, G., Bialek, W.: Information Processing in Living Systems. arXiv:1412.8752v1 [q-bio. QM], 2014

Trietsch, D., Mazmanyan, L., Gevorgyan, L., Baker, K.R.: Modeling activity times by the Parkinson distribution with a lognormal core: theory and validation. Eur. J. Oper. Res. **216**(2), 445–452 (2012)

Yassine, A., Joglekar, N., Eppinger, S.D., Whitney, D.: Information hiding in product development: the design churn effect. Res. Eng. Des. **14**(3), 145–161 (2003)

Chapter 6
Conclusions and Outlook

In this book, we have presented theoretical and empirical analyses of the dynamics and emergent complexity of cooperative work in product development projects. To do so, we have mainly focused on projects that are organized according to the management concept of concurrent engineering. Concurrent engineering offers a systematic approach to the integrated, concurrent design of products and their related processes, including manufacture and support (Winner et al. 1988). Designed to encourage developers to consider all elements of the product life cycle from the outset, it requires intensive cooperation between and within teams. We opted for a model-driven approach and formulated various mathematical models to analyze cooperative work in these kinds of open organizational systems. These models are based on the fundamental work of Smith and Eppinger (1997, 1998) and Yassine et al. (2003) on a deterministic product development flow and also take account of the important developments by Huberman and Wilkinson (2005) on the theory of stochastic processes. We preferred statistical models, because they can account for unpredictable performance fluctuations in the iteration process due to limited information-processing capacities and the intrinsic mechanisms of error correction in product development. Further, from an ergonomics perspective, unpredictable performance fluctuations can be seen as essential components of creative activities in design work and are therefore basic ingredients of success that should not be limited in their reach and capacity to benefit the whole process. These fluctuations are inherent to open organizational systems and are especially prevalent in concurrent engineering because of the multitude of interfaces between mechanical, electrical and electronic modules and the high level of integration of technical knowledge and methodological approaches to problem-solving, all of which are aspects crucial to the success of a parallel and highly iterative execution of the work processes.

To gain a deeper understanding of the dynamics of cooperative work, we did not simply consider a basic model class in which the project state under uncertainty is represented by a linear combination of observable random vectors, but also a broader class with latent variables which cannot be directly observed but must be

© Springer International Publishing Switzerland 2016
C. Schlick, B. Demissie, *Product Development Projects*, Understanding Complex Systems, DOI 10.1007/978-3-319-21717-8_6

inferred through a causal model from the estimated project state. As in the works of Smith and Eppinger (1997, 1998) and Yassine et al. (2003), the project state was expressed as the amount of work remaining for all tasks at time step t. The work remaining can be measured by the time left to finalize a specific design, by the definable labor units required to complete a particular development activity or component of the work breakdown structure, by the number of engineering drawings requiring completion before design release, by the number of engineering design studies required before design release, or by the number of issues that still need to be addressed/resolved before design release (Yassine et al. 2003). The basic model class comprises vector autoregression models of finite order which can capture the apparent cooperative processing of the development tasks with short iteration length in a given project phase. Despite their simple structure and low logical depth, these models can simulate a surprisingly diverse set of patterns of cooperative task processing in product development projects. Furthermore, they can be used to explain the "problem-solving oscillations" (Mihm et al. 2003) specified in the introductory chapter by making only a few, very reasonable assumptions about the problem-solving processes in the iteration process. According to the deterministic and stochastic parts of the state equations, the irregular oscillations between being on, ahead of, and behind schedule can be interpreted as excited performance fluctuations (Schlick et al. 2008; Schlick et al. 2013). For first-order models the excitation can occur because of the interrelationship between the dependency structure encoded in the work transformation matrix A_0 and the forcing matrix K. These mechanisms were explicitly revealed in the spectral basis (see e.g. Eqs. 43 and 47 in conjunction with Eqs. 258 and 262). An augmented state-space formulation also makes it possible to model and simulate more complex autoregressive processes with periodically correlated components. These so-called periodic vector autoregressive stochastic processes can be used to model hierarchical coordination structures in large-scale product development projects and to simulate the long-term effect of intentionally withholding the release of design information for a certain period of time and not immediately disseminating it to lower hierarchical levels. To go beyond vector autoregressive processes, we also considered the theoretically interesting class of linear dynamical systems with additive Gaussian noise. As mentioned above, the state of the project cannot be directly observed in these system models with latent variables but is inferred through a causal model. This makes it possible to distinguish between a "hidden" state process of cooperative development and the observation process in the product development flow. The internal configuration is not entirely accessible to the project manager but must be estimated on the basis of repeated readings from dedicated performance measurement instruments. This fundamental degree of uncertainty in the project state and its evolution can lead to a non-negligible fraction of long-term correlations between development activities and therefore significantly increase emergent complexity. In addition to the statistical models of cooperative work, the corresponding least squares and maximum likelihood estimation methods were introduced to demonstrate how the parameters can be efficiently estimated from time series of task processing in industrial product development environments. To validate the basic vector autoregression and linear dynamical system models with field data, a case study was carried out at a German industrial company that develops sensor technologies for the automotive industry. The validation results show that in terms of the standards of organizational modeling and simulation, the overall predictive accuracy of the parameterized models is high. As is to be expected, the more complex linear dynamical system model leads to slightly better predictions. Moreover, for periodic vector autoregressive stochastic processes the accuracy of the introduced least squares estimation technique was investigated in a simulation study in order to completely control the confounding factors. We formulated a periodic vector autoregression model that connects the dynamics of module design and integration in a vehicle door development project with component development. This served as a reference model to simulate task processing and generate time series for different lengths of work remaining. We then "reconstructed" the reference model representation purely from data. To simplify the analysis, we focused on the door module of the vehicle door subsystem. Here too, we were able to show that in terms of the standards of organizational modeling and simulation, the identifiability of the reference model of cooperative work was high and the uncertainty in parameter estimation was low. The results were highly consistent and replicable.

Furthermore, and most importantly from a scientific point of view, the complexity framework, consisting of theories and measures developed in organizational theory, systematic engineering design and basic scientific research on complex systems, was reviewed in great detail and applied to project management as far as possible. To evaluate emergent complexity of cooperative work in product development projects, an information-theory measure from basic scientific research—termed "effective measure complexity"—was chosen because it can be derived from first principles and therefore offers high construct validity. Effective measure complexity quantifies the mutual information between the infinite past and future histories of a stochastic process. According to this principle, this measure is of particular interest for evaluating the time-dependent complexity of cooperative development processes and identifying the essential interactions between activities. Effective measure complexity and the underlying complexity theory can be traced back to the theoretical physicist Grassberger (1986), whose seminal work has been completely overlooked in organization theory and engineering management literature. It is important to point out that effective measure complexity is not limited to a specific class of statistical models of cooperative work: if the data is generated by a process in a specific class but with unknown parameter values, we can calculate the effective measure complexity explicitly, as we did. It is also possible, however, to quantify the complexity of processes that fall outside the conventional models. The formulated vector autoregression models provided the mathematical foundation for the calculation of several closed-form solutions of effective measure complexity in the original state space, solutions that allow an explicit complexity evaluation based on the model's parameters. For first-order models we also carried out a transformation into the spectral basis to obtain additional, more expressive solutions in matrix form. In the spectral basis, the essential parameters driving emergent complexity, which are surprisingly few in number, were identified and the effects of cooperative relationships were directly interpreted. The essential parameters include the eigenvalues of the work transformation matrix as a dynamical operator of the vector autoregression model and the correlation coefficients between components of unpredictable performance fluctuations. In this context, the closed-form solution from Eq. 262 is especially interesting for a complexity analysis with respect to first-order processes, as it significantly reduces the effective number of parameters without blurring the essential spatiotemporal structures that shape emergent complexity in the product development projects modeled. Through a simple rewriting of the state equation as a first-order recurrence relation it was also possible to calculate the effective measure complexity of processes that are generated by higher-order autoregressive models. Furthermore, different types of closed-form solutions of effective measure complexity in the original state-space coordinates were calculated for linear dynamical systems with additive Gaussian noise. Because linear dynamical systems with additive Gaussian noise are very common in mechanical, electrical and control engineering, these solutions are not only interesting for evaluating emergent complexity in open organizational systems; they can also be used to analyze, design and control purely technical systems. Due to the comparatively complicated structure of linear dynamical systems the derivation of the closed-form solutions was much more involved mathematically speaking. The most sophisticated solution is based on infinite dimensional matrices and was presented in Eq. 291 in explicit form. A similar result has been obtained in a more general context by de Cock (2002), whose seminal work also made it possible to express effective measure complexity in a structurally rich but much simpler implicit form. The implicit form is significantly easier to interpret because its parameters can be derived from the solutions of fundamental equations. This most intuitive solution is provided in Eq. 299. Both the explicit and implicit formulations of the solutions lead to consistent and robust numerical complexity results, as we have shown in the sensitivity analysis of Section 2.11. Hence, both approaches can be very helpful for evaluating strong emergence in terms of mutual information communicated from the infinite past to the infinite future by the stochastic process as a model of cooperative work in product development projects.

The closed-form solutions obtained show that effective measure complexity is non-negative. The detailed discussions in Section 4.1 made it clear that this measure has four especially favorable properties in the application domain of project management:

1. It is small for project phases in which tasks can be processed independently and it assigns larger complexity values to intuitively more complex work processes with the same dominant eigenvalue of the corresponding dynamical operator but stronger task coupling due to intense cooperation. The effective measure complexity equals zero if the process observed is completely temporally uncorrelated and therefore no meaningful informational dependency structure exists between the processed tasks, which can be used to make good predictions (see Section 4.1). Moreover, if the vector of work remaining over all tasks can be

divided into two subsets which are completely uncorrelated and therefore represent completely independent work streams in the product development flow, the effective measure complexity of the whole process equals the sum of the complexities of each subprocess resulting from the division (see Section 4.1). The importance of the nature, quantity and magnitude of organizational subtasks and subtask interactions is also pointed out in the theoretical and empirical analyses of Tatikonda and Rosenthal (2000). Interestingly, the empirical studies of Höltä-Otto and Magee (2006) show that estimation of effort in product development projects is primarily based on the scale and stretch of the project and not on interactions. This is due to the fact that the balancing or reinforcing effects of concurrent interactions in open organizational systems are very difficult for project managers to anticipate. Accordingly, the measure can contribute to achieving more reliable estimates of time and effort. The dependencies between tasks were also mentioned as complexity-contributing elements in four out of six cases in the empirical analysis put forward by Bosch-Rekveldt et al. (2011). Mulenburg (2008) considers the number and types of task-based interactions between actors in projects of different types as one of six main sources of complexity, while Summers and Shah (2010) consider "complexity as coupling" to be one of three main aspects of design complexity. Both the number of activities in a project and their interconnectedness are complexity-shaping factors in the UCP and NTCP models developed by Shenhar and colleagues (Shenhar and Dvir 2007; Shenhar 1998; Shenhar and Dvir 1996).

2. The effective measure complexity tends to assign larger complexity values to project phases with more tasks if the intensity of cooperative relationships is similar, and is thus sensitive to the dimensionality of the state space of the process. The measure is also a strictly increasing function of each of the canonical correlations (see Section 4.1 in conjunction with Section 4.1.3). The complexity-reinforcing effects of the "size" of a project are also stressed in Mihm et al. (2003), Huberman and Wilkinson (2005), Suh (2005), Höltä-Otto and Magee (2006), Mihm and Loch (2006), Shenhar and Dvir (2007), Mulenburg (2008), Hass (2009), Summers and Shah (2010), and Bosch-Rekveldt et al. (2011). Alternatively, one can divide the effective measure complexity by the dimension of the state space and compare processes with different dimensionalities.
3. The measure can evaluate both weak and strong emergence in an uncertain product development environment. According to Chalmers (2002), weak emergence means that there is in principle no choice of outcome. As such, the outcome can be anticipated without a detailed inspection of particular instances of task processing. Given the state equation, there are entirely reproducible features of its subsequent evolution that inevitably emerge over time, such as reaching a steady state. In light of our approach, a simple technique for evaluating weak emergence in the class of vector autoregression models is the eigenvalue analysis of the dynamical operator (assuming that for higher-order models the state equation was rewritten as a first-order recurrence relation and the combined dynamical operator can be evaluated, see Section 4.1.6). It is

evident that effective measure complexity indicates the same bound of asymptotic stability as does a classic eigenvalue analysis by assigning infinite complexity values: if the dominant eigenvalue has modulus less than one, the infinite sum in Eq. 247 converges, and the project will converge toward the asymptote of "no remaining work;" on the other hand, if the dominant eigenvalue has modulus greater than one, the sum diverges, and the work remaining grows over all given limits. The emergence of complexity is termed strong if the patterns of cooperative task processing can only be reliably forecasted by observing the distant past of each particular development task instance and with relevant knowledge of the prior history of the interacting processes (Chalmers 2002). In management literature this phenomenon is also known as "path dependence" (Maylor et al. 2008). Relevant information on the prior history is extracted through Bialek's predictive information (cf. Eq. 226) and utilized in the different closed-form solutions in the limit of an infinite block length. The importance of the factor "uncertainty" in the scope and methods of a project in conjunction with "stability of project environment" is also pointed out in the TOE framework of Bosch-Rekveldt et al. (2011). Mulenburg (2008) identifies uncertainty in the sense of the inability to pre-evaluate actions as one of the six main sources of complexity in projects, listing the unpredictability of the project state and events as another main source. It is readily apparent that for a first-order vector autoregression model of cooperative work, for instance, these sources are clearly separated through the definition of the effective measure complexity from Eq. 246: the inherent inability to pre-evaluate the consequence of actions concerning the amount of work is expressed by the determinant of the covariance matrix C of the intrinsic prediction error. The unpredictability of the iteration process, which can only be "explained away" through an increasingly detailed inspection of the particular instances of task processing is represented by the ratio of generalized variances $\mathrm{Det}[\Sigma]$ and $\mathrm{Det}[C]$. Suh's information axiom (2005) addresses both size and uncertainty, while Hass (2009) focuses on the stability of the requirements, as well as the clarity of the problem and its solution. The simulation study contributed by Lebcir (2011) shows that development time significantly increases when project uncertainty is changed from low to reference level.

4. The measure is independent of the basis in which the state vectors are represented. It is invariant under arbitrary reparameterizations based on smooth and uniquely invertible maps (Kraskov et al. 2004) and is therefore independent of the project manager's subjective choice of measurement instrument. As such, it can contribute to reducing ambiguity in projects by promoting awareness of causality and persistent dynamical dependency structures (sensu Mulenburg 2008). To the best of our knowledge, this fundamental objectivity is a unique property that other metrics do not possess. Moreover, this invariance property is mandatory, given that the likelihood function $\mathcal{L}(\theta|\{y_t\}_0^T)$ of a sequence of fixed observations $\{y_t\}_0^T$ of a linear dynamical system possesses the same invariance property, as outlined in Section 2.10.

Finally, the theoretical complexity analyses were elucidated in practical terms using three validation studies. We investigated the validity of closed-form solutions for effective measure complexity that were obtained for vector autoregression models as the basic mathematical representations of cooperative work in product development projects. With respect to classic validity theory, the focus was on the criterion-based conception of validity. We evaluated two types of criterion-related validity: concurrent validity and predictive validity. In the first study, we concentrated on concurrent validity. To do so, we expanded the theoretical foundations on model selection, primarily building upon the work of Li and Xie (1996) on the principle of minimal mutual information. Although this principle was developed independently from the work of Grassberger (1986) and others, it is very closely related, as the mutual information communicated from the infinite past to the infinite future by the stochastic process is evaluated to select the class of models under certain constraints. Furthermore, this principle allowed us to formulate a complexity-based model selection criterion (termed the mutual information criterion) that could be directly applied to identify an optimal model order within the class of vector autoregression models. The studies of concurrent validity were based on two Monte Carlo experiments, which shared the same overall objective but used different parametric model forms. The overall objective was to compare the accuracy of the mutual information criterion with standard criteria like the (original and bias corrected) Akaike information criterion and the Schwarz-Bayes criterion. It was hypothesized that model selection based on effective measure complexity makes it possible to select the true model order with high accuracy and that the histogram distributions of the selected model orders are not significantly different from the distributions obtained under the alternative criteria. The parametric model forms were not only derived on the basis of field data from the previously mentioned PD project at the small industrial company, but were also synthetically generated to allow a systematic comparison of concurrent validity of the different model selection criteria (cf. Lütkepohl 1985). The results of the Monte Carlo experiments unambiguously show that the mutual information criterion is not only very effective for making model selection decisions in specific PD environments, but also appears to offer a highly accurate universal quantity for model selection in the class of vector autoregression models. Further, additional analytical and numerical considerations have shown that the formulation of the criterion can easily be generalized to the class of linear dynamical systems, and that using the criterion for model selection within this broader class also allows us to select the true model order with a high degree of accuracy. The significance of the criterion's effectiveness at universally penalizing unnecessarily complex models should not be underestimated. It not only shows that vector autoregression and linear dynamical system models as specific instances of homogeneous recurrence relations can be used to validate a quantitative theory of emergent complexity in open organizational systems, but also demonstrates that theoretical knowledge can be transferred and a systematic method provided to find optimal parametric representations for different classes of systems that are completely independent of the systems' representation of its organization. The second and third validation studies focused on predictive validity. In the second study our goal was to optimize organization design, and concretely to minimize emergent complexity by selecting the optimal staffing of three concurrent engineering teams with developers who have different levels of productivity (attributable to their skills, abilities or access to information) in a simulated product development project. We hypothesized that for significant individual differences "productivity balancing" at the team level leads to minimal emergent complexity. The aim of the third study was to optimize the period for minimal emergent complexity in which information about integration and tests of geometric/topological entities is deliberately withheld by subsystem-level teams and not released to component-level teams. This type of non-cooperative behavior is justified by the aim to improve the implementation of the product architecture and reduce coordination efforts. In both the second and third studies we formulated unconstrained as well as constrained optimization problems and solved them on the basis of the analytically obtained complexity solutions. We consistently modeled and analyzed small but complete product development projects in both studies where all tasks have to be processed completely in parallel and are fully interdependent, as this approach allowed us to demonstrate the introduced concepts in a holistic manner and simplified the interpretation of the results. Furthermore, we carried out Monte Carlo experiments to investigate the influence of emergent complexity in the sense of effective measure complexity on means and standard deviations of the project duration and the effort. In our cases the term effort referred to the total amount of work done in the iteration process to complete the project as a whole. In the constrained optimization problems we considered the mean total amount of work as the externally set constraint and concentrated on the interrelationship between emergent complexity and project duration. In practical terms, the most important finding of the second validation study was that an assignment of team members with "balanced" productivity at the team level and therefore maximum diversity within teams produces minimal emergent complexity. Moreover, this organization design led to the lowest means and standard deviations of the project duration in the Monte Carlo experiments. These results are independent of whether a constraint was put on the mean total amount of work or not. When the mean total amount of work was considered as a constraint in the Monte Carlo experiments, the standard deviation of the total amount of work that was calculated for the optimal project organizational design was also smallest. Hence, the results consistently supported the formulated "productivity balancing" and the complementary "team diversity" hypothesis. From a theoretical perspective another interesting finding of the second validation study was that the larger the emergent complexity, the more the evolution toward a stable solution of the project duration at any particular instance can differ from the average unperturbed process. Hence, projects that in the absence of unpredictable performance fluctuations would converge smoothly to the desired goal of zero work remaining for all tasks can deviate significantly from this path. When emergent complexity is low, convergence is smooth, and both the project duration and the total amount of work are distributed approximately log-normally with small variance. Above certain complexity thresholds, we observed that the distributions undergo a transition to a long-tailed log-normal form. The appearance of a long-tailed log-normal form implies that while there may be work processes where time to completion seems to be compatible with the expected duration of the project, on other occasions it will be significantly longer, with the concomitant aggravation that the project manager is unable to predict when such long delays will arise (Huberman and Wilkinson 2005). One of the most important findings of the third validation study was that in unconstrained optimization, the smallest complexity values are assigned to periodically correlated work processes with minimum correlation length. For the organizational conditions investigated we must therefore recommend minimizing the period in which information about the system design and integration testing is deliberately withheld by subsystem-level teams. By minimizing the length of this period, the emergent complexity can be kept as low as possible. This recommendation is also fully supported by the results of the Monte Carlo experiments because they show that, as the release period increases, the means and standard deviations of the project duration and the total amount of work also increase significantly. In the case of constrained optimization, specifying a constraint on the mean total amount of work leads to complexity values that strictly decrease as the release period increases and are therefore in contrast to the solution of the unconstrained optimization problem. However, the results of the Monte Carlo experiments again support the predictive validity of the complexity measure, as the means and standard deviations of the project duration and the total effort expended strictly decrease as well. The differences in project duration are much smaller but still statistically significant. Maximizing the period in which information is deliberately withheld by subsystem-level teams can only be recommended if the work processes can be organized in such a way that the fractions of work that are put on hold at each short iteration can be reduced by at least 25%. In real product development environments this can usually only be achieved through substantially intensified communication between developers or with an exorbitant number of standards on how a work process is to be executed. Not only do these organization-level interventions often result in excessive work stress; they can also have adverse effects on creative solutions and innovative approaches and are therefore not desirable. The combined computational analyses in the second and third validation study show that the effective measure complexity is not only a highly satisfactory quantity in theory, but also a good predictor of the mean and standard deviation of the project duration for different staffings of concurrent engineering teams, for different release periods of design information and for different formulations of the optimization problem – regardless of whether the mean total amount of work is unconstrained or constrained. The predictive accuracy is higher for the unconstrained optimization of the release period. In summary, inconsistencies in performance predictions or disordinal interactions between the independent variables did not occur, proving the predictive validity of the closed-form solutions for the investigated class of vector autoregression models. The studies also showed that the developed statistical models of cooperative work in conjunction with the application of the complexity theory can lead to interesting and useful results in applied organizational simulation and optimization.

The information-theory approach to evaluating emergent complexity in product development projects in conjunction with the state-space representations of cooperative work still needs to be worked out in more detail in the future. A first step in this direction would be to compute the effective measure complexity of processes that are generated by the statistical project model developed by Huberman and Wilkinson (2005), which incorporates multiplicative instead of additive noise to model non-predictable performance fluctuations (cf. Eq. 8). The multiplicative approach to modeling performance variability is interesting not only because it can reproduce the critical effects of large design teams and highly interdependent tasks with conceptually convincing and practically useful conclusions about the cooperative problem-solving processes, but also because of its theoretically reasonable assumption that the autonomous task processing rates and the intensity of cooperative relationships are subject to random influences. However, to the best of our knowledge, the Huberman-Wilkinson model has not been supported by any empirical evidence and therefore it remains an open question whether it can produce better managerial decisions than our additive approach. Furthermore, since there is no sufficient reason to believe that this is the only theoretically reasonable assumption, let alone whether this assumption would be empirically especially relevant for the product development flow, we believe that our simpler autoregressive models of cooperative work already offer a good explanation of the cited problem-solving oscillations and high predictive validity. As mentioned before and shown in detail in Section 4.1.2, within this basic class of models one can explain problem-solving oscillations by the interference pattern between all design and performance fluctuation modes that can exacerbate performance fluctuations. It may be possible to calculate the effective measure complexity related to the Huberman-Wilkinson model by following three main steps: First, calculate the covariance matrix in the steady state. Second, show that performance variability in the steady state is governed by a log-normal distribution. The simulation results indicate that the distribution in the steady state may be log-normal (Huberman and Wilkinson 2005). Third, calculate the dynamic entropies of the state variables and derive the corresponding measure. Unfortunately, these three calculations are extremely involved, mathematically speaking. A natural extension on the basis of linear state-space models would be to incorporate both multiplicative and additive performance fluctuations (Arnold and Wishtutz 1982). However, under this premise the parameter estimation from the small sample size typical for organizational modeling is difficult and can be unreliable.

In the long term, we plan to conduct an external validation study of the complexity theory and the related measures with experienced project managers in industry. By "measures" we mean not only the effective measure complexity which has been heavily stressed in this book, but also the persistent mutual information that was recently proposed by Ball et al. (2010, see definition in Eq. 229). The closed-form solutions that were presented in Section 4.1 and the theoretical analysis in Section 4.1.6 have shown that the persistent mutual information can be considered as a complexity measure in its own right and can lead to meaningful complexity values even if there are zeros on the unit circle, which can easily cause the effective measure complexity to be infinite (Li 2006). It is hypothesized that the effective measure complexity and the persistent mutual information are also externally valid complexity metrics, which can be used for the simulation-supported optimization of organizational structures and processes in projects of different scale and scope (cf. Sections 5.2 and 5.3). Furthermore, they also have the potential to capture the implicit knowledge of project managers based on the nature, quantity and magnitude of concurrent tasks and can therefore serve as a good foundation for developing effective managerial interventions to cope with emergent complexity in the product development flow. In general, we believe that the effective measure complexity and the persistent mutual information provide valuable information enabling the project manager and concurrent engineering teams to better organize their work and to improve coordination processes. The results of the optimization study on project organization in Section 5.2 may also have practical implications. Within the modeling framework developed for evaluating complexity we found some experimental evidence that the self-developed concept of productivity balancing can lead to a reduction in the finishing time in product development projects and its variance. The concept also has the potential to simultaneously reduce the total effort expended. As we pointed out in Section 5.2, the general idea behind the concept of productivity balancing is not to standardize the work and complete the tasks at a constant rate, which would severely limit the developers' scope of action and place limits on their creative thinking. Rather, the goal is to implement the basic principle of diversity management and find optimal (or near optimal) assignments of team members with different levels of performance so as to maximize the potential advantages of diversity and keep its potential disadvantages down to a tolerable level (sensu Cox 1994). This approach makes it possible to carry out the project work effectively and efficiently without design churns (Yassine et al. 2003) or other critical emergent phenomena of complex sociotechnical systems that often lead to unacceptably high levels of stress and excessive workloads. Productivity balancing promotes the continuous development of human knowledge, skills and abilities through effective cooperation and intensive communication. As such, it can be regarded as a holistic concept for the goal-oriented design of interactions between individuals, tasks and products/services that views performance fluctuations as an opportunity to innovate and learn. Thanks to its participatory design and evaluation of these interactions it also aims to increase awareness of characteristic emergent phenomena in open organizational systems and to leverage from them the greatest advantage for the individuals and work teams. A theoretically possible but still highly speculative effect of leverage is that under certain circumstances unpredictable performance fluctuations in conjunction with team diversity can accelerate the completion of the development tasks: the mean project duration obtained in the Monte Carlo experiments is shorter than the duration without performance variability. However, the standard deviation usually grows simultaneously with the acceleration factor and therefore the risk of falling behind the schedule tends to increase. Furthermore, there is a certain degree of risk that the resulting work intensity coupled with multitasking could have a negative effect on well-being. Using analytical methods, simulation models or empirical studies, we need to find the optimal project organization arrangement in which either this trade-off is non-existent or its potential negative consequences on individuals and teamwork can be significantly reduced. The theoretical framework developed here would seem to offer a stable foundation for analyzing the novel concept of productivity balancing in more detail in the future and for developing human-centered solutions for cooperative work in product development projects.

References

Arnold, L., Wishtutz, V.: Stationary solutions of linear systems with additive and multiplicative noise. Stochastica **7**(1), 131–155 (1982)

Ball, R.C., Diakonova, M., MacKay, R.S. (2010). Quantifying emergence in terms of persistent mutual information. arXiv:1003.3028v2 [nlin.AO]

Bosch-Rekveldt, M., Jongkind, Y., Mooi, H., Bakker, H., Verbraeck, A.: Grasping project complexity in large engineering projects: The TOE (Technical, Organizational and Environmental) framework. Int. J. Project Manag. **29**(6), 728–739 (2011)

Chalmers, D.J.: Strong and weak emergence. In: Clayton, P., Davies, P. (eds.) The Re-Emergence of Emergence, pp. 244–256. Oxford University Press, Oxford (2002)

Cox, T.: Cultural Diversity in Organizations: Theory, Research and Practice. Berrett-Koehler Publishers, San Francisco, CA (1994)

de Cock, K.D.: Principal angles in system theory, information theory and signal processing. Ph.D. thesis, Katholieke Universiteit Leuven. (2002)

Grassberger, P.: Toward a quantitative theory of self-generated complexity. Int. J. Theor. Phys. **25** (9), 907–938 (1986)

Hass, K.B.: Managing Complex Projects. A New Model. Management Concepts. Leesburg Pike, PA (2009)

Hölttä-Otto, K., Magee, C.L.: Estimating factors affecting project task size in product development—An empirical study. IEEE Trans. Eng. Manag. **53**(1), 86–94 (2006)

Huberman, B.A., Wilkinson, D.M.: Performance variability and project dynamics. Comput. Math. Organ. Theor **11**(4), 307–332 (2005)

Kraskov, A., Stögbauer, H., Grassberger, P. (2004). Estimating mutual information. arXiv:cond-mat/0305641v1 [cond-mat.stat-mech]

Lebcir, M.R. (2011). Impact of project complexity factors on new product development cycle time. University of Hertfordshire Business School Working Paper. https://uhra.herts.ac.uk/dspace/handle/2299/5549

Li, L.: Some notes on mutual information between past and future. J. Time Ser. Anal. **27**(2), 309–322 (2006)

Li, L., Xie, Z.: Model selection and order determination for time series by information between the past and the future. J. Time Ser. Anal. **17**(1), 65–84 (1996)

Lütkepohl, H.: Comparison of criteria for estimating the order of a vector autoregressive process. J. Time Ser. Anal. **6**, 35–52 (1985)

Maylor, H., Vidgen, R., Carver, S.: Managerial complexity in project-based operations: A grounded model and its implications for practice. Project Manag. J. **39**(1), 15–26 (2008)

Mihm, J., Loch, C.: Spiraling out of control: Problem-solving dynamics in complex distributed engineering projects. In: Braha, D., Minai, A.A., Bar-Yam, Y. (eds.) Complex engineered systems: science meets technology, pp. 141–158. Springer, Berlin (2006)

Mihm, J., Loch, C., Huchzermeier, A.: Problem-solving oscillations in complex engineering. Manag.. Sci. **46**(6), 733–750 (2003)

Mulenburg, J.: What does complexity have to do with it? Complexity and the management of projects. In: Proceedings of the 2008 NASA Project Management Challenge Conference. (2008)

Schlick, C.M., Duckwitz, S., Gärtner, T., Schmidt, T. A complexity measure for concurrent engineering projects based on the DSM. In: Proceedings of the 10th International DSM Conference, pp. 219–230. (2008)

Schlick, C.M., Duckwitz, S., Schneider, S.: Project dynamics and emergent complexity. Comput. Math. Organ. Theor. **19**(4), 480–515 (2013)

Shenhar, A.J.: From theory to practice: toward a typology of project management styles. IEEE Trans. Eng. Manag. **45**(1), 33–48 (1998)

Shenhar, A.J., Dvir, D.: Toward a typological theory of project management. Res. Pol. **25**(4), 607–632 (1996)

Shenhar, A.J., Dvir, D.: Reinventing Project Management: The Diamond Approach to Successful Growth and Innovation. Harvard Business School Press, Boston, MA (2007)

Smith, R.P., Eppinger, S.D.: Identifying controlling features of engineering design iteration. Manag. Sci. **43**(3), 276–293 (1997)

Smith, R.P., Eppinger, S.D.: Deciding between sequential and concurrent tasks in engineering design. Concurrent Eng.: Res Appl **6**(1), 15–25 (1998)

Suh, N.P.: Complexity—Theory and Applications. Oxford University Press, Oxford (2005)

Summers, J.D., Shah, J.J.: Mechanical engineering design complexity metrics: Size, coupling, and solvability. J. Mech. Des. **132**(2), 1–11 (2010)

Tatikonda, M.V., Rosenthal, S.R.: Technology novelty, project complexity and product development project execution success. IEEE Trans. Eng. Manag. **47**, 74–87 (2000)

Winner, R.I., Pennell, J.P., Bertrand, H.E., Slusarezuk, M.M.: The role of concurrent engineering in weapons system acquisition. In: IDA-Report R-338, Institute for Defense Analyses; Alexandria, VA, (1988)

Yassine, A., Joglekar, N., Eppinger, S.D., Whitney, D.: Information hiding in product development: The design churn effect. Res. Eng. Des. **14**(3), 145–161 (2003)

GPSR Compliance
The European Union's (EU) General Product Safety Regulation (GPSR) is a set of rules that requires consumer products to be safe and our obligations to ensure this.

If you have any concerns about our products, you can contact us on

ProductSafety@springernature.com

In case Publisher is established outside the EU, the EU authorized representative is:

Springer Nature Customer Service Center GmbH
Europaplatz 3
69115 Heidelberg, Germany

www.ingramcontent.com/pod-product-compliance
Lightning Source LLC
Chambersburg PA
CBHW071415230926
51846CB00001B/92
* 9 7 8 3 3 1 9 3 7 3 5 4 6 *